DESCRIPTIVE SENSORY ANALYSIS IN PRACTICE

PUBLICATIONS IN FOOD SCIENCE AND NUTRITION

Books

NUTRACEUTICALS: DESIGNER FOODS III, P.A. Lachance
DESCRIPTIVE SENSORY ANALYSIS IN PRACTICE, M.C. Gacula, Jr.
APPETITE FOR LIFE: AN AUTOBIOGRAPHY, S.A. Goldblith
HACCP: MICROBIOLOGICAL SAFETY OF MEAT, J.J. Sheridan et al.
OF MICROBES AND MOLECULES: FOOD TECHNOLOGY AT M.I.T., S.A. Goldblith
MEAT PRESERVATION: PREVENTING LOSSES AND ASSURING SAFETY,
  R.G. Cassens
S.C. PRESCOTT, PIONEER FOOD TECHNOLOGIST, S.A. Goldblith
FOOD CONCEPTS AND PRODUCTS: JUST-IN-TIME DEVELOPMENT, H.R. Moskowitz
MICROWAVE FOODS: NEW PRODUCT DEVELOPMENT, R.V. Decareau
DESIGN AND ANALYSIS OF SENSORY OPTIMIZATION, M.C. Gacula, Jr.
NUTRIENT ADDITIONS TO FOOD, J.C. Bauernfeind and P.A. Lachance
NITRITE-CURED MEAT, R.G. Cassens
POTENTIAL FOR NUTRITIONAL MODULATION OF AGING, D.K. Ingram et al.
CONTROLLED/MODIFIED ATMOSPHERE/VACUUM PACKAGING, A.L. Brody
NUTRITIONAL STATUS ASSESSMENT OF THE INDIVIDUAL, G.E. Livingston
QUALITY ASSURANCE OF FOODS, J.E. Stauffer
THE SCIENCE OF MEAT AND MEAT PRODUCTS, 3RD ED., J.F. Price and
  B.S. Schweigert
HANDBOOK OF FOOD COLORANT PATENTS, F.J. Francis
ROLE OF CHEMISTRY IN PROCESSED FOODS, O.R. Fennema et al.
NEW DIRECTIONS FOR PRODUCT TESTING OF FOODS, H.R. Moskowitz
ENVIRONMENTAL ASPECTS OF CANCER: ROLE OF FOODS, E.L. Wynder et al.
FOOD PRODUCT DEVELOPMENT AND DIETARY GUIDELINES, G.E. Livingston,
  R.J. Moshy and C.M. Chang
SHELF-LIFE DATING OF FOODS, T.P. Labuza
ANTINUTRIENTS AND NATURAL TOXICANTS IN FOOD, R.L. Ory
UTILIZATION OF PROTEIN RESOURCES, D.W. Stanley et al.
POSTHARVEST BIOLOGY AND BIOTECHNOLOGY, H.O. Hultin and M. Milner

Journals

JOURNAL OF FOOD LIPIDS, F. Shahidi
JOURNAL OF RAPID METHODS AND AUTOMATION IN MICROBIOLOGY,
  D.Y.C. Fung and M.C. Goldschmidt
JOURNAL OF MUSCLE FOODS, N.G. Marriott, G.J. Flick, Jr. and J.R. Claus
JOURNAL OF SENSORY STUDIES, M.C. Gacula, Jr.
JOURNAL OF FOODSERVICE SYSTEMS, C.A. Sawyer
JOURNAL OF FOOD BIOCHEMISTRY, N.F. Haard, H. Swaisgood and B. Wasserman
JOURNAL OF FOOD PROCESS ENGINEERING, D.R. Heldman and R.P. Singh
JOURNAL OF FOOD PROCESSING AND PRESERVATION, D.B. Lund
JOURNAL OF FOOD QUALITY, J.J. Powers
JOURNAL OF FOOD SAFETY, T.J. Montville and D.G. Hoover
JOURNAL OF TEXTURE STUDIES, M.C. Bourne and M.A. Rao

Newsletters

MICROWAVES AND FOOD, R.V. Decareau
FOOD INDUSTRY REPORT, G.C. Melson
FOOD, NUTRITION AND HEALTH, P.A. Lachance and M.C. Fisher

DESCRIPTIVE SENSORY ANALYSIS IN PRACTICE

Edited by

M.C. Gacula, Jr., Ph.D.

Gacula Associates
Scottsdale, Arizona

**FOOD & NUTRITION PRESS, INC.
TRUMBULL, CONNECTICUT 06611 USA**

Copyright © 1997 by
FOOD & NUTRITION PRESS, INC.
6527 Main Street
Trumbull, Connecticut 06611 USA

All rights reserved. No part of this publication may be reproduced, stored in a retrieval system or transmitted in any form or by any means: electronic, electrostatic, magnetic tape, mechanical, photocopying, recording or otherwise, without permission in writing from the publisher.

Library of Congress Catalog Card Number: 97-060573
ISBN: 0-917678-37-0

DEDICATION

To my lovely daughters

Karen, Lisa, and Elena

PREFACE

In the last two decades, the need by Product Research and Development personnel to define what is in a product in terms of sensory impressions, rather than by instrumental measures, has been met by Descriptive Sensory Analysis techniques. This is an important development in sensory science because only human beings can accurately describe and identify the sensory properties of products and materials related to the basic senses of taste, smell, touch, and sight. By descriptive analysis, the senses that are perceived are quantified and related to product acceptance/preference, which is the ultimate goal in product development.

The motivation of this book comes from the need to collect in one document published materials dealing with the technical developments and applications of descriptive analysis to various types of products and materials, such as, dairy, meats, alcoholic beverages, textile materials, and other general applications. Each chapter in this book contains a wealth of materials on the various applications of descriptive analysis–its sensory philosophy, its statistical philosophy, and test execution which provides the readers a wide spectrum of the uses of descriptive analysis, and an opportunity to improve the current descriptive analysis techniques. Although there is no specific article in the book that deals with personal care products (soap, lotion, shampoo, conditioner, toothbrush, and shaving materials) and household products, it is an established fact that descriptive analysis has been widely and successfully used for these types of products.

The availability of many statistical software packages greatly enhanced the implementation of descriptive analysis techniques. In this book, the following packages were used to illustrate various techniques of data analyses:

SAS/STAT, a registered trademark of SAS Institute, Inc.
STATISTIX, a registered trademark of Analytical Software
DESIGN-EXPERT, a registered trademark of Stat-Ease, Inc.
DESIGN-EASE, a registered trademark of Stat-Ease, Inc.
Microsoft EXCEL, a registered trademark of the Microsoft Corporation

The author is indebted to all authors and publishers for their kind permission to reprint original papers in this book. The technical assistance of John Ose is gratefully acknowledged. I thank Food & Nutrition Press for publishing this book, and in particular, Jennifer Schuchman whose diligent work kept the publication process in order.

MAXIMO C. GACULA, JR.

CONTENTS

CHAPTER		PAGE
1.0	INTRODUCTION, Gacula, M.C.	1
	1.1 DESCRIPTIVE SENSORY ANALYSIS METHODS, Gacula, M.C.	5
	1.2 FLAVOR PROFILES: A NEW APPROACH TO FLAVOR PROBLEMS, Cairncross, S.E. and Sjöström, L.B.	15
	1.3 SENSORY EVALUATION BY QUANTITATIVE DESCRIPTIVE ANALYSIS, Stone, H., Sidel, J., Oliver, S., Woolsey, A. and Singleton, R.C.	23
	1.4 THE SELECTION AND USE OF JUDGES FOR DESCRIPTIVE PANELS, Zook, K. and Wessman, C.	35
	1.5 EXAMINING METHODS TO TEST FACTOR PATTERNS FOR CONCORDANCE, Fischman, E.I., Shinholser, K.J. and Powers, J.J.	51
	1.6 QUANTITATIVE DESCRIPTIVE ANALYSIS, Stone, H., Sidel, J.L. and Bloomquist, J.	63
	1.7 IMPORTANCE OF REFERENCE STANDARDS IN TRAINING PANELISTS, Rainey, B.A.	71
	1.8 THE IMPORTANCE OF LANGUAGE IN DESCRIBING PERCEPTIONS, Civille, G.V. and Lawless, H.T.	77
	1.9 ILLUSTRATIVE EXAMPLES OF PRINCIPAL COMPONENTS ANALYSIS, Federer, W.T., McCulloch, C.E. and Miles-McDermott, N.J.	91
	1.10 REDUCING THE NOISE CONTAINED IN DESCRIPTIVE SENSORY DATA, Bett, K.L., Shaffer, G.P., Vercellotti, J.R., Sanders, T.H. and Blankenship, P.D.	109
	1.11 EXPERTS VERSUS CONSUMERS: A COMPARISON, Moskowitz, H.R.	127
2.0	DAIRY PRODUCTS, Gacula, M.C.	147
	2.1 SENSORY ASPECTS OF MATURATION OF CHEDDAR CHEESE BY DESCRIPTIVE ANALYSIS, Piggott, J.R. and Mowat, R.G.	149

- 2.2 SENSORY PROFILING OF DULCE DE LECHE, A DAIRY BASED CONFECTIONARY PRODUCT, *Hough, G., Bratchell, N. and Macdougall, D.B.* 163
- 2.3 SENSORY PROPERTIES OF FERMENTED MILKS: OBJECTIVE REDUCTION OF AN EXTENSIVE SENSORY VOCABULARY, *Hunter, E.A. and Muir, D.D.* 185
- 2.4 MEASURING SOURCES OF ERROR IN SENSORY TEXTURE PROFILING OF ICE CREAM, *King, B.M. and Arents, P.* 201
- 2.5 EFFECTS OF STARTER CULTURES ON SENSORY PROPERTIES OF SET-STYLE YOGHURT DETERMINED BY QUANTITATIVE DESCRIPTIVE ANALYSIS, *Rohm, H., Kovac, A. and Kneifel, W.* 219
- 2.6 THE USE OF STANDARDIZED FLAVOR LANGUAGES AND QUANTITATIVE FLAVOR PROFILING TECHNIQUE FOR FLAVORED DAIRY PRODUCTS, *Stampanoni, C.R.* 235

3.0 MEATS, *Gacula, M.C.* 253

- 3.1 DEVELOPMENT OF A TEXTURE PROFILE PANEL FOR EVALUATING RESTRUCTURED BEEF STEAKS VARYING IN MEAT PARTICLE SIZE, *Berry, B.W. and Civille, G.V.* 255
- 3.2 A STANDARDIZED LEXICON OF MEAT WOF DESCRIPTORS, *Johnsen, P.B. and Civille, G.V.* 267
- 3.3 DEVELOPMENT OF CHICKEN FLAVOR DESCRIPTIVE ATTRIBUTE TERMS AIDED BY MULTIVARIATE STATISTICAL PROCEDURES, *Lyon, B.G.* 275
- 3.4 A TECHNIQUE FOR THE QUANTITATIVE SENSORY EVALUATION OF FARM-RAISED CATFISH, *Johnsen, P.B. and Kelly, C.A.* 289

4.0 ALCOHOLIC BEVERAGES, *Gacula, M.C.* 301

- 4.1 SENSORY PROFILING OF BEER BY A MODIFIED QDA METHOD, *Mecredy, J.M., Sonnemann, J.C. and Lehmann, S.J.* 303
- 4.2 FACTOR ANALYSIS APPLIED TO WINE DESCRIPTORS, *Wu, L.S., Bargmann, R.E. and Powers, J.J.* 313

	4.3	DESCRIPTIVE ANALYSIS AND QUALITY RATINGS OF 1976 WINES FROM FOUR BORDEAUX COMMUNES, *Noble, A.C., Williams, A.A. and Langron, S.P.*	335
	4.4	SENSORY PANEL TRAINING AND SCREENING FOR DESCRIPTIVE ANALYSIS OF THE AROMA OF PINOT NOIR WINE FERMENTED BY SEVERAL STRAINS OF MALOLACTIC BACTERIA, *McDaniel, M., Henderson, L.A., Watson, Jr., B.T. and Heatherbell, D.*	351
	4.5	DESCRIPTIVE ANALYSIS OF PINOT NOIR WINES FROM CARNEROS, NAPA AND SONOMA, *Guinard, J.X. and Cliff, M.A.*	371
	4.6	DESCRIPTIVE ANALYSIS FOR WINE QUALITY EXPERTS DETERMINING APPELLATIONS BY CHARDONNAY WINE AROMA, *McCloskey, L.P., Sylvan, M. and Arrhenius, S.P.*	383
5.0	TEXTILE MATERIALS, *Gacula, M.C.*		403
	5.1	THE JUDGMENT OF HARSHNESS OF FABRICS, *Bogaty, H., Hollies, N.R.S. and Harris, M.*	405
	5.2	MEASUREMENT OF FABRIC AESTHETICS ANALYSIS OF AESTHETIC COMPONENTS, *Brand, R.H.*	417
	5.3	DEVELOPMENT OF TERMINOLOGY TO DESCRIBE THE HANDFEEL PROPERTIES OF PAPER AND FABRICS, *Civille, G.V. and Dus, C.A.*	443
6.0	GENERAL APPLICATIONS, *Gacula, M.C.*		457
	6.1	THE USE OF FREE-CHOICE PROFILING FOR THE EVALUATION OF COMMERCIAL PORTS, *Williams, A.A. and Langron, S.P.*	459
	6.2	A COMPARISON OF THE AROMAS OF SIX COFFEES CHARACTERISED BY CONVENTIONAL PROFILING, FREE-CHOICE PROFILING AND SIMILARITY SCALING METHODS, *Williams, A.A. and Arnold, G.M.*	477
	6.3	EVALUATION AND APPLICATIONS OF ODOR PROFILING, *Jeltema, M.A. and Southwick, E.W.*	493

6.4 COMPONENT AND FACTOR ANALYSIS APPLIED TO DESCRIPTORS FOR TEA SWEETENED WITH SUCROSE AND WITH SACCHARIN, Rogers, N.M., Bargman, R.E. and Powers, J.J. 507

6.5 INTENSITY VARIATION DESCRIPTIVE METHODOLOGY: DEVELOPMENT AND APPLICATION OF A NEW SENSORY EVALUATION TECHNIQUE, Gordin, H.H. 519

6.6 DEVELOPMENT OF A LEXICON FOR THE DESCRIPTION OF PEANUT FLAVOR, Johnsen, P.B., Civille, G.V., Vercellotti, J.R., Sanders, T.H. and Dus, C.A. 533

6.7 SENSORY MEASUREMENT OF FOOD TEXTURE BY FREE-CHOICE PROFILING, Marshall, R.J. and Kirby, S.P.J. 543

6.8 TASTE DESCRIPTIVE ANALYSIS: CONCEPT FORMATION, ALIGNMENT AND APPROPRIATENESS, O'Mahony, M., Rothman, L., Ellison, T., Shaw, D., and Buteau, L. 561

6.9 CONTROL CHART TECHNIQUE: A FEASIBLE APPROACH TO MEASUREMENT OF PANELIST PERFORMANCE IN PRODUCT PROFILE DEVELOPMENT, Gatchalian, M.M., de Leon, S.Y. and Yano, T. 595

6.10 COMPARISON OF THREE DESCRIPTIVE ANALYSIS SCALING METHODS FOR THE SENSORY EVALUATION OF NOODLES, Galvez, F.C.F. and Ressurreccion, A.V.A. 613

6.11 A COMPARISON OF FREE-CHOICE PROFILING AND THE REPERTORY GRID METHOD IN THE FLAVOR PROFILING OF CIDER, Piggott, J.R. and Watson, M.P. 627

6.12 DESCRIPTIVE ANALYSIS OF ORAL PUNGENCY, Cliff, M. and Heymann, H. 641

6.13 A COMPARISON OF DESCRIPTIVE ANALYSIS OF VANILLA BY TWO INDEPENDENTLY TRAINED PANELS, Heymann, H. 653

6.14 MULTIVARIATE ANALYSIS OF CONVENTIONAL PROFILING DATA: A COMPARISON OF A BRITISH AND A NORWEGIAN TRAINED PANEL, Risvik, E., Colwill, J.S., McEwan, J.A. and Lyon, D.H. 665

7.0   COMPUTER SOFTWARE, *Gacula, M.C.* 687

     7.1   SOFTWARE PACKAGES, *Gacula, M.C.* 689

INDEX 711

CHAPTER 1.0

INTRODUCTION

In the early 1950s, the Arthur D. Little company pioneered the Flavor Profile method (Cairncross *et al.* 1950; Caul *et al.* 1957, 1958; see Chapter 1.2) which became the foundation of the current descriptive sensory analysis techniques. Descriptive analysis is a sensory technique used to obtain an objective description of the sensory properties of various types of products and materials. Since the development of the Flavor Profile, new methods have evolved, such as the Quantitative Descriptive Analysis (Stone *et al.* 1992, see Chapters 1.3 and 1.6), the Spectrum Descriptive Analysis Method (Civille *et al.* 1991; Meilgaard *et al.* 1991), and the Free-Choice Profiling (see Chapter 6.7). The use of these methods in the sensory evaluation of various types of products is well-documented in both academic research and industrial work. A detailed discussion of the Flavor Profile, Quantitative Descriptive Analysis, and the Spectrum Descriptive Analysis Method is contained in an ASTM publication (Hootman 1992). Also discussed in this publication is the Texture Profile Method (Brandt *et al.* 1963) written by Munoz *et al.* (1992), which is not covered in this book because the documents pertaining to this subject could make a book by itself.

Since its development, descriptive analysis has been successfully used in quality control to maintain sensory quality characteristics of products, in comparison of product prototypes, in understanding consumer responses in relation to product sensory attributes, in exploring the marketplace by sensory mapping so that gaps and opportunities in the map can be examined for possible development of new products, and in product matching, useful for claims substantiation and product improvement.

The success of the use of descriptive analysis depends on four factors: the training and experience of the judges, the panel leader, the sensory execution, and a long-term commitment by company management. Training is product-dependent because the sensory attributes vary among products, i.e., attributes for lotion products are different than those of wines. The length of training also depends on the product; some products require longer training than others. An experienced judge, by virtue of product exposure and product usage, should not be considered a trained judge, because they were not taught in scaling procedures, attribute definition, and other aspects of product-related training. The ideal situation is the existence of experienced and trained judges in an organization. The panel leader or program administrator has a critical role in the establishment and maintenance of a descriptive analysis panel, particularly in maintaining motivation of panel members. Sensory execution would include the choice of reference standards, conduct of the test, and test design. These factors are exemplified in several chapters of the book as applied to various types of products and experimental conditions. The last factor, management commitment, is the prime mover for a successful sensory program in both academia and industry. Development of a descriptive analysis program, as everyone knows, requires time and a special physical facility that requires capital investment.

Consumer testing is generally expensive compared to a descriptive analysis, hence, in product development, descriptive analysis is done first to screen and eliminate prototypes that do not meet the prescribed sensory criteria. A research guidance panel type of study is conducted to determine consumer liking for these prototypes, and the resulting data are correlated with the data from descriptive analysis.

The product development process is more effective when prototypes have undergone thorough descriptive analyses before subjecting the product to a marketing consumer test, such as a central location test (CLT). It is important in this type of application that results from descriptive analysis must be predictive of consumer test results, hence, the development of descriptive analysis must be consumer-oriented. However, there are products that cannot be packaged for laboratory testing because of the expense involved. In this case a surrogate package is used during sensory evaluation that simulates consumer use of the product. Remember that the ultimate goal is to produce a robust prototype from descriptive analysis.

An important application of descriptive analysis is in sensory evaluation of samples from formula optimization studies that utilize the principles of design of experiments (DOE). The use of DOE in product development is highly recommended because it is more efficient and in the long run, less costly than the traditional one-variable at a time approach. Although, the initial number of samples (design points) to be evaluated is larger than the traditional approach, the repetition of the study would be unlikely; it is more efficient in the sense that the effects of more than one ingredient in the formulation can be studied simultaneously. Tables 1.0.1 and 1.0.2 show the number of samples in a mixture experimental design according to the number of ingredients to be studied in the formula. In using mixture designs, it is important to know that the response to be measured is dependent on ingredient proportion rather than amount, otherwise the mixture design will not apply and the response surface design should be used. Examples of responses that depend on the amount of ingredient in the formula are fertilizer experiments and the level of salts in an antiperspirant formula. In sensory optimization studies it is highly recommended that a control sample should be included in the experiment.

With a properly designed study, sensory attributes from descriptive analysis can be simultaneously optimized to obtain a number of optimal formulas for testing by the research guidance panel. Several DOE useful in formula optimization work are given in Gacula and Singh (1984) and Gacula (1993). DESIGN-EXPERT and DESIGN-EASE (Stat-Ease, Inc) are software packages that can generate experimental designs based on the objectives and types of studies. These are illustrated in Chapter 7.

TABLE 1.0.1
A THREE-INGREDIENT MIXTURE DESIGN

Design points	Ingredient A	Ingredient B	Ingredient C
1	1	0	0
2	0	1	0
3	0	0	1
4	.5	.5	0
5	.5	0	.5
6	0	.5	.5
7	.33	.33	.33

Total number of samples = 7 + Control = 8. Coded ingredient levels are shown.

TABLE 1.0.2
A TWO-INGREDIENT MIXTURE DESIGN

Design points	Ingredient A	Ingredient B
1	1	0
2	.67	.33
3	.5	.5
4	.33	.67
5	0	1

Total number of samples = 5 + Control = 6. Coded ingredient levels are shown.

REFERENCES

BRANDT, M.A., SKINNER, E.Z. and COLEMAN, J.A. 1963. Texture Profile Method. J. Food Sci. *28*, 404-409.

CAIRNCROSS, S.E. and SJÖSTROM, L.B. 1950. Flavor Profiles—A new approach to flavor problems. Food Technol. *4*, 308-311.

CAUL, J.F. 1957. The Profile Method of flavor analysis. Advances in Food Research *7*, 1-40.

CAUL, J.F., CAIRNCROSS, S.E. and SJÖSTROM, L.B. 1958. The Flavor Profile in review. Flavor Research and Food Acceptance, Arthur D. Little, Inc., Reinhold Publishing Co., New York.

CIVILLE, G.V. and DUS, C.A. 1991. Evaluating tactile properties of skincare products: A descriptive analysis technique. Cosmetic and Toiletries *106*, 83-88.

GACULA, JR., M.C. and SINGH, J. 1984. *Statistical Methods in Food and Consumer Research*. Academic Press, San Diego, CA.

GACULA, JR., M.C. 1993. *Design and Analysis of Sensory Optimization*. Food & Nutrition Press, Trumbull, CT.

HOOTMAN, R.C. (ed.). Manual on Descriptive Analysis Testing for Sensory Evaluation. ASTM Manual Series: MNL 13, ASTM, Philadelphia, PA.

MUNOZ, A.M., SZCZESNIAK, A., EINSTEIN, M.A. and SCHWARTZ, N.O. 1992. The Texture Profile. In *Manual on Descriptive Analysis Testing for Sensory Evaluation*, (R.C. Hootman, ed.), ASTM Manual Series: MNL 13. ASTM, Philadelphia, PA.

MEILGAARD, M., CIVILLE, G.V. and CARR, B.T. 1991. Sensory Evaluation Techniques. CRC Press, Boca Raton, FL.

STAT-EASE, Inc. Minneapolis, MN.

STONE, H. 1992. Quantitative Descriptive Analysis (QDA). In *Manual on Descriptive Analysis Testing for Sensory Evaluation*, (R.C. Hootman, ed.), ASTM Manual Series: MNL 13. ASTM, Philadelphia, PA.

CHAPTER 1.1

DESCRIPTIVE SENSORY ANALYSIS METHODS

In this book, three methods will be briefly presented because several works have already been published describing the details and use of these methods. There are similarities among the methods, but they differ in sensory philosophy, length of training, presentation of results, and sensory scales. The three methods are thoroughly described in a publication edited by Hootman (1992) and in other publications by Heymann *et al.* (1993), Powers (1988), and Einstein (1991). Unlike the cited publications, detailed applications of these methods in various experimental situations are reported in various chapters of the book.

Flavor Profile and Profile Attribute Analysis

The Flavor Profile Method (FP), developed by the Arthur D. Little, Inc., in 1949, was the first technique to assess the flavor and aroma impressions of food products. An extension of the Flavor Profile is the Profile Attribute Analysis (PAA), which incorporates numerical aspects of sensory description. As a result, standard statistical methods, such as analysis of variance, factor analysis, principal component analysis, and others are used to analyze the data. A detailed discussion of both FP and PAA are given by Neilson *et al.* (1988).

Quantitative Descriptive Analysis

The Quantitative Descriptive Analysis (QDA) was developed by the Tragon Corporation in the mid-1970s to address the problem of quantifying sensory description. As a means of quantifying sensory perception, an unstructured line scale is used that approaches a continuous scale, an important property that permits the use of standard statistical procedures. The spider plot, that characterizes QDA, is used as a graphical tool for presenting the results. Plotting can be accomplished by using the Microsoft Excel. See Chapters 1.3 and 1.6 for the original articles pertaining to QDA and that by Zook and Pearce (1988).

The Spectrum Descriptive Analysis Method

This method was developed by the Spectrum, Inc. in the late 1970s. Like the QDA and PAA, it also utilizes statistics to analyze the data obtained from a line scale anchored on both ends. Bar charts are used to portray the data, and again can be accomplished by the Microsoft Excel. See Meilgaard *et al.* (1991) for a thorough description and applications of the Spectrum Descriptive Analysis Method. Refer also to a paper by Civille *et al.* (1991).

Variants of Descriptive Analysis

The Free-Choice Profiling is a popular method which, unlike the traditional methods, uses untrained judges for evaluating products. A special type of statistical analysis is used known as Procrustes analysis that accounts for the effect of using untrained judges. See Chapters 6.1, 6.2, 6.7, and 6.11 for its application. Another variation is the Repertory Grid given in Chapter 6.11 and that for tobacco evaluation reported by Gordin (1987) given in Chapter 6.5.

Overview of Statistical Analyses

The most popular statistical methods are analysis of variance, factor analysis, principal component analysis, and regression analysis. The applications of these methods are described in various chapters of the book. The analysis of variance is a well-known method that breaks down the total variation into several sources. It is mainly used in hypothesis testing, i.e., test of significant difference between products, test of significant difference between panelists, etc. Another application of the analysis of variance is in the estimation of variance components. In this application, one desires to determine the percentage contribution of each source of variation to the total variability. This application is illustrated by Finkey *et al.* (1987) and in Gacula and Singh (1984). A comprehensive discussion of data relationships between descriptive analysis and consumer testing is given in an ASTM publication edited by Munoz (1997).

Regression and Correlation Analyses. The initial analyses in relating data obtained by descriptive analysis and consumer testing are regression and correlation analyses. Since different panels are used on both data sets, the input data are product means for each attribute. The data structure is shown in Table 1.1.1. It is desirable that many products with varying degrees of attribute intensities should be used. The variation in intensities will provide a better definition of attribute relationships. If the range of variation is not sufficient, misleading results may occur. The initial analysis is a simple linear regression,

$$Y_{ij} = B_o + B_1 X_{ij} + E_{ijm} \quad \begin{matrix} i = 1,2, .., \text{kth product} \\ j = 1,2, .., \text{nth attribute} \\ m = 1,2, .., \text{lth rating} \end{matrix} \quad \text{(Eq. 1.1.1)}$$

where Y_{i1} is the overall liking mean score, B_o is the intercept, X_{ij} is the kth sensory attribute, and E_{ijm} is random error. The plot between Y_{ij} and X_{ij} provides an initial view of the relationship, how the products are positioned against attribute X_{ij}. The STATISTIX software, among others, can be used to provide the scatterplot of the mean scores with the regression line superimposed (Fig. 1.1.1). In this example, the overall liking for the product increases with increasing score intensity of attribute X8.

TABLE 1.1.1.
LAYOUT OF OBSERVATIONS FOR REGRESSION ANALYSIS
OF DATA OBTAINED FROM DESCRIPTIVE ANALYSIS AND CONSUMER TEST:
5 PRODUCTS, 6 ATTRIBUTES (X1-X6)

Products	X1	X2	X3	X4	X5	X6
1	y11	y12	y13	y14	y15	y16
2	y21	y22	y23	y24	y25	y26
3	y31	y32	y33	y34	y35	y36
4	y41	y42	y43	y44	y45	y46
5	y51	y52	y53	y54	y55	y56

Note: y11, . ., y56 are mean scores.

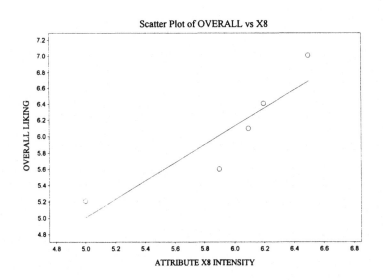

FIG. 1.1.1. SCATTERPLOT OF PRODUCT MEANS (○) USING THE STATISTIX SOFTWARE
Products plotted are with increasing amount of flavor additive.

The next step is to look at all the sensory attributes at the same time by running a stepwise regression analysis. The result of the analysis will provide the important attributes that predict overall liking. There are choices in the model to use in stepwise regression analysis. One must consult a statistician regarding the choice of the model. An example of a stepwise regression analysis is given in Chapter 7.

Factor Analysis and Principal Components. Descriptive sensory and consumer data are characterized by the presence of correlations among attributes in a given product. This correlation arises from the process of sensory evaluation which involves a "memory capacity," i.e., context effect, during evaluation of samples or products, as opposed to the use of a mechanical instrument. In addition, there is an intrinsic relationship in most sensory attributes of products due to synergistic or antagonistic effect of various ingredients used in product formulation.

To make use of the correlation among observations, multivariate statistical procedures are used in addition to univariate methods. Factor analysis and principal component analysis are the most common multivariate procedures used in the industry and academia for the analysis of this type of data. Another procedure gaining popularity is the Procrustes analysis described in Chapters 6.1, 6.2, 6.7, and 6.11. Multivariate methods have been known for a long time and books that vary in statistical complexity are widely available (Anderson 1958; Harman 1976; Morrison 1967; Afifi and Azen 1979). Briefly, let us discuss the methods of statistical analyses.

Factor analysis and principal component analysis are similar in many ways, the major similarities being that both methods make use of the correlation (variance-covariance) among attributes, and both methods have the objectives of reducing the number of attributes into a new set of attributes, the so-called factors or components. The reduction is expected to retain as much information in the original variables or attributes as possible. The resultant components, which are now a linear combination of the original attributes, are uncorrelated (statistically orthogonal). For example, there may be 20 original attributes used in rating the products; by using a principal component or factor analysis, the original number of attributes of 20 may be reduced, say, to five components. That is the data or the products can now be represented by these five components instead of 20, making the relationships among products and among attributes easily visualized and more manageable. Then the five components are given a hypothetical descriptive name in relation to the sensory and physical characteristics of the products.

The first principal component, PC1, accounts for the largest variance in the data; PC2, which is uncorrelated with PC1, the second largest; PC3, which is uncorrelated to PC1 and PC2, the third largest, and so on in a decreasing variance order. Suppose that PC1 consists of the following attributes: gentle to the gum, cleans teeth, bristle density, bristle stiffness. A descriptive name may be a mouthfeel component.

An important use of principal components is in the comparison of products based on principal component scores. This is accomplished by statistical conversion of the original ratings into principal component scores associated with each product. The PC scores can be used to correlate with consumer liking to aid product formulations and/or product improvements. Statistically, the use of PC scores in multiple regression analysis is not biased by collinearity because the principal components are uncorrelated. Furthermore, a multiple comparison tests of PC scores can be done to provide a separation of the products, the separation of which is based on the integrated sensory dimension—the principal components. This application is illustrated in Chapter 7.

When there are no significant differences among product means, a principal component analysis defines the overall sensory dimensions of the data. When products are similar, the plot of the products, for example PC1 and PC2 would cluster around the (0,0) coordinate (Fig. 1.1.2). This type of analysis is useful in a study dealing with ingredient substitution and/or ingredient change in a formula; the analysis provides assurance that the overall sensory properties of the products did not change by the ingredient substitution. The traditional method of analysis has been the use of difference tests. The PC analysis is also useful in product matching studies. When products are dissimilar in many sensory characteristics, the products on the plot would scatter (Fig. 1.1.3), hence there is no match; on the contrary, the plot in Fig. 1.1.2 would indicate a reasonable match among products.

One of the differences between factor analysis and principal component analysis is in the model. For the principal component analysis, the model is

$$\begin{bmatrix} Y_1 \\ Y_2 \\ \cdot \\ \cdot \\ \cdot \\ Y_p \end{bmatrix} = \begin{bmatrix} A_{11} & A_{12} & \ldots & A_{1p} \\ A_{21} & A_{22} & \ldots & A_{2p} \\ \cdot & & & \\ \cdot & & \cdot & \\ \cdot & & & \\ A_{p1} & & & A_{pp} \end{bmatrix} \begin{bmatrix} X_1 \\ X_2 \\ \cdot \\ \cdot \\ \cdot \\ X_p \end{bmatrix} \quad \text{(Eq. 1.1.2)}$$

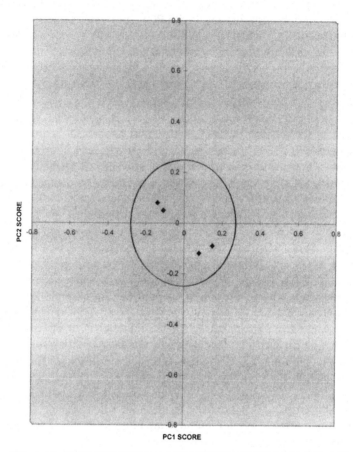

FIG. 1.1.2. PLOT OF PRINCIPAL COMPONENT SCORES PC1 AND PC2 FOR PRODUCTS (♦) WHICH SHOW NO SIGNIFICANT DIFFERENCES

where Y_i, $i = 1, 2, .., p$, are the linear combinations of the original attributes; thus, Y_1 is the first principal component, Y_2 is the second principal component, and so on; the estimates of Y_i are uncorrelated; X_p is the observed sensory ratings of the attributes. The term A_{pp} is the coefficient (mathematically known as eigenvector) that needs to be obtained by solving Eq. (1.1.3). In matrix notation, the model is

$$Y = AX \qquad \text{(Eq. 1.1.3)}$$

where Y is a p × 1 matrix, A is a p × p matrix, and X is a p × 1 matrix. As one can see, principal components Y_i are statistical functions of the observed

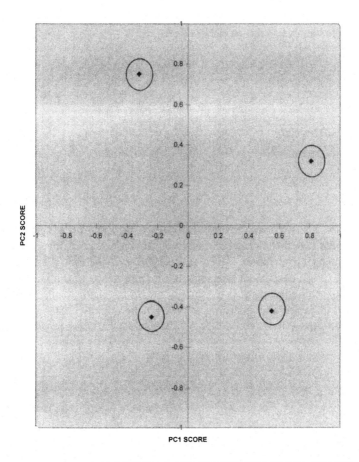

FIG. 1.1.3. PLOT OF PRINCIPAL COMPONENT SCORES PC1 AND PC2 FOR PRODUCTS (♦) WHICH SHOW SOME SIGNIFICANT DIFFERENCES

variables X_p. Thus, the equation for the first principal component is

$$Y_1 = A_{11}X_1 + A_{12}X_2 + \ldots + A_{1p}X_p \qquad \text{(Eq. 1.1.4)}$$

and for the second principal component,

$$Y_2 = A_{21}X_1 + A_{22}X_2 + \ldots + A_{2p}X_p$$

and so on for the remaining components. Substitution of standardized ratings into the above equations produces the principal component scores. There is software available to solve Eq. (1.1.3) and the SAS software (SAS 1990) is used in this book (see Chapter 7).

In factor analysis, the model is

$$\begin{bmatrix} X_1 \\ X_2 \\ \cdot \\ \cdot \\ \cdot \\ X_p \end{bmatrix} = \begin{bmatrix} A_{11} & A_{12} & \ldots & A_{1m} \\ A_{21} & A_{22} & \ldots & A_{2m} \\ \cdot & & \cdot & \\ \cdot & & \cdot & \\ \cdot & & \cdot & \\ A_{p1} & & & A_{pm} \end{bmatrix} \begin{bmatrix} Y_1 \\ Y_2 \\ \cdot \\ \cdot \\ \cdot \\ Y_m \end{bmatrix} + \begin{bmatrix} E_1 \\ E_2 \\ \cdot \\ \cdot \\ \cdot \\ E_p \end{bmatrix} \quad \text{(Eq. 1.1.5)}$$

where X_p, $i = 1, 2, .., p$, are the observed ratings for p sensory attributes, Y_j, $j = 1, 2, .., m$, are called the principal/common factors extracted from the original number of sensory attributes, given as rotated factor pattern in a SAS output. The term A_{pm}, is a coefficient that reflects the importance of the ith attribute on the jth factor; this coefficient is commonly known as factor loadings. The term E is the error component unaccounted for by the common factors. In matrix notation, the model is

$$X = AY + E \quad \text{(Eq. 1.1.6)}$$

where X is a p × 1 matrix, A is a p × p matrix, Y is a p × 1 matrix, and E is also a p × 1 matrix. Factor scores are obtained by substituting the ratings into the resultant common factor equation given as standardized scoring coefficients in a SAS output. Again, the SAS software will be used to evaluate Eq. (1.1.6).

Another difference is that in factor analysis it is assumed that some underlying factors which are smaller than the number of the observed variables (m < p), are responsible for the correlations among the observed variables. The SCREE plot in SAS provides the appropriate number of underlying factors for inclusion (see Chapter 7). In the principal component analysis, this assumption is not made, instead the total variation in the data is exhaustively divided into component parts; that is why the error term E is not shown in Eq. (1.1.3). The choice between principal component and factor analysis depends on the purpose of the statistical evaluation. It is not an easy choice because of their similarities; the results of statistical analyses by both methods may differ in some degree due to the type of mathematical rotations used in the analysis. In sensory evaluation, it is a common practice to combine sensory attributes into integrated or composite attributes (underlying factors), hence the factor analysis may be the appropriate choice. On the other hand, the principal component analysis can also be used by specifying in the SAS code the number of components to be included in the analysis.

REFERENCES

AFIFI, A.A. and AZEN, S.P. 1979. *Statistical Analysis: A Computer Oriented Approach*. Academic Press, San Diego, CA.

ANDERSON, T.W. 1958. *An Introduction to Multivariate Statistical Analysis*. John Wiley & Sons, New York.

CIVILLE, G.V. and DUS, C.A. 1991. Evaluating tactile properties of skincare products: A descriptive analysis technique. Cosmetic and Toiletries *106*, 83-88.

EINSTEIN, M.A. 1991. Descriptive Techniques and their Hybridization. In *Sensory Science Theory and Applications in Foods*, (H.T. Lawless and B.P. Klein, eds.), Marcel Dekker, New York.

FINKEY, M.B., GACULA, JR., M.C. and MILLER, J.K. 1987. Site to site variation in the soap chamber test. J. Sensory Studies *2*, 293-300.

GACULA, JR., M.C. and SINGH, J. 1984. *Statistical Methods in Food and Consumer Research*. Academic Press, San Diego, CA.

HARMAN, H.H. 1967. *Modern Factor Analysis*. The University of Chicago Press, Chicago, IL.

HEYMANN, H., HOLT, D.L. and CLIFF, M.A. 1993. Measurement of Flavor by Sensory Descriptive Techniques. In *Flavor Science — Sensible Principles and Techniques*, (T.E. Acree and R. Teranishi, eds.), ACS, Washington, DC.

HOOTMAN, R.C. (ed.). Manual on Descriptive Analysis Testing for Sensory Evaluation. ASTM Manual Series: MNL 13. ASTM, Philadelphia, PA.

MEILGAARD, M., CIVILLE, G.V. and CARR, B.T. 1991. *Sensory Evaluation Techniques*. CRC Press, Boca Raton, FL.

MORRISON, D.F. 1967. *Multivariate Statistical Methods*. McGraw-Hill, New York.

MUNOZ, A.M. (ed.). 1997. Relating Consumer, Descriptive, and Laboratory Data to Better Understand Consumer Responses. ASTM Manual 30. ASTM, West Conshohocken, PA.

NEILSON, A.J., FERGUSON, V.B. and KENDALL, D.A. 1988. Profile Methods: Flavor Profile and Profile Attribute Analysis. In *Applied Sensory Analysis of Foods*, (H. Moskowitz, ed.), CRC Press, Boca Raton, FL.

POWERS, J.J. 1988. Current Practices and Application of Descriptive Methods. In *Sensory Analysis of Foods*, (J.R. Piggott, ed.), Elsevier Applied Science, London.

SAS. 1990. SAS/STAT User's Guide, Ver. 6, 4th Ed. SAS Institute, Inc., Cary, NC.

ZOOK, K.L. and PEARCE, J.H. 1988. Quantitative Descriptive Analysis. In *Applied Sensory Analysis of Foods*, (H.R. Moskowitz, ed.), CRC Press, Boca Raton, FL.

CHAPTER 1.2

FLAVOR PROFILES
A NEW APPROACH TO FLAVOR PROBLEMS

S.E. CAIRNCROSS and L.B. SJÖSTRÖM

*At time of original publication,
all authors were at Arthur D. Little, Inc.
Cambridge, Massachusetts.*

The Flavor Profile presents a descriptive analysis of flavor expressing in common language terms the characteristic notes of both aroma and flavor, their order of appearance and intensities, and the amplitudes of total aroma and total flavor. The method enables a panel to evaluate the small differences which occur in the usual steps of flavor improvement. The concept furnishes insight to the philosophy of seasoning.

INTRODUCTION

The Flavor Profile represents a method of flavor analysis which makes it possible to indicate degrees of difference between samples on the basis of the intensity of individual character notes, the degree of blending and the overall amplitude. In the study of flavor improvement, laboratory panels often operate on the basis of preference. Secondarily, descriptive terms are used to indicate points of difference, but the interpretation of such observations as a guide to product improvement is usually difficult, especially in a prolonged series and in a series where small differences exist. In reviewing the efforts of the flavor group in these laboratories, in this direction, as well as those of others, it is apparent that this type of approach makes progress by minor degrees of improvement difficult and at times appears to lead in circles. The need has been felt for a more objective method of judging products which does not depend upon personal preference and which integrates the points of difference in such a way that products can be judged separately as well as in groups.

Sensory studies of the effect of monosodium glutamate on food flavor sharply focused attention on the inadequacy of any known method of expressing the phenomenon of the blending of flavor. Similarly, this work brought a great deal of attention to bear on the mechanism of seasoning, particularly with respect to the augmentation of certain character notes and the suppression of others. In the course of this work, a philosophy of seasoning has been developed which attempts to describe the general mechanism of the seasoning and flavoring of foods. In order to be able to analyze and describe the complex effects of various seasoning agents, it was decided to formalize a method of expression and description which was being used by the panels in a general way. This complete description of aromatic and flavor characteristics has been given the name of "Flavor Profile". The "Flavor Profile", which is a sort of flavor spectrum, may be expressed in either diagrammatic or tabular form. In the latter case it resembles a system of grading with a semiquantitative indication of the variation in the principal character notes. While the system has not yet been developed into a fully quantitative procedure, whenever necessary, measurement of the degree of intensity of any one factor can readily be reduced to quantitative evaluation by the use of outside standards. In any given series of products being examined for the first time, it is customary to use a rather complete tabular description, but eventually it is possible to concentrate upon only the principal points. Saltiness, sweetness and certain aromatic character notes may be judged individually, and by matching samples with outside standards these can be given fairly reproducible quantitative ratings.

It is the authors' belief that this method is the most flexible and adaptable method at present available for the operation of small research panels, particularly as a guide to flavor improvement studies. This procedure has been found useful in work on such diversified materials as: frozen fruits, dairy products, chewing gum, vitamin tablets, coffee, rum, beer, and soft drinks. In all of these products it is possible for a small panel to render descriptive analyses which are readily readable, understood and applied by both production and management. Although the method is still in a formative phase of development, it is felt that there should be no further delay in presenting it to other workers in the field who might wish to experiment with its applications.

In this presentation of the Flavor Profile it is desirable to stress three basic considerations of panel operation: training of panel members, development of descriptive terms and interpretation of results.

TRAINING

Frequently it is common practice in flavor development work to expend much effort in attacking a problem and then to squeeze the evaluation panels into a crowded schedule, drafting panel members from occupations that they may consider more important. In flavor work at the Arthur D. Little, Inc. laboratories equal weight is given to the developmental and evaluation procedure, emphasizing scheduling and adequate time for panel sessions. This promotes full concentration, fosters interest and avoids the viewpoint that panel sessions are merely interruptions of regular work.

Since flavor analyses are no different from other laboratory analyses in requiring time for developing technique, persons who are to become members of a flavor panel are allowed ample opportunity to learn the fundamentals of taste testing. They sit in on panel sessions and are introduced to the ramifications of many flavor problems, in course acquiring the necessary experience and confidence. Likewise, every panel member is allotted time to become acquainted with each new flavor problem under consideration. He is encouraged to study the material by himself and later compare findings with fellow members. At the end of this acquaintance period all panel members together in seminar fashion prepare the Flavor Profile for the material which they have studied.

TERMINOLOGY OF QUALITATIVE ASPECTS

Since each material possesses flavor and aroma characteristics which in type and intensity distinguish it from any other material, great emphasis is therefore given to exact descriptive terminology. The objective should be to use terms that most nearly characterize each component note.

The establishment of descriptive terminology in each flavor study begins when the panel members individually examine the material. While principal attention is given to aroma and flavor, other sensory phenomena may also be observed. In these laboratories aroma is defined as the odor impressions perceived in the nose by sniffing; for example, burnt, fragrant, pungent, sharp, and acidic. Taste is used to refer only to the basic factors perceived in the mouth: sweet, sour, salty, and bitter. Flavor or flavor by mouth includes those impressions received when the substance is taken into the mouth: the basic taste factors, odor, and feeling sensations, and often aftereffects.

It is customary to use associative terms such as "eggy", "rubbery", "cabbage-like" and "skunky" for the naturally-occurring odors due to sulfides and organic sulfur compounds. However, direct association with a definite chemical is attempted whenever possible; for example, phenylacetic acid has been the reference odor for a principal aromatic note in both beer and honey. An attempt is made to analyze the more important factors specifically, for instance, in the case of astringency, location of the effect is cited, as dry mouth, cheek puckering, tongue coating, and tooth roughing.

Useful descriptive terminology for a given product can usually be established after several panel sessions. The principal components of aroma and flavor that have been noted during the preliminary study are introduced and checked at round table discussions during which all panel members come to understand the meaning of each descriptive term, and adopt common terminology. It is understood that the panel be fairly homogeneous as to personality and training in order to encourage complete freedom of expression during all discussions. Every effort is made to establish the ideal conditions for round table discussion, avoiding pressure and domination.

While determining the nature of the principal components of flavor and odor, the tasters endeavor to establish the order of appearance and intensities of the individual notes as well as the overall amplitude of aroma and flavor. As will be shown later, the order of appearance of flavor notes in beer begins with salt, sweet, sour, and fruity.

QUANTITATIVE

Intensity of individual character notes is judged initially by an arbitrary scale based upon the recognition threshold, using the following designations: not detectable, just detectable, slightly strong, moderately strong, and strong.

Amplitude is regarded as the summation of all intensities and is applied to the overall impression of aroma body (total aroma) and flavor body (total flavor). In the case of a series, the sample having the least body is arbitrarily given unit amplitude, and the remainder rated accordingly.

To summarize, the Flavor Profile presents a descriptive analysis of flavor expressing in common language terms the characteristic notes of both aroma and flavor, their order of appearance and intensities, and the amplitudes of total aroma and flavor.

From this tabulation a response sheet is prepared which is intentionally simple, for elaborate rating sheets have been found to divert the attention of the analyst and complicate interpretation of results.

Figure 1 shows the response sheet developed for beer by the Flavor Profile method during an experimental program centering on beer flavor. A sheet like this is presented to each panel member who individually examines the sample at hand and records his findings. First he checks the amplitude of aroma and then studies aroma, looking for component notes indicated on the left of the sheet, and records their intensities in the parallel column in the middle. Other notes perceived can be recorded. Similarly he studies the flavor by mouth. Frequently in a flavor problem it is necessary to follow one characteristic intensively. For example, in Fig. 1 there is plotted a curve which indicates the amplitude of flavor with respect to its duration. In beer and soft drinks as well, the rate at which maximum intensity is reached and rate of disappearance of total flavor apparently are important factors in product appeal.

EXAMINATIONS OF BEER

Test No._____ Date_____ Signature_____

Aroma	Intensity	Amplitude
Hop Fragrance		
Estery		
Fruity		
Yeast		
Sour		
Malt		
Hop Resin		
Phenylacetic Acid		
Gassy (Illuminating)		
Skunk		
Others		

Flavor By Mouth	Intensity	Amplitude
Salt		
Sweet		
Sour		
Fruity		
Bitter		
Yeast		
Malt		
Phenylacetic Acid		
Skunk		
Metallic		
Astringent (Dry Throat)		
Others		

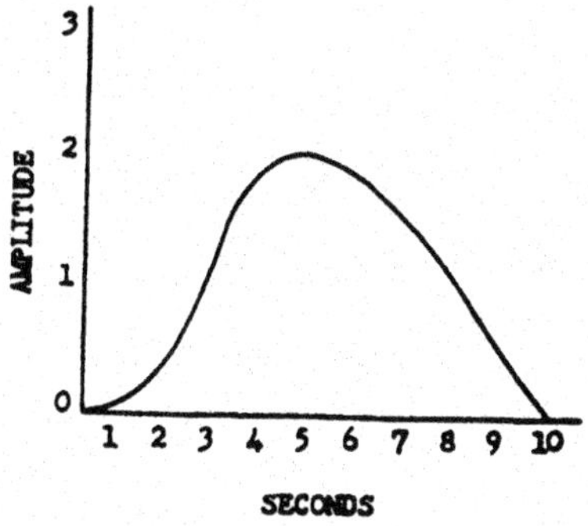

FIG. 1. RESPONSE SHEET DEVELOPED FOR THE EXAMINATION OF BEER

INTERPRETATION OF RESULTS

Upon completion of the individual examinations the findings are discussed in open panel, one person in the group acting as moderator and recorder. This discussion at the end of the closed panel is a unique feature of the Flavor Profile method, stimulating panel members to increase their acuteness and reliability, also fostering interest and self-confidence. Here, the moderator resolves differences existing as to the intensity of notes by resubmission of samples. Thus, a single composite Flavor Profile is obtained.

To determine the effect of an experimental modification of the sample, it is only necessary to compare its composite profile with that of the original sample.

With regard to size of panel, we have used groups of six to ten people and find that six is a practical panel for most purposes. The panels at the Flavor Laboratories of Arthur D. Little, Inc. are generally made up of chemists, but there is no reason why any intelligent non-technical group could not operate according to the same procedure, using descriptive terms common to their own experience.

APPLICATION OF FLAVOR PROFILES

In general, studies of flavor and aroma may be classified as corrective, formulative and analytical. The flavor profile method has been found particularly valuable in steering a program of flavor correction for foods and pharmaceuticals. For example, in exploring certain pharmaceuticals the two outstanding notes were a rubbery, thiol note and a sour-yeasty note. The profile method showed that some additives were capable of suppressing one factor and not the other, and some covered both. These findings enable one to indicate small increments of improvement effected by different additives in such a way that combinations can be developed to include several different partial effects in a cumulative fashion.

The effect of partials as encountered in formulation studies illustrates the authors' concept of flavor construction. In developing a soft drink flavor one objective is to produce a high-body drink with distinctive interest factors. This may be achieved by the careful selection and blending of ten to twenty ingredients each of which, while not identifiable, blends with the others and contributes to the total flavor. Then several compatible aromatic flavorants are allowed to project beyond the flavor body, adding interest, identity and character.

If one were to attempt to diagram the flavor using the Flavor Profile, a semicircle of fairly large amplitude could represent the complex flavor body, and beyond would project the several outstanding notes. The majority of the ingredients would be at near-threshold level where they are not readily identifiable. Each by itself would be of little consequence in producing character, but by the addition of many partials an interesting and blended complex is built.

Figure 2 is a diagram of the profiles for both the aroma and the flavor of a carbonated beverage, the amplitudes being represented by the areas of the semicircles. Thresholds are indicated by the perimeters of the semicircles, and the individual notes perceived are designated by the solid lines that originate at the center. Some of the lines reach just to the perimeter. These notes are just at recognition level. The distance to which the other lines extend beyond the threshold border indicates detectable intensity of the notes they represent.

Not all of the formula ingredients of the beverage are identifiable, for the majority go to make up the blended body; however, it is probable that several are represented by one line. For example, in the aroma diagram, fragrance, as one of the perceptible notes, might be attributable to unrecognizable quantities of several different aromatics.

FIG. 2. DIAGRAMMATIC REPRESENTATION OF THE AROMA AND FLAVOR PROFILES OF A CARBONATED BEVERAGE

For the past three years these laboratories have been studying the effect of monosodium glutamate on a variety of foods. Certain of the typical seasoning properties of this salt (increase of blending and increased amplitude) are shown in Figure 3, which is the diagrammatic flavor profile for boiled, seasoned, fresh summer squash with and without 0.5 percent added monosodium glutamate.

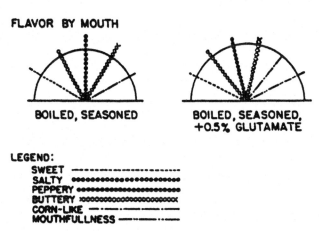

FIG. 3. DIAGRAMMATIC REPRESENTATION OF THE FLAVOR PROFILES OF SEASONED SUMMER SQUASH

The analytical panel found that first sweetness was perceptible in the boiled squash, seasoned with 0.5 percent salt, 0.05 percent pepper and 1 percent butter. After sweetness they tasted salt, a moderate amount of pepper and butter, and finally the squash itself which was found to resemble sweet corn. In the sample containing the added glutamate, seasoned and tasted at the same time, the outstanding quality was the blended character of the component notes, which is indicated by the augmented size of the semi-circle and the reduction of the outstanding notes. The order of perception of the flavor components was not altered but another factor, mouthfullness, was added.

Diagrammatic representations such as these profile drawings are useful in gaining the understanding of persons not versed in flavor techniques, for they are visual summaries and express fundamental concepts of flavor. These together with the tabular data sheets are valuable in providing management groups with a greater understanding of their own flavor problems and of the alternatives presented by research and production.

Inherent in any successful system of seasoning and flavoring is the building of an interesting complex of flavor. This is accomplished by the increase of blending, the building of greater amplitude, and the addition of interest factors. Such an approach allows for suppression of undesirable notes and augmentation of the desirable ones. This concept supplies a working scheme and philosophy to be followed in all problems of flavoring and seasoning. The Flavor Profile method is a means of indicating degrees of success in the development and control of optimum flavor.

CHAPTER 1.3

SENSORY EVALUATION BY QUANTITATIVE DESCRIPTIVE ANALYSIS

HERBERT STONE, JOEL SIDEL, SHIRLEY OLIVER, ANNETTE WOOLSEY
and RICHARD C. SINGLETON

*At time of original publication,
all authors were affiliated with
the Dept. of Food and Plant Sciences
Stanford Research Institute
Menlo Park, CA 94025.*

In the evolution of sensory evaluation techniques, there has been a strong orientation toward the stimulus and its properties as well as the receptor and its properties. To a considerable extent, this has meant a lack of appreciation for the behavioral aspects of what is perceived. In this article, we discuss a technique for characterizing the perceived sensory attributes of a product in quantitative terms. The focus of this technique is on the psychophysical aspects of perception and the application of an interval scaling technique to the problem of flavor characterization.

THE TRADITIONAL VIEW

Characterization of the perceived flavor of a food is a complex task (Amerine *et al.* 1965). Traditionally, each company had in its employ one or more individuals who were considered "experts." Having tasted, smelled, touched, etc., a material along with many other similar materials, this individual established an impressive "catalog" of information as to the ultimate quality of this material and thus became the judge of "standard of excellence."

The judgment of the expert (or collective judgments of selected experts) was relied on for all manner of decisions, including which of several ingredients or products was most appropriate for the marketplace. This traditional situation is shown in Figure 1.

Undoubtedly, an individual who has spent many years using his senses (and who also has a good memory) can make worthwhile contributions to his company's success, but there are limits to one person's abilities. Today's

*Reprinted with permission of the Institute of Food Technologists, Chicago, Illinois.
©Copyright 1974. Originally published in Food Technology 8, 24-32.*

markets are multinational, more competitive, and certainly more complex than they were twenty or thirty years ago. It is estimated that only 1 in 100 new products succeeds in the marketplace, and still more items never leave the company's internal screening system.

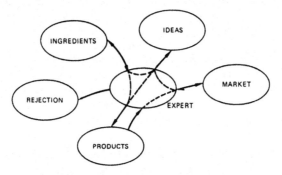

FIG. 1. IN THE TRADITIONAL, INDUSTRIAL ENVIRONMENT, THE EXPERT SERVED IN A PIVOTAL POSITION, MAKING DECISIONS AS TO WHAT INGREDIENTS AND PRODUCTS WERE MOST APPROPRIATE FOR THE MARKETPLACE

AN ALTERNATIVE TO THE EXPERT

Figure 2 is a schematic diagram more representative of the marketplace today. We are now dealing with a greatly expanded information base, infinitely more complex than in the previous diagram, and a situation that makes it practically impossible for the expert to function successfully in his typical, all-inclusive role.

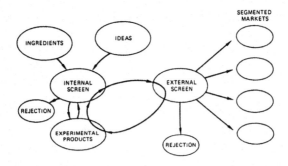

FIG. 2. IN THE CONTEMPORARY MARKETPLACE, WHICH IS SEGMENTED, MORE COMPLEX, AND MORE COMPETITIVE, THE EXPERT CONTINUES TO FUNCTION WITHIN THE INTERNAL SCREENING ENVIRONMENT, BUT OTHER FUNCTIONS ARE HANDLED BY SENSORY AND CONSUMER PANELS

Many companies and industry associations have recognized this problem and as an alternative have developed specialized scoring systems and panels of trained judges for specific products. Implied in these standardized procedures and panels was the assumption that the consumer would recognize these "standards of excellence" and purchase those products that best reflected the "standards."

Most of these methods involved category scales in which the individual used either words or numbers (the categories) to characterize the specific sensory attributes of a product. In many respects, these were attempts to move away from total reliance on the expert and his strictly qualitative judgments and to introduce a greater degree of objectivity to the measurement process.

About 1949, the Arthur D. Little Company proposed The Flavor Profile as a means of dealing with the complex world of food flavor (Anonymous 1963). The technique acquired a considerable following because it represented an alternative to existing techniques which were considered less than adequate at that time.

The Flavor Profile method called attention to the fact that a person can be trained to perceive and recognize individual flavor characteristics and, with the use of appropriate training aids, can reach an agreement with his or her fellow panel members on a total product impression, followed by detectable factors, their intensities, and their order of detection. It replaced the individual expert with a "group expert" that provided a collective response.

IMPROVED TECHNIQUE DEVELOPED

Believing that knowledge about human perception and quantification of sensory measurement should be applied to the problem of flavor evaluation, we at Stanford Research Institute have developed what we consider to be an improved sensory evaluation technique. Our research focused on subject selection and training, the judgmental process, the scaling technique, the collection of repeated judgments from each subject, and the data processing and handling system (Garner 1951; Garner and Halse 1960; Green and Swets 1966; Tanner and Swets 1954).

Selection of the most appropriate technique for measuring the responses of an individual judge to a stimulus (the perceived sensory characteristics of an ingredient or a product) was considered basic to the development of the descriptive technique. The choice of scale was influenced by the early work of Baten (1946) and a subsequent study by Hall (1958). Baten showed that a scale word-anchored only at the ends yields greater product differences than the typical category scale. Hall used a technique that was a variation of the Flavor Profile procedure. After round-table discussions, the panelists were provided with a rating form listing each attribute along with a scale in the form of an elongated rectangle with words at each end. The subjects' score was then converted to a numerical rating for subsequent analysis. He reported that this method provided a continuous rather than stepwise scale, and that a scale without numbers eliminated possible bias for particular numbers. This latter point was an empirical observation that Stevens (1957) had shown to be correct.

Work on scaling by Stevens and Galanter (1957) and Ekman and Sjöberg (1965) also had an important influence on the development of our scaling methodology. Stevens clearly showed that it was possible for individuals to make direct assessments of subjective magnitude for perceived sensory attributes. The use of category scales for judgments of relative intensities was discarded because of difficulties inherent in such a scale. For example, it is assumed that each word phrase of a category scale has the same meaning to each judge, and that the psychological interval between each category is equivalent. Experimentally, these criteria have been found to operate to varying degrees and not to be at all easy to control. This does not mean, however, that category scales can never be used — numerous appropriate and successful examples can be cited (Stevens 1957; Eisler 1963) to defend their use under certain circumstances.

EQUAL-INTERVAL SCALE CHOSEN

Using these results as background, we developed a scaling method that incorporated the apparent advantages of these previous studies adjusted to meet the more practical requirements of complex products (most scaling research is based on a single stimulus — e.g., weight or loudness — rather than a mixture of fragrance chemicals or flavor ingredients). The scale is an interval scale with the following features: a line of specific length — 6 inches — with anchor points ½ inch from each end, usually but not necessarily a third anchor at the midpoint; and usually one word or expression at each anchor. The subject has the task of placing a vertical mark across the line at that point which best reflects the magnitude of his or her perceived intensity of that attribute. An example of the scale and the specific instructions for the subject are shown in Figure 3.

A recent paper by Weiss (1972) described a scaling method called "graphic rating" that is similar to our method except that his subjects are provided with reference stimuli at each test session. With our scale method, subjects receive instruction on the range of intensities during training, but during product testing, the anchor words may serve the same purpose. In a study of the grayness of color chips, Weiss reported that the graphic rating method yielded data that followed a straight line. These results supported our approach that the scaling technique yielded data appropriate to the statistical model and were consistent

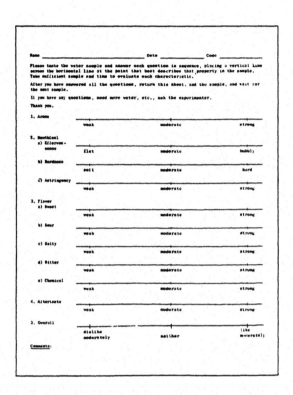

FIG. 3. EXAMPLE OF A SCORECARD USING THE INTERVAL SCALE; PERCEIVED ATTRIBUTES AND WORDS TO DESCRIBE THEM ARE DERIVED BY THE PANEL; LINE SCORES, IN TENTHS OF AN INCH, ARE CONVERTED TO NUMBERS BY USE OF A TEMPLATE

with what would be predicted with an equal-interval scale. Analyses of data from several experiments using our scaling method supported our continued use of this equal-interval scale. A detailed discussion of this aspect of our work will be published soon (Singleton *et al.* 1974).

Finally, we observed that use of this scaling technique was consistent with the analysis of variance model. This enabled us to develop a special computer program to determine subject performance, repeatability, and related analyses in addition to the basic issue of product differences.

WHAT QDA IS

In this technique — which we call "Quantitative Descriptive Analysis" (QDA) — trained individuals identify and quantify, in order of occurrence, the sensory properties of a product or an ingredient. These data enable us to develop the appropriate product multidimensional models in a quantitative form that is readily understood in both marketing and R&D environments. The technique has also been used successfully to develop data about concepts and idealized products before actual developmental efforts are initiated. The basic features of the technique are as follows:

·**Introspection Is Relied On** to develop perceived sensory attributes, with formal statistical testing for reliability.
·**Development of the Language** is a group process, with the panel administrator providing leadership and guidance but not actively participating in product evaluation.
·**Subject Selection is** based on performance with test products, without using model systems.
·**Repeated Judgments** (as many as 12-16) are collected from each subject to monitor individual and panel performance.
·**Subject Replacement** requires less than a month, usually not more than 20 hours of training.
·**All Data** are collected individually in isolated booths, using coded samples.
·**Interval Scales** are used to measure perceived intensities of individual attributes.
·**Analysis of Variance** (one- and two-way) is used to analyze individual and panel performance data.
·**Correlation Coefficients** are used to determine relationships among the various scales.
·**Principle Component Analysis,** factor analysis, etc., are used to determine primary sensory variables and eliminate redundancy.
·**A Multidimensional Model** can be developed and its relationship established with consumer responses and other external measures.

ANALYZING SUBJECT PERFORMANCE

As noted above, the collection of repeated judgments from each subject is basic to the method. These data are analyzed to enable the experimenter to evaluate:

·Subject Performance: Are individual subjects responding consistently, or is more training required? Are individual subjects adequately discriminating differences among products?

·Scale Performance: Are individual scales producing consistent results? Are individual scales adequately discriminating differences?

·Product Performance: To what extent do products differ on the specific attributes judged?

The computer program for analyzing the sensory data treats each rating scale as an independent unit. The data are analyzed with a two-way analysis of variance model, showing both subject and product effects. In the one-way analysis of variance, the error term is listed subject by subject. The corresponding probability levels are summarized in a separate table of subject performance (Table 1).

During training, subjects are encouraged toward uniformity in the spread of observations for differing products. Significant differences in subject means are to be expected, but are not of direct importance, i.e., adding a constant value does not alter the product or subject-product interaction results. It is not-critical which segment of the scale is used by a judge, only that performance be consistent. Figure 4 shows schematically how four different judges might use the scale.

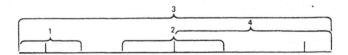

FIG. 4. WAYS IN WHICH FOUR SUBJECTS MIGHT USE THE INTERVAL SCALE.
During training, subjects are encouraged to make use of the entire scale,
but in practice the portion of the scale used may vary from
subject to subject but should be internally consistent.

The analysis enables the panel administrator to select the most accurate and consistent subjects, and to identify those individuals who require more training (and the specific sensory attributes for this training) and those judges to be excused from further testing. The range in performance levels is shown in Table 1.

TABLE 1.

SUBJECT PERFORMANCE: Probability levels for individual subjects across all products by individual scales derived form the one-way analysis of variance for the individual subject. A value of 0.50 or greater is indicative of no contribution to sample discrimination for that subject.

Scale	1	2	3	4	5	6	7	8	9	10	Panel
AAA	0.04	0.53	0.39	0.07	0.14	0.34	0.00	0.11	0.54	0.34	0.000
BBB	0.00	0.00	0.59	0.03	0.00	0.07	0.00	0.00	0.00	0.00	0.000
CCC	0.00	0.11	0.04	0.78	0.00	0.00	0.00	0.00	0.47	0.00	0.000
DDD	0.00	0.64	0.06	0.00	0.02	0.00	0.00	0.00	0.02	0.00	0.000
EEE	0.00	0.31	0.61	0.14	0.73	0.45	0.31	0.39	0.09	0.75	0.114
FFF	0.07	0.13	0.03	0.81	0.51	0.96	0.20	0.32	0.03	0.20	0.012
GGG	0.00	0.14	0.73	0.50	0.00	0.00	0.00	0.00	0.00	0.00	0.000
HHH	0.10	0.11	0.03	0.61	0.29	0.02	0.00	0.04	0.07	0.20	0.000
III	0.00	0.02	0.17	0.00	0.00	0.00	0.00	0.00	0.00	0.00	0.000
JJJ	0.01	0.00	0.00	0.00	0.00	0.00	0.00	0.00	0.00	0.00	0.000
KKK	0.00	0.00	0.01	0.15	0.00	0.00	0.00	0.00	0.34	0.04	0.000
LLL	0.15	0.01	0.00	0.00	0.00	0.00	0.00	0.02	0.00	0.00	0.000
MMM	0.05	0.01	0.58	0.30	0.17	0.00	0.00	0.02	0.55	0.53	0.000
NNN	0.03	0.20	0.02	0.16	0.52	0.44	0.00	0.05	0.06	0.78	0.001
OOO	0.00	0.06	0.54	0.73	0.00	0.00	0.01	0.00	0.00	0.00	0.000
PPP	0.00	0.00	0.00	0.13	0.00	0.00	0.00	0.00	0.00	0.00	0.000
No. of attributes missed (out of 15)	0	2	5	5	3	1	0	0	2	3	

Another factor considered in subject selection and training is the degree of agreement between the subject and the panel as a whole. As a measure of this factor, the interaction sum of squares is estimated subject by subject and expressed as F-ratios; the F ratio for the entire panel is then the average of the F-ratios for the individual subjects. A high F-ratio for an individual subject is indicative of disagreement (in scores assigned to samples) with the panel as a whole and warrants a review of that subject's performance for that particular attribute.

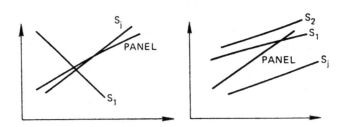

FIG. 5. POSSIBLE FORMS OF INTERACTION. Interaction may be considered the relationship between products and subjects when effects are not additive. For example, subject 1 on the left perceives the products differently from the remainder of the panel (typical form of interaction).
In the figure on the right, subjects 1 and 2 also exhibit interaction, but this interaction is attributable to the relatively small differences in perceived intensity among the samples.

By interaction, we refer to the problem of judge-byproduct confusion in scoring, as shown in Figure 5. We accept some confusion in scoring, especially when samples are not very different from each other; however, to score one or more attributes in a series of samples in the reverse order from all other judges warrants further attention. The judge's response, in fact, may reflect a real difference in perception of the product that may not be as readily seen without the test for interaction.

DETECTING PRODUCT DIFFERENCES

Having achieved the appropriate performance level, it is reasonable to proceed to the question of whether there are product differences. Table 2 is a printout of one attribute, including the scale-by-scale results tabulated individually, with product mean values, the overall mean, standard deviation, F-ratio, and appropriate probability values. Reported in Table 3 are the analysis of variance results for this attribute, the main effects being tested against both error and interaction mean square. When interactions are significant, the test against interaction is used, and it can be quickly determined whether significant differences among products are due to artifacts or to real differences among the products. In this particular situation, the ANOVA data showed that there were significant differences among the four samples and no significant interaction effect.

TABLE 2.
MEAN CHARACTERISTICS FOR CHARACTERISTIC CCCC: Computer printout for an attribute in which the display includes product mean values, overall means, standard derivations, F-ratios, probabilities, interaction F-ratios, subject performance, probabilities, and t-value differences for product comparisons.

Sample	Panel	1	2	3	4	5	6
AAAA	19.806	12.58	21.25	25.75	31.00	13.08	15.17
BBBB	19.069	10.25	13.42	30.33	33.75	18.08	8.58
CCCC	22.458	16.08	19.00	30.42	34.25	14.00	21.00
DDDD	29.222	16.83	24.08	37.42	36.08	33.33	27.58
Mean	22.639	13.94	19.44	30.98	33.77	19.62	18.08
S. dev.	12.764	10.86	8.37	9.23	19.40	12.49	13.11
N	288	48	48	48	48	48	48
F-ratio	9.449	0.96	3.50	3.26	0.14	6.78	4.60
Prob.	0.000	0.42	0.02	0.03	0.93	0.00	0.01
Interaction:							
F-ratio	1.227	0.62	0.93	0.43	0.92	2.90	1.57
Prob.	0.251	0.54	0.40	0.65	0.40	0.06	0.21
T-value differences of panel means, D.F. = 264							
DDDD	0.00						
CCCC	4.50	0.00					
AAAA	6.26	1.76	0.00				
BBBB	6.75	2.25	0.49	0.00			

TABLE 3.
ANALYSIS OF VARIANCE FOR SENSORY DATA FOR CHARACTERISTIC C

	D.F.	Sum Sq.	Mean Sq.	Vs. error F-ratio	Vs. error Prob.	Vs. interaction F-ratio	Vs. interaction Prob.
Sample	3	4618.194	1539.398	9.449	0.000	7.700	0.002
Subject	5	14845.444	2969.089	18.225	0.000	14.851	0.000
Interaction	15	2998.806	199.920	1.227	0.251		
Error	264	43008.000	162.909				

Having established that the mean values for the products were different, it is important to know which values were different from each other. Differences in t-values for panel means are tabulated for convenience in making a Duncan multiple range test without the need for further arithmetic. When there is a significant interaction equal to or greater than 10%, the t-values are based on the interaction, and the degrees of freedom are concomitantly reduced.

DEVELOPING A MODEL

With the completion of these analyses, attention can be given to the correlations of the various rating scales shown in Table 4. These data enable us to determine what relationships exist between the various scales, and to minimize attribute redundancies. From these data and the relative intensity values, we can then develop the appropriate multidimensional model, as well as hypothesize and test relationships to chemical, physical, and consumer attributes.

TABLE 4.
CORRELATION COEFFICIENTS FOR EACH SCALE WITH ALL OTHER SCALES.

Scale	1	2	3	4	5	6	7	8	9	10	11	12	13	14	15	16
1	100															
2	−5	100														
3	−66	−63	100													
4	−33	−84	91	100												
5	88	−44	−34	−1	100											
6	29	−88	46	79	51	100										
7	−69	−51	98	88	−47	42	100									
8	85	−5	−44	−8	64	47	−37	100								
9	−64	−54	92	89	−45	52	97	−28	100							
10	−85	−38	95	76	−63	22	97	−58	93	100						
11	−65	−50	79	78	−48	48	85	−31	94	85	100					
12	43	87	−84	−87	1	−62	−74	41	−74	−72	−73	100				
13	77	43	−89	−79	60	−36	−94	43	−98	−95	−96	71	100			
14	13	−62	54	78	21	86	59	52	66	38	53	−41	−50	100		
15	−46	70	4	−31	−66	−72	12	−36	−3	23	−19	44	−4	−31	100	
16	−8	−72	48	72	8	82	50	18	67	40	79	−66	−63	68	−71	100

One example of a model is shown in Figure 6, where obvious differences were found between the test and market leader products. The angular differences relate directly to the correlation coefficients, and the distance on the line from the center for an attribute is the mean value assigned by the panel. Probably the most important point here is the realization that, with data obtained by this scaling technique, we can proceed to almost any type of multidimensional model, using data for which statistical reliability is found.

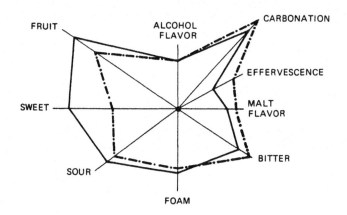

FIG. 6. QUANTITATIVE DESCRIPTIVE ANALYSIS OF TWO COMPETITIVE PRODUCTS; the distance from the center is the mean value for that attribute, and the angles between the outer lines are derived from the correlation coefficients.

When concepts are converted into protocepts or prototypes, a QDA panel can measure, quantitatively, the degree to which these match the concept. The technologist can then prepare products that closely agree with the concept.

Knowing which attributes are related to each other, as well as to specific physical and chemical characteristics, is extremely important in quality assurance and in designing questionnaires for home-use testing. Large-scale testing by some of our research sponsors with thousands of consumers has yielded information consistent with what has been found using small QDA panels. This does not obviate the need for consumer testing, but it does provide management with additional information about the possible outcome of subsequent consumer testing, as well as expanded understanding of its products and their competition.

AN EVOLUTIONARY PROCESS

We consider the QDA technique to be a part of the evolutionary process in sensory evaluation. Over the past five years, many aspects of the technique — especially training, information feedback to subjects, and data presentation — have undergone revision and improvement. However, the fundamental principles of scaling — i.e., the measurement process — remains unchanged, as does the basic need for obtaining repeated judgments from individual subjects in a controlled test environment.

As we seek more specific kinds of information about products and the reaction of the consumer to these products, we will need to utilize techniques such as QDA to do better than in the past.

REFERENCES

AMERINE, M.A., PANGBORN, R.M. and ROESSLER, E.B. 1965. *Principles of Sensory Evaluation of Food*. Academic Press, New York.

ANONYMOUS. 1963. The Flavor Profile. Arthur D. Little, Cambridge, Mass.

BATEN, W.D. 1946. Organoleptic tests pertaining to apples and pears. Food Res. 11: 84.

EISLER, H. 1963. Magnitude scales, category scales and Fechnerian integration. Psychol. Rev. 70: 243.

EKMAN, G. and SJÖBERG, L. 1965. Scaling. Ann. Rev. Psychol. 16: 451.

GARNER, W.R. 1960. Rating scales, discriminability, and information transmission. Psychol. Rev. 67: 343.

GARNER, W.R. and HAKE, H.W. 1951. The amount of information in absolute judgments. Psychol. Rev. 58: 446.

GREEN, D.M. and SWETS, J.A. 1966. *Signal Detection Theory and Psychophysics*. John Wiley & Sons, New York.

HALL, R.L. 1958. Flavor study approaches at McCormick & Co., Inc. In *Flavor Research and Food Acceptance*, Arthur D. Little, p. 224. Reinhold, New York.

SINGLETON, R.C., SIDEL, J. and STONE, H. 1974. Scaling and data processing with an equal-interval scale. In press.

STEVENS, S.S. 1957. On the psychophysical law. Psychol. Rev. 64: 153.

STEVENS, S.S. and GALANTER, G.H. 1957. Ratio scales and category scales for a dozen perceptual continua. J. Exp. Psychol. 54: 377.

TANNER, W.P. and SWETS, J.A. 1954. A decision-making theory of visual detection. Psychol. Rev. 61: 401.

WEISS, D.J. 1972. Averaging: An empirical validity criterion for magnitude estimation. Perception and Psychol. Phys. 12: 385.

CHAPTER 1.4

THE SELECTION AND USE OF JUDGES FOR DESCRIPTIVE PANELS

KATHERINE ZOOK and COLLEEN WESSMAN

*At time of original publication,
all authors were affiliated with
the Product Evaluation Dept.
The Quaker Oats Research Laboratory
Barrington, IL.*

Sensory descriptive data from experienced individuals have been used as an adjunct to laboratory taste tests at Quaker Oats for some period of time. For the past 2½-3 years, however, a more tightly controlled method of handling descriptive sensory data, the Quantitative Descriptive Analysis (QDA), has been continuously in use. During that time, ten major descriptive panels have been developed to meet different objectives. All of these groups have had two things in common: They have been selected by a series of triangle difference tests, and they have received an intensive period of training before beginning their descriptive work.

The purpose of this article is to examine this method of selecting and training as a means of producing good descriptive judges and to look at some of the types of questions which have been answered with QDA data from such judges.

BASICS OF THE QDA METHOD

The QDA method of descriptive analysis was described in detail by its developers (Stone *et al.* 1974) several months after it was implemented by them at Quaker Oats. Early usage with beer at the Schlitz Beer Co. was described at the same time by Mecredy *et al.* (1974).

The method as used at Quaker Oats followed this sequence of events:

Screening of 24-36 prospective judges by means of 12 triangle difference tests, each administered twice.

Selection of 10-12 of the most discriminating judges. Availability for the time involved in training was a prime requisite for judge use.

Training for around 10 hr usually 1 hr/day. During this period, the terminology was developed to describe the appearance, flavor, and texture of the products. Judges were provided with a broad assortment of training products as they modified and perfected the evaluation sheets.

Reprinted with permission of the Institute of Food Technologists, Chicago, Illinois.
©Copyright 1977. Originally published in Food Technology **8**, 56-61.

A Series of 4 replicated judgments on training products. After grading and statistical analysis, one or more correction sessions were conducted to clarify any confusion in the use of terms. The basic unit of evaluation in the QDA method is an unstructured line 6 in. long anchored ½ in. from either end by pairs of terms (Fig. 1). The judge evaluates the intensity of each sensory attribute by placing a vertical line across the unstructured line. This is translated into a score from 0 to 60 for statistical analysis. All data reported in this article are in terms of this scale, which is unstructured when the judge uses it and is assigned numerical values later.

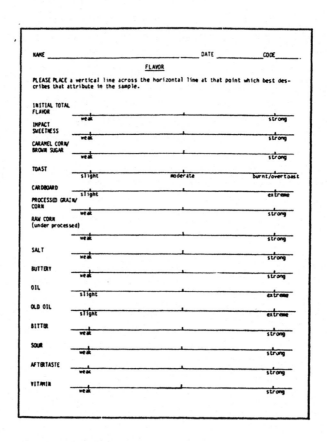

FIG. 1. SCORE SHEET FOR QUANTITATIVE DESCRIPTIVE ANALYSIS; the judge evaluates the intensity of each sensory attribute by placing a vertical line across the unstructured line. This intensity is later translated into a numerical score for statistical analysis by use of a template which divides the line into values ranging from 0 at the left end to 60 at the right end.

Replicated Judgments, using 10-12 replications on the first use of the panel immediately after training. With repeated usage of the group, 4-6 replications with 8 judges have been found to produce reliable data.

Analysis of Variance of each sensory attribute separately (Table 1), together with development of correlations between attributes and between attributes and an overall acceptance term.

TABLE 1.
AVERAGE INTENSITIES OF VARIOUS ATTRIBUTES FOR THE FLAVOR OF A SINGLE SAMPLE OF CAP'N CRUNCH® READY-TO-EAT CEREAL, DRY

Flavor attribute	Average intensity*
Intensity of flavor	29.69
Sweetness	32.14
Toast	19.95
Caramel corn/brown sugar	16.28
Processed grain/corn	20.05
Raw corn (underprocessed)	10.88
Buttery	12.60
Salt	11.99
Sour	12.05
Bitter	7.22
Oil	11.07
Old oil	9.26
Cardboard	14.36
Vitamin	12.06
Aftertaste	27.68
Fresh/stale	32.05

*11 judges × 10 replications = 110 evaluations; intensities are from analysis of variance of panel as a whole; analysis is also made on the data of each individual judge, to monitor judge performance.

Examination and Interpretation of data.

Preparation of QDA Configurations describing products, and visual or written presentation. A typical QDA configuration (Fig. 2) has lines radiating outward from a center point. Each line represents a particular descriptive term, and the average intensity for that term is plotted on that line (the center represents an intensity of 0, and the outer point a value of 35-60). Connecting the average intensities for all the terms provides a product profile.

Judges who made up the trained QDA groups were all employees of the Quaker Oats Research and Development Laboratory. A total of 279 screenings were conducted, from which 110 judges were selected for membership in the groups. About half of the judges were in food-related jobs where they might have experience in tasting food—the other half were in nonfood-related positions, but were found to be able to discriminate well by the triangle testing procedure. All of the groups developed were mixtures of the two types, as shown in Table 2.

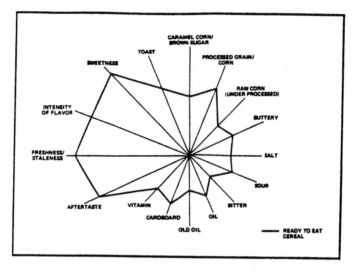

FIG. 2. A TYPICAL QDA CONFIGURATION FOR THE FLAVOR OF A
READY-TO-EAT CEREAL, DRY

The average intensities for the various attributes are graphed on lines radiating outward from a value of 0 at the center point to a value of 35 at the outer perimeter. For example, the average intensity of sweetness is located at a point representing 32.14.

TABLE 2.
MAKEUP OF QDA DESCRIPTIVE PANELS

Product evaluated	Food-related personnel	Non-food-related personnel	Total
Breakfast item	8	2	10
Natural Cereal®	5	6	11
Syrup	7	4	11
Pizza	3	8	11
Life® cereal	6	5	11
Shredded Wheat	3	7	10
Cap'n Crunch® cereal	5	6	11
Frozen pancakes	7	5	12
Oatmeal cookies	3	9	12
Instant grits	8	3	11
Total	55 (50%)	55 (50%)	110 (100%)

PRACTICAL CONSIDERATIONS

There are a number of considerations in selecting and using judges for QDA, or any other type of continuing descriptive work, which have nothing to do with the technical aspects of selecting a good judge, but are critical to the success of the program:

(1) First, and most important, the whole program must have upper management support.

(2) This must be communicated to middle management personnel who will be giving permission to release employees for the time required for the training and development of the configurations. It may be necessary to hold brief seminars or educational meetings to acquaint these people with examples of what QDA can do and how it is accomplished.

(3) The time requirements must be clearly spelled out to both the participant and his immediate superior before participation begins. Securing the panelist for one hour per day during the training period is the crucial factor, the screening and repeated evaluations requiring much less time.

(4) If two people from a small department both qualify, it is wiser to select only one for training. Thus, when any one group meets, participation is spread among a number of departments and no one department is suddenly denuded of a large part of its work force.

(5) After management support, motivation and interest on the part of the panel member is the single most important factor in obtaining on-time, conscientious presence at all panel sessions. This is enhanced by giving the panel some idea of the purpose or importance of the project to the company. There is a very fine line between providing the panel with information on the overall aims of the work and giving clues which may bias their descriptive work.

(6) As a morale booster, luncheon together at a restaurant of their choice for the group after the completion of a set or series of samples has proved to be effective not only as a reward but as a means of building "esprit de corps".

(7) Occasionally, one person will qualify and serve on more than one group, but this has not been encouraged because of the potential conflicting demands of the two groups.

TECHNICAL ABILITIES NEEDED

The qualifications for expert descriptive judges have been thoroughly described in various papers. Some of the more important are:

Taste Acuity or ability to duplicate a difference judgment (Amerine *et al.* 1965). Normal gustatory and olfactory ability is presupposed.

Ability to Deal Analytically with a complex test situation. Girardot *et al.* (1951) believed that it was impossible to test independently for all the factors underlying this unitary skill but that the test situation must be set up to require acts of discrimination and judgment, such as would be used later in the experiments. (These skills include flavor memory, ability to deal logically with flavor perceptions, and general adjustment to the test situation).

Good Health and freedom from allergy, frequent head colds, and sickness (Amerine *et al.* 1965).

Stable Personality, neither overly passive nor overly dominant (Caul 1957). From some recent studies with personality trait scaling (Henderson and Vaisey 1970), it has been suggested that individuals would continue to operate at a higher level of performance if they score high in the "need for achievement". Our observation is that an individual who is not easily distracted from the task at hand, who can perform no matter what the emotional upsets in his life, usually makes an excellent judge.

Ability to Verbalize or describe what they taste.

Interest and Motivation.

Availability.

SCREENING BY TRIANGLE TEST

The triangle test is not a new test, nor is its use as a means of selecting judges new. This difference test, first suggested by Bengtsson (1943), was used by Helm and Trolle (1946) as a method for selecting expert beer tasters. Peryam and Swartz (1950) reported that the triangle test had been developed independently in the laboratories of Joseph E. Seagram and Sons in 1941. Amerine *et al.* (1965) states that "because of its extensive application the test has been the most thoroughly studied and criticized of all test designs."

There are several advantages in screening with triangle tests:

(1) It is obvious that those who do best on the series of triangle tests have the underlying abilities to make flavor and texture judgments on that particular product.

(2) The results of the triangle tests are objective. Panelists know that if they are selected for training, they have a similar level of discriminatory ability to the remainder of the panel.

(3) Panel members work directly with the product under consideration. Included in the triangle testing can be some of the same variables which will be present in the test products.

(4) The preparation of the products for the triangle tests by the panel leader gives this individual a great deal of experience in what the product and its variations are like.

(5) By the time the participants have completed the triangle tests, they also have a background of tasting experience with the product even before they are trained.

(6) If special products are prepared for the triangle tests (such as a special plant run of undercooked or overcooked product), they also make excellent training products for later use.

However, screening by triangle test also has several disadvantages:

(1) It does not tell you anything about the ability of the candidate to verbalize what he tastes or what kind of a personality he will have for group interaction.

(2) The triangle tests themselves must be quite carefully prepared. If you have set up a test set to measure a certain flavor difference, you do not want the judge to make a correct pairing from some appearance difference inadvertently present in the samples. Masking of appearance by reduced or minimal lighting is very often necessary. Colored lights which may mask color often leave other clues which the participant can pick up.

(3) The general plan of arranging the tests is to give the easier ones first and the more difficult ones later in the series. It is very difficult to get a perfect progression from easy to difficult. However, even without this, the judge gets exposure to a broad range of tasks.

(4) To be fair to the candidates, tests must be given twice at the same sitting; once with a set containing two A samples and once with a set containing two B samples. This doubles the number of test sets used for screening.

RESULTS OF SCREENING

The samples for the 12 triangle tests were all pre-screened before the tests were administered, and an attempt was made to start with fairly easy tests and make them progressively more difficult. The aim was to expose the judges to a broad range of discriminatory tasks and to include in the series some of the types of variables which would later be described by the panel. Tests were administered twice, once with two A samples and once with two B samples, because it was felt that certain combinations might be easier to pair correctly than others. Table 3 shows a sample set of 12 triangle tests as administered for Cap'n Crunch ready-to-eat cereal, from which a group of 11 of the most discriminating judges were selected for training.

An overview of the 10 different groups screened by triangle tests shows that:

·Not all people are equally discriminating on all products. Thus, it proved worthwhile to test a significant percentage of candidates on more than one product type (Table 4).

·Some products were more difficult to discriminate with (pair correctly) than others. This is shown in Table 5, which gives the record of the 10-12 most discriminating members who were chosen for training on each product type. For 7 out of the 10 products, panelists averaged better than 75% correct identifications; for 2 products, the level was about 64%; for one, slightly lower.

TABLE 3.

RESULTS OF SCREENING BY TRIANGLE DIFFERENCE TEST, USING VARIATIONS OF CAP'N CRUNCH READY-TO-EAT CEREAL, DRY, WHEN TESTED USING TWO A SAMPLES AND TWO B SAMPLES (X INDICATES AN INCORRECT IDENTIFICATION)

	Day 1		Day 2		Day 3		Day 4		Day 5		Day 6		Day 7		Day 8		Day 9		Day 10		Day 11		Day 12		
	Control vs Low sweetness		Control vs High sweetness		Control vs No salt		Normal vs Under-cook		Control vs No oil		Normal vs Under-dried		Normal vs Over-dried		Control vs High salt		Normal vs Process change		Plant 1 vs Plant 2		Lab product vs Plant 1		Lab product vs Plant 2		Total incorrect out of 24 for each judge
Judge	with 2As	2Bs	with 2As	2Bs	with 2As	2Bs	with 2As	2Bs	with 2As	2Bs	with 2As	2Bs	with 2As	2Bs	with 2As	2Bs	with 2As	2Bs	with 2As	2Bs	with 2As	2Bs	with 2As	2Bs	
1				X	X	X			X		X		X		X		X	X			X		X		11
2						X							X	X											3
3				X	X	X			X	X	X		X	X		X				X					10
4			X	X					X	X			X		X			X							7
5	X			X	X				X		X		X	X						X					8
6			X	X						X			X	X	X	X		X	X						9
7	X		X	X		X				X				X						X		X			8
8									X	X			X	X	X	X	X	X	X	X			X	X	12
9	X		X	X		X				X			X	X	X		X		X	X	X		X		13
10		X								X	X	X	X	X	X				X	X	X	X			10
11			X		X	X				X		X	X		X				X		X		X	X	11
12									X	X					X	X									4
13				X		X			X							X									4
14		X			X	X							X	X										X	6
15			X						X					X	X	X									5
16	X		X	X	X	X					X		X	X	X	X		X	X	X	X		X		15
17		X	X	X					X	X					X	X	X	X	X	X	X		X		13
18	X		X	X					X					X					X	X	X		X		9
19		X	X			X			X					X		X									6
20	X	X		X	X									X	X	X		X		X					9
21	X				X	X				X	X				X		X				X			X	9
Total incorrect for total panel of 21 judges	7	5	10	12	8	11	0	0	11	11	6	2	12	14	13	10	5	7	8	10	8	2	7	4	

TABLE 4.
NUMBER AND PERCENTAGE OF CANDIDATES SCREENED BY TRIANGLE TEST
ON ONE OR MORE TYPES OF PRODUCTS;
170 persons participated in 279 screening tests to obtain 110 judges.

No. of times tested, each time on a different type of product	No. of persons tested	% of persons tested
1	97	57.1
2	51	30.0
3	12	7.1
4	6	3.5
5	2	1.2
7	1	0.1

TABLE 5.
AVERAGE LEVEL OF DISCRIMINATION FOR JUDGES SELECTED
FOR QDA PANELS

Product	No. of persons screened	No. of judges selected	% correct identifications in 24 tests for judges selected
Breakfast item	33	10	71.3
Natural Cereal*	35	11	79.6
Syrup	30	11	76.3
Pizza	31	11	64.6
Life* cereal	28	11	63.8
Shredded Wheat cereal	26	10	58.7
Cap'n Crunch* cereal	21	11	72.5
Frozen pancakes	25	12	71.7
Oatmeal cookies	23	12	78.3
Instant grits	27	11	85.8

·The level of discrimination attained by the groups tested appeared to vary within: (a) The complexity of the product-pizza contained many ingredients which often masked the real variable. (b) The variability of the product-shredded wheat biscuits showed unavoidable differences between individual units. (c) The difficulty of the test sets—although the same general plan was set up for all products, the tests on Life cereal were based on small variations in plant-made product which proved extremely difficult to identify.

·The level of discrimination attained for all individuals on all products in 279 screenings (Table 6) showed that: (a) 28.7% of those tested were very discriminating, picking the odd sample 75% of the time. Another 24.7% picked the odd sample 66.7% of the time. (b) 36.6% of those tested discriminated at a level in which they identified 50-62% of samples correctly. (c) Around 10% of the group did little better at discriminating than the level which could be attained by guessing on a triangle test (33⅓ %correct).

Thus, generally speaking, if a level of discrimination of 67-75% correct identifications is desired, about 2-3 times as many candidates would have to be screened as would be selected for final training.

TABLE 6.
LEVEL OF DISCRIMINATION (% CORRECT IDENTIFICATIONS)
Shown by candidates for expert QDA descriptive panels in screening by triangle difference tests on various food products.

Product	No. of judges who paired correctly at the following levels of discrimination (% correct identifications):								No. of judges	
	75% or better	71%	67%	62%	58%	54%	50%	Less than 50%	Screened	Trained
Breakfast item	10	8	4	3	5	0	1	2	33	10
Natural Cereal*	17	2	3	3	6	3	0	1	35	11
Syrup	7	9	2	4	1	1	5	1	30	11
Pizza	2	1	3	3	2	8	4	8	31	11
Life* cereal	1	2	5	0	5	4	7	4	28	11
Shredded Wheat cereal	1	0	3	2	4	8	3	5	26	10
Cap'n Crunch* cereal	5	2	2	4	1	1	3	3	21	11
Frozen pancakes	5	9	4	1	1	1	1	3	25	12
Oatmeal cookies	11	4	2	1	1	2	1	1	23	12
Instant Grits	21	2	2	1	1	0	0	0	27	11
Total	80 (28.7%)	39 (24.7%)	30	22 (17.6%)	27	28 (19.0%)	25	28 (10%)	279	110

TRAINING OF JUDGES IMPORTANT

The importance of the training which is conducted with the judges after they are selected cannot be emphasized too much. For the training to be successful, an experienced and perceptive leader is necessary and several basic points must be observed:

·The panel leader does not take part in the descriptions of the products. His or her role is to keep the group functioning, provide standards and training samples as needed, prepare trial score sheets from the terms suggested, think of ways to clarify confusion, and test and monitor the judges.

·The terminology used to describe the products comes from the panel members themselves. All must understand and feel comfortable with the descriptive terms to be able to use them effectively in grading. This is why it is so important that members not miss sessions.

·Members of the group must feel on an even footing so that all will make contributions to the general pool of knowledge about the sensory characteristics of the product.

·The ingredients from which the product is made are not identified at first but only after a need for them is indicated, since they could possibly influence the expectations of the judges in their first impressions of the product.

·The physical surroundings of the group are important; these include adequate privacy, ventilation, and lighting and a conference table around which all may sit to take part.

USES FOR QDA DATA

What are some of the uses of an expert panel developing descriptive data by the QDA method? There are undoubtedly many uses for a trained panel, but I will confine myself to examples of projects on which we have accumulated data and have results:

(1) As an Aid in Product Development:
a. To describe the sensory characteristics of competitive products and one or more experimental prototypes (Fig. 3 and 4).

b. To document that the scaled-up product is like that originally developed.

c. To optimize a formula by describing several levels of several variables in more complex designs.

d. To provide language for description of an ideal or favorite product by consumers.

e. To describe changes in storage.

(2) As an Aid in the Maintenance or Improvement of an Established Product. Here the QDA data furnished by a descriptive panel are used as an analytical tool and can function by themselves or in conjunction with objective measurements, or even consumer acceptance data. The following are several examples of these kinds of applications relating to typical problems which arise in continued manufacture of a product:

a. Can the variability be reduced? In one case, it was desired to reduce color variability of a cereal by setting color ranges on a colorimeter. Here, QDA data described the changes which took place so that realistic cut-off ranges could be set (Fig. 5). After the first description, product was made again, described a second time by QDA, and also submitted to children, the consumers of this cereal. Note how closely the results of the second QDA description duplicated those of the first, and also how the children reacted to the products described.

b. Can the manufacturing process be changed, for reasons of improvement or cost cutting? Descriptive data have been successfully used in changing extrusion conditions for a ready-to-eat cereal, as one example.

c. Can an ingredient be changed? Figure 6 shows that a desired change in processing of a particular ingredient could be used with negligible results on the product. Figure 7 shows that a 25% reduction in level of one ingredient will probably cause problems with the product at the low end of the permitted pouch weight; the projected decrease in the ingredient was dropped.

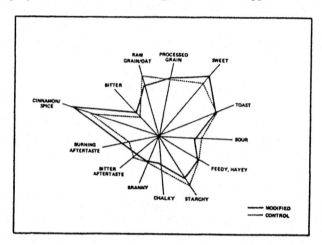

FIG. 3. QDA CONFIGURATIONS FOR A PROTOTYPE CINNAMON-FLAVORED CEREAL AND A MODIFICATION

Increasing the "cinnamon spice" and "sweetness" slightly resulted in a product with more impact and less "starchy" and "raw grain/oat" character. This modification was found to be significantly more acceptable in consumer testing and went on to become a successful market introduction. This is an example of how small but significant changes in key sensory attributes can be crucial to a product's success.

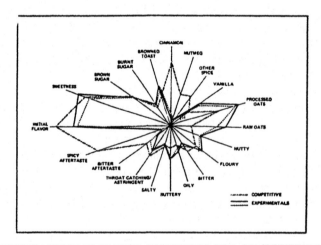

FIG. 4. CONFIGURATIONS OF FLAVOR ATTRIBUTES OF A COMPETITIVE BRAND OF OATMEAL COOKIES AND TWO EXPERIMENTAL PROTOTYPES
(SCALE=0 AT CENTER TO 40 AT PERIMETER)

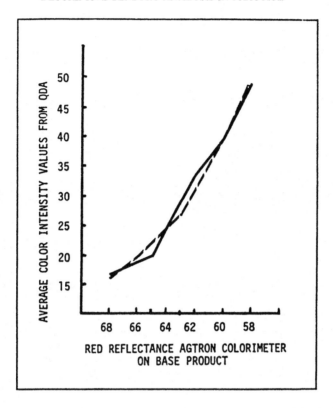

FIG. 5. QDA DATA CAN BE USED TO ESTABLISH ACCEPTABLE RANGES OF OBJECTIVE DATA

For example, QDA averages for color correlate with color readings from an Agtron colorimeter. Results are shown for two separate QDA descriptions. When product from the second was tested among 162 children (consumers of the cereal), the children significantly preferred product from the light end of the scale (Agtron value 68) and rejected the product from the dark end (Agtron value 58).

(3) As a Diagnostic Tool when a product is slowly losing its accustomed share of the market. A detailed sensory description, together with that of the competitive product is the first step in determining whether the problem is in the sensory area or in some other area, such as marketing. Figure 8 indicates that a shredded cereal biscuit possesses negative notes that the competitive product doesn't. Packaging and processing can now be examined to find the source of the critical off-notes.

(4) As a Quality Control Measure. This depends upon the development and maintenance of a master configuration or QDA profile of the product against which later profiles of the product can be compared to measure "drift". This presupposes that the samples described are truly representative of most of the product that is being made at the point in time when the measurement is taken

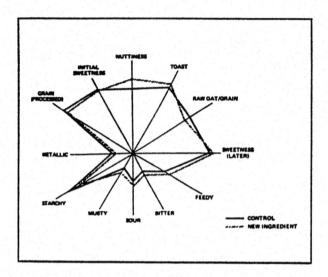

FIG. 6. EFFECT OF INGREDIENT REPLACEMENT CAN BE DETERMINED BY QDA
Here, a sample of ready-to-eat oat cereal with milk was compared with another sample containing an ingredient prepared by a new process. The almost identical configurations (not significantly different) indicate that little flavor change need be expected if the process change is implemented.

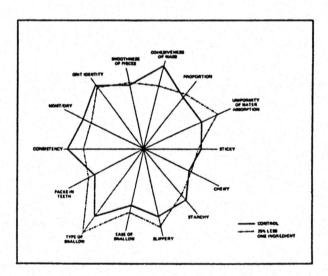

FIG. 7. EFFECT OF REDUCTION IN LEVEL OF ONE INGREDIENT CAN ALSO BE DETERMINED BY QDA
Here, a comparison of a control sample of an instant cereal and a sample containing 25% less of one ingredient shows that the reduction would significantly alter the consistency and cohesiveness of the product.

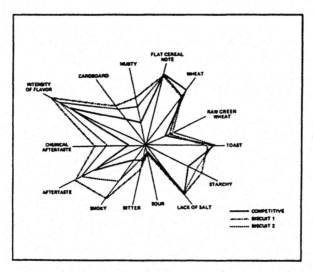

FIG. 8. AS A DIAGNOSTIC TOOL, QDA CAN INDICATE WHY A PARTICULAR PRODUCT MIGHT BE LESS ACCEPTABLE THAN A COMPETITIVE PRODUCT

Here, an unwrapped shredded wheat biscuit (Biscuit 1) and a wrapped biscuit (Biscuit 2) are compared to a competitive product. The comparison shows that the wrapped and unwrapped biscuits have negative notes that may be due to packaging or processing.

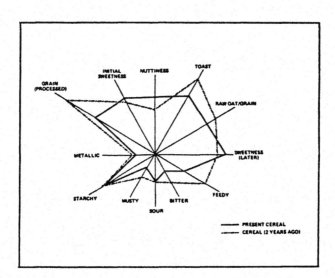

FIG. 9. AS A QUALITY CONTROL TOOL, QDA CAN DETERMINE WHETHER A PRODUCT HAS CHANGED IN CHARACTER WITH TIME

Here, a profile of a dry ready-to-eat cereal is compared to a profile of the same cereal drawn two years previous. Note the increase in "nuttiness" (a positive attribute) and decrease in "feediness" (a negative attribute), as well as other changes.

or that there is a historical knowledge of the amount of normal variability in manufacture. Figure 9 shows a ready-to-eat cereal as presently manufactured vs the same product from two years previous.

These are but a few illustrations of how the data from trained descriptive judges, when used in conjunction with statistical analysis to judge variability and reliability, can answer questions and give information.

The question inevitably arises as to whether sensory descriptive data can be used to predict acceptance. The experience with the 10 groups over a period of 2½-3 years suggests that detailed sensory data can be used very effectively in combination with acceptance information obtained from larger or consumer groups.

In summary, each of the 10 descriptive sensory groups discussed in this article was developed by (1) screening 21-33 individuals to obtain the 10-12 most discriminating and (2) giving these individuals intensive training. They all went on to produce sensory descriptive data which met the standards of the particular technique for sensory descriptions which we were using (QDA). This was true even though there was some variation in the level of discrimination attained in the original screening on the various products.

REFERENCES

AMERINE, M.A., PANGBORN, R.M. and ROESSLER, E.B. 1965. *Principles of Sensory Evaluation of Food*. Academic Press, New York.

BENGTSSON, K. 1943. Provsmakning som analysmetod. Statistisk behandling av resultaten. Svenska Bryggureforen Manadsblad 58.

CAUL, J.F. 1957a. The profile method of flavor analysis. Adv. Food Res. 7:1.

GIRARDOT, N.F., PERYAM, D.R. and SHAPIRO, R. 1952. Selection of sensory testing panels. Food Technol. 6: 140.

HELM, E. and TROLLE, B. 1946. Selection of a taste panel. Wallerstein Lab. Commun. 9: 181.

HENDERSON, D. and VAISEY, M. 1970. Personality traits related to performance in a repeated sensory task. J. Food Sci. 35: 407.

MECREDY, J.M., SONNEMANN, J.C. and LEHMAN, S.J. 1974. Sensory profiling of beer by a modified QDA method. Food Technol. 28: 36.

PERYAM, D.R. and SWARTZ, V.W. 1950. Measurement of sensory differences. Food Technol. 4: 390.

STONE, H., SIDEL, J., OLIVER, S., WOOLSEY, A. and SINGLETON, R.C. 1974. Sensory evaluation by Quantitative Descriptive Analysis. Food Technol. 28: 24.

CHAPTER 1.5

EXAMINING METHODS TO TEST FACTOR PATTERNS FOR CONCORDANCE

ELIZABETH I. FISCHMAN, KATHLEEN J. SHINHOLSER
and JOHN J. POWERS

*At the time of original publication,
all authors were affiliated with
the Department of Food Science
University of Georgia, Athens, GA 30602.*

ABSTRACT

To test whether the individual factor patterns of panelists and the panel factor pattern agreed, five grades of tea of the same type, ranging from low to high quality were studied. When a likelihood-ratio (LR) test and cluster analysis (CA) were applied to the 55-term correlation matrices of the seven individual panelists, according to the LR test only two of the matrices were homogeneous, but the CA test grouped all the matrices into the same cluster. The study demonstrated that the correlation matrices of the individual panelists or factor patterns themselves should be examined for agreement among panelists when factor analysis is used.

INTRODUCTION

Most sensory technologists accept as a matter of course that data sets should he examined to learn if product-panelist interaction exists. The same has generally not been true as to factor analysis (FA). With increasing frequency, FA is being applied to sensory results (Harries *et al.* 1972; Wu *et al.* 1977; Powers *et al.* 1977; Lyon 1980; Kwan and Kowalski 1980; Galt and McLeod 1983; McLellan *et al.* 1983, 1984), but only Harries *et al.* (1972) and Powers *et al.* (1985) appear to have evaluated concurrence among the panelists as to their individual factor patterns. The matter is important, for a factor pattern derived from panel results pooled might be merely the mean of a group of diverse patterns and thus misleading. Harries *et al.* (1972) compared by inspection the first three eigenvalues for the responses of six panelists to note how well they agreed with the eigenvalues for the panel results pooled. They agreed quite well. Powers *et al.* (1985) used the DISCRIM program of the SAS package (1982) to learn whether there was homogeneity between the variance-covariance matrices of the individual panelists. All but two of the matrices differed significantly.

Means of examining factor patterns for significance have been described by Bartlett (1950), Burt (1952), Rao (1955), Lawley (1956), Anderson (1958), Lawley and Maxwell (1971) and Arnold (1981), but programs to make the analysis are not generally available. The purpose of this study was to learn whether programs available in common statistical packages would permit sensory technologists to make, as a routine matter, tests to ascertain whether a factor pattern probably represents real underlying structural order or is but the mean of a group of divergent responses.

MATERIALS AND METHODS

The tea used had been procured for us by McCormick & Co. for a prior study (Godwin 1984). An expert tea buyer had chosen five grades of tea, ranging from low to high quality, of the same type and fanning grade. Immediately upon receipt, the five lots had been subpackaged in No. 303 cans, flushed with N_2, sealed and stored at -10C. Although the two panels described below evaluated the tea at times 6 months apart, it was assumed that the tea would not be a cause of any major experimental error because of the way it was stored. Later statistical analysis showed that to be true and the panel results were then pooled for further analysis.

The tea infusion was prepared by adding 400 mL freshly boiled water to 6 g tea in a flask, followed by steeping for 4 min. At 30 s intervals, the flasks were gently swirled to facilitate leaching. At the end of the steeping period, the infusion was filtered through four layers of cheesecloth. To standardize the contact time, the infusions were collected for 30 s only. Each of the five lots of tea was started 1 min apart. Once the infusion was collected, the flask containing the tea was held in a water bath at 85C to maintain a constant temperature. Tea infusions were served to the panelists approximately 5 min after preparation.

To ensure that each sample was presented to the panelists at the same temperature, a single sample was dispensed to the panelist upon request. Four cups coded with a 3-digit number and randomized as to order were placed in each booth beforehand and the panelists were instructed to request samples according to the codes on the cups, progressing from left to right.

Matzoh crackers were provided for mouth-clearing if the panelist so wished. The panelists were instructed that if they used the crackers, they had to use them between each sample. Water was also provided. The panelists were required to rinse their mouths with water between samples.

Panelist Selection

Six formal training sessions were conducted to train 15 experienced sensory panelists to evaluate tea. Five grades of tea used for the main trial and four other types of tea, somewhat similar, were used so that the panel candidates would have an idea of the range of characteristics to be encountered. Ultimately from among the trainees, five individuals were chosen for the first panel and six for the second. Some terms were provided to the panelists initially; they added two, "oregano-like" and "citrus-like" upon their own volition. The set of terms finally employed included: the desirability of flavor and aroma, and the intensities of briskness, grass-like, oregano-like, bitterness, floral-note, hay-like, metallic, sweetness and citrus-like. A 15 cm unstructured scale was used. During the training period the trainees received five types or grades of tea. At the end of each session, the identities were revealed and various facets of the evaluation were discussed in an effort to bring the panelists as much into agreement with each other as was possible. When the main trials commenced, four samples were provided at each session; no information was given about these samples or the panelists' performance until the conclusion of the study.

Statistical Analysis

The panelists' results were first analyzed by one-way analysis of variance (ANOVA) to ascertain the effectiveness of each panelist for each attribute. Panelists were accepted who were significant at $p < 0.50$ for 8 of the 11 descriptors, both trials pooled. After pooling there were 10 replications for Panel 1, 15 for Panel 2. Two-way ANOVA was used to detect product-panelist interaction. Its application showed that significant product-panelist interaction existed for only two of the terms; nonetheless discriminant analysis and cluster analysis were also applied to the panel results to learn how pervasive product-panelist interaction was.

The simple correlations between each panelist's responses for each combination of the 11 descriptors were examined by the DISCRIM program of SAS (1985) to secure an estimate of the degree of homogeneity between correlation matrices. The output from the DISCRIM program cannot be used directly; a correction in degrees of freedom has to be made to account for each "within" correlation matrix (Powers *et al.* 1985). Bartlett's sphericity test (Bartlett 1950) was applied to each matrix to test for randomness (Knapp and Swoyer 1967; Tobias and Carlson 1969). The VARCLUS program of SAS (1985) was used to cluster the panelists according to their correlation matrices. The CANDISC program of SAS (1985) was also employed to observe interspatial distances between panelists. Both the SAS and BMDP (1979) programs for principal-component and for the maximum-likelihood method of FA were utilized inasmuch as there are some differences in the output and for some purposes the factor patterns of one program were more useful than the other. Kendall's tau procedure was used to ascertain the correlation between factors produced by the panel and by different subgroups of the panel.

RESULTS

Out of the 11 panelists chosen for the main trials, the results of three were deleted because they did not attain significant differences for 8 of the 11 attributes examined. The results of one panelist were deleted because he refused to take part in the follow-up trial 6 months later.

The only two terms for which significant product-panelist interaction existed were "floral" and "hay-like." When discriminant analysis was used to classify the panelists, success ranged from 38 to 60% indicating that product-panelist interaction was somewhat more prevalent than significance for two out of 11 attributes would suggest. When Ward's minimum variance cluster analysis was applied to the scale values, panelist 6 broke off immediately, but no further segregation of the panelists occurred until the semi-partial R^2 value had fallen to 0.27. Had panelist 6 been eliminated at this stage, her results would have had to be restored later, for cluster analysis applied to the correlation matrices resulted in an entirely different conclusion being arrived at as to agreement among the panelists.

The correlation matrix of each panelist was subjected to Bartlett's sphericity test to determine whether the correlation matrix was significantly different from an identity matrix. At the two different trial periods and for the two different panels, only once was a panelist's correlation matrix determined to be nonsignificant. When the VARCLUS program was applied to the correlation matrices, only one cluster of seven panelists resulted. The program had to be forced to form two or more clusters for the reason explained below.

Likelihood-ratio Test

Except for the correlation matrices of two of the panelists, the DISCRIM program demonstrated that heterogeneity existed between the matrices of all the other panelists. That result agreed exactly with that of Powers *et al.* (1985) who found the variance-covariance matrices of only two panelists to be homogeneous. Some of the x^2 values associated with the likelihood-ratio test are listed in Table 1.

TABLE 1.
LIKELIHOOD-RATIO TEST FOR HOMOGENEITY OF CORRELATION
MATRICES OF INDIVIDUAL PANELISTS

Panelists	x^2 Observed	x^2 Tabular $p = 0.10$	df
7 panelists pooled	1064.18	363.30	330
2,7,11	185.86	129.37	110
2,7	38.32	68.78	55
9,10	184.59	68.78	55
5,6	187.83	68.78	55

The two multivariate methods of analysis thus gave conflicting answers. The likelihood-ratio test indicated that the correlation matrices of individual panelists were nearly all heterogeneous. The result is understandable. Only rarely are panel members' results for all attributes free of product-panelist interaction. A few discrepancies can cause matrices to be heterogeneous. In producing 55 simple correlation coefficients from the scale values of 11 descriptors, any discrepancy in scaling is correspondingly magnified. The likelihood ratio test thus said heterogeneity abounded; clustering said the correlation matrices were so much alike that all the panelists belonged in the same cluster. For practical reasons, a course has to be struck somewhere between the very demanding requirement of the likelihood-ratio test that differences be almost nil and the clustering process which is designed to accommodate some variation in location in multidimensional space. The problem was to select panelists whose correlation matrices were similar enough to yield comparable factor patterns even though the likelihood-ratio test would still indicate heterogeneity was significant. The CANDISC program of SAS (1985) was initially utilized. The correlation matrices of the panelists for each tea were examined separately. An indication was thus obtained as to those panelists who tended to be a part of the same group notwithstanding relations among the panelists varied according to the particular tea being examined. A disadvantage to using the CANDISC procedure is that the final decision is subjective. The VARCLUS program was therefore returned to except it was forced to form two and three clusters. Panelists who were grouped together in both clusters were considered to be responding sufficiently alike.

Table 2 shows the results when the Varclus program was forced to generate two and three clusters. The correlation matrices of panelists 2, 5, 6, and 7 resulted in their being included in the dominant cluster upon both occasions with the matrix of panelist 11 included within the dominant cluster when only two clusters were formed. The CANDISC procedure, having generally led to panelists 11 being grouped with 2 and 7, he was therefore also included with panelists 2, 5, 6, and 7. The factor pattern based upon the results of these five panelists was quite similar to that of the panel results pooled (Fig. 1), indicating that the panel results were the consequence of agreement among a majority of the panelists.

TABLE 2.
VARCLUS PROGRAM (OBLIQUE PRINCIPAL COMPONENT) FORCED TO FORM TWO AND THREE CLUSTERS BASED UPON INDIVIDUAL PANELIST'S CORRELATION MATRICES

Cluster groupings	Panelists[a]
Original cluster	2 5 6 7 11 9 10
Two cluster stage	2 5 6 7 11 9 10
Three cluster stage	2 5 6 7 10 11 9

[a] Panelists underscored by the same line were assigned to the same cluster.

It might be argued that when the results of 5 out of 7 panelists are used, the patterns for the subgroup and the panel would generally be similar whether or not there was agreement within the subgroup. To show that two of the panelists likewise yielded a pattern comparable to that of the panel. Table 3 lists the pattern results for the panel and for panelists 5 and 6 combined. There are some differences, but not many. If a corresponding value is not listed, that is because the attribute's correlation with the factor was non-significant. Table 4 lists Kendall's tau correlation values when the corresponding factors were compared. For the three factors and the communality estimates, the correlations were all significant.

Three-factor patterns were derived for Fig. 1 because graphic presentation naturally cannot go beyond three dimensions. To facilitate visual inspection, Table 3 likewise lists only three factors. In some instances, four- and five-factor patterns were more appropriate. The forcing of clusters, more and more compact, still applied. If however the number of factors appropriate for the subgroup is different from that of the panel, overall correlation naturally is poor. In fact, numerical analysis is generally not at all necessary in that case.

TABLE 3.
FACTOR REFERENCE STRUCTURE (SEMIPARTIAL CORRELATIONS), SEVEN PANELISTS POOLED AND PANELISTS 5 AND 6 COMBINED, TEA LOT 1

Attribute	Factor 1		Factor 2		Factor 3	
	Pooled panelists	Panelists 5 & 6	Pooled panelists	Panelists 5 & 6	Pooled panelists	Panelists 5 & 6
Flavor	56	64	59			
Aroma	68	62				
Briskness		79				
Oregano-like		72				
Bitter	65	70				
Floral	66	79				
Metallic	43	53				
Grass-like			87	92		
Hay-like			83	84	44	
Sweetness					72	78
Citrus-like					83	83

TABLE 4.
CORRELATIONS OF FACTOR PATTERN FOR PANEL (7 MEMBERS) AND PATTERN YIELDED BY PANELISTS 5 AND 6, COMBINED, TEA LOT 1

	Variables			
Variables	R11	R12	R13	R14
R21	0.51 (0.029)			
R22		0.75 (0.001)		
R23			0.49 (0.036)	
R24				0.78 (0.0008)

On their faces the patterns are not in accord with each other. In other instances, correlation was poor, but for an entirely different reason. Sometimes all the factors would be well correlated, but the best correlation for a variable would not be with its corresponding factor in the other pattern but with a different factor. When oblique factors were calculated and the Procrustean transformation (SAS 1985) is a part of the calculation process, some shifting of the factors between position is understandable. The intent of a Procrustean transformation is to harmonize as much as possible the configurations of the different panelists making up a set. In rotating 5 configurations as compared with 7 to effect maximum harmony within a set, a variable may shift from one factor to another. The problems above are pointed out because forcing to form clusters more and more compact does not necessarily lead to factor patterns being more and more alike. Our panel of 11 was initially culled so as to retain only the most consistent and discriminating panelists and those who were not contributing substantially to product-panelist interaction; nonetheless additional culling was sometimes needed to secure a "panel", the factor pattern of which could be demonstrated to result from substantial agreement among the individual members rather than being mere averaging of a mixture of similar and dissimilar patterns.

One observation which clearly came out of the study is that various products or grades should be examined separately. Ideally, the investigator would like to have a factor pattern which is so general it encompasses all variations of the same commodity. Cluster analysis applied to the five grades of tea showed that Products 1 and 5 each formed a class. Products 2, 3 and 4 formed a third class though upon further clustering it ultimately could be split apart. The factor patterns for Products 1 and 5 were different as was that for products 2, 3 and 4 pooled. Before any firm decisions are made as to the factor pattern existing for a given food, the products should be grouped into any classes known in advance or subjected to principal-component analysis or cluster analysis to learn if separate classes possibly exist. That rubric is in line with Cattell's (1965) recommendation. If different classes exist, separate factor analyses should be calculated for each class to demonstrate whether the patterns are alike or different. For any given commodity there is a core pattern, but variations of the pattern should be expected if the characteristics of the product vary as they did here for the different grades of tea. This seems so obvious as not to merit mention. Frequently however factor patterns are presented for a product as if the pattern were a general characteristic for all variations of that product when in fact each grade or species may have its own pattern. While the ultimate objective of FA is to reduce a set of entities to a lesser number of factors, the products and the performance of the panelists should be examined separately since they may be confounded in some fashion.

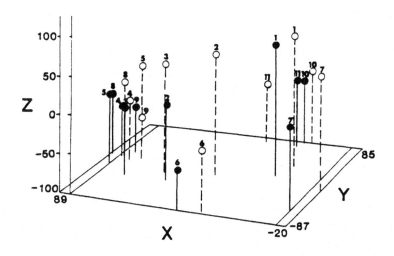

FIG. 1. FACTOR PATTERNS OF PANEL MEMBERS POOLED (7 MEMBERS) AND OF PANELISTS 5 AND 6 COMBINED, TEA, PRODUCT 5
• = pattern for panelists 5 and 6 combined; o = pooled pattern;
the numbers, in the same order, stand for the terms listed in Table 1.

SUMMARY

This study shows that the factor patterns of individual panelists and their patterns for individual products should be examined to learn whether the panelists' responses are sufficiently concordant to suggest that an underlying structural order exists. Patterns may appear to be logical, but before any import is attributed to the pattern, the responses of individual panelists should first be examined to learn if they agree reasonably well. "Reasonably well" is not used as weasel words, but rather in a practical sense. Some heterogeneity is to be expected. The degree of heterogeneity needs to be ascertained so that the investigator has an indication as to whether the factor pattern discerned borders on randomness due to the panelists responding in different ways or there is sufficient agreement among the panelists to indicate that relations among the attributes are being reflected by the panelists' scale values.

The methods examined here, while unquestionably not as rigorous as the tests devised by various statisticians (see citation in the introduction), have the merit of being procedures found in most computer packages and thus available for routine use by sensory technologists.

The study demonstrated that panelists shown to be consistent and discriminating in their responses and exhibiting little product-panelist interaction as judged by analysis of their original scale values may nonetheless be the cause of interaction when their results are subjected to FA. Unless a reasonable percent of the panelists are shown to yield comparable factor patterns, the trustworthiness of the panel's pattern is considerably in doubt.

REFERENCES

ANDERSON, T.W. 1958. *An Introduction to Multivariate Statistical Analysis*, p. 247. Chapman & Hall Ltd., London.

ARNOLD, S.F. 1981. *The Theory of Linear Models and Multivariate Analysis*. John Wiley & Sons, New York.

BARTLETT, M.S. 1950. Tests of significance in factor analysis. British J. Stat. Psychol. *111*, 77.

BMDP. 1979. *BMDP-79 Biomedical Computer Programs P-Series*. University of California Press, Berkeley, CA.

BRACH, N.M. 1982. Interrelationships among sensory attributes of tea. Ph.D. dissertation; Univ. of Georgia, Athens, GA.

BURT, C. 1952. Tests of significance in factor analysis. British J. Stat. Psychol. V: 109.

CATTELL, RAYMOND, B. 1965. Factor analysis: an introduction to essentials. II. The role of factor analysis in research. Biometrics *21*, 405.

GALT, A.M. and MacLEOD, G. 1983. A research note: the application of factor analysis to cooked beef aroma descriptors. J. Food Sci. *48*, 1354.

GODWIN, D.R. 1984. Relationships between sensory response and chemical composition of tea. Ph.D. dissertation, University of Georgia, Athens, GA.

HARRIES, J.M., RHODES, D.N. and CHRYSTALL, B.B. 1972. Meat texture. 1. Subjective assessment of the texture of cooked beef. J. Texture Studies *3*, 101.
KNAPP, T.R. and SWOYER, V.H. 1967. Some empirical results concerning the power of Bartlett's test of the significance of a correlation matrix. Am. Educ. Res. Journ. *4*, 13.
KWAN, W. and KOWALSKI, B.R. 1980. Data analysis of sensory scores. Evaluations of panelists and wine score cards. J. Food Sci. *45*, 213.
LAWLEY, D.N. 1956. Tests of significance for the latent roots of variance and correlation matrices. Biometrika *43*, 128.
LAWLEY, D.N. and MAXWELL, A.E. 1971. *Factor Analysis as a Statistical Method*. Butterworth & Co., London.
LYON, B.G. 1980. Sensory profiling of canned boned chicken: sensory evaluation procedures and data analysis J. Food Sci. *45*, 1341.
MCLELLAN, M.R., CASH, J.N. and GRAY, J.I. 1983. Characterization of the aroma of raw carrots (Daucus carota L.) with the use of factor analysis. J. Food Sci. *48*, 71.
MCLELLAN, M.R., LIND, L.R. and KIME, R.W. 1984. Determination of sensory components accounting for intervarietal in apple sauce and slices using factor analysis. J. Food Sci. *49*, 751.
POWERS, J.J. 1984. Current practice and applications of descriptive analysis. In *Sensory Analysis in Foods*, (J.R. Piggot, ed.), p. 179. Applied Sci. Publishers, London.
POWERS, J.J., GODWIN, D.R. and BARGMANN, R.E. 1977. Relation between sensory and objective measurements for quality evaluation of green beans. In *Flavor Quality: Objective Measurements*, (R.A. Scanlan, ed.), p. 51. ACS Symposium Series No. 51, Am. Chem. Soc., Washington, DC.
POWERS, J.J., SHINHOLSER, K. and GODWIN, D.R. 1985. Evaluating assessors' performance and panel homogeneity using univariate and multivariate statistical analysis. In *Progress in Flavour Research 1984*, (J. Adda, ed.), p. 202. Elsevier Sci. Publishers, Amsterdam.
RAO, C.R. 1955. Estimation and tests of significance in factor analysis. Psychometrika. *20*, 93.
SAS. 1982. SAS User's Guide: Statistics. SAS Institute, Inc., Box 8000, Cary, NC 27511.
SAS. 1985. SAS User's Guide: Statistics. Version 5 Edition. SAS Institute Inc., Cary, NC.
TOBIAS, SIGMUND and CARLSON, JAMES E. 1969. Brief Report: Bartlett's test of sphericity and chance findings in factor analysis. Multivariate Behavioral Research *4*, 375.
WU, L.S., BARGMANN, R.E. and POWERS, J.J. 1977. Factor analysis applied to wine descriptors. J. Food Sci. *42*, 944.

CHAPTER 1.6

QUANTITATIVE DESCRIPTIVE ANALYSIS

HERBERT STONE, PH.D., JOEL L. SIDEL
and JEAN BLOOMQUIST

At time of original publication,
all authors were affiliated with
the Tragon Corporation
Palo Alto, California.

In 1974 after more than five years of study we introduced a new quantitative approach to descriptive analysis called QDA (1). Until that time, interest was very high in procedures that yielded descriptions of what people perceived, but the delivery was not always adequate (2,3,4). Basically a descriptive procedure was needed that was applicable to all products, food and nonfood, and was quantitative, as an alternative to available qualitative procedures.

Descriptive methods are defined as those methods that provide a word description for a product or set of products. For many years very few formal approaches to descriptive analysis existed; Flavor Profile, a method developed by Arthur D. Little, Inc., was one of the few approaches attempting to satisfy this need (5). Many companies, however, remained unconvinced and continued to rely on the individual expert or a small group of experts to make decisions regarding a product's sensory properties, storage stability, etc. Early attempts at quantification resulted in some success although attempts were limited to such methods as Texture Profile developed at the General Foods Company (6).

Progress toward greater quantification was minimal until the introduction of the QDA system of descriptive analysis (1). At that time we believed that sufficient advances had been made in scaling techniques, as well as in small group dynamics, to realize a quantitative approach to descriptive procedures. Building on what was already known, we devised a system that would satisfy the requirements of sensory professionals and function in a business environment. We determined that a quantitative approach to descriptive analysis should:

Reprinted with permission of the American Association of Cereal Chemists, St. Paul, MN.
©Copyright 1980. Originally published in Cereal Foods World 25(10), 642-644.

- Be capable of responding to all the sensory characteristics of a product, i.e., appearance, aroma, flavor, feel, etc.;
- Have quantitative procedures for determining individual and panel reliability;
- Have face validity;
- Require not more than six to 10 panelists per test;
- Be independent of the individual panelist (no experts);
- Be capable of being used with any product;
- Have a language development procedure that is easily learned and free from leader influence
- Have a language verification procedure;
- Have a data processing system that allows development of sensory derived diagrams for products;
- Be reasonably rapid.

These requirements were met, and, in the process, the QDA system evolved into a much more important management, as well as sensory tool than was appreciated at the outset.

QDA BACKGROUND

We believed a descriptive procedure must be applicable to all products with no limitations in terms of the type of product or sensory property. For example, one cannot realistically expect to have a useful method if it cannot be used to evaluate a fragrance, a fabric, a food, a hair conditioner, etc. Any product that has sensory properties should have a sensory description. To accomplish this, one cannot expect a panelist to ignore certain properties; for example, a panelist should not be asked to ignore a product's color and only evaluate its flavor. Although it is true that a panelist can score a product on a dimension-by-dimension basis, this does not mean that the perceptual process functions on a solely independent basis. Sensations enter and travel the nervous system via distinctly different pathways; in the brain considerable sensory and cognitive integration occurs and the responses are primarily interdependent. Consequently, a descriptive analysis of a product should be a total sensory picture (with no limitations).

The QDA system, because it is quantitative, requires replicate responses from the individual panelists. This allows constant monitoring of judge reliability and within sample variation. The procedure also requires that each panelist evaluate products independent of the other panelists. This too was a calculated departure from the qualitative approaches in which panel members collectively reached a group decision. Such a departure was desirable in light of the negative effects of group dynamics and the desire to quantify responses.

The system also needed to be capable of dealing with a set of products. Human beings achieve a high degree of accuracy and reliability when making relative judgements but are very poor at making absolute judgements. Thus, a single QDA session typically involved as few as three and as many as five or six products. Initial concern that an individual would be unable to evaluate so many products was very quickly proven to be of no consequence with controlled time intervals and rinse activities between samples. In fact, the presence of multiple samples provided additional information and helped characterize product similarities and differences.

The decision to have a system independent of specific panelists reflected an understanding of human behavior. Individuals do not maintain their sensory skills unless they use them regularly; if they are used too much, boredom and/or fatigue will reduce their effectiveness. Individuals have different levels of sensory skills that vary depending on the type of product. Promotions, retirement, and changing company affiliations contribute to the dynamics of the system, independent of panel skill. Finally, individuals familiar with a particular product or privy to experimental conditions may contribute biased responses and thus cannot always serve as subjects. These considerations necessitated development of a screening procedure that provided a group of individuals from which a panel could be selected, depending on the specific test product and objectives.

We considered reliability and validity measures to be essential to the overall usefulness of the system. We monitored subject reliability by evaluating replicate responses within each study, as an integral component of the data analysis. We monitored the reliability of the system as a whole by checking that different panels evaluating the same set of products made similar assessments. Tests involving systematic product variable changes also strengthened the validity of the system. The validity of product information needed to be addressed as well; for example, products are perceived as different, market shares are different, and the expectation is that their descriptions also will be different. In fact, the descriptions will be different.

The ability to visually display the sensory characteristics of a set of products, in quantitative terms, represented a significant achievement for sensory evaluation. Figure 1 illustrates a typical use for descriptive analysis and, in this instance, a quantitative form, QDA. It shows the effect of modifications to processing conditions on the sensory characteristics of a grain-flavored, formulated breakfast cereal. If the processing changes are a part of energy conservation, then it is necessary to relate these results to consumer acceptance and, if necessary, to determine the processing condition that yields the least product flavor change relative to the maximum energy savings. Additional examples of visual displays reflecting different applications for QDA are shown in an article by Zook and Wessman (7).

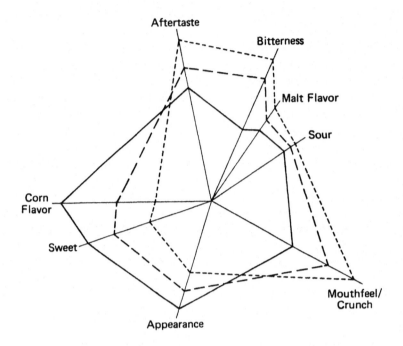

FIG. 1. QDA PICTURE SHOWING THE RELATIONSHIPS BETWEEN THE SPECIFIC
CHARACTERISTICS PERCEIVED BY THE PANELISTS
The relative intensity for each characteristic is depicted by the length of the line from the center.

CURRENT DEVELOPMENTS

As indicated in the original QDA publication (1), the technique represented an evolutionary process, an evolution that has continued to the present. Although the principles on which the method was founded were not altered, technological modifications have been made.

Additional support for using graphic line scales came from basic researchers in human measurement, especially from the work of Anderson (8,9) and from applied sensory evaluation laboratories. We improved the graphic line scales used in QDA by eliminating the middle anchor point (and its word anchor). Our current QDA data are derived from graphic line scales anchored only ½ in. from each end (of the 6-in. line).

Initially, we took a very conservative approach on replicate responses and obtained as many as eight to 12 responses from each panelist for each product.

Now we find that most products require about four responses per panelist to enable product differences and similarities to be discussed with confidence.

A procedure for pooling the responses for statistical analysis and, finally, for presenting the results in a format that could be easily communicated was also necessary. The actual data base from a study could involve as many as 10,000 entries. It became obvious to QDA users that a specialized system was needed to enter the responses (the data base) that was both rapid and error free. Such a system was developed that incorporated a digitizer with visual display, a storage system for holding information until needed, a microprocessor for information control, a printer, and interface with a main frame computer. This system more closely links the sensory professional with the data.

We also made improvements in the software for data analysis. We increased, from 25 to 50, the number of dimensions that could be analyzed at a time, and permitted up to 25 products in a test. This increase in the number of dimensions was necessary for products having large numbers of sensory characteristics, an increasingly common occurrence with technologically complex products. Another change was a summary table for individual subject F-ratios and probability values for interaction. We modified the formula for computing the individual interaction F-ratios for increased accuracy. We also provided a table of frequency of mean ranks for each characteristic to assist in distinguishing between crossover and magnitude of interactions, as described by Stone *et al.* (1). These modifications assisted the sensory professional in understanding the data.

We also made modifications in organizing the visual displays of the QDA data. Whereas originally a single diagram was used to display the correlations among dimensions and sample mean values, these two kinds of information are now displayed separately. This change does not alter the information; rather it provides a more comprehensible sensory picture of even the most complex consumer product. The primary objective of the QDA diagram is to communicate test results in a form that is readily understood with minimum discussion. This modification represented an improvement in communicating the information, a factor of extreme importance in depicting the relationships between products.

On the negative side, the popularity and acceptance of the QDA system has led some investigators to describe results of their studies as QDA data. What those investigators have failed to realize is that using a graphic line scale, analyzing data with the analysis of variance, or depicting data on circular plots does not in itself constitute a QDA. QDA is a total system covering sample selection, judge screening, language development, testing, and data. The data are only as good as the other components of the system, and QDA prescribes and carefully controls those components.

Applications of QDA

At the outset, the QDA system was considered to be a sensory evaluation tool of particular benefit to research and development. As more companies became familiar with the system, the applications were expanded. Production and quality control found the method to be especially useful as a means for checking drift in the formulation (ingredients) and in processing (operator/equipment). Market research used the method to monitor competition and to ascertain whether recent changes in share of market were due to product changes or to other factors. Market research also used the method as a product language that was understood by the consumer and that could, in turn, be used in advertising. Thus it became apparent that the system provided a basis on which sensory evaluation communicated (and operated) outside the research and development structure. A direct link was being established between production, quality control, research and development, market research, and the consumer.

Thus, what started out as a new method of quantifying the characteristics of a set of products gradually evolved into an integrated testing and evaluation system. This system provided a more precise picture of the similarities and differences of a set of products, and a direct link between consumer responses and these similarities and differences. Because it is a dynamic system, further progress continues to be made.

LITERATURE CITED

1. STONE, H., J. SIDEL, S. OLIVER, A. WOOLSEY and R.C. SINGLETON. Sensory evaluation by quantitative descriptive analysis. Food Technol. 28(11):24, 1974.
2. AMERINE, M.A., R.M. PANGBORN, and E.B. ROESSLER. Principles of Sensory Evaluation of Food. Academic Press, New York, 1965. p. 377.
3. HALL, R.L. Flavor study approaches at McCormick & Co., Inc. In *Flavor Research and Food Acceptance*. A.D. Little, Reinhold, New York, 1958. p. 224.
4. JONES, F.N. Prerequisites for test environment. In *Flavor Research and Food Acceptance*. A.D. Little, Reinhold, New York, 1958. p. 107.
5. CAIRNCROSS, S.E. and L.B. SJÖSTROM. Flavor profiles--a new approach to flavor problems. Food Technol. 4:308, 1950.
6. BRANDT, M., E. SKINNER and J. COLEMAN. Texture profile method.

J. Food Sci. 28:404, 1963.
7. ZOOK, K. and C. WESSMAN. The selection and use of judges for descriptive panels. Food Technol. 31(11):56, 1977.
8. ANDERSON, N.H. Functional measurement and psychophysical judgement. Psychol. Rev. 77(3): 153, 1970.
9. ANDERSON, N.H. Algebraic models in perception. In (Caterette, E.C. and M.P. Freidman, eds.) *Handbook of Perception*, Vol.2, Academic Press, New York, 1974. p. 215.

This publication was excerpted, in part, from the book Sensory Evaluation Practices.

CHAPTER 1.7

IMPORTANCE OF REFERENCE STANDARDS IN TRAINING PANELISTS

BARBARA A. RAINEY

*At the time of original publication,
the author was affiliated with
Sensory Evaluation Consultant
P.O. Box 622, Manteca, CA 95336.*

ABSTRACT

When training a sensory evaluation panel, reference standards play an important role in developing appropriate terminology, establishing intensity ranges and showing the action of an ingredient. In addition, use of reference standards reduces the amount of training time while providing documentation for terminology. Product characteristics can be demonstrated through reference standards for application in plant quality assurance programs as well as for project planning in new product development, product maintenance (i.e., shelf-life), product improvement and cost reduction programs.

INTRODUCTION

Reference standards are useful tools in the training of a sensory evaluation panel because they: (1) help panelists develop terminology to properly describe products, (2) help determine intensities; anchor end points, (3) show action of an ingredient and the interaction of ingredients, (4) shorten training time, (5) document terminology, (6) identify important product characteristics for the plant quality assurance program, (7) provide useful discussion tools to be used with the project team for planning new product development, product maintenance, product improvement and cost reduction programs.

While some authors discuss reference standards during their explanation on panel training (Abo Gnah and Harris 1985; Caul 1957; Civille and Szczesniak 1973; Cross *et al.* 1978; Stone and Sidel 1985; Swartz and Furia 1977), the majority do not. The importance of reference standards is either assumed and not enough information is given to understand what reference materials were utilized; or, reference standards were given cursory attention or never even used during panel training. This article was written to examine the above points of the importance of reference standards in the hopes of encouraging more attention and practice being given to the use of reference standards when training a sensory evaluation panel.

A description of a reference standard is in order. A reference standard is any chemical, spice, ingredient, or product which can be used to characterize or identify an attribute or attribute intensity found in whatever class of products (whether food or nonfood, such as hot dogs or floor wax) is being evaluated by the trained panel. Generally, the less complex the reference standard, the better, as there is less chance of confusion on the part of the panel as to what exactly the reference standard is helping describe in the actual products. An ideal reference standard is one which is simple, reproducible, identifies only one term and can be diluted without changing character (such as a sucrose solution for the term "sweet"). Terms often can be referenced by more than one standard, such as using ferrous sulfate and a 3-day-old open can of pineapple juice to reference the term "metallic". Obviously, all attributes are not so easily described by an ideal reference standard. Sometimes, a simple chemical or a single ingredient does not fully or properly describe what the panelists are finding in the product(s). In this case, several products may need to be brought in as reference standards to show that each has a similar characteristic, perhaps at differing intensities, but the same characteristic nevertheless which is found in the training (and obviously potential test) products — an example of this characteristic might be "earthy". This same technique is used to demonstrate the possible intensity range of a given characteristic. Also, while every attribute that the panel evaluates should have a written description, it is not always possible to find a reference standard for each attribute; however, this should be the exception rather than the rule. Any manner of producing or gathering a reference standard to help the panel with their task is appropriate.

Terminology Development

The most important aspect of using reference standards in training a panel is to help panelists develop the terminology they use to describe the products. Reference standards are brought in after the orientation session(s) and are generally suggested by the panel members themselves or may be brought in by the panel leader or moderator based on the descriptive terms used by the panel. Reference standards can help make the task of developing the terminology not only easier but much more accurate. Each training session should have a goal, one or two specific characteristics on which to focus attention, with appropriate reference standards brought in for discussion. This technique of breaking the larger task of describing all the sensations of a product into smaller tasks of describing just one or two sensations at each training session, provides for less confusion on the part of the panelist and increased motivation by having achieved noticeable progress during each session. Let's take the case of a cherry-flavored yogurt. Specific training sessions would focus on the terms the panelists are using to describe the fruit flavor. These terms include cherry, strawberry, fruity, artificial cherry, and orange. The panel leader/moderator would bring in reference standards to demonstrate these different fruit flavors to aid the discussion. Additional reference standards may need to be brought in, for instance, it could be that the cherry standard was a fresh bing cherry which described some of the products, but another product seemed to have more of a maraschino cherry flavor, in which case a reference standard of maraschino cherries would be brought in the next session. The panel would continue to work on describing the fruit flavor until the entire panel was satisfied with the terminology and the reference standards. As another example, cinnamon, clove, nutmeg and an "unidentified" spice are mentioned as possible attributes in a breakfast roll. The next step would be to bring in these and similar spices to show what the spices are actually like. Spices may be blended, diluted and/or incorporated into a basic roll to help identify the proper terms. In another case, panelists may only be able to say "this reminds me of the sourness found in yogurt rather than the sourness found in oranges". The panel leader/moderator would then bring in reference standards, such as lactic acid and citric acid solutions, which would demonstrate these types of sourness as a means of discussing and clarifying this term. Reference standards can help alleviate questionable attributes where panel members are not in agreement when rating products — the best place to start when there is disagreement is to be sure panelists have a clear understanding of the terminology and this is where reference standards are invaluable. When a portion of the panelists feel strongly that a characteristic is in the product(s) and the remainder of the panel cannot find this characteristic, those panelists work with the panel leader/moderator to find a reference standard for the whole panel; this should help to resolve any difference of opinion (sensitivity, of course, differs among people but "seeing" what this characteristic is through the use of a reference standard can often make the remainder of the panel aware of this characteristic).

Determine Intensities; Anchor End Points

References standards can also be used to help determine intensities and anchor the end points of the attribute scales. It is important to note here that reference standards establish relative rather than absolute intensities, however, without reference standards, intensities are entirely subjective — just how strong is "strong"? For any given attribute, the panelist can call any low level of intensity by the name "low" or "slight" but what helps anchor the term "low" or "slight" is a specific reference standard (possibly a diluted solution) that shows this lower level versus a stronger level which then can be used to anchor the other ends of the scale. For instance, for the term "pungency", full strength vinegar would represent the high range while a dilute vinegar (1 part vinegar, 3 parts water) would represent the low range. This anchoring of scales is particularly important in descriptive work where products are not directly compared or where a consensus is reached, such as in flavor profile analysis, where agreeing on low and high ranges of each attribute and corresponding reference standards during the orientation or training sessions will provide more consistency during the evaluation sessions.

Actions of Ingredients

Another use of reference standards during the training period is to show the action of specific ingredients or the interaction of ingredients. Products can be produced with and without a particular ingredient to see what this ingredient actually does for the product. Salt is a good example of this concept. Products made with and without salt are usually quite different — salt blends the components while often contributing just a very low, if any, salty taste. The interaction of ingredients can be studied to further the knowledge of the panelists and increase their awareness of the scope of potential products that the panel might evaluate once trained.

Shorten Training Time

An added benefit of using reference standards is that the training time is shortened by reducing the amount of time spent in developing the appropriate terminology and getting panelists to agree. Panelists "groping" for words to describe sensations will find it less frustrating and will learn faster if reference standards are supplied, by the panelists and the panel leader/moderator, as the terms are being identified and discussed. In addition, new panelists (or entire panels) can be added to the panel much more quickly by having the reference standards there (which have already been determined by the current panel) while discussing the attribute scales (terms). However, be careful not to unduly influence the panelists. Let the panelists try to describe the products and then bring in reference standards rather than bringing in reference standards and saying "here is what you'll find in the product".

If it has been a while since the trained panel has met, or in the case of multiproduct panels where the panel has not seen the product line recently, reviewing the reference standards and the terms prior to actual test sessions will help get the trained panel to the point they were when they last evaluated the product. Orienting the panel in this manner will make you more confident that any changes noted in product testing are more likely to have occurred because of changes in the product themselves rather than a change in the panelists' perceptions of the products.

Document Reference Standards

Having documented reference standards along with the written definition of each term will provide a more reliable description of these terms for future use. Reference standards should be documented as to what they are (chemical, brand name, product ingredients, etc.), how they were prepared (full strength, diluted solution, incorporated into a product, etc.) and how they were presented to the panel (room temperature, covered container, etc.). Even a term so simple as sweet should be documented as to it referring to a sucrose sweet with low and high ranges referenced by solution strength. This information should be recorded in the project file as well as in a continuing reference file for use on other projects. A word of caution, a reference standard may drift over time if a product, rather than a chemical, is used as the reference standard — the majority of products change slightly through the years and a "new & improved" product may no longer exhibit the characteristic that was the reference standard for the panel's descriptor.

QA Program

Reference standards can also be used in plant quality assurance programs, if the trained panel can identify the main difference(s) between a more preferred and a less preferred product (preference determined by affective testing not the trained panel), this can be a checking point for the quality assurance department. For instance, a sauce which should have a caramelized note becomes less preferred by the consumer when this note becomes burnt sugar. The trained panel can identify reference standards for the caramelized note and for the point at which it becomes burnt sugar — the QA department then would check the production line product against the reference standards for this particular characteristic rather than needing to do an entire product profile at each pull time.

Discussion Aids

Once the panel has established reference standards which help define the descriptive terms of the products, these reference standards can be used to demonstrate the characteristics of the test products to the project team for use in planning new product development, product maintenance (i.e. shelf-life), product improvement and cost reduction programs. For instance, with a competitive product survey of a processed meat product, the descriptive panel has evaluated four products and the results of the study are being presented to marketing, market research and R&D. One of the major differences among the products is the type and intensity of smoke found in the products. To highlight the product differences, reference standards of several varieties and intensities of liquid smoke and natural hickory chips can be used as demonstrations to show how the four competitive products differ in this characteristic, thus aiding the discussion and "next-step" planning.

For all of the above reasons, reference standards should be employed during panel training to make the best use of the panel's and panel leader/moderator's time while providing for more reliable and consistent results.

REFERENCES

ABO GNAH, Y.S. and HARRIS, N.D. 1985. Determination of musty odor compounds produced by Streptomyces griseus and Streptomyces odorifer. J. Food Science 50, 132-135.

CAUL, J.F. 1957. The profile method of flavor analysis. Adv. in Food Res. 7, 1-40.

CIVILLE, G.V. and SZCZESNIAK, A.S. 1973. Guidelines for training a texture profile panel. J. Texture Studies 4(2), 204.

CROSS, H.R., MOEN, R. and STANFIELD, M.S. 1978. Training and testing of judges for sensory analysis of meat quality. Food Technol. 32, 48-54.

STONE, H. and SIDEL, J.L. 1985. Sensory Evaluation Practices, pp. 205-206, Academic Press, Orlando, Florida.

SWARTZ, M.L. and FURIA, T.E. 1977. Special sensory panels for screening new synthetic sweeteners. Food Technol. 31, 51-55, 67.

CHAPTER 1.8

THE IMPORTANCE OF LANGUAGE IN DESCRIBING PERCEPTIONS

GAIL VANCE CIVILLE and HARRY T. LAWLESS

*At the time of original publication,
author Civille was affiliated with
Sensory Spectrum, Inc.*

and

*author Lawless was affiliated with
S.C. Johnson & Son, Inc.*

ABSTRACT

A set of terms in descriptive analysis should enable differentiation among products, specification of the sensory properties of the products, and sufficient characterization to permit its recognition or identification. Meeting these criteria enables the use of descriptive data in several applications: (1) interpretation of other sensory data, (2) correlation with instrumental measures, (3) quality monitoring, and (4) product development and maintenance. Several aspects of the specific terminology and the way it is taught are important in effective descriptive analysis. Terms should be uncorrelated with one another. Terms should be related to physical or chemical references. Training a panel requires the establishment of a system in which relevant dimensions can be abstracted and attended to. In this paper, specific problems of descriptive analysis training such as the choice of reference sets and the boundaries of terms are discussed.

INTRODUCTION

One major goal of sensory evaluation in general and of descriptive analysis in particular is the objective description of a product in terms of perceived sensory attributes. To the extent that these descriptions are necessarily verbal, language plays a central role in determining the accuracy and potential benefits of a given evaluation. A primary issue in descriptive analysis is the choice of terms to be used for any particular product or category of products. This issue has received little attention in spite of its importance.

Different approaches are taken in industrial product evaluation in choosing qualitative descriptors. For example, terms may be chosen and defined by underlying physical or chemical properties of the product, as opposed to consumer-directed or consumer-oriented language. At one end of this continuum is the Texture Profile method which derives sensory texture terms from objective texture characteristics based on rheological principles (Brandt et al. 1963; Szczesniak et al. 1963). On the other extreme, some forms of descriptive analysis allow judges to develop terminology similar to common consumer vocabulary.

Regardless of the orientation toward the selection of descriptive terms, clearly, language can play an important role in influencing our perceptions. This can work for the better, in the direction of greater accuracy and objectivity, or for the worse, in biasing the panelists to perceive attributes that may not be present, or even to fail to differentiate fundamental attributes that should be perceptually separate. Over the years psychologists and anthropologists have discussed the interplay of language and perception. Benjamin Whorf and his followers took one extreme point of view in saying that language both reflects and determines the way in which we perceive the world (Whorf 1956). For example, eskimos have several words in their vocabulary to describe and differentiate among types and conditions of snow. From another perspective, Gibson (1966) argued that perception is largely determined by the information and structure offered by stimulation from the environment. A related position, taken by Eleanor Rosch, Donald Homa and others, holds that we learn to abstract patterns of correlated sensory characteristics in forming concepts and categories (Rosch et al. 1976; Homa 1978). This process influences the basic classes with which we most commonly refer to objects. Patterns such as "four-legged, furry, and tail-wagging" together signify "dog."

In experimental psychology, several areas of research relate to terminology development. Early in this century, Gestalt psychologists demonstrated that people are often not analytical in their perceptions (Garrett 1930). Rather, they tend to synthesize or integrate whole patterns of stimulation as single figures; they are automatically influenced by the context within which a stimulus is viewed. Thus, the fractionation of a perception into its component sensory impressions may be very difficult. Research in concept learning addressed the ways in which attributes were identified to classify similar objects together and the rules for combining those attributes in determining class membership (Dominowski 1974). More recently, literature on category development also discussed the ways in which objects are grouped (Homa and Chambliss 1975; Rosch and Mervis 1975; Rosch et al. 1976).

Of these three major areas of perceptual and cognitive investigation, the category formation literature is most relevant to our questions of terminology development, since these studies have focussed on identification of critical attributes and factors affecting the rate of learning of new categories. The concept formation literature has been more concerned with the purely cognitive activity of abstracting logical relationships (e.g. Bourne 1970), and in many cases the rule for using certain sensory terms are less than logical and explicit. The Gestalt tradition simply offers a warning that our perceptions (and therefore our sensory terms) are often integrative, and the analytical process of applying descriptive terminology may be a rocky road at times. In the discussion that follows, we outline several principles of terminology development which are supported by the psychological literature.

Language uses words to represent concepts, objects, or attributes. A central theme in our thinking has been that language does not coerce perception but permits observers to tune their perception to certain differences rather than others. Words and gestures develop into code for given concepts and then direct the observer to selected parts of the environment (Gibson 1966). Gibson also suggests that the organs of perception are oriented so that the whole system of input and output resonates to the external information. Words per se may also act as memory links to aid in the retrieval or establishment of a mental image. There is some recent thought that in olfaction at least, images may function as a template to match to incoming information, thus aiding identification (Freeman 1983).

Criteria for Systems of Terminology

From a psychological perspective a descriptive term does not reproduce the perception it represents, but, as outlined by Harper *et al.* (1974), it can and must permit (1) differentiation from similar sensations, (2) identification of the object it describes, and (3) recognition of the object by others seeing the term.

Differentiation. The act of description often begins with various classifications of a stimulus into familiar categories. This process of categorization is so fundamental to human perception that the classification of objects may even precede the naming of the class or category (Rosch *et al.* 1976). Classification uses the information which is common to the members of a class, as well as the information which differentiates that class from others. Objects classified in the same category usually possess significant numbers of common attributes. As observers compare, match and or distinguish objects they take greater notice of critical differences among category features and less notice of irrelevant characteristics (Gibson and Gibson 1955). Furthermore, the greater the importance of the differentiation to the observer, the greater is the diversification and refinement of categorization. Even skiers feel the need to use several terms to describe the various ski conditions.

In addition to being able to differentiate objects, any classification system must employ terminology which permits the distinguishing sensory properties to be completely or nearly completely specified. Terminology used to classify should imply the broad similarities among members of that class. Structure in the environment often determines "natural" categories, which are reduced to behaviorally and cognitively simple and useful factors (Gibson 1966; Rosch *et al.* 1976). These natural categories (table, nutty, shoes, citrus) yield the most information with the least verbal complexity.

Identification. Identification (naming) a stimulus which is coded, unlabeled or otherwise disguised is most difficult in olfaction. Even when the name of a substance is known and the odor is familiar a searching process may occur, which even after a reasonable time span fails to elicit the required word (Harper *et al.* 1974). This phenomenon of smelling a familiar fragrance, but finding an annoying mnemonic gap from the sensation to the name, was once called the "tip of the nose" phenomenon (Lawless and Engen 1977). The presence of a list of words to jog the odor memory enhances the identification rate of odor impressions, since it provides a bridge across the olfactory-verbal gap that most people experience (Cain 1979; Lawless 1985). The memory code for odors may incorporate semantic information, which enhances performance on later recognition (Walk and Johns 1984). An important part of terminology development and descriptive training, then, is providing either an explicit or an internalized list of attributes which may be expected in any given type of product. These expectations may then act as memory aids. For example, expecting a certain type of Bordeaux wine to have notes of raspberry, currants, and cherry may enhance the descriptive panelists' accurate identification of those attributes if they are indeed present.

Recognition. Globally, a good description of an object would allow you to recognize that object after someone had described it to you. Such matching of descriptions to products has been used to check the validity of descriptive systems in the absence of other criteria for what makes a description "good" (Lehrer 1983; Lawless 1984). This recognition process establishes the communication value of a description and depends upon a commonly understood (shared) set of terms.

Individually, these terms must enable recognition by others as to the nature of the perceived attribute or categorization, and assumes an ability to recognize those contributing features which distinguish one class or category from another. Terminology should enable those distinguishing features to be isolated and described. Some terms may be more informative than others. Rosch *et al.* (1976) describe cue validity in terms of having many attributes common to members of the category and few attributes shared with members of other categories.

Validation by Convergence. If a classification systems permits recognition among different groups of observers, i.e., if it has the communication value discussed above, the terminology system not only implies a shared language and shared understanding, but that system may also be reflecting some underlying structure in the environment and the resonance of the sensory system to that structure. For example, when different scientists in different laboratories converge on similar descriptive systems, one can infer a high degree of validity in the system.

An example of such convergence was recently observed in fragrance description systems. To establish a set of odor classes for fragrance evaluation of consumer products, a list of ten general fragrance descriptors was selected during panel training at S.C. Johnson, based on our experience with fragrance and flavor evaluations. These major fragrance categories are listed in Table 1.

TABLE 1.
FRAGRANCE TERMS DERIVED BY EXPERIENCE

WOODY, NUTTY
MINTY
FLORAL
FRUITY (NOT CITRUS)
CITRUS
SWEET (e.g., VANILLA, MALTOL)
SPICE
GREEN
HERBAL

In research conducted at Philip Morris and reported by Melissa Jeltema and Rhett Southwich at the Sensory Evaluation Symposium on November 21, 1985, (sponsored by Western New York Section of IFT), a panel of 20 judges evaluated thirty-five odorants from the ASTM Atlas of Odor Character Profiles (Dravniecks 1985). Panelists chose from one or more of the 146 descriptors used in the odor profile text. Factor analysis of these terms reduced them to a list of 16 factors shown in Table 2 which are closely aligned to the list on Table 1.

The initial part of the lists are quite similar. Apparent points of divergence are actually predictable. For example, the coconut/almond, and anise/caraway characteristics in Table 2 were points of much discussion among the panels using the words in Table 1. Panelists argued that the almond and coconut odors could fall into either nutty or sweet categories and that fragrances with anise notes could fall into either sweet or spice. For the terms listed at the bottom of Table 2, the panelists who formulated Table 1 were provided with a catchall category of "other" for some off-notes that were rarely encountered in their products. This category requires panelists to specify the descriptor other than the specific major classes listed.

TABLE 2.
FRAGRANCE TERMS DERIVED BY FACTOR ANALYSIS

WOODY
NUTTY
COOL, MINTY
FRUITY, NONCITRUS
CITRUS
BROWN (e.g., VANILLA, MOLASSES)
SPICY
GREEN
COCONUT, ALMOND
CARAWAY, ANISE
ANIMAL, FOUL
SOLVENT
BURNT
SULFIDIC
RUBBER

There appears to be strong evidence for some underlying structure in odor classification since two very similar systems were derived independently from different odor evaluation methods and different stimulus sets from panelists working in different industries. Dominowski (1974) defined a "class concept" as a list of relevant attributes plus the rule applied to be attributes to determine category membership. When the proper term can elicit the same class concept in two or more groups of subjects the selected classes appear to reflect some underlying structure.

Criteria for Specific Terms

In constructing a word set which adequately describes a product's sensory properties the key aspects of the words which researchers should consider are (1) that the terms should be orthogonal, (2) that they should be based on underlying structure if it is known, (3) that the terms should be based on a broad reference set, (4) that they should be precisely defined, and finally, that they should be "primary" rather than "integrated." These individual aspects are elaborated below.

The terms should be *orthogonal*, that is, uncorrelated with each other. Redundancy in terms confuses both the panelist and the researchers or managers who are trying to interpret the sensory data. For example, in texture work, panelists are often initially confused by hardness and denseness terms, because many products increase in hardness as they become more compact. By constructing a correlating graph, as shown in Fig. 1, and by discussing products which differ widely and sometimes independently in these attributes, a texture panel leader can show the lack of correlation between the two texture terms. Such demonstrations are important since the terms are positively correlated in the (naive) impressions of untrained observers.

Descriptive terminology should be *related to the underlying natural structure*, if it is known. Errors in identification of natural categories often stem from ignorance or inattentiveness to attributes, and ignorance or exaggeration of the underlying structure (Rosch *et al.* 1976). In devising a classification or terminology system subjects must be made aware of the underlying physical or rheological principles for texture terms, chemical principles for aroma or flavor terms, and color and geometrical principles for appearance terms. Because there are terms in the vocabulary that are used too broadly or applied imprecisely, clarification by an individual who understands the structure which underlies the categories can aid learning. The potential of such an individual to influence the resulting classification or terminology system requires that this teacher or trainer have a clear working knowledge of the underlying structure and the various objects or exemplars within each class.

The understanding of the underlying structure is influenced by the *frame of reference* of the perceiver. Choice of the reference set is critical in affecting the proper assignment or words. Gibson and Gibson (1955) stressed the need for a broad category experience to enhance knowledge of the structure, discrimination among objects, and future uses of classification. The fact that categorization is more accurate when it is initially abstracted from a broad array of examples has been repeatedly verified (Homa *et al.* 1973; Homa and Vosburgh 1976; Homa 1978; Homa and Cultice 1984).

In abstracting and subsequently understanding the concept of blue, the use of only one reference or exemplar, such as a royal blue, impairs the comprehension of the range or scope of the class or category which the reference is meant to represent. Selection of a broad array of blue objects or color standards better defines the entire frame of reference for the observer, and thus enables improved discrimination, identification, and recognition in future applications of the class or term. The use of several reference samples to represent one concept, which itself is almost always a range of objects rather than a singular phenomenon, enhances the initial learning and future use of term to describe that concept.

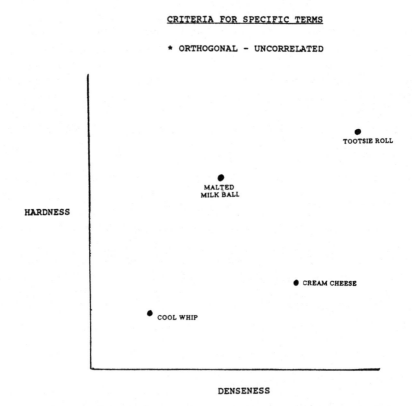

FIG. 1. TO DEMONSTRATE THE INDEPENDENCE OF THE TERMS "HARDNESS"AND "DENSENESS", PRODUCTS WHICH ARE HARD BUT NOT DENSE (eg., MALTED MILK BALL) AND DENSE BUT NOT HARD (CREAM CHEESE) CAN BE POINTED OUT TO PANELISTS IN TRAINING

In addition, the understanding of the boundaries of a class or term greatly influences the proper application of the term and the use of rating scales to indicate the amount of that attribute that is perceived. In the color section shown in Fig. 2, the clear definition of the term red greatly affects responses to the question: What percent of the bar is red? Subjects defining red, as "exclusively red" report 25-40%. Subjects defining red as "mostly red" report 40-50%. And, subjects defining red as "any redness at all" report 50-60%. With olfactory descriptions in particular, such as in fragrance and volatile flavor characteristics, there is a high degree of overlapping of terms and a blurring of boundaries. Panelists often have difficulty cutting up the pie appropriately and assigning intensity values to reflect the relevant strength and proportion of each aroma attribute to the whole.

CRITERIA FOR SPECIFIC TERMS

*FRAME OF REFERENCE

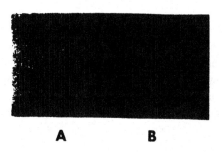

ESTIMATE THE % REDNESS

A - Red area
B - Blue area

FIG. 2. PROBLEMS MAY ARISE WHEN PANELISTS SET DIFFERENT CRITERIA
FOR WHICH QUALITIES THEY WILL INCLUDE IN A CATEGORY
Depending upon your definition of "redness," the amount
of red that is rated in this stimulus will vary.

Another factor which strongly influences a terminology system is the quality of training, which can affect both the rate and amount of learning. Referring specifically to fragrance training, Jellinek (1964) recommended teachers "with various aids to memory", knowledge and experience, which enables them to interpret "versatile associations" from subjects. She also asserted that rapid progress can be made under such a teacher. Responses to the odor of oil of bitter almond often include descriptors such as almond, cherry, cough drops, Amaretto, and Danish pastry. The experienced teacher or panel leader can recognize the commonality in these responses to the underlying benzaldehyde character, which is also a major component in cherry pits, wild cherry flavors, and almond oils. Appropriate and timely feedback is crucial in learning complex systems of classification. Homa and Cultice (1984) showed that appropriate feedback contributed substantially to learning and that without feedback learning is often reduced because the observer must extract the structure of the categories and relationships among the stimuli without help.

In addition to appropriate reference sets and feedback, discussion of the terms and their underlying structure by the panelists themselves can also aid learning. In research on concept learning, group discussion has been shown to facilitate learning (Loughlin and Dougherty 1967). The benefits involve the ability of individuals to monitor and check the inferences and reasoning of the others in the group, which then speeds the process of concept attainment by all members.

The precision in definition of the terms is related to the above mentioned understanding of the underlying natural, chemical or physical structure, the proper determination of terminology boundaries, and the proper abstraction of concepts by the group. Without such considerations the identification of attribute words and the precise characteristics which define those words can be sources of error. Words can facilitate and even direct the generation of a search image causing "ideational excitement" (James 1913) which helps the observer to perceive a particular attribute or object more readily, given training about what to look for. Experts who are knowledgeable about the structure and detail of the environment may see certain attributes and therefore use words which are ignored by the average person.

Observers may also fail to identify a given attribute as being relevant to a category of objects, because that attribute or term fails to differentiate among the members of the category. For example, descriptive panelists may not use the term "dense" to describe a series of cream cheeses, because all the objects in the set are quite dense. Likewise, a group working with hard candies may not describe the set members as sweet, because all are similarly very sweet. The experienced teacher or leader should decide if the term may still be necessary to describe the objects in the set and differentiate them from objects in other sets, although it may not differentiate within the set.

The *use of primary terms*, those which are elemental, rather than combinations of several terms, reduces confusion among the observers and the end users of the data. Combination or integrated terms, such as creamy, soft, refreshing, and clean, may be measurable in large groups of consumers who generally originate such terms. In fact, it is a principle or cognitive development that the inexperienced observers in a new problem solving task will often represent stimuli holistically, rather than breaking them into their analytical parts (Sternberg 1986; Smith and Kemler 1977). While such integrated terms may be useful for advertising purposes, they provide little actionable information to product formulators who need objectively anchored and elemental sensory information in order to adjust ingredients.

Furthermore, trained observers operating in small groups often encounter difficulty with integrated terms because each individual may weight the underlying elemental attributes (component attributes that contribute to the integrated term) differently. For example, if "creamy" is a function of smoothness, viscosity, and fatty mouthfeel, the observer who gives greater weight to increases in viscosity will rate the product creaminess different than the observer responding strongly to change in smoothness. Attempts should be made to evaluate the individual attributes (smooth, fatty, viscous) and then construct a model correlating the consumer's perception of the integrated term ("creamy") with the analytical component attributes.

Applications/Use of Terminology

The principles outlined above are useful in developing sensory terminology which can be applied to a broad array of industrial product evaluation needs. The precise and reliable use of words to describe the sensory attributes of products and materials has many benefits. When panelists have constructed a list of specific product attributes, which can cover appearance, flavor, texture, aroma, and sound, these can be used to document and define products in the following situations:

When *other sensory data* yield confusing or incomplete results, descriptors can be used to characterize the product attributes and elucidate the source of anomaly. When a product does poorly in a consumer test, descriptive words can help detect problems, such as the presence of off-flavors or a lack of appropriate or expected crispness. Such descriptors are often helpful in the construction of consumer questionnaires when consumer attribute responses are employed. The descriptors document which attributes discriminate among the test products and which attributes are non-discriminators. The consumer research can then be directed to the perceivable attribute differences.

The use of *instrumental data* cannot provide direct appearance, flavor, aroma, or texture information. Only human subjects can measure sensory attributes and are, therefore, needed to correlate with and explain the physical and chemical measurements recorded with instrumentation. The detailed description of the chemical and physical sensory attributes, their precise definitions, and the exact procedures used during evaluation can often provide the starting place for the analytical chemist and rheologist in establishing protocols and direction for the instrumental analyses.

In *product development, product maintenance, and quality monitoring*, the use of qualitative documentation provides a means to measure and monitor changes in products from a control or target due to changes in age, storage, packaging, ingredients, processing, and formulation. Such tracking is widely used in the food industry to document research and manufacturing changes reliably.

A better understanding of the psychological literature with respect to classification systems, terminology development and usage, and aids to learning should help sensory analysts to provide better systems for descriptive analysis and other attribute evaluations. The use of precise terminology, which is clearly defined and referenced, and which is based on the underlying structure in the environment should yield sensory attributes that can provide valuable information in addressing business decisions related to aroma, flavor and texture of consumer products.

REFERENCES

BOURNE, L.E. 1976. Knowing and using concepts. Psych. Rev. *77*, 546.

BRANDT, M.A., SKINNER, E.Z. and COLEMAN, J.A. 1963. Texture profile method. J. Food Sci. *28*, 404.

CAIN, W.S. 1979. To know with the nose: keys to odor identification. Science *203*, 467.

DOMINOWSKI, R.L. 1974. How do people discover concepts? In *Theories in Cognitive Psychology*, (R.L. Solso, ed.), Lawrence Erlbaum Associates, Potomac, MD.

DRAVNIEKS, A. 1985. *Atlas of Odor Character Profiles*. American Society for Testing and Materials, Philadelphia, PA.

FREEMAN, W.J. 1983. The physiological basis of mental images. Biol. Psychiat. *18*, 1107.

GARRETT, H. 1930. *Great Experiments in Psychology*. The Century Co., New York.

GIBSON, J.J. 1966. *The Senses Considered as Perceptual Systems*. Houghton Mifflin, Boston, MA.

GIBSON, J. and GIBSON, E. 1955. Perceptual learning: differentiation or enrichment? Psych. Rev. *62*, 32.

HARPER, R., BATE SMITH, E.C. and LAND, D.G. 1974. *Odour Description and Odour Classification: A Multidisciplinary Examination*. American Elsevier Publishing Co., New York.

HOMA, D. 1978. Abstraction of ill-defined form. J. Exp. Psychol.: Human Learning and Memory *4*, 407.

HOMA, D. and CHAMBLISS, D. 1975. The relative contributions of common and distinctive information on the abstraction from ill-defined categories. J. Exp. Psychol.: Human Learning and Memory *104*, 351.

HOMA, D. and CULTICE, J. 1984. Role of feedback, category size, and stimulus distortion on the acquisition and utilization of ill-defined categories. J. Exp. Psychol.: Learning, Memory and Cognition. *10*, 83.

HOMA, D. and LITTLE, J. 1985. The abstraction and long-term retention of ill-defined categories by children. Bull. Psychon. Soc. *23*, 325.

HOMA, D. and VOSBURGH, R. 1976. Category breadth and the abstraction of prototypical information. J. Exp. Psychol.: Human Learning and Memory *2*, 322.

HOMA, D., CROSS, J., CORNELL, D., GOLDMAN, D. and SCHWARTZ, S. 1973. Prototype abstraction and classification of new instances as a function of number of instances defining the prototype. J. Exp. Psychol. *101*, 116.

JAMES, W. 1913. *Psychology*. Henry Holt & Co., New York.

JELLINEK, G. 1964. Introduction to and critical review of modern methods of sensory analysis with special emphasis on descriptive sensory analysis. J. Nutr. Dieta. (Dehli) *1*, 219.

JELTEMA, M.A. and SOUTHWICK, E.W. 1986. Evaluation and applications of odor profiling. J. Sensory Studies *1*(2), 123-136.

LAUGHLIN, P. and DOHERTY, M.A. 1967. Discussion vs. memory in cooperative group concept attainment. J. Ed. Psychol. *58*, 123.

LAWLESS, H. 1984. Flavor description of white wine by "expert" and non-expert wine consumers. J. Food Sci. *49*, 120.

LAWLESS, H. 1985. Psychological perspectives on wine tasting and recognition of volatile flavors. Ch. 8. In *Alcoholic Beverages* (G.G. Birch and M.G. Lindley, eds.), p. 97. Elsevier Applied Science Publishers, London.

LAWLESS, H. and ENGEN, T. 1977. Associations to odors: Interference mnemonics and verbal labeling. J. Exp. Psychol. Human Learning and Memory *3*, 52.

LEHRER, A. 1983. *Wine and Conversation*. University of Indiana Press, Bloomington, IN.

ROSCH, E. and MERVIS, C.B. 1975. Family resemblances: studies in the internal structure of categories. Cog. Psychol. *7*, 573.

ROSCH, E., MERVIS, C.B., GRAY, W.D., JOHNSON, D.M. and BOYES-BRAEM, P. Basic objects in natural categories. Cog. Psychol. *8*, 382.

SMITH, L.B. and KEMLER, D.G. 1977. Developmental trends in free classification: Evidence for a new conceptualization of perceptual development. J. Exp. Child Psychol. *24*, 279.

SZCZESNIAK, A.S., BRANDT, M.A. and FRIEDMAN, H.H. 1963. Development of standard rating scales for mechanical parameters of texture and correlation between the objective and sensory methods of texture evaluation. J. Food Sci. *28*, 397.

STERNBERG, R.J. 1986. Inside intelligence. Am. Scientist. *74*, 137.

WALK, H.A. and JOHNS, E.A. Interference and facilitation in short-term memory for odors. Percep. Psychophys. *36*, 508.

WHORF, B.J. 1956. *Language, Thought and Reality*. MIT Press and John Wiley & Sons, New York.

CHAPTER 1.9

ILLUSTRATIVE EXAMPLES OF PRINCIPAL COMPONENTS ANALYSIS

W.T. FEDERER, C.E. McCULLOCH
and N.J. MILES-McDERMOTT

*At the time of original publication,
all authors were affiliated with
Biometrics Unit, Cornell University
Ithaca, New York 14853-7801.*

ABSTRACT

In order to provide a deeper understanding of the workings of principal components, four data sets were constructed by taking linear combinations of values of two uncorrelated variables to form the X-variates for the principal component analysis. These examples are believed useful for aiding researchers in the interpretation of data and they highlight some of the properties and limitations of principal components analyses.

INTRODUCTION

Principal components is a form of multivariate statistical analysis and is one method of studying the correlation or covariance structure in a set of measurements on m variables for n individuals. For example, a data set may consist of n = 260 samples and m = 15 different fatty acid variables. It may be advantageous to study the structure of the 15 fatty acid variables since some or all of the variables may be measuring the same response. One simple method of studying the correlation structure is to compute the m(m-1)/2 pairwise correlations and note which correlations are close to unity. When a group of variables are all highly intercorrelated, one may be selected for use and the others discarded or the sum of all the variables can be used. When the structure is more complex, the method of principal components analysis (PCA) becomes useful.

In order to use and interpret a principal components analysis there needs to be some practical meaning associated with the various principal components. In this paper, we describe the basic features of principal components and examine some constructed examples to illustrate the interpretations that are possible.

Reprinted with permission of Food & Nutrition Press, Inc., Trumbull, Connecticut.
©Copyright 1987. Originally published in Journal of Sensory Studies 2 (1987) 37-54.

BASIC FEATURES OF PRINCIPAL COMPONENTS ANALYSIS

PCA can be performed on either the variances and covariances among the m variables or their correlations. One should always check which is being used in a particular computer package program. First we will consider analyses using the matrix of variances and covariances. A PCA generates m new variables, the principal components (PCs), by forming linear combinations of the original variables, $X = (X_1, X_2, ..., X_m)$, as follows:

$$PC_1 = b_{11}X_1 + b_{12}X_2 + ... + b_{1m}X_m = Xb_1$$
$$PC_2 = b_{21}X_1 + b_{22}X_2 + ... + b_{2m}X_m = Xb_2$$

$$PC_m = b_{m1}X_1 + b_{m2}X_2 + ... + b_{mm}X_m = Xb_m.$$

In matrix notation,

$$P = (PC_1, PC_2, ..., PC_m) = X(b_1, b_2, ..., b_m) = XB.$$

The rationale in the selection of the coefficients, b_{ij}, that define the linear combinations that are the PC_1 is to try to capture as much of the variation in the original variables with as few PCs as possible. Since the variances of a linear combination of the Xs can be made arbitrarily large by selecting very large coefficients, the b_{ij} are constrained by convention so that the sum of squares of the coefficients for any PC is unity:

$$\sum_{j=1}^{m} b_{ij}^2 = 1 \qquad i=1,2,...,m.$$

Under this constraint, the b_{1j} in PC_1 are chosen so that PC_1 has maximal variance among all PCs.

If we denote the variance of X_i by s_j^2 and if we define the total variance, T, as $T = \sum_{i=1}^{m} s_j^2$, then the proportion of the variance in the original variables that is captured in PC_1 can be quantified as $var(PC_1)/T$. In selecting the coefficients for PC_2, they are further constrained by the requirement that PC_2 be uncorrelated with PC_1. Subject to this constraint and the constraint that the squared coefficients sum to one, the coefficients b_{2j} are selected so as to maximize $var(PC_2)$. Further coefficients and PCs are selected in a similar manner, by requiring that a PC be uncorrelated with all PCs previously selected and then selecting the coefficients to maximize variance. In this manner, all the PCs are constructed so that they are uncorrelated and so that the first few PCs capture as much variance as possible. The coefficients also have the following interpretation which helps to relate the PCs back to the original variables. The correlation between the ith PC and the jth variable is

$$b_{ij} \sqrt{\mathrm{var}(PC_i)}/s_j$$

After all m PCs have been constructed, the following identity holds:
$$\mathrm{var}(PC_1) + \mathrm{var}(PC_2) + \ldots + \mathrm{var}(PC_m) = T = \sum_{i=1}^{m} s_i^2$$
This equation has the interpretation that the PCs divide up the total variance of the Xs completely. It may happen that one or more of the last few PCs have variance zero. In such a case, all the variation in the data can be captured by fewer than m variables. Actually, a much stronger result is also true; the PCs can also be used to reproduce the actual values of the Xs, not just their variance. We will demonstrate this more explicitly later.

The above properties of PCA are related to a matrix analysis of the variance-covariance matrix of the Xs, S_x. Let D be a diagonal matrix with entries being the eigenvalues, λ_i, of S_x arranged in order from largest to smallest. Then the following properties hold:

(1) $\lambda_i = \mathrm{var}(PC_i)$

(2) $\mathrm{trace}(S) = \sum_{i=1}^{m} s_i^2 = T = \sum_{i=1}^{m} \lambda_i = \sum_{i=1}^{m} \mathrm{var}(PC_i)$

(3) $\mathrm{corr}(PC_i, X_j) = \dfrac{b_{ij} \sqrt{\lambda_i}}{s_j}$

(4) $S_x = B'DB$,

where B is the matrix of the coefficients of PCs defined earlier.

The statements made above are for the case when the analysis is performed on the variance-covariance matrix of the Xs. The correlation matrix could also be used, which is equivalent to performing a PCA on the variance-covariance matrix of the standardized variables,

$$Y_{ij} = \frac{X_{ij} - \bar{X}_i}{s_i}$$

PCs using the correlation matrix is different in these respects:

(1) The total "variance" is m, the number of variables. (It is not truly variance anymore.)

(2) The correlation between PC_i and X_j is given by $b_{ij} \sqrt{\mathrm{var}(PC_i)} = b_{ij} \sqrt{\lambda_i}$. Thus PC_i is most highly correlated with the X_j having the largest coefficient in PC_i in absolute value.

The experimenter must choose whether to use standardized (PCA on a correlation matrix) or unstandardized coefficients (PCA on a variance-covariance matrix). The latter is used when the variables are measured on a comparable basis. This usually means that the variables must be in the same units and have roughly comparable variances. If the variables are measured in different units then the analysis will usually be performed on the standardized scale, otherwise the analysis may only reflect the different scales of measurement. For example, if a number of fatty acid analyses are made, but the variances, s_i^2, and means, $\bar{X}_i$, are obtained on different bases and by different methods, then standardized variables would be used (PCA on the correlation matrix). A note of caution in using standardized variables is that this makes the ranges of all variables comparable and this may not be what an investigator desires.

To illustrate some of the above ideas, a number of examples have been constructed and these are described in the next section. In each case, two variables, Z_1 and Z_2, which are uncorrelated, are used to construct X_i. Thus, all the variance can be captured with two variables and hence only two of the PCs will have nonzero variances. In matrix analysis terms, only two eigenvalues will be nonzero. An important thing to note is that, in general, PCA *will not recover the original variables Z_1 and Z_2*. Both standardized and nonstandardized computations will be made.

To compute residuals after computing a particular PC, we proceed as follows. The residuals after computing PC_1 are:

$$\mathbf{X} - \mathbf{X}b_1 b_1' = \hat{e}_1;$$

those residuals after fitting both the first and second PCs are:

$$\mathbf{X} - \mathbf{X}b_1 b_1' - \mathbf{X}b_2 b_2' = \hat{e}_{12};$$

and so forth. The fitting of residuals is also a convenient way of envisioning the extraction of further PCs. PC_2 can be calculated by extracting the *first* PC from the residuals, e_1. That is, it can be calculated by forming the variance-covariance matrix of the residuals and performing PCA on it.

The computations to calculate the residuals are performed on each of the n observations consisting of m variables. A study of residuals may be helpful in finding outliers and patterns of observations.

EXAMPLES

Throughout the examples we will use the variables Z_1 and Z_2 (with $n = 11$) from which we will construct $X_1, X_2, \ldots, X_m$. We will perform PCA on the Xs.

Thus, in our constructed examples, there will only really be two underlying variables.

Values of Z_1 and Z_2

Z_1	-5	-4	-3	-2	-1	0	1	2	3	4	5
Z_2	15	6	-1	-6	-9	-10	-9	-6	-1	6	15

Notice that Z_1 exhibits a linear trend through the 11 samples and Z_2 exhibits a quadratic trend. They are also chosen to have mean zero and be uncorrelated. Z_1 and Z_2 have the following variance-covariance matrix (a variance-covariance matrix has the variance for the ith variable in the ith row and i^{th} column and the covariance between the ith variable and the jth variable in the ith row and j^{th} column).

Variance-covariance matrix of Z_1 and Z_2

$$\begin{bmatrix} 11 & 0 \\ 0 & 85.8 \end{bmatrix}$$

Thus the variance of Z_1 is 11 and the covariance between Z_1 and Z_2 is zero. Also, the total variance is 11 + 85.8 = 96.8.

Example 1: In this first example we analyze Z_1 and Z_2 as if they were the data. Thus $X_1 = Z_1$ and $X_2 = Z_2$ and m = 2. If PCA is performed on the variance-covariance matrix then the output is as follows:

	EIGENVALUE	DIFFERENCE	PROPORTION	CUMULATIVE
PRIN1	85.80000[3]	74.80000	0.88636	0.88636
PRIN2	11.00000[4]		0.11364	1.00000

EIGENVECTORS

	PRIN 1	PRIN2
X1	0.000000[1]	1.000000[2]
X2	1.000000	0.000000

The above output was produced using SAS/PRINCOMP, Version 82.3. Note that PRIN1 = PC_1 in the previous notation. We can interpret the results as follows:

(1) The first principal component is $PC_1 = 0 \cdot X_1 + 1 \cdot X_2 = X_2$

(2) $PC_2 = 1 \cdot X_1 + 0 \cdot X_2 = X_1$

(3) $Var(PC_1)$ = eigenvalue = 85.8 = $Var(X_2)$

(4) $Var(PC_2)$ = eigenvalue = 11.0 = $Var(X_1)$

The PCs will be the same as the Xs whenever the Xs are uncorrelated. Since X_2 has the larger variance it becomes the first principal component.

If PCA is performed on the correlation matrix we get slightly different results.

Correlation Matrix of Z_1 and Z_2

$$\begin{bmatrix} 1 & 0 \\ 0 & 1 \end{bmatrix}$$

A correlation matrix always has unities along its diagonal and the correlation between the ith variable and the jth variable in the ith row and j^{th} column. PCA would yield the following output:

	EIGENVALUE	DIFFERENCE	PROPORTION	CUMULATIVE
PRIN1	1.000000	0.000000	0.500000	0.500000
PRIN2	1.000000		0.500000	1.000000

EIGENVECTORS

	PRIN 1	PRIN2
X1	1.000000	0.000000
X2	0.000000	1.000000

The principal components are again the Xs themselves, but both eigenvalues (var(PC)s) are unity since the variables have been standardized first.

Example 2: Let $X_1 = Z_1$, $X_2 = 2Z_1$ and $X_3 = Z_2$. The summary statistics are given below.

	X1	X2	X3
MEAN	0.000000	0.000000	0.000000
ST DEV	3.316625	6.63325	9.262829

CORRELATIONS

	X1	X2	X3
X1	1.0000	1.0000	0.0000
X2	1.0000	1.0000	0.0000
X3	0.0000	0.0000	1.0000

COVARIANCES

	X1	X2	X3
X1	11	22	0
X2	22	44	0
X3	0	0	85.8

TOTAL VARIANCE = 140.8

Note that on the above SAS output, CORRELATIONS means correlation matrix and COVARIANCES means variance-covariance matrix. If the analysis is performed on the variance-covariance matrix the results are shown in Table 1.

TABLE 1.
SAS OUTPUT FROM PCA OF DATA FROM EXAMPLE 2 (VARIANCE-COVARIANCE)

	EIGENVALUE	DIFFERENCE	PROPORTION	CUMULATIVE
PRIN1	85.80000	30.80000	0.60938	0.60938
PRIN2	55.00000	55.00000	0.39062	1.00000
PRIN3	0.00000		0.00000	1.00000

EIGENVECTORS

	PRIN1	PRIN2	PRIN3
X1	0.000000	0.447214	0.894427
X2	0.000000	0.894427	-.447214
X3	1.000000	0.000000	0.000000

OBS	X1	X2	X3	PRIN1	PRIN2	PRIN3
1	-5	-10	15	15	-11.180	0
2	-4	-8	6	6	-8.944	0
3	-3	-6	-1	-1	-6.708	0
4	-2	-4	-6	-6	-4.472	0
5	-1	-2	-9	-9	-2.236	0
6	0	0	-10	-10	0.000	0
7	1	2	-9	-9	2.236	0
8	2	4	-6	-6	4.472	0
9	3	6	-1	-1	6.708	0
10	4	8	6	6	8.944	0
11	5	10	15	15	11.180	0

The values under PRIN1 and PRIN2 are calculated by evaluating the PCs for each sample. For example,

$$-11.180 = .447214\ X_1 + .894427\ X_2 + 0X_3$$

$$= .447214(-5) + .894427(-10) + 0(15)$$

Analyzing the correlation matrix gives the results in Table 2.

TABLE 2.
SAS OUTPUT FROM PCA OF DATA FOR EXAMPLE 2 (CORRELATION)

	EIGENVALUE	DIFFERENCE	PROPORTION	CUMULATIVE
PRIN1	2.000000	1.000000	0.666667	0.666667
PRIN2	1.000000	1.000000	0.333333	1.000000
PRIN3	0.000000		0.000000	1.000000

EIGENVECTORS

	PRIN1	PRIN2	PRIN3
X1	0.707107	0.000000	-.707107
X2	0.707107	0.000000	0.707107
Z	0.000000	1.000000	0.000000

OBS	X1	X2	X3	PRIN1	PRIN2	PRIN3
1	-5	-10	15	-2.1320	1.6194	0
2	-4	-8	6	-1.7056	0.6478	0
3	-3	-6	-1	-1.2792	-0.1080	0
4	-2	-4	-6	-0.8528	-0.6478	0
5	-1	-2	-9	-0.4264	-0.9716	0
6	0	0	-10	0.0000	-1.0796	0
7	1	2	-9	0.4264	-0.9716	0
8	2	4	-6	0.8528	-0.6478	0
9	3	6	-1	1.2792	-0.1080	0
10	4	8	6	1.7056	0.6478	0
11	5	10	15	2.1320	1.6194	0

The values for PRIN1, PRIN2 and PRIN3 are calculated in a fashion similar to the variance-covariance analysis, but using the standardized variables. For example,

$$-2.1320 = .707107\left(\frac{X_1-\bar{X}_1}{s_1}\right) + .707107\left(\frac{X_2-\bar{X}_2}{s_2}\right) + 0\left(\frac{X_3-\bar{X}_3}{s_3}\right)$$

$$= .707107\left(\frac{-5-0}{3.316625}\right) + .70107\left(\frac{-10-0}{6.63325}\right) + 0\left(\frac{15-0}{9.262829}\right)$$

There are several items to note in these analyses:

(1) There are only two nonzero eigenvalues since given X_1 and X_3, X_2 is computed from X_1.

(2) X_3 is its own principal component since it is uncorrelated with all the other variables.

(3) The sum of the eigenvalues is the sum of the variances, i.e.,

$$11 + 44 + 85.8 = 55 + 85.8 = 140.8$$

and

$$1 + 1 + 1 = 2 + 1 = 3.$$

(4) For the variance-covariance analysis, the ratio of the coefficients of X_1 and X_2 in PC_2 is the same as the ratio of the variables themselves (since $X_2 = 2X_1$).

(5) Since there are only two nonzero eigenvalues, only two of the PCs have nonzero variances (are nonconstant).

(6) The coefficients help to relate the variables and the PCs. In the variance-covariance analysis,

$$\text{Corr}(PC_2, X_1) = \frac{(\text{coefficient of } X_1 \text{ in } PC_2) \sqrt{\text{var}(PC_2)}}{\sqrt{\text{var}(X_1)}}$$

$$= \frac{b_{21}\sqrt{\lambda_2}}{s_1}$$

$$= \frac{.447214\sqrt{55}}{3.16625}$$

$$= 1.$$

In the correlation analysis,

$$\text{Corr}(PC_1, X_1) = b_{11}\sqrt{\lambda_1}$$
$$= .707107\sqrt{2}$$
$$= 1.$$

Thus, in both these cases, the variable is perfectly correlated with the PC.

(7) The Xs can be reconstructed exactly from the PCs with nonzero eigenvalues. For example, in the variance-covariance analysis, X_3 is clearly given by PC_1. X_1 and X_2 can be recovered via the formulas

$$X_1 = PC_2/\sqrt{5}$$
$$X_2 = 2 \cdot PC_2/\sqrt{5}.$$

As a numerical example,

$$-5 = -11.180/\sqrt{5}.$$

Example 3. For Example 3 we use $X_1 = Z_1$, $X_2 = 2(Z_1 + 5)$, $X_3 = 3(Z_1 + 5)$ and $X_4 = Z_2$. Thus X_1, X_2 and X_3 are all created from Z_1. The data and summary statistics are given in Table 3.

TABLE 3.
SAS OUTPUT OF DATA, MEANS, STANDARD DEVIATIONS (ST DEV) CORRELATION MATRIX, AND VARIANCE-COVARIANCE MATRIX FOR EXAMPLE 3

OBS	X1	X2	X3	X4
1	-5	0	0	15
2	-4	2	3	6
3	-3	4	6	-1
4	-2	6	9	-6
5	-1	8	12	-9
6	0	10	15	-10
7	1	12	18	-9
8	2	14	21	-6
9	3	16	24	-1
10	4	18	27	6
11	5	20	30	15

	X1	X2	X3	X4
MEAN	0.000000	10.00000	15.00000	0.00000
ST DEV	3.316625	6.63325	9.94987	9.62823

CORRELATION MATRIX

	X1	X2	X3	X4
X1	1.0000	1.0000	1.0000	0.0000
X2	1.0000	1.0000	1.0000	0.0000
X3	1.0000	1.0000	1.0000	0.0000
X4	0.0000	0.0000	0.0000	1.0000

COVARIANCES

	X1	X2	X3	X4
X1	11	22	33	0
X2	22	44	66	0
X3	33	66	99	0
X4	0	0	0	85.8

TOTAL VARIANCE = 239.8 = 11 + 44 + 99 + 85.8

The analyses for the variance-covariance matrix (unstandardized analysis) and correlation matrix (standardized analysis) are given in Table 4.

TABLE 4.
SAS OUTPUT FROM PCA OF DATA FOR EXAMPLE 3

VARIANCE-COVARIANCE ANALYSIS

	EIGENVALUE	DIFFERENCE	PROPORTION	CUMULATIVE
PRIN1	154.0000	68.2000	0.6422	0.6422
PRIN2	85.8000	85.8000	0.3578	1.0000
PRIN3	0.0000	0.0000	0.0000	1.0000
PRIN4	0.0000		0.0000	1.0000

EIGENVECTORS

	PRIN1	PRIN2	PRIN3	PRIN4
X1	0.267261	0.000000	0.358569	0.894427
X2	0.534522	0.000000	0.717137	-.447214
X3	0.801784	0.000000	-.597614	0.000000
X4	0.000000	1.000000	0.000000	0.000000

OBS	X1	X2	X3	X4	PRIN1	PRIN2	PRIN3	PRIN4
1	-5	0	0	15	-1.336	15	-1.793	-4.472
2	-4	2	3	6	2.405	6	-1.793	-4.472
3	-3	4	6	-1	6.147	-1	-1.793	-4.472
4	-2	6	9	-6	9.889	-6	-1.793	-4.472
5	-1	8	12	-9	13.630	-9	-1.793	-4.472
6	0	10	15	-10	17.372	-10	-1.793	-4.472
7	1	12	18	-9	21.114	-9	-1.793	-4.472
8	2	14	21	-6	24.855	-6	-1.793	-4.472
9	3	16	24	-1	28.597	-1	-1.793	-4.472
10	4	18	27	6	32.339	6	-1.793	-4.472
11	5	20	30	15	36.080	15	-1.793	-4.472

CORRELATION ANALYSIS

	EIGENVALUE	DIFFERENCE	PROPORTION	CUMULATIVE
PRIN1	3.000000	2.000000	0.750000	0.750000
PRIN2	1.000000	1.000000	0.250000	1.000000
PRIN3	0.000000	0.000000	0.000000	1.000000
PRIN4	-.000000		-.000000	1.000000

EIGENVECTORS

	PRIN1	PRIN2	PRIN3	PRIN4
X1	0.577350	-.000000	0.408248	-.707107
X2	0.577350	0.000000	0.408248	0.707107
Z	-.000000	1.000000	-.000000	0.000000
X3	0.577350	0.000000	-.816497	0.000000

OBS	X1	X2	X3	X4	PRIN1	PRIN2	PRIN3	PRIN4
1	-5	0	0	15	-2.6112	1.6194	0	0
2	-4	2	3	6	-2.0889	0.6378	0	0
3	-3	4	6	-1	-1.5667	-0.1080	0	0
4	-2	6	9	-6	-1.0445	-0.6478	0	0
5	-1	8	12	-9	-0.5222	-0.9716	0	0
6	0	10	15	-10	0.0000	-1.0796	0	0
7	1	12	18	-9	0.5222	-0.9716	0	0
8	2	14	21	-6	1.0445	-0.6478	0	0
9	3	16	24	-1	1.5667	-0.1080	0	0
10	4	18	27	6	2.0889	0.6478	0	0
11	5	20	30	15	2.6112	1.6194	0	0

For the variance-covariance analysis, the coefficients in PC_1 are in the same ratio as their relationship to Z_1. In the correlation analysis X_1, X_2 and X_3 have equal coefficients. In both analyses, as expected, the total variance is equal to the sum of the variances for the PCs. In both cases two PCs, PC_3 and PC_4, have zero variance; in the correlation analysis the PCs are identically zero but in the variance-covariance analysis they have a constant value.

Example 4: In this example we take more complicated combinations of Z_1 and Z_2. Let

$$X_1 = Z_1$$
$$X_2 = 2Z_1$$
$$X_3 = 3Z_1$$
$$X_4 = Z_1/2 + Z_2$$
$$X_5 = Z_1/4 + Z_2$$
$$X_6 = Z_1/8 + Z_2$$
$$X_7 = Z_2$$

Note that X_1, X_2 and X_3 are collinear (they all have correlation unity) and X_4, X_5, X_6 and X_7 have steadily decreasing correlations with X_1. The data and data summaries are given in Table 5.

TABLE 5.
SAS OUTPUT OF DATA, MEANS, AND STANDARD DEVIATION
FOR DATA FROM EXAMPLE 4

OBS	X1	X2	X3	X4	X5	X6	X7
1	-5.000	-10.000	-15.000	12.500	13.750	14.375	15.000
2	-4.000	-8.000	-12.000	4.000	5.000	5.500	6.000
3	-3.000	-6.000	-9.000	-2.500	-1.750	-1.375	-1.000
4	-2.000	-4.000	-6.000	-7.000	-6.500	-6.250	-6.000
5	-1.000	-2.000	-3.000	-9.500	-9.250	-9.125	-9.000
6	0.000	0.000	0.000	-10.000	-10.000	-10.000	-10.000
7	1.000	2.000	3.000	-8.500	-8.755	-8.875	-9.000
8	2.000	4.000	6.000	-5.000	-5.500	-5.750	-6.000
9	3.000	6.000	9.000	0.500	-0.250	-0.625	-1.000
10	4.000	8.000	12.000	8.000	7.000	6.500	6.000
11	5.000	10.000	15.000	17.500	16.250	15.625	15.000

	X1	X2	X3	X4	X5	X6	X7
MEAN	0.00000	0.00000	0.00000	0.00000	0.00000	0.00000	0.00000
ST DEV	3.31662	6.63325	9.94987	9.41010	9.29987	9.27210	9.26283

The PCAs for the variance-covariance and correlation matrices are in Table 6. We note several things:

(1) In both analyses there are only two eigenvalues that are nonzero indicating that only two variables are needed. This is not readily apparent from the correlation or variance-covariance matrix.
(2) In PC_1, PC_2, PC_3, PC_4, and PC_6 where the standardized X_1, X_2 and X_3 are the same, they have the same coefficients.
(3) Neither PCA recovers Z_1 and Z_2. The PCs with nonzero variances have elements of both Z_1 and Z_2 in them, i.e., neither PC_1 or PC_2 is perfectly correlated with one of the Zs.

If one computes residuals for Example 4, there will be no residuals after the second PC since the first two PCs explain all the variance. The residuals from PC_1, and a PCA for the residuals, are given in Table 7. The SAS program for their calculations is given in Appendix 1. Several things are noteworthy:

(1) The variance-covariance matrix for $ê_1$ has one nonzero eigenvalue.
(2) PC_1 from the analysis is the same as PC_2 from the original analysis.
(3) The total variance is the original total variance minus the variance of the original PC_1.

SUMMARY AND DISCUSSION

PCA provides a method of extracting structure from the variance-covariance or correlation matrix. If a multivariate data set is actually constructed in a linear fashion from fewer variables, then PCA will discover that structure. PCA constructs linear combinations of the original data, X, with maximal variance, $P = XB$. This relationship can be inverted, $PB^{-1} = X$, to recover the Xs from the PCs (actually only those PCs with nonzero eigenvalues are needed - see Example 2). Though PCA will often help discover structure in a data set, it does have limitations. It will not necessarily recover the exact underlying variables, even if they were uncorrelated (Example 4). Also, by its construction, PCA is limited to searching for linear structure in the Xs.

Although the motivation for a PCA is to explain variance with a linear combination of the m variables, there are other things that can be accomplished with a PCA. For example, residuals may be computed (see Example 4). These can be studied to investigate their distribution, or to find patterns or outliers. Linear combinations of variables can also be eliminated if one or more eigenvalues are near zero. A study of collinearity or near collinearity among the variables may be important.

TABLE 6.
SAS OUTPUT OF PCAs FOR DATA FROM EXAMPLE 4

COVARIANCES

	X1	X2	X3	X4	X5	X6	X7
X1	11.00000	22.00000	33.00000	5.50000	2.75000	1.37500	0.00000
X2	22.00000	44.00000	66.00000	11.00000	5.50000	2.75000	0.00000
X3	33.00000	66.00000	99.00000	16.50000	8.25000	4.12500	0.00000
X4	5.50000	11.00000	16.50000	88.55000	87.17500	86.48750	85.80000
X5	2.75000	5.50000	8.25000	87.17500	86.48750	86.14375	85.80000
X6	1.37500	2.75000	4.12500	86.48750	86.14375	85.97188	85.80000
X7	0.00000	0.00000	0.00000	85.80000	85.80000	85.80000	85.80000

	EIGENVALUE	DIFFERENCE	PROPORTION	CUMULATIVE
PRIN1	347.01507	193.22077	0.69291	0.69291
PRIN2	153.79430	347.01507	0.30709	1.00000
PRIN3	0.00000	0.00000	0.00000	1.00000
PRIN4	0.00000	0.00000	0.00000	1.00000
PRIN5	0.00000	0.00000	0.00000	1.00000
PRIN6	0.00000	0.00000	0.00000	1.00000
PRIN7	0.00000	0.00000	0.00000	1.00000

EIGENVECTORS

	PRIN1	PRIN2	PRIN3	PRIN4	PRIN5	PRIN6	PRIN7
X1	0.02502	0.26479	0.96362	0.01243	0.00000	0.01806	0.01450
X2	0.05004	0.52957	-0.14825	0.02485	0.83205	0.03613	0.02899
X3	0.07505	0.79436	-0.22237	0.03728	-0.55470	0.05419	0.04349
X4	0.50482	0.02744	0.00000	-0.70228	0.00000	-0.29061	-0.40835
X5	0.49856	-0.03876	0.00000	0.71035	0.00000	-0.28719	-0.40354
X6	0.49544	-0.07186	0.00000	-0.00326	0.00000	-0.28623	0.81697
X7	0.49231	-0.10495	0.00000	-0.00481	0.00000	0.86404	-0.00508

OBS	PRIN1	PRIN2	PRIN3	PRIN4	PRIN5	PRIN6	PRIN7
1	25.921	-21.332	0	0	0	0	0
2	8.790	-15.937	0	0	0	0	0
3	-4.359	-10.918	0	0	0	0	0
4	-13.525	-6.275	0	0	0	0	0
5	-18.709	-2.009	0	0	0	0	0
6	-19.911	1.881	0	0	0	0	0
7	-17.131	5.395	0	0	0	0	0
8	-10.368	8.533	0	0	0	0	0
9	0.377	11.294	0	0	0	0	0
10	15.104	13.679	0	0	0	0	0
11	33.813	15.688	0	0	0	0	0

TABLE 6. (*Continued*)

CORRELATIONS

	X1	X2	X3	X4	X5	X6	X7
X1	1.00000	1.00000	1.00000	0.17623	0.08916	0.04471	0.00000
X2	1.00000	1.00000	1.00000	0.17623	0.08916	0.04471	0.00000
X3	1.00000	1.00000	1.00000	0.17623	0.08916	0.04471	0.00000
X4	0.17623	0.17623	0.17623	1.00000	0.99614	0.99124	0.98435
X5	0.08916	0.08916	0.08916	0.99614	1.00000	0.99901	0.99602
X6	0.04471	0.04471	0.04471	0.99124	0.99901	1.00000	0.99900
X7	0.00000	0.00000	0.00000	0.98435	0.99602	0.99900	1.00000

	EIGENVALUE	DIFFERENCE	PROPORTION	CUMULATIVE
PRIN1	4.05217	1.10433	0.57888	0.57888
PRIN2	2.94783	2.94783	0.42112	1.00000
PRIN3	0.00000	1.00000	0.00000	1.00000
PRIN4	0.00000	1.00000	0.00000	1.00000
PRIN5	0.00000	1.00000	0.00000	1.00000
PRIN6	0.00000	1.00000	0.00000	1.00000
PRIN7	0.00000	0.00000	0.00000	1.00000

EIGENVECTORS

	PRIN1	PRIN2	PRIN3	PRIN4	PRIN5	PRIN6	PRIN7
X1	0.14429	0.55733	-0.01047	-0.03822	-0.40825	0.01819	0.70711
X2	0.14429	0.55733	-0.01047	-0.03822	-0.40825	0.01819	-0.70711
X3	0.14429	0.55733	-0.01047	-0.03822	0.81650	0.01819	0.00000
X4	0.49334	-0.06831	-0.00052	0.86715	0.00000	0.00121	0.00000
X5	0.48633	-0.11881	0.70763	-0.28505	0.00000	-0.40912	0.00000
X6	0.48133	-0.14409	-0.70635	-0.28505	0.00000	-0.40912	0.00000
X7	0.47535	-0.16918	0.00134	-0.28490	0.00000	0.81501	0.00000

OBS	PRIN1	PRIN2	PRIN3	PRIN4	PRIN5	PRIN6	PRIN7
1	2.238	-3.284	0	0	0	0	0
2	0.543	-2.304	0	0	0	0	0
3	-0.737	-1.432	0	0	0	0	0
4	-1.600	-0.668	0	0	0	0	0
5	-2.048	-0.011	0	0	0	0	0
6	-2.080	0.538	0	0	0	0	0
7	-1.695	0.980	0	0	0	0	0
8	-0.895	1.314	0	0	0	0	0
9	0.321	1.540	0	0	0	0	0
10	1.953	1.658	0	0	0	0	0
11	4.001	1.669	0	0	0	0	0

TABLE 7.
RESIDUALS AFTER PC$_1$ FOR DATA FROM EXAMPLE 4 AND THE SAS OUTPUT USING THESE RESIDUALS AS DATA INSTEAD OF ORIGINAL DATA

E1	COL1	COL2	COL3	COL4	COL5	COL6	COL7
ROW1	-5.64854	-11.2971	-16.9454	-0.585338	0.826926	1.5328	2.23893
ROW2	-4.21992	-8.43984	-12.6597	-0.617727	0.617727	1.14515	1.67266
ROW3	-2.89094	-5.78188	-8.67287	-0.29958	0.423133	0.784534	1.14589
ROW4	-1.6616	-3.3232	-4.98493	-0.172188	0.243144	0.450945	0.658611
ROW5	-0.531891	-1.06378	-1.59586	-0.0551207	0.144385	0.210825	0.210825
ROW6	0.498181	0.996361	1.49434	0.0516225	-0.0730223	-0.135146	-0.197468
ROW7	1.42862	2.85723	4.28568	0.148041	-0.209199	-0.387647	-0.566267
ROW8	2.25942	4.51883	6.77814	0.234135	-0.33077	-0.61312	-0.895572
ROW9	2.99058	5.98116	8.97174	0.309905	-0.437738	-0.811563	-1.18538
ROW10	3.62211	7.24421	10.8665	0.37535	-0.530101	-0.982977	-1.4357
ROW11	4.154	8.30799	12.4623	0.430471	-0.607859	-1.12736	-1.64653

SIMPLE STATISTICS

	COL1	COL2	COL3	COL4	COL5	COL6	COL7
MEAN	0.000000	0.000000	0.000000	0.0000000	0.0000000	0.000000	0.888888
ST DEV	3.283719	6.567437	9.851156	0.3402822	0.4806475	0.8911123	1.301577

COVARIANCES

	COL1	COL2	COL3	COL4	COL5	COL6	COL7
COL1	10.783	21.566	32.348	1.1174	-1.578	-2.926	-4.274
COL2	21.566	43.131	64.697	2.2348	-3.157	-5.852	-8.548
COL3	32.348	64.697	97.045	3.3522	-4.735	-8.778	-12.82
COL4	1.1174	2.2348	3.3522	0.11579	-0.1636	-0.3032	-0.4429
COL5	-1.578	-3.157	-4.735	0.23102	0.23102	0.42831	0.6256
COL6	-2.926	-5.852	-8.778	-0.3032	0.42831	0.79408	1.1599
COL7	-4.274	-8.548	-12.82	-0.4429	0.6256	1.1599	1.6941

TOTAL VARIANCE = 153.7943

	EIGENVALUE	DIFFERENCE	PROPORTION	CUMULATIVE
PRIN1	153.7943	153.7943	1.0000	1.0000
PRIN2	0.0000	0.0000	0.0000	1.0000
PRIN3	0.0000	0.0000	0.0000	1.0000
PRIN4	0.0000	0.0000	0.0000	1.0000
PRIN5	-0.0000	0.0000	-0.0000	1.0000
PRIN6	-0.0000	0.0000	-0.0000	1.0000
PRIN7	-0.0000	0.0000	-0.0000	1.0000

EIGENVECTORS

	PRIN1	PRIN2	PRIN3	PRIN4	PRIN5	PRIN6	PRIN7
COL1	0.264786	-0.251156	-0.019310	0.836265	0.064150	0.138641	0.379158
COL2	0.529573	-0.502312	0.013628	-0.392969	-0.048768	0.556175	-0.030259
COL3	0.794359	0.446382	0.095146	-0.059164	0.003596	-0.395920	0.020439
COL4	0.027439	-0.010654	0.013988	0.007718	0.973360	0.042589	-0.222788
COL5	-0.038758	0.6520600	0.258748	0.120481	-0.036856	0.699033	-0.042929
COL6	-0.071856	-0.216434	0.960808	-0.007869	-0.000523	-0.018358	0.895849
COL7	-0.104954	0.114924	-0.009529	-0.357909	0.211436	0.018358	0.895849

There does not seem to be any systematic way to use PCA as a variable selection tool. Since PCA works by calculating linear combinations of all the variables, usually a PC will have a contribution from each of the original variables. Thus, even in cases in which we need only consider a few of the PCs, each one of them often includes each of the original variables with a substantial coefficient.

Readers interested in further reading about PCA should consult a book on multivariate statistical methods. Some good references are Harris (1975), Morrison (1976), and Johnson and Wichern (1982). Federer *et al.* (1986) have a technical report which describes in detail output from statistical packages that perform PCA.

ACKNOWLEDGMENT

Partly supported by the U.S. Army Research Office through the Mathematical Sciences Institute of Cornell University.

APPENDIX I

```
DATA ONE;
INPUT Z1 Z2;
X1=Z1;
X2=2*Z1;
X3=3*Z1;
X4=(Z1/2)+Z2;
X5=(Z1/4)+Z2;
X6=(Z1/8)+Z2;
X7=Z2;
DROP Z1 Z2;
CARDS;
-5  15
-4   6
-3  -1
-2  -6
-1  -9
 0 -10
 1  -9
 2  -6
 3  -1
 4   6
 5  15
PROC MATRIX PRINT;
V1=.02502/.05004/.07505/.50482/.49856/.49544/.49231;
FETCH X DATA=ONE;
```

```
E1=X-((V1*V1')*X')';
OUTPUT E1 OUT=TWO;
PROC PRINCOMP COV DATA=TWO;
PROC MATRIX PRINT;
V1=.02502/.05004/.07505/.50482/.49856/.49544/.49231;
V2=.26479/.529573/.794359/.027439/-.038757/-.071855/-.104954;
FETCH X DATA=ONE;
E2=X-(((V1*V1')*V1')*X')+((V2*V2')*X'))';
OUTPUT E2 OUT=FOUR;
PROC PRINCOMP COV DATA=FOUR;
```

REFERENCES

FEDERER, W.T., MCCULLOCH, C.E. and MILES-MCDERMOTT, N. 1986. Illustrative Examples of Principal Components Analysis Using SAS/PRINCOMP. BU-918-M in the Biometrics Unit Series.

HARRIS, R.J. 1975. *A Primer of Multivariate Statistics*. Academic Press, New York.

JOHNSON, R.A. and WICHERN, D.W. 1982. *Applied Multivariate Analysis*. Prentice Hall, Englewood Cliffs, New Jersey.

MORRISON, D.F. 1976. *Multivariate Statistical Methods* (2nd Ed.). McGraw-Hill Book Co., New York.

CHAPTER 1.10

REDUCING THE NOISE CONTAINED IN DESCRIPTIVE SENSORY DATA

K.L. BETT, G.P. SHAFFER, J.R. VERCELLOTTI,
T.H. SANDERS and P.D. BLANKENSHIP

*At the time of original publication,
authors Bett and Bercellotti were affiliated with
USDA-ARS-SRRC, New Orleans, LA,*

*author Shaffer was affiliated with
Southeastern Louisiana University
Dept. of Biological Science
Hammond, LA*

and

*authors Sanders and Blankenship
were affiliated with
USDA-ARS-NPRC, Dawson, GA.*

ABSTRACT

A data reduction protocol was designed to minimize distortion inherent in sensory data. Following removal of nonexistent attributes and treatment levels, extreme value analysis and distribution comparisons combined with graphical representations, facilitated elimination of inconsistent (with respect to overall consensus) panelists. Application of a calibration factor showed super-responsive panelists (those with intensity values consistently higher than other panelists) were among the most accurate and thus were retained in spite of their tendency to produce extreme value data. Panelists that consistently produced a narrow variance around the overall mean and rarely produced extreme values were classified as noncommittal and removed. Analysis of variance calls for a split plot design; blocks (sessions) and treatments in main plot, and panelists in subplot. In general, the subplot can be ignored. These methods are suggested for evaluating panelists' training needs; and for eliminating data that distorts the statistical analysis.

INTRODUCTION

Descriptive sensory analysis is a very important and widely accepted research tool in all aspects of the food industry. It can be used to describe the flavor changes caused by a new treatment or process. Generally, a group of 8-20 highly trained panelists rate the intensities of the flavor characteristics of the sample in question (Meilgaard et al. 1987). Several types of problems can exist with descriptive panel data. Most deal with panelist responses, such as misjudged scores, super-responsive panelists, nondiscerners for one or more attributes and a noncommittal (narrow variance within panelist) panelist.

In the original Profile Method (Caul 1957) panelists would discuss their evaluations and make adjustments when a sample was misjudged. Many descriptive panels do not presently do this because of economic and time constraints; or because of innovative methodology. Super-responsive panelists use intensity scales correctly, but they have low thresholds for some attributes. Therefore, their responses are sufficiently greater than the other panelists. Nondiscerners have high thresholds for a particular attribute and, therefore, do not detect it, even though the majority of the panel perceived it. Noncommittal panelists may not perceive the attribute, but for fear of being wrong they respond with a nonzero intensity, and do this consistently for all samples. Another problem that skews a data set is treatment levels that contain an abundance of nondetectable intensities, resulting in a stacked frequency distribution at or near zero. Many of these problems can be dealt with by remedial training for future sample evaluations. This study describes methods for identifying these problems and minimizing noise in the data set. Consequently, these noise reduction techniques aid the researcher in finding actual differences between treatments.

Several methods have been published to identify specific sensory panel problems. A number of methods in the literature utilize statistical analysis, such as analysis of variance or multivariate analysis techniques (Powers 1984; Kwan and Kowalski 1980; Oreskovich et al. 1991; Marshal and Kirby 1988; Lundahl and McDaniel 1990; Naes and Solheim 1991) to identify noise in descriptive data sets. These methods give the investigator a good indication of which panelists may be outliers and which panelists group together in the way they evaluated a sample. Miller (1988) suggests using a shift or scale factor that results in each panelist having the same overall mean and standard deviation. This method corrects problems encountered in changes in panel memberships over long periods of time. Pokorny et al. (1982) reported on Neumann's repeatability index (Neumann and Molnar 1974), which assesses a panelist's ability to repeat their evaluations of duplicate samples. Neumann's performance index compares individual panelist's scores with those of the group mean. Neumann's methods are designed to eliminate panelists that deviate widely in their scores.

Over the past few years noise problems in sensory data have been prevalent in several of our descriptive flavor data bases in which either beef, peanuts, or catfish were subjected to a variety of factorial treatment arrangements. O'Mahony (1982) discussed some of these problems and suggested caution in the use of parametric statistics. We attempted to develop a general procedure that reduces the distortion in the data. An unintended, but beneficial, result of the procedure is that the reduced data are more likely to be appropriate for parametric analysis. The noise reduction methods presented here give details of trends in the data set that may hinder statistical analysis. They identify inaccurate judgements so they can be considered for deletion. These methods aid in evaluating judges, in knowing which judges need further training and can be used for eliminating judges or attributes from an analysis.

METHODS

Data Base

The descriptive sensory data were collected at the U.S. Department of Agriculture, Agricultural Research Service, Southern Regional Research Center (SRRC), New Orleans, LA. The data base used for this study (collected November 1986-March 1987) concerned the effects of curing temperature treatments on the flavor of peanuts of different maturities (Sanders *et al.* 1987, 1989a). Methods for maturity determination, size selection, roasting and sensory analysis have been described by Sanders *et al.* (1989a, b). Three curing treatments and five maturities were completely cross-classified in a 3 × 5 factorial treatment arrangement replicated six times (sessions) for each of the 13 panelists. This yielded 1170 experimental observations with a control peanut randomly presented 18 times to each of the 13 panelists for a total of 1404 potential observations. Maturity was determined by hull mesocarp colors designated as black, brown, orange B, orange A, and yellow (mature and immature). The three curing temperature treatments were ambient air, ambient +8.4C and ambient +16.8C (forced air during postharvest bulk drying). The panel consisted of SRRC employees who were trained in Spectrum descriptive methods (Meilgaard *et al.* 1987). The panel had recently finished their training when these data were collected. The flavor characteristics rated by the panel were similar to those described by Johnson *et al.* (1988), except for the addition of the attribute 'fruity/fermented' (Sanders *et al.* 1989b). The intensities of the flavor characteristics were evaluated on a 15-point scale that is anchored according to the Spectrum method using intensity references that are universal for all foods (Meilgaard *et al.* 1987).

TABLE 1.
INDICATORS USED TO IMPLICATE SUPER-RESPONSIVE, NONDISCERNING, AND NONCOMMITTAL PANELISTS[a]

Panelist	% in Class[b]			Mean	Standard Deviation	Dissimilarity Index[c]
	I	L	U			
A	68.50	0.00	31.48	4.44	1.39	47
B	86.96	0.00	13.04	4.06	1.32	27
C	98.15	0.00	1.85	2.81	1.19	44
D	100.00	0.00	0.00	2.52	0.51	32
E	100.00	0.00	0.00	2.50	0.38	39
F	99.07	0.93	0.00	2.41	0.58	30
G	100.00	0.00	0.00	2.25	0.43	22
H	100.00	0.00	0.00	2.20	0.56	28
I	100.00	0.00	0.00	1.83	0.41	17
J	99.07	0.93	0.00	1.75	0.60	32
K	100.00	0.00	0.00	1.48	0.67	38
L	60.06	34.86	5.08	1.33	1.18	56
M	55.56	44.44	0.00	1.14	1.15	133

[a]Example from flavor attribute astringent.
[b]Shown are percentage of observations within the 95% confidence interval (I), percentage beyond the lower (L) 95% and upper (U) 95% bound, means and standard deviation of each panelist, and the dissimilarity index.
[c]The absolute value of the distance between the overall finger print and that for each panelist.

Treatment Levels

Treatment level omissions occurred when an attribute was detectable over only a subset of the designated treatment levels. Treatment levels where the attribute was undetectable (and presumably could not occur) produced an overabundance of zeros in the data set. In the extreme case, frequency plots of this nature contained a mode of zero. Treatment levels were omitted until the distribution as a whole approximated normality. When panelist and/or treatment level omissions were extreme, it was necessary to eliminate entire attributes from the analysis.

Tabular Procedures

Using a combination of the MEANS, SORT, and RANK procedures of SAS (1985), tables were constructed to identify panelists that were either superresponsive, nondiscerning, or noncommittal (Table 1). For each attribute, each observation was labelled either 'I', 'U' or 'L', depending on whether it was within the 95% confidence interval of the mean (I), beyond the upper 95% confidence bound (U), or beyond the lower 95% bound (L). Observations were

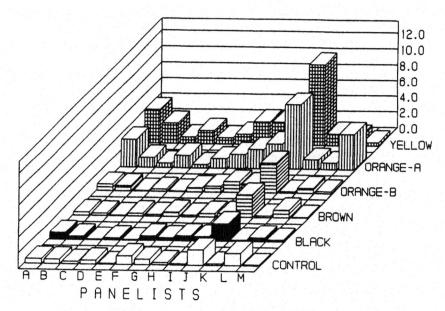

FIG. 1. EXAMPLE OF THREE DIMENSIONAL GRAPH WITH PANELISTS REPRESENTED ON THE X-AXIS, PEANUT MATURITY ON Y-AXIS, AND RAW MEANS OF THE DEPENDENT VARIABLE FRUITY/FERMENTED INTENSITY OF THE Z-AXIS

listed by attribute and panelists; for a particular panelist, an abundance of either I's or U's indicated either normal or super-responsive tendencies, respectively, whereas *many* L's or a combination of U's and L's indicated nondiscerning tendency. Also in the table is the means of each panelist ranked from low to high and their associated standard deviations. Variance of the means located near the overall mean with very small standard deviations were tested against the variance of the overall distribution using a Chi-square test. A significant Chi-square test implicated noncommittal panelists. The dissimilarity index in Table 1 is explained under the graphical procedures.

Graphical Procedures

Two graphical representations were used to augment the tabular information and served to further verify inconsistent panelists. First, three dimensional graphs were constructed for each attribute with all panelists represented on the x-axis and peanut maturity (the most pronounced of the two treatments) represented along the y-axis (Fig. 1). In general, the treatment expected to contain the most behavior (with respect to the dependent variable) constitutes the y-axis. If the interaction between treatments is expected to contain the most behavior, then cross-classified treatment combinations constitute the y-axis.

Plots of this nature using the raw data overemphasized the behavior of super-responsive panelists and deemphasized (flattened) the behavior of the other panelist responses. Consequently, a simple transformation was performed to calibrate responses of the individual panelists to a common 'benchmark'. This transformation is for exploring the data only. Once the problems are identified and eliminated, the original data are used for the experimental analysis, minus inconsistent panelists. The steps involved in this operation were computed separately for each attribute: (1) compute the overall mean, ignoring panelists, (2) compute the mean for each panelist, (3) obtain a correction factor for each panelist by forming a ratio of the overall mean to the within-panelist mean, and (4) multiply each attribute-by-panelist observation by the appropriate constant ratio. For each attribute, this operation had the effect of sliding the observations within each panelist toward the overall mean without affecting the behavior of the data across treatment levels (addressed in detail in RESULTS and DISCUSSION) while maintaining intensity units.

The calibrated data were also subjected to a second graphical procedure devised to construct "fingerprints" of the overall response of each attribute across the treatment gradient, along with that for each panelist. The calibrated observations were input to the G3GRID procedure of SAS (1985) to create a detailed grid of the original 15 treatment combinations. These data were then input to the G3D procedure of SAS to obtain contour plots such as in Fig. 2. In addition, for each point on the grid, the absolute value of the distance between the overall fingerprint and that for each panelist was computed and summed to obtain a dissimilarity index. By using the calibrated data, this dissimilarity index measured the difference in behavior between the overall and by-panelist fingerprints and was not influenced by errors due to over- or underrating tendencies. The appropriate test criteria is the paired-t test on the distribution of differences of the treatment combination.

Once the individual indices were computed, the data sets were merged together by panelist and one table for each attribute (e.g., Table 1) was produced using a series of programming statements that resulted in output lines. Panelists were removed from subsequent analysis if they were noncommittal (had nearly average means and narrow standard deviations and produced a flat response across all treatments), or nondiscerning (produced either a preponderance of L's, or a combination of U's and L's in the extreme value analysis, displayed a flat or incongruous response across treatments, and produced a high dissimilarity (index).

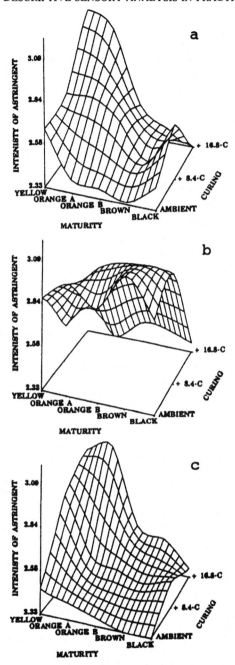

FIG. 2. INTENSITY RATINGS FOR THE COLOR BY TEMPERATURE RESPONSE SURFACE OF THE ATTRIBUTABLE ASTRINGENT
(a) The average of all panelists, (b) panelist E displaying an inconsistent pattern, and (c) panelist M displaying a consistent pattern.

In summary, the analysis proceeded in the following order: (1) removal of nonexistent attributes and treatment levels; (2) computation and tabulation of numerical indices; (3) production of graphics; (4) removal of inconsistent panelists; (5) analysis of variance and reproduction of graphics on the reduced data set.

RESULTS

By-attribute frequency plots indicated certain attributes contained an overabundance of data at or near zero (Fig. 3a). This type of stacking resulted from the presence of certain maturities of peanuts that contained undetectable amounts of certain attributes. For example, maturity levels black and brown for the attribute fruit/fermented contained low ratings across all panelists. As a result, the overall mean of the frequency distribution ($\overline{Y} = 0.82$, Fig. 3a) did not reflect central tendency. After elimination of maturities black and brown, the resulting distribution (Fig. 3b) better approximated normality ($\overline{Y} = 1.37$) and the residuals from analysis of variance were normally distributed.

When the extreme value, variance comparison, and dissimilarity information were examined for each attribute, decisions to omit inconsistent panelists were greatly facilitated. For example, the results shown in Table 1 for the attribute 'astringent' indicated that for 9 of the 13 panelists (C-K), distributions met the 95% confidence criterion, contained relatively commensurate dissimilarity indices and did not produce variances significantly different than the overall variance. Two panelists (A and B) rated over 13% of the observations as extreme values and all were overestimated. This is over five times that which would be expected by chance alone. These panelists were deemed superresponsive. The two remaining panelists (L and M) rated from one third to almost half of the observations beyond the lower confidence bound (10 to 20 times that expected by chance alone) and produced the highest dissimilarity indices; these panelists were deemed nondiscerners. For astringent, none of the panelists were noncommittal. However, noncommittal panelists were extreme for certain attributes; one panelist rated 100% of the observations identically (and near the overall mean) for the attribute dark roasted, yielding a standard deviation of zero. Interestingly, none of the panelists were habitually inconsistent across attributes. In fact, of the 12 attributes retained, no panelist was inconsistent on greater than three and no attribute had fewer than two inconsistent panelists. Six of the 13 panelists were nondiscerners for at least one attribute and yet super-responsive for at least one other attribute. Importantly, these results suggested that any overall measure devised to omit individual panelists from the entire analysis (such as the sum of the dissimilarity indices across attributes) should be avoided.

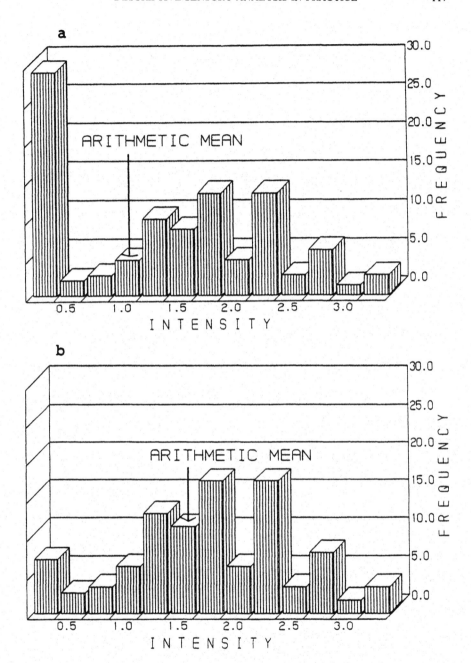

FIG. 3. FREQUENCY PLOT OF FRUITY/FERMENTED DATA ILLUSTRATING
(a) the stacking effect caused by an abundance of treatments with zero intensity, and
(b) the same data following removal of two treatments (maturities).

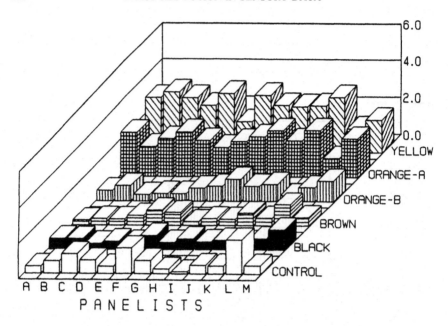

FIG. 4. INTENSITY MEANS OF THE ATTRIBUTE FRUITY/FERMENTED
ACROSS MATURITY LEVELS AND PANELISTS
FOLLOWING APPLICATION OF CALIBRATION FACTOR

After applying the calibration factor, the overall trend of each attribute (across the treatment gradient) was much more transparent than the same trend prior to transformation. For example, compare the ratings of fruity/fermented prior to calibration in Fig. 1 with those following calibration in Fig. 4. The large majority of panelists in Fig. 4 perceived a dramatic increase in 'fruity/fermented' for the two most immature peanut groups and an overall trend of increasing fruity/fermented with decreasing maturity. The nondiscerning tendency of panelist L was clear in Fig. 4, yet almost imperceptible in Fig. 1. This was largely the result of super-responsive panelist J whose tendency to overemphasize intensities made the scores of several panelists appear flat across the treatment gradient. It was noteworthy that the intensity ratings of the super-responsive panelist produced a behavior exemplifying the consensus; considering this panelist an outlier (as we had in our original attempt at noise reduction) would have resulted in a loss of consistent data.

The transformation was also useful in identification of attributes that display little variation (i.e., appear flat) over the treatment gradient. Such attributes carry little information (Shannon 1948) and could be eliminated. The attribute salty displayed this general lack of behavior (Fig. 5). Sporadic behavior was also easily detected following calibration. For example, the attribute painty in Fig. 6 displayed no broadly consistent pattern across maturities. This type of

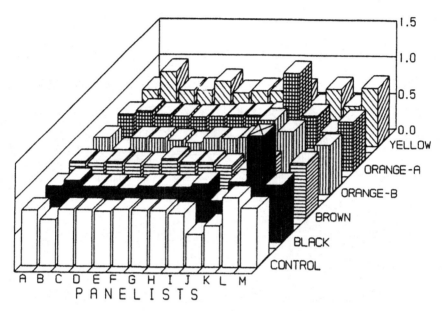

FIG. 5. INTENSITY MEANS OF THE ATTRIBUTE SALTY ACROSS MATURITY LEVELS AND PANELISTS FOLLOWING APPLICATION OF CALIBRATION FACTOR

attribute was more difficult to evaluate than one producing no behavior, because it was tempting to search for a subset of panelists showing similar behavior (addressed in discussion).

A more specific visualization of the relationship of each attribute across the treatment gradient was attainable through inspection of the fingerprint plots (see Fig. 2a, b, and c, for example). Severely distorted fingerprints were easily detectable when the plot average across panelists (e.g., Fig. 2a) was placed alongside the by-panelists plot (e.g., Fig. 2b,c). For example, in Fig. 2c, this panelist had a fingerprint resembling the mean fingerprint (Fig. 2a), whereas the panelist in Fig. 2b was clearly inconsistent.

Analysis of Variance

The most commonly used design for analysis of variance of sensory data is the randomized block design, blocking on the panelist effect (or the double block design blocking on panelist and session). Blocking on panelist is inappropriate

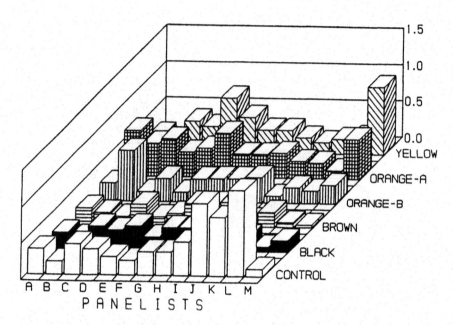

FIG. 6. INTENSITY MEANS OF THE ATTRIBUTE PAINTY ACROSS MATURITY AND PANELISTS FOLLOWING APPLICATION OF CALIBRATION FACTOR

because generally panelists are presented with uniform subsamples of each experimental unit. Therefore, panelists approximate repeated measures rather than true replications and thus a type of split plot design (Cochran and Cox 1957) is appropriate. Table 2 presents an example of this design for the attribute astringent. The analysis of variance is performed on the raw data and not on the transformed data.

It is noteworthy that, following the noise reduction procedure, none of the effects in the subplot (panelist and panelist interactions) are of interest; the sums of squares attributable to the panelist and panelist-by-treatment interaction effects have already been interpreted. This is particularly fortuitous because the error term of the subplot often does not meet the criteria for parametric statistics.

The simple randomized block design shown in Table 2b was obtained by averaging over the remaining panelists. It approximates the main plot of the split plot (compare Table 2a and 2b). The statistical power of the two designs is identical (though they differ slightly in Table 2 due to 75 missing observations) and analysis of several data sets reveals that the distribution of the residual

TABLE 2.
ANALYSIS OF VARIANCE COMPARING STRIP-SPLIT PLOT
AND RANDOMIZED BLOCK DESIGN[a]

A) SOURCE	df	SS	F value	P>F
Session	5	28.384	6.40[b]	0.0001
Color	4	25.119	7.08[b]	0.0001
Temp	2	9.331	5.26[b]	0.0076
Color*Temp	8	15.572	2.19[b]	0.0391
Error A	65	57.655		
Panel	10	560.106	17.91[c]	0.0001
Color*Panel	40	74.524	2.18[c]	0.0001
Temp*Panel	20	47.457	2.78[c]	0.0001
Color*Temp*Panel	80	70.657	0.85[c]	0.8106
Error B	680	703.311		
Corrected Total	914	1580.577		

B) Source	df	SS	F value	P>F
Session	5	2.771	6.62	0.0001
Color	4	2.334	6.99	0.0001
Temp	2	0.854	5.10	0.0088
Color*Temp	8	1.479	2.21	0.0380
Residual Error	65	5.444		
Corrected Total	84	12.882		

[a]Example from flavor attribute astringent with data averaged over functional panelists.
[b]Tested with the pairwise and three-way pooled block interaction terms (Error A).
[c]Tested with pooled residual error term (Error B).

error generally meets the criteria for parametric analysis. Since only the main plot of the split plot is of interest, we suggest the simple randomized block design. Of course, since the design matrix for each attribute is identical and certain attributes are expected to behave similarly or inversely, the analysis should be conducted as a multivariate analysis of variance (MANOVA) rather than as separate univariate ANOVA's.

DISCUSSION

The operations described herein were devised to reduce noise expressed as: (1) a stacking of zeros that result from inclusion of treatment levels that do not exist for certain attributes, (2) nondiscerning panelists (3) noncommittal panelists, and (4) attributes devoid of information or containing highly incongruous patterns. We also suggest an analysis of variance approach once the data are reduced that incorporates graphics for ease in interpretation of treatment effects.

Treatment Levels

By-attribute frequency distribution plots indicated that certain variables contained an overabundance of observations at or near zero. This stacking occurred because certain treatment levels did not contain detectable amounts of certain attributes. We assumed that these treatment levels did not (under these conditions) contain detectable amounts of these attributes and therefore did not constitute true samples. This interpretation is synonymous to ordination studies where certain extremes of a gradient are removed because of life history constraints (e.g., omissions of elevations that are subaqueous, since pine trees cannot survive underwater). Removing treatment levels of this nature does not violate the random sampling criteria for parametric analysis because these are not meaningful samples. Removal would enhance the ability of the analysis to find differences between samples that exhibit the sensory attribute, and should only be done at the discretion of the investigator.

Detection of Inconsistent Panelists

Elimination of inconsistent panelists inclines the data toward true behavior. It is in essence equivalent to omitting data collected with instrumentation that was later found faulty. Noncommittal panelists consistently (and incorrectly) rated the majority of responses near the overall mean. Consequently, these panelists rarely produced outlier data, yet inherently produced non-homogeneous variance.

Calibration Factor. For the graphic representations of the data, a transformation was employed to make the ratings of different panelists commensurate. The calibration factor we arrived at had absolutely no effect on the behavior of the data across treatment levels, because each observation within each panelist was multiplied by a constant. But, why not subtract by a constant (i.e., the overall mean) rather than multiply, since addition or subtraction of a constant leaves the variances unchanged, whereas multiplication by a constant changes the variance by the constant squared? The answer is twofold: first, the variances of the transformed population are not used, and secondly, multiplication by the calibration factor maintains the units of the intensity ratings.

Removal of Attributes. Elimination of entire attributes appear to constitute a severe loss of carefully gathered data. Yet, if responses for an attribute are flat across treatment levels, then the attribute carries no information in terms of Shannon's (1948) measure of information. This type of response is characteristic of attributes that are subordinate to the more dominant attributes and, therefore, difficult to detect. Subordinate attributes may also result in inconsistent ratings across panelists and treatment levels for the attribute 'painty' in Fig. 6. This type of attribute is more difficult for the investigator to evaluate because it is tempting to search for a subset of panelists (no matter how small) that show similar (and sensible) behavior. For example, in Fig. 6 panelists E, F and M appeared to display similar behavior. However, when the majority of a trained panel does not perceive a note (as in this case) then any use of that attribute must be done with extreme caution. We do not object to exploratory analysis of this nature, so long as it is well-documented and the investigator realizes she/he is capitalizing on chance. If such analyses are to be performed, one can protect against Type I error (claiming a difference when none exists) by decreasing the alpha level (e.g., from say $\alpha = 0.05$ to $\alpha = 0.01$) required for a significant difference. The most conservative action is to omit the entire attribute from the inference phase.

Analysis of Variance

As mentioned previously, blocking on panelists is inappropriate because panelists more closely approximate sampling units or repeated measures than true experimental units (i.e., panelists are pseudoreplicates). If the panelist effect and panelist interactions are to be included in the analysis of variance, a split plot design is appropriate (e.g., Table 2a). However, for three reasons we chose to average over the panelists directly and use a simple randomized block design, blocking on session. First, results from several of our data sets revealed that the subplot residual error often violated distributional assumptions. Second, after examination of the various indicators discussed herein, the panelist and panelist interactions are of little value. Significant differences due to inconsistent panelists have been removed and those due to super-responsive panelists are known. In addition the panelist-by-treatment interactions have already been plotted (e.g., in Fig. 4-6). Furthermore, in terms of using the subplot information as a training tool to improve panelists' future judgments, panelists find graphical indicators far more enlightening. Third, the main plot effects and residual error are identical to the simple randomized block design with the data averaged over panelist, and the residual error from this averaged data often meets the criteria for parametric analysis (i.e., see-Central Limit Theorem). The authors did this work with a program written in SAS (SAS 1985, 1988). Since this original work was done, the program has been rewritten in a more efficient manner that has shortened the analysis time.

ACKNOWLEDGMENTS

The authors are especially grateful to Alejandra Munoz, Deborah Boykin, Dallas Johnson, and Bryan Vinyard for advice and discussion of sensory design problems, and Deborah Boykin, Peter C. Dickinson, Bryan Vinyard and one anonymous reviewer for their helpful comments on an earlier version of this manuscript. Authors Bett and Shaffer developed the statistical protocols, author Vercellotti contributed to the conceptualization and resolution of the problem and, with authors Sanders and Blankenship, initiated and conducted the original experiment (Sanders *et al.* 1989a) which provided the data used in this manuscript.

REFERENCES

CAUL, J.F. 1957. The profile method of flavor analysis. Adv. Food Res. *7*, 1.

COCHRAN, W.G. and COX, G.M. 1957. *Experimental Designs*, 2nd ed., John Wiley & Sons, New York.

JOHNSEN, P.B., CIVILLE, G.V., VERCELLOTTI, J.R., SANDERS, T.H. and DUS, C.A. 1988. Development of a lexicon for the description of peanut flavor. J. Sensory Studies *3*, 9.

KWAN, W.O. and KOWALSKI, B.R. 1980. Data analysis of sensory score. Evaluations of panelists and wine score cards. J. Food Sci. *45*, 213.

LUNDAHL, D.S. and McDANIEL, M.R. 1990. Use of contrasts for the evaluation of panel inconsistency. J. Sensory Studies *5*, 265.

MARSHALL, R.J. and KIRBY, S.P.J. 1988. Sensory measurement of food texture by free-choice profiling. J. Sensory Studies *3*, 63.

MEILGAARD, M., CIVILLE, G.V. and CARR, B.T. 1987. Sensory Evaluation Techniques, Vol. 1 and 2, CRC Press, Boca Raton, FL.

MILLER, A.J. 1988. Adjusting taste scores for variations in use of scales. J. Sensory Studies *2*, 231.

NAES, T. and SOLHEIM, R. 1991. Detection and interpretation of variation within and between assessors in sensory profiling. J. Sensory Studies *6*, 159.

NEUMANN, R. and MOLNAR, P. 1974. Planning of the quality of foods. Libensmittel-Industrie *21*, 347.

O'MAHONY, M. 1982. Some assumptions and difficulties with common statistics for sensory analysis. Food Technol. *36*(11), 75.

ORESKOVICH, D.C., KLEIN, B.P. and SUTHERLAND, J.W. 1991. Procrustes Analysis and its applications to free-choice and other sensory profiling. In *Sensory Science Theory and Applications in Foods*, (H.T. Lawless and B.P. Klein, eds.), Marcel Dekker, New York.

POKORNY, J., MARCIN, A., PAVLIS, J. and DAVIDEK, J. 1982. Testing of the performance of assessors in the evaluation of sensory profiles. Scientific Papers Prague Inst. Chem. Technol. *E*(53), 25.

POWERS, J. 1984. Using general statistical programs to evaluate sensory data. Food Technol. *38*(6), 74-84.

SAS 1985. *SAS User's Guide; Basics*, SAS Institute, Cary, NC.

SAS 1988. *SAS/STAT User's Guide*, SAS Institute, Cary, NC.

SANDERS, T.H., BLANKENSHIP, P.D., VERCELLOTTI, J.R., CRIPPEN, K.L. and CIVILLE, G.V. 1989a. Interaction of maturity and curing temperature on descriptive flavor of peanuts. J. Food Sci. *54*, 1066.

SANDERS, T.H., VERCELLOTTI, J.R., BLANKENSHIP, P.D. and CIVILLE, G.V. 1987. Interaction of maturity and curing temperature on peanut flavor. Paper No. 41, presented at 194th Amer. Chem. Soc. Meeting, New Orleans, LA, Aug. 30-Sept. 4.

SANDERS, T.H., VERCELLOTTI, J.R., CRIPPEN, K.L. and CIVILLE, G.V. 1989b. Effect of maturity on roast color and descriptive flavor of peanuts. J. Food Sci. *54*, 475.

SHANNON, C.E. 1948. A mathematical theory of communication. Bell Syst. Technol. J. *27*, 379.

CHAPTER 1.11

EXPERTS VERSUS CONSUMERS: A COMPARISON

HOWARD R. MOSKOWITZ

*At the time of original publication,
the author was affiliated with
Moskowitz Jacobs Inc.
1025 Westchester Ave.
White Plains, New York 10604.*

ABSTRACT

This paper compares experts and consumers to determine the degree to which they agree with each other on ratings of 37 sauce products, using the same sensory attributes. The paper also assesses the degree to which sensory attribute ratings correlate with objective physical measures. The ratings of experts (1-9 scale) and consumers (0-100 scale) agree quite highly, as shown by the high correlation between the two panels across the 37 products. The paper refutes the notion that consumers are incapable of validly rating the sensory aspects of products. The paper therefore presents the case for using consumers to assess the sensory characteristics of products.

INTRODUCTION

Importance of Experts in the History of Sensory Analysis

Any history of sensory analysis will quickly reveal that experts have played an important part in the evaluation of foods and drinks. Indeed, the early history of sensory analysis often relied upon experts, such as brewmasters and winetasters, to guide development and quality assurance. More recently experts have played an important role because they comprise individuals who have been tested for sensory acuity, and have undergone extensive training in descriptive analysis, including training in the recognition/description of attributes and the use of scales. These experts provide reliable ratings for the descriptive characteristics of products, and often the data provided by such experts is the only information available to developers because instrumental measures cannot be programmed to pick up the sensory nuances of a product that a human being can.

Expert panelists have been recommended for and used in product development and quality control. For the most part these expert panelists have been developed and used by R&D groups, rather than by marketing groups. Over the years the use of such experts has grown, beginning with the introduction of experts some fifty years ago in the food industry. As Moskowitz (1983) pointed out, the heyday and hegemony of experts extended from food to drink, from perfumes to other consumer items. These experts were either home-grown or received formal training (especially after the Flavor Profile was developed by the Arthur D. Little Co. of Cambridge, Mass) (Caul 1957). As time progressed, a variety of commercial approaches sprang up to create experts out of the typical consumer. Thus individuals could become experts in evaluating texture by learning the texture profile (Szczesniak *et al.* 1963), or become experts in flavor by learning either the Flavor Profile, the Quantitative Descriptive Analysis Method (Stone *et al.* 1974), or the Sensory Spectrum (Meilgaard *et al.* 1987). No effort is made to include individuals with vastly superior sensory capacities, although there are screening tests to ensure that the individuals would possess at least the requisite sensory acuity.

Researchers espousing the use of experts, rather than consumers, for profiling the product characteristics of foods used arguments varying from improved sensitivity to product variables (through development of a special descriptive language and training on reference products), to cost efficiencies (expert panelists often work at the test location, and can be run in experiments at vastly reduced costs). Most of the effort is focused on training the consumers to describe the characteristics of the products. Occasionally expert panelists are used to guide new product development, but new product developments is left to consumer researchers rather than to those working with experts. Today, expert panels are used for routine quality control, where daily testing is necessary to maintain product quality. Typically, the use of expert panelists is justified by their ability to notice differences between product-run, and their ability to spot sensory defects before the batch goes out to the market.

At the same time as sensory analysis was celebrating the expert, psychophysicists in experimental psychology were investigating the ability of untrained individuals to scale the sensory intensities of stimuli, ranging from the size of geometric shapes to the loudness of sounds, brightness of lights, and strengths of stimuli that were tasted or smelled (Stevens 1953, 1975). The results of these studies were remarkable. They showed that untrained panelists could easily scale perceived intensity, and that when the physical stimulus was systematically varied the ratings correlated highly with the physical measures. Using the method of magnitude estimation as a scaling procedure Stevens showed that the ratings for intensity (S) generated a power function (Vs physical intensity I) of the form: $S = kI^n$. The exponent n was reproducible from study to study. Stevens averred (1969, personal communication) that a well instructed panelist could perform almost as well as an expert. Thus the psychophysical traditional stands in direct contrast to the food science tradition. Psychophysicists believe that it is the experiment, not the panelist, which when properly set up reveals the key sensory aspects of the product. To a psychophysicist the panelists are interchangeable. To the food scientist steeped in the tradition of expert panelists it is the trained expert panelist who brings the value to the study. The product and the experimental design of the study are important, but the trained expert panelist is the key factor.

Given this divergence between those who believe only in experts as capable of assessing the sensory stimuli and those who believe in consumers as being able to scale intensities when well instructed, can we determine the degree to which the two viewpoints agree versus disagree? The psychophysical literature provides a mass of literally hundreds (if not thousands) of refereed scientific papers showing that untrained (but well instructed) consumers can validly scale the sensory intensity of stimuli, and that their ratings track known physical variations. In contrast, in the main sensory professionals holding for the use of experts do not provide a similar mass of papers, but rather show profiles provided by expert panelists. These papers do not demonstrate the superiority of expert panel data over consumer data, but rather only argue (generally without corroborating data) that consumers cannot validly profile the sensory aspects of stimuli, but rather only can rate degree of liking. The best way to decide between these two radically different opinions is to study how experts and consumers scale the same stimuli.

The current study compares the data obtained from a panel of 12 experts with a data obtained from a consumer panel of 225 individuals. The category is a flavored gravy/sauce product. The original study was designed to evaluate all key competitors in the market, in order to identify new opportunity areas. To this end the study considered a limited set of 37 different products from more than 150 available across the US and Europe. When both panelists evaluate the same set of products it is possible to identify areas of agreement and disagreement between experts and consumers. This study parallels the work done on breads some sixteen years ago (Moskowitz *et al.* 1979) where interest focused on the relation between experts and consumers in a food category. The current study is much broader, however, dealing as it does with in-market products of a more diverse sensory nature.

METHOD

Stimuli

The stimuli comprised 37 different, commercially available sauces/gravies that would be used either with meat or with pasta. All products were of potential dual use. These products were selected from a larger array of 155 different products evaluated by the sensory assessment group at the manufacturer. These 155 products were profiled on a variety of sensory attributes. Based upon the profiles, the sensory evaluation group selected 37 most different products, using a clustering program (Systat 1994). The clustering program was used to collapse the 155 different products into 37 clusters. (The number 37 for the total set of products was based upon the potential maximum number of products that could be tested by consumers within the corporate budget).

Test Procedure — Experts

Twelve experts rated each of the 37 products twice, rating 15 samples per day. The experts used an attribute list that was developed to cover the sensory nuances. The list had been in use for three years prior to the study. These attributes covered appearance, aroma, taste/flavor and texture/mouthfeel. The experts used an anchored 9 point scale with which they were familiar and on which they had been trained.

Test Procedure — Consumers

Each of the 225 consumers (equally distributed across four markets) rated a randomized set of 10 products during each four hour test session. Each consumer participated for two days, so that the panelist rated a total of 20 of the products. For purposes of product control the 37 products were blocked so that during a single test period only five products were produced. There were ten such blocks each day. Blocking reduced the order bias. The composition of the blocks and the order of blocks were changed from market to market to ensure that the same products would not appear in each block across markets.

Statistical Analysis

The analysis of this data proceeds in four steps as follows:

(1) Assess the degree to which experts versus consumers space their ratings on the scale. That is, do experts show greater discrimination than consumers by locating the 37 products further apart from each other on the scale?

(2) Correlate the ratings assigned by experts and those assigned by consumers. The correlation coefficient assumes a linear relation between the two variables. To what degree do the same attributes correlate when used by experts versus consumers?

(3) Factor analyze the attributes to determine the underlying basic factors. Factor analysis identifies underlying basic dimensions. What is the factor

structure for experts rating these products versus for consumers?

(4) Create an integrative model, allowing the researcher to estimate attribute levels for a consumer sensory profile, given attribute levels for an expert sensory profile (and vice versa).

RESULTS

Analysis 1 — Range and Variability of Means

This first analysis looks at the standard deviations of the mean ratings for experts and consumers on comparable attributes. Since the experts used a 9 point scale and the consumers used a 0-100 point scale there must first be a transformation to render the two scales of equal length. This transformation turns a 9 point rating into a rating of 100 and a 1 point rating into a rating of 0. The transformation is as follows:

$$\text{Transformed Expert Panel} = (12.5)(\text{Expert Rating} - 5) + 50$$

Table 1 presents the standard deviations for the consumer data, for the expert data (transformed). Figure 1 presents the scatterplot for the different attributes. Each point in Fig. 1 corresponds to a sensory attribute. Figure 1 reveals a strong linear relation with a slope near 1 for the standard deviation of experts plotted against the standard deviation of consumers. This suggests that the two groups space their ratings equally. Therefore we should expect no greater discrimination based upon mean data for experts versus for consumers. For texture the experts use a slightly broader range on the scale than do consumers. For appearance and for taste/flavor the experts and the consumers use similar ranges, although sometimes experts use a larger range for an attribute, whereas other times consumers use a larger range. Thus for this first analysis the data suggest that experts and consumers show similar, albeit not identical abilities to discriminate among products, based upon the scatter of products on the scale. The advantage of the expert appears in texture, but the advantage is not dramatic.

Analysis 2 — Correlations Between Similar Attributes

The correlation analysis assumes a linear relation between two variables. Since both experts and consumers rate the same products on simple sensory scales we expect to see a positive correlation. With 37 products the magnitude of the correlation becomes a telling statistic. Table 2 presents the correlation analysis. For many attributes (but not all) there is a very high and significant correlation between the two sets of measures. The key area where this relation

TABLE 1.
COMPARISON OF STANDARD DEVIATIONS FOR CONSUMER AND EXPERT PANEL RATINGS ON 37 PRODUCTS*

Sense	Attribute	Consumer	Expert	Difference
Appearance	Flecks	15.9	13.2	2.7
Appearance	Amount Tomato Pieces	17.5	16.3	1.2
Appearance	Size Tomato Pieces	14.9	15.7	-.8
Appearance	Size Vegetable Pieces	16.1	17.2	-1.1
Appearance	Brown	11.7	15.7	-4.0
Flavor	Sour	11.9	4.9	7.0
Flavor	Black Pepper Flavor	15.8	9.2	6.6
Flavor	After taste	14.0	7.8	6.2
Flavor	Flavor	11.5	6.6	4.9
Flavor	Green Pepper Flavor	15.4	10.7	4.7
Flavor	Tomato Flavor	9.8	5.5	4.3
Flavor	Onion Flavor	9.3	5.6	3.6
Flavor	Mushroom Flavor	11.5	9.3	2.2
Flavor	Garlic Flavor	9.1	7.0	2.1
Flavor	Aroma	7.4	6.5	0.9
Flavor	Meat Flavor	11.0	10.5	0.5
Flavor	Sweet Flavor	9.8	10.0	-0.2
Flavor	Oily Flavor	6.4	7.2	-0.8
Flavor	Vegetable Flavor	12.5	14.0	-1.5
Flavor	Salty Flavor	3.5	5.6	-2.1
Flavor	Cheese Flavor	4.7	7.7	-3.0
Texture	Crispy Vegetable	12.2	16.4	-4.2
Texture	Thickness	9.6	14.0	-4.4
Texture	Oily Mouthfeel	5.4	13.4	-8.0

*Ratings for experts transformed to a 0-100 point scale to facilitate comparison of standard deviations

falls down is in the basic tastes (salty, sour). For other, more unusual flavor attributes, including black pepper, the correlation is very high. It may well be that for the more common tastes consumers and experts differ because experts have been taught to recognize nuances in these common tastes that consumers do not. For more unusual attributes, such as black pepper or garlic, respectively, there is not as much time for the concepts of these flavors to diverge in the minds of experts and consumers, respectively. Hence these attributes show high correlation between consumers and experts. Attributes where there is less agreement, however, include tomato flavor, oil flavor, oily texture, and thickness, as well as sour and bitter tastes.

Standard Error Across Products

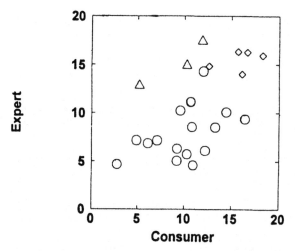

FIG. 1. STANDARD ERRORS ACROSS THE 37 PRODUCTS TESTED
IN THE STUDY
The figure compares the standard errors for consumers versus for experts,
after the scales used by experts have been adjusted to become a
0-100 point scale. The shapes refer to different sensory attributes.
Diamonds = visual attributes; Triangles = texture attributes;
Hexagons = flavor attributes

It is important to note that the attributes showing less agreement between consumers and experts comprise relatively simple attributes, commonly used (except for tomato flavor). Attributes where there is greater agreement often comprise terms such as black pepper or garlic. This result is counter-intuitive. One would expect that after training the experts would perform differently on the specific attributes such as garlic and black pepper, because these attributes are unusual. Yet it is the more ordinary and general attributes on which the consumers and experts disagree.

Analysis 3 — Factor Structure of Experts Versus Consumers

Factor analysis of the expert ratings and the consumer ratings each show four major factors when the analysis is done by the principal components method, and the roots extracted using a stringent criterion (eigenvalue > 2). However, unlike the correlation analysis above, factor analysis shows that the experts have a far more meaningful, logical set of factors than do consumers.

TABLE 2.
SIMPLE CORRELATION BETWEEN CONSUMER RATING ATTRIBUTES
AND PARALLEL EXPERT RATING ATTRIBUTES

	Correlation
Appearance Attributes	
Smooth	0.95
Brown Color	0.92
Amount of tomato pieces	0.91
Size of tomato pieces	0.89
Amount of Flecks	0.88
Size of vegetable pieces	0.74
Aroma Attributes	
Aroma Intensity	0.68
Flavor Attributes	
Herb Flavor	0.91
Meat Flavor	0.89
Black Pepper Flavor	0.88
Garlic Flavor	0.88
Cheese Flavor	0.87
Flavor Intensity	0.86
Mushroom Flavor	0.86
Vegetable Flavor	0.86
Sweet Taste	0.86
Green Pepper Flavor	0.83
Onion Flavor	0.76
Aftertaste	0.63
Tomato Flavor	0.60
Oily Flavor	0.53
Sour Taste	0.43
Salty Taste	0.37
Texture Attributes	
Thickness	0.59
Oily Texture	0.49
Crispness Of Vegetables	0.79

Experts separate their factors into more meaningful, coherent groups than do consumers.

Table 3A presents the factor structure for experts:

Factor 1 = Primarily flavor and tomato (e.g., flavor, spices, etc.)
Factor 2 = Vegetables (size, texture, flavor)
Factor 3 = Sweet flavor, tomato flavor Vs aftertaste
Factor 4 = Appearance/texture (darkness, consistency)

Experts also have other attributes such as meat flavor, cheese flavor, etc., which do not load on any of the four axes, and may require separate dimensions to be captured.

Table 3B presents the factor structure for consumers. The factors are less reasonable, and show a great deal more confusion among attributes:

Factor 1 = Vegetables (texture, appearance, flavor), along with many other attributes such as tomato pieces, thickness, etc. Similar to Factor 2 for experts
Factor 2 = Taste as well as oiliness
Factor 3 = Meat flavor, brown color
Factor 4 = Salt taste

From the data in Tables 3A and 3B we conclude, therefore, that the structure for consumers suggests a melange with many unrelated attributes thrown together. In contrast, for experts the factor structure is more simple. It is from the factor analysis that we begin to see differences between consumers and experts in the way attributes are used.

Analysis 4 — An Integrative Model Relating Experts and Consumers

One of the most recurring problems in product testing and sensory evaluation has been to relate the data from experts to the data from consumers in order to use the data, rather than simply to establish that a relation exists. Many companies invest a great deal of money to create trained panels. Often these panels are used as simple quality control systems, with the expert panelist acting as a machine. These quality control panels are critical to insure ongoing product quality for attributes that an instrument cannot easily measure.

Moskowitz (1994) presented a method to relate profiles from one domain (e.g., experts) to profiles from another domain. The method requires that the same products be assessed by panelists or instruments from both domains. The six steps of the procedure follow:

TABLE 3A.
INDIVIDUAL FACTOR SPACE — EXPERTS

	Factor 1E	Factor 2E	Factor 3E	Factor 4E
Flavor - Intensity	0.85	-0.02	0.30	0.04
Flavor - Spices	0.81	0.16	0.23	0.14
Flavor - Onion	0.80	-0.07	-0.10	0.09
Appearance -Tomato Pieces	0.75	0.25	-0.32	-0.28
Flavor - Garlic	0.73	-0.14	0.09	-0.01
Appearance - Amount of Flecks	0.71	0.32	-0.25	0.30
Aroma - Intensity	0.71	0.00	0.35	0.03
Appearance - Smooth	0.70	0.55	-0.04	-0.12
Appearance - Size Of Tomato Pieces	0.65	0.27	-0.45	-0.29
Flavor - Black Pepper Flavor	0.60	0.09	0.55	-0.18
Appearance - Size of Vegetable Pieces	0.17	0.93	-0.12	-0.02
Texture- Texture Of Vegetable Pieces	0.24	0.89	0.00	-0.07
Flavor- Vegetable Flavor	0.16	0.86	0.14	0.10
Flavor - Green Pepper Flavor	0.08	0.73	0.20	-0.02
Flavor -Mushroom Flavor	-0.18	0.69	-0.25	0.10
Flavor - Aftertaste	0.21	0.01	0.76	0.29
Flavor - Tomato Flavor	-0.11	0.24	-0.75	-0.15
Flavor - Sweet	0.11	-0.12	-0.78	0.26
Appearance - Consistency	-0.03	0.14	-0.04	0.89
Appearance - Dark	0.14	0.08	0.05	0.85
Flavor - Sour	-0.19	-0.12	0.23	0.47
Flavor - Meat Flavor	0.08	-0.41	0.30	0.40
Flavor - Salty	0.37	-0.40	-0.28	-0.09
Flavor - Cheese	-0.19	-0.24	-0.34	-0.12
Appearance- Oily Appearance	0.42	-0.38	0.26	-0.37

(1) Array the full data profile in a rectangular matrix, with the columns corresponding to attributes, and the rows corresponding to the products.

(2) Isolate the sensory attributes from consumers and experts, as well as the instrumental measures. It is important here to use only those attributes which describe the products (either from a sensory or a physical aspect), and not to use attributes which reflect liking or image.

(3) For the variables identified in Step 2 above, perform a principal components analysis, extract a reasonable number of orthogonal dimensions using a modest criterion (e.g., eigenvalue > 2), rotate the solution (e.g., using quart-

TABLE 3B.
INDIVIDUAL FACTOR SPACE — CONSUMERS

	Factor 1C	Factor 2C	Factor 3C	Factor 4C
Texture - Crispness Of Vegetables	0.92	-0.19	-0.03	-0.19
Flavor - Vegetable	0.89	-0.06	-0.04	-0.17
Appearance - Chunky	0.88	0.02	0.07	-0.38
Appearance - Size Of Vegetable Pieces	0.87	0.01	-0.08	-0.34
Appearance - Flecks	0.85	-0.26	0.09	0.08
Flavor - Herbs	0.84	0.31	0.06	0.29
Flavor - Onion	0.84	0.16	0.01	0.06
Appearance - Thick	0.82	-0.10	0.32	0.29
Flavor - Pepper	0.77	0.25	-0.02	-0.01
Texture - Thick	0.77	-0.10	0.39	0.30
Aroma - Intensity	0.75	0.36	0.15	-0.16
Appearance - Amount Of Tomato Pieces	0.73	-0.02	-0.37	-0.41
Flavor - Black Pepper	0.70	0.59	0.04	0.27
Flavor - Intensity	0.69	0.57	0.11	0.31
Appearance - Size Of Tomato Pieces	0.68	-0.08	-0.43	-0.45
Flavor - Garlic	0.67	0.34	0.00	0.26
Appearance - Oily	-0.56	0.13	0.47	-0.34
Flavor - Sour	0.03	0.82	-0.04	0.22
Flavor - Aftertaste	0.52	0.79	0.06	0.24
Flavor - Oily	-0.38	0.63	0.46	-0.27
Texture - Oily	-0.41	0.56	0.47	-0.24
Flavor - Meat	0.02	0.01	0.90	0.12
Appearance - Brown	0.28	-0.06	0.75	0.25
Flavor -Mushroom	0.45	-0.45	0.17	-0.04
Flavor - Salt	-0.06	0.17	0.11	0.78
Flavor - Cheese	-0.18	-0.57	-0.03	0.06
Flavor - Sweet	-0.22	-0.81	-0.03	0.04
Flavor - Tomato	-0.29	-0.31	-0.66	0.14

imax rotation), and finally save the factor scores. Typically this procedure generates a far more limited number of factors (e.g., 2-7, depending upon the number of stimuli, their diversity, and the diversity of the attributes used to measure the products). When the expert panel and consumer panel data are combined to create one large data set, the factor analysis generates 6 factors, $F_1..F_6$.

(4) Using multiple regression, let the factor scores be independent variables, and all rating attributes be dependent variables, respectively (whether or not these rating attributes had been previously used to create the factor scores). For each attribute in turn, relate the attribute rating to the factor scores by means of a quadratic function, forcing in linear and quadratic terms, and allowing additional significant) pairwise interaction terms to enter the equation. The regression equation is written as follows:

$$\text{Attribute Rating} = k_0 + k_1(F_1) + k_2(F_2).. + k_{11}(F_1*F_1) + k_{22}(F_2*F_2).. + \text{significant cross terms such as } k_{12}(F_1*F_2) \text{ etc.}$$

Table 4A shows the degree of fit of the factor-based equations to the attributes.

(5) The equations created in Step 4 relate the factor scores to rating attributes. For any combination of factor scores one can estimate the profile of rating attributes by solving the different equations. Each equation generates an expected level of a rating attribute. With M rating equations, therefore, we have M numerical estimates.

(6) Turn the problem around. This time choose a subset Q of the rating attributes to act as goals. Vary the factor scores (viz., the independent variables) within the range tested until the expected profile of the subset Q is as close as possible to the goals specified. This set of factor scores is therefore the set which generates the match. Now solve all of the M equations, given the set of factor scores which had produced the fit. The result is an estimate of the full profile of attributes corresponding to the subset Q of the ratings which had acted as goals.

Table 4B presents parameters of some of the equations. Note that the fits are quite good, indicating that the factor scores can act as independent variables. Table 4C shows predicted expert panel data given consumer data, and the reverse (predicted consumer panel data given expert data).

DISCUSSION

Although the relation between expert panel data and consumer panel data has long been a bone of contention among researchers, the current data for a sauce/gravy set suggest that the two types of panelists show similar types of results when instructed to evaluate the same set of products on similar attributes. The correlations between the two data sets are high, albeit certainly not perfect. Furthermore, both experts and consumers show similar but again not identical factor structures, although the expert factor structure is clearer.

TABLE 4A.
GOODNESS OF FIT OF REGRESSION EQUATIONS RELATING FACTOR SCORES TO ATTRIBUTES

Attribute	Con Mul/R	Con Adj/R	Con SE	Exp Mul/R	Exp Adj/R	Exp SE
Darkness of Color	0.97	0.95	2.49	0.96	0.94	0.31
Oily appearance	0.91	0.86	2.32			
Chunky appearance	0.99	0.98	3.26	0.98	0.96	0.34
Thick appearance	0.98	0.97	2.05			
Aroma	0.94	0.89	2.41	0.90	0.82	0.22
Amount Of Flecks	0.97	0.95	3.41	0.97	0.94	0.25
Amount Tomato Pieces	0.99	0.99	1.55	0.99	0.98	0.19
Size Tomato Pieces	0.99	0.98	2.34	0.98	0.97	0.21
Size Vegetable Pieces	0.95	0.92	4.50	0.97	0.95	0.32
Flavor	0.99	0.98	1.83	0.99	0.97	0.09
Tomato Flavor	0.98	0.97	1.68	0.98	0.96	0.09
Meat Flavor	0.99	0.98	1.41	0.97	0.95	0.19
Mushroom Flavor	0.93	0.90	3.69	0.94	0.91	0.22
Onion Flavor	0.94	0.91	2.77	0.91	0.87	0.16
Green Pepper Flavor	0.94	0.91	4.63	0.91	0.86	0.32
Vegetable Flavor	0.94	0.92	3.61	0.95	0.92	0.31
Herb Flavor	0.99	0.99	1.57			
Spice Flavor	0.98	0.96	0.16			
Oregano	0.86	0.74	0.29			
Basil Flavor	0.91	0.83	0.23			
Pepper Burn	0.99	0.98	0.18			
Black Pepper Flavor	0.99	0.99	1.73	0.99	0.98	0.11
Garlic Flavor	0.98	0.96	1.85	0.91	0.85	0.21
Cheese Flavor	0.98	0.97	0.87	0.96	0.94	0.15
Salt	0.96	0.92	0.93	0.95	0.92	0.13
Sweet	0.98	0.97	1.82	0.98	0.97	0.14
Aftertaste	0.99	0.98	1.76	0.91	0.86	0.24
Sour	0.97	0.95	2.57	0.82	0.75	0.19
Oily Flavor	0.99	0.98	0.89	0.93	0.88	0.20
Crispness of Vegetables	0.98	0.98	1.80	0.96	0.94	0.32
Oily Mouthfeel	0.98	0.96	1.03	0.96	0.92	0.28
Thickness Of Texture	0.96	0.93	2.52	0.96	0.94	0.28

Mul/R = Pearson R (multiple linear regression)
Adj/R = Adjusted Pearson R (for number of predictors)
S.E. = Standard error of regression
Con = Consumers
Exp = Experts

The approaches here show that it is possible to integrate expert and consumer data together into a single coherent database, provided the same samples are used for both. The integrative model using factor scores as predictors enables the researcher to traverse between the domain of experts and the domain of consumers. Rather than searching for a 1:1 relation between data from these two domains the integrative model allows for the estimation of profiles in one domain from profiles in the other domain.

TABLE 4B.
COEFFICIENTS OF EQUATIONS RELATING FACTOR SCORES TO SENSORY
ATTRIBUTES: COMPARISON OF CONSUMER EQUATIONS (CON) AND
EXPERT EQUATIONS (EXP)

	CON Dark	EXP Dark	CON Aroma	EXP Aroma	CON Tomato Flavor	EXP Tomato Flavor
Multiple R^2	0.96	0.92	0.89	0.80	0.95	0.84
Adjusted R	0.94	0.87	0.82	0.64	0.92	0.75
St.Err. Reg	2.93	0.46	3.12	0.31	2.79	0.22
CONSTANT	49.51	4.43	49.21	5.14	49.65	5.11
F_1	0.27	-0.09	5.04	0.24	-3.19	0.04
F_2	2.38	0.17	7.02	0.53	-3.48	-0.20
F_3	-4.00	-0.50	-3.52	-0.40	9.81	0.34
F_4	6.31	0.88	-0.51	0.06	1.14	0.10
F_5	-1.25	-0.14	0.60	-0.08	1.06	0.19
F_6	2.65	0.18	-0.92	0.01	-2.57	-0.10
$(F_1)^2$	0.04	0.27	-2.20	-0.14	1.10	0.11
$(F_2)^2$	-0.43	0.03	-0.95	-0.16	0.41	-0.02
$(F_3)^2$	0.63	-0.01	-0.14	-0.09	1.00	0.01
$(F_4)^2$	0.02	-0.16	-0.04	-0.02	-1.01	-0.06
$(F_5)^2$	-0.03	0.21	0.20	-0.02	0.59	0.04
$(F_6)^2$	1.05	0.12	0.91	0.12	2.38	0.04
F_1F_2	0	0	0	0.12	0	0.11
F_1F_3	-3.24	0	0	0	0	0
F_1F_4	0	-0.21	0	0	0	0
F_1F_5	0	0	0	0	0	0
F_1F_6	0	0	0	0	-3.31	0
F_2F_3	0	0	0	0.25	0	0
F_2F_4	0	0.19	0	-0.09	0	0
F_2F_5	0	0.26	0	-0.16	0	0.091
F_2F_6	0	0	0	0	0	0
F_3F_4	-4.51	-0.43	0.08	0	0	0
F_3F_5	0	0	0	0	0	0
F_3F_6	0	0	1.28	0	0	-0.15
F_4F_5	0	0	0	0	1.26	0
F_4F_6	0	0.21	-1.34	0	-2.27	0
F_5F_6	0	0	0	0.28	0	0

The Role of Experts Versus the Role of Consumers

Those who aver that experts are the only panelist who can validly profile the sensory characteristics of products may be incorrect. Consumers can also profile these characteristics. Quite possibly the profiles will differ, and perhaps the cognitive organization of the profiles will differ from experts to consumers, but the reverse engineering procedures (Table 4C) allow the researcher to estimate

TABLE 4C.
REVERSE ENGINEERING PROFILES USING FACTOR EQUATIONS

Goal Profile Obtained:	Consumers		Experts	
Estimated Profile For:	Experts		Consumers	
F1	1.47		1.69	
F2	0.9		1.38	
F3	0.91		0.45	
F4	-2.66		0.14	
F5	-0.3		0.03	
F6	-0.52		-0.64	
	Estimate	Goal	Estimate	Goal
C - Brown	38	50	48	
C -Chunky	65	50	82	
C - Aroma	48	50	59	
C - Flecks	50	50	62	
C -Amount Of Tomato Pieces	50	50	58	
C- Size Of Tomato Pieces	43	50	43	
C - Size Of Vegetable Pieces	51	50	62	
C - Strength Of Flavor	62	50	72	
C -Tomato Flavor	46	50	55	
C - Meat Flavor	1		8	
C - Mushroom Flavor	24		0	
C - Onion Flavor	33		45	
C -Green Pepper	47		63	
C - Vegetable Flavor	40		57	
C - Black Pepper Flavor	47		36	
C - Garlic Flavor	30		46	
C - Cheese Flavor	7		7	
C - Salty	26		25	
C - Sweet	19		25	
C - Aftertaste	63		57	
C - Sour	34		37	
C - Oily Flavor	21		22	
C - Crisp Vegetable	58		66	
C - Oily In Mouth	21		25	
C - Thick Texture	38		65	
Obtained Profile From:	Consumers		Experts	
Estimated Profile For:	Experts		Consumers	
E - Color	2.7		5.1	5.0
E - Smooth	4.8		7.3	5.0
E - Aroma	5.6		5.8	5.0
E - Amount Of Flecks	2.6		4.6	5.0
E - Amount Of Tomato Pieces	3.5		4.5	5.0
E - Size Of Tomato Pieces	3.9		5.0	5.0
E - Size Of Vegetable Pieces	4.0		5.0	5.0
E - Strength Of Flavor	6.1		6.4	5.0
E - Tomato Intensity	5.0		5.7	5.0
E - Meat Flavor	0.6		1.5	
E - Mushroom Flavor	2.3		0.1	
E - Onion Flavor	2.6		3.5	
E - Green Pepper Flavor	2.6		3.8	
E - Vegetable Flavor	4.2		5.0	
E - Black Pepper Flavor	4.6		3.5	
E - Garlic Flavor	3.0		3.0	
E - Cheese Flavor	1.0		1.4	
E - Salt	4.6		3.8	
E - Sweet	2.6		3.6	
E - Aftertaste	2.5		3.4	
E - Sour	3.9		4.2	
E - Oily Flavor	2.0		1.8	
E - Texture Vegetable Pieces	4.3		4.4	
E - Oily	4.5		3.1	
E - Consistency	2.4		6.3	

the likely expert panel profile given a consumer panel profile, and vice versa. Thus the sets of profiles can be related to each other, or at least approximated, so that given one profile the researcher can estimate the other.

If it is possible, then, to interrelate experts and consumers in this fashion, then the next question deals with the role of the expert versus the role of the consumer. Clearly it is consumers who are the most appropriate individuals to profile liking and image characteristics of products, and not experts who should do this profiling. If, in addition, consumers can validly profile the characteristics of the product, then what is the role of the expert? It is clearly not to be the primarily profiler of sensory characteristics because the consumer data can serve just as well. Experts therefore become surrogate consumers on the one hand (since experts are cheaper to run than consumers), or surrogate machines on the other (because experts are trained to assign reproducible, valid ratings to products). Experts provide another, possibly cheaper form of attribute profiles, which can be mathematically transformed to consumer profiles. Furthermore the language used by the expert and the consumer need not be the same. Each group can use its own language. The mathematical modeling and the reverse engineering process transforms data from one domain to another.

APPENDIX

Validating the Reverse Engineering Model

To validate the reverse engineering model we must determine how well a profile from one set of attributes (e.g., consumer ratings) generates a profile from another set of attributes (e.g., experts). We can perform this validation for each of the products tested. This appendix shows summary results from five different products in this study. Each product is treated separately. The validation steps follow:

(1) *Input Data*: Each product comprises 25 attributes from experts and 25 attributes from consumers.

(2) *Assess Each Product Separately*: Begin with one product (e.g., product 101).

(3) *Define The Goal*: The consumer profile comprises the goal to be matched. The objective is to determine the profile of factor scores which, when expanded into the consumer attributes, generates a fit as close as possible to the goal that was specified.

(4) *Match The Goal, Retain The Factor Scores Creating The Match*: The goal fitting program estimates the consumer profile (which has been optimized to match the goal expert profile in step 3). Retain the factor scores.

(5) *Estimate The Profile For Experts, Given These Factor Scores*: Using the model, and the factor scores which created the match (step 4, e.g., for consumers), now estimate the profile for the other group of panelists (viz., experts), using these factor scores and the model. Compute the expected expert panel profile corresponding to the factor scores.

(6) *Compare Expected And Obtained Profiles*: Subtract the expected panel profile (estimated by the reverse engineering model) from the actual expert panel profile (obtained from the experts themselves and used in this study). This is the signed difference. Also compute the absolute value of the signed difference. This is the absolute difference.

(7) *Compute Relevant Statistics*: Compute the average signed difference, the average absolute difference, the standard deviation of the signed difference, and the standard deviation of the absolute difference.

TABLE 5.
SUMMARY TABLE SHOWING DIFFERENCES BETWEEN ACTUAL AND EXPECTED RATINGS FOR CONSUMER PROFILES GIVEN EXPERT PROFILES, AND FOR EXPERT PROFILES GIVEN CONSUMER PROFILES

Product	Difference Statistic Across Attributes	Consumer* Mean	Consumer Standard Deviation	Expert** Mean	Expert Standard Deviation
101	Signed	-0.4	2.7	0.0	0.3
101	Absolute	1.9	1.9	0.3	0.2
107	Signed	-0.3	1.0	0.0	0.1
107	Absolute	0.7	0.7	0.1	0.1
113	Signed	0.9	1.9	0.2	0.4
113	Absolute	1.5	1.5	0.4	0.3
120	Signed	-1.6	5.5	0.1	0.4
120	Absolute	2.9	4.9	0.3	0.2
129	Signed	0.0	2.8	-0.1	0.4
129	Absolute	2.0	2.0	0.3	0.2

*Consumer = comparison of attribute profile generated for the product by consumers versus the profile estimated for the product by using the expert profile + reverse engineering model for the same product.
**Expert = comparison of attribute profile generated for the product by experts versus the profile estimated for the product by using the consumer profile + reverse engineering model for the same product.

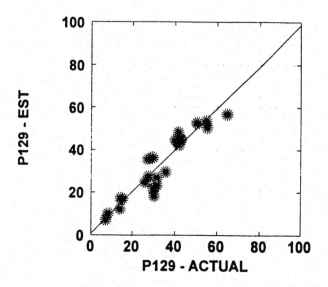

FIG. 2A. PRODUCT 129 (CONSUMER PROFILE DATA)

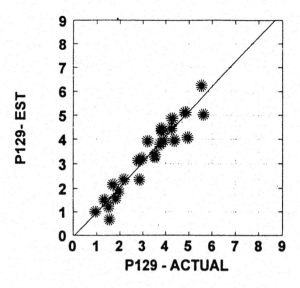

FIG. 2B. PRODUCT 129 (EXPERT PROFILE DATA)

Table 5 shows that these differences are quite small relative to the range of the scale that could be used. Figure 2A shows the plot of the expected consumer attribute ratings for product 129 versus the obtained (actual) ratings for product 129. Figure 2B shows the same type of plot, this time however for the expert panel data. Similar plots can be drawn for any product. The results from these analyses for the five products used in the validation show linear relations with slope near 1.0 for estimated attribute profiles versus actual attribute profiles. In all cases the relation between expected and obtained is linear, with correlation higher than 0.95. The figures suggest that on an attribute by attribute basis the reverse engineering model can estimate profile data from one set (e.g., consumer), given profile data from another set (e.g., expert).

REFERENCES

CAUL, J.F. 1957. The profile method of flavor analysis. *Advances in Food Research*, 1-40.

MEILGAARD, M., CIVILLE, G.V. and CARR, B.T. 1987. *Sensory Evaluation Techniques*. pp. 127-134, CRC Press, Boca Raton, FL.

MOSKOWITZ, H.R. 1983. *Product Testing and Sensory Evaluation Of Foods: Marketing and R&D Approaches*. Food & Nutrition Press, Trumbull, CT.

MOSKOWITZ, H.R. 1994. *Food Concepts and Products: Just In Time Development*. Food & Nutrition Press, Trumbull, CT.

MOSKOWITZ, H.R., KAPSALIS, J.G., CARDELLO, A., FISHKEN, D., MALLER, G. and SEGARS, R. 1979. Determining relationships among objective, expert and consumer measures of texture. Food Technol. *33*, 84-88.

STEVENS, S.S. 1953. On the brightness of lights and the loudness of sounds. Sci. *118*, 576.

STEVENS, S.S. 1969. Personal communication.

STEVENS, S.S. 1975. *Psychophysics: An Introduction To Its Perceptual, Neural and Social Prospects*. John Wiley & Sons, New York.

STONE, H., SIDEL, J.L., OLIVER, S., WOOLSEY, A. and SINGELTON, R. 1974. Sensory evaluation by quantitative descriptive analysis. Food Technol. *28*, 24-34.

SYSTAT, 1994. *Systat, The System For Statistics*, Evanston, IL.

SZCZESNIAK, A.S., BRANDT, M.A. and FRIEDMAN, H.H. 1963. Development of standard rating scales for mechanical properties of texture and correlation between the objective and sensory methods of texture evaluation. J. Food Sci. *28*, 397-403.

CHAPTER 2.0

DAIRY PRODUCTS

The use of the principles of sensory evaluation in animal and dairy sciences was seen in the 1950s, when scientists in these fields evaluated fresh meats and milk by-products using a quality grading system. For example, in the area of animal breeding, offsprings were graded on conformation, carcass quality, and other desirable physical characteristics, i.e., symmetry and balance in the animal as a whole, and grading continues on to their offsprings. Offsprings with good ratings are allowed to breed and those that failed due to poor ratings are slaughtered. The effectiveness of a quality grading system has been demonstrated through physical and carcass quality improvements of their offsprings. The graders are considered expert with no formal training—all based on experience, where agreement between graders is improved by verbal discussion and repeated participation.

In recent years, quality ratings on dairy products have been greatly improved using sensory science principles (Bodyfelt *et al.* 1988). Chapters 2.1 through 2.6 provide specific applications of descriptive analysis.

REFERENCE

BODYFELT, F.W., TOBIAS, J. and TROUT, G.M. 1988. Sensory Evaluation of Dairy Products. Chapman & Hall, New York.

CHAPTER 2.1

SENSORY ASPECTS OF MATURATION OF CHEDDAR CHEESE BY DESCRIPTIVE ANALYSIS

J.R. PIGGOTT and R.G. MOWAT

*At the time of original publication,
all authors were affiliated with
the University of Strathclyde
Department of Bioscience and Biotechnology
131 Albion Street
Glasgow G1 1 SD, Scotland.*

ABSTRACT

Previous research on the flavor development of Cheddar cheese has largely concentrated on the chemical changes that occur, and less work has been published on the sensory aspects of cheese maturation. Cheddar cheese was produced under controlled conditions and a quantitative descriptive analysis procedure was developed and used to assess development of sensory properties during maturation. The method was successful and weekly changes were recorded during maturation. The sensory data also grouped cheeses according to commercial classifications such as Vintage, Mature and Medium, and showed how these were related to a reduced fat cheese. The quantitative descriptive analysis procedure could be used in conjunction with other chemical and physical methods to help in gaining a full understanding of the maturation process in Cheddar cheese.

INTRODUCTION

World cheese production continues to increase steadily, from 11.5 Mt in 1980 to over 14 Mt in 1988. UK production has increased over the same period from 0.24 to 0.29 Mt, and 91% of the 0.25 Mt consumed in 1984 was hard cheese including Cheddar (FAO 1988; IDF 1986). The production process for Cheddar was developed in England during the 16th Century (Scott 1986), and although the principal steps in the manufacturing method still stand, as new varieties of Cheddar develop, changes in the recipe and new methods are introduced. There has been some interest recently in reduced fat cheeses, to reduce dietary intake of saturated fats, but Cheddar cheese variations are restricted in the UK by legislation (S.I. 1970, 1974). Cheddar cheese must contain no less than 48% milk fat in the dry matter and no more than 39% moisture. A typical Cheddar contains around 37% moisture, 26% protein, 35.5% fat and 1.7% salt (Egan et al. 1987).

Traditionally much of the sensory analysis carried out of maturing and matured Cheddar cheeses has been by small groups of experts or by individual cheese graders. The aims of grading schemes are generally to classify the potential of a cheese to develop a satisfactory character during maturation, and to maintain quality at the point of sale. A grading system generally starts off with a maximum score and points are deducted for defects depending on the rating and intensity of the defect compared to overall quality (Bodyfelt et al. 1988; Manning et al. 1984).

There are however many examples of the use of panels of untrained and trained assessors in recent cheese research. Webster and Frye (1987) used a consumer panel to establish the feasibility of selling cheese after 6 months of maturation instead of the more expensive 9 month storage time. Banks and Muir (1984) used a trained panel of ten members with several years experience and a structured scoring system to evaluate quality and intensity of flavor along with cheese body. Colwill (1989) showed how a trained panel could be used in conjunction with a larger untrained consumer panel to identify desirable qualities that could be emphasized in cheese to meet consumer preferences. McEwan et al. (1989) carried out Free-Choice and conventional profiling on seven varieties of Cheddar cheese, and concluded that the results were very similar and possibly the less time consuming procedure of FCP was an appropriate technique to use. They demonstrated differences between classifications of matured cheeses, and found that a range of commercial cheeses was well described by two major sources of variation, intensity of flavor and a textural variable. They also showed that consumer preferences could be related to sensory properties of the cheeses, but had insufficient data to build a predictive model, and in any case considered only differences between different classifications of cheese and not variations within a maturity classification.

Chemical measures have been widely researched in attempts to find instrumental correlates of cheese flavor and quality (Aston et al. 1983a; Burton 1989; Hill and Ferrier 1989; Kristoffersen 1967; Kristoffersen 1973; Manning et al. 1984), but there has been no systematic study of the sensory changes occurring during maturation, and the chemical basis of Cheddar cheese flavor is still a subject of debate. Flavor components in finished cheese can arise from: (a) microbial metabolism; (b) breakdown of milk proteins; (c) degradation of milk lipids. In enzyme-accelerated ripening of cheese, many authors (e.g., Sood and Kosikowski 1979; Aston et al. 1983b) have concluded that proteolysis in treated cheese has a direct relationship with maturity. However, lipase action and fatty acid breakdown is also known to be required to develop a full Cheddar flavor (Arbige *et al.* 1986).

The aims of the work described here were to develop a descriptive sensory analysis procedure that could be used to follow the sensory changes over the early stages of maturation of Cheddar cheese; to establish the differences that occur between varieties of commercially available Cheddar cheese; and to compare normal maturation and an accelerated maturation produced by the addition of excess rennet and starter inoculum.

MATERIALS AND METHODS

Materials

Two different sources of cheese were used: 13 commercially available Cheddar or similar cheeses were purchased locally (Table 1), and experimental cheeses, produced according to the method of Banks and Muir (1984), were supplied by the Hannah Research Institute (HRI), Ayr, Scotland (Table 2). Two batches of cheese were produced simultaneously, and therefore in identical environmental conditions, from 45 L of heat-treated milk (72C for 16 s), with varying levels of rennet and direct vat inoculum (DVI) starter. The curd was pressed into blocks of approximately 4 kg, and divided into 8 portions (ca. 500 g) which were individually vacuum packed.

TABLE 1.
COMMERCIAL CHEESE VARIETIES USED

Abbreviation	Cheese variety
O	Orkney
IMi	Irish Mild (Kerrygold)
C	Canadian Mature
IEM	Irish Extra Mature (Kerrygold)
NZV	New Zealand Vintage (Anchor)
A	Asda* Value Pack
NZM	New Zealand Mature (Anchor)
EMi	English Mild (St. Ivel)
K	Kraft Cheese Slices
LF	English Low Fat (St. Ivel Delight)
SS	Squidgy Squad (Asda* child brand)
NZJ	New Zealand Mature (Anchor Jubilee)
SMe	Scottish Medium

* UK retail chain

TABLE 2.
PRODUCTION AND STORAGE OF CHEESE SAMPLES BY HANNAH
RESEARCH INSTITUTE, AYR, SCOTLAND

Formulation*	Maturation Temperature (C)	Duration of Storage weeks
1	10	2
1	10	3
1	10	4
1	10	7
1	10	12
1	10	16
1	10	36
1	10	40
1	15	3
1	15	6
1	15	7
1	10	2
1	10	4
1	10	5
1	10	7
1	15	5
1	15	6
1	15	7

*Formulation 1 = 13.5 g rennet and 2.0 g direct vat inoculum (DVI) starter in 45 L milk; formulation 2 = 20.3 g rennet and 2.2 g DVI starter in 45 L milk.

All cheese was stored at 10C for the first 14 days to stabilize the blocks and prevent accelerated growth of non-starter bacteria (Law 1984a). After two weeks, 4 of the portions from each block were stored at 10C, and the remaining 4 at 15C ± 2C to accelerate the ripening procedure and possible development of off-flavors (Law 1984b). The HRI also supplied portions of cheese produced 3 and 9 months previously and stored at 10C.

Sensory Analysis

Prospective assessors (22 in total) attended the initial training session, and the selection criteria of availability, interest and motivation and good health reduced the final number to 13, who were all either students or staff of the Department and had some previous experience with sensory analysis. Training and vocabulary development were carried out over four sessions; two for vocabulary development and two for familiarization and clarification of the panel's use of the descriptors.

The assessors were asked to suggest descriptive terms while they were tasting 16 varieties of cheese. A wide range of cheeses was presented, including Cheddar cheese spreads, processed cheese slices, a variety of Cheddars, Double Gloucester, Derby, Leicester and other hard pressed cheeses. The descriptive terms suggested were reduced to 24 (Table 3) following the guidelines of Piggott (1991), but at this time no reference materials were provided. Of these terms, 9 had also been employed by McEwan *et al.* (1989) during their analysis of seven commercial cheeses, as shown in Table 3. Each assessor was then required to familiarize himself with the vocabulary and with the use of the scale, where 0 represented "not present" and 5 represented "as much as possible".

In order to prompt comments from the assessors, a questionnaire was circulated at the end of the third training session. This gave a clear indication of where the assessors were experiencing difficulties. The necessary changes were made to the score sheet before the final training session after which all assessors reported themselves to be satisfied with it.

TABLE 3.
TERMS USED FOR DESCRIPTIVE ANALYSIS OF CHEDDAR CHEESE

Descriptor	Abbreviation
White to Orange	col
Milky*	mi
Buttery	bu
Cheesy	ch
Mouldy	mo
Rancid	ra
Pungent	pu
Sour* (Aroma)	So, ar
Sweet (Aroma)	Sw, ar
Salty* (Taste)	Sa, ta
Sour* (Taste)	So, ta
Bitter (Taste)	bi
Processed	pr
Strength*	st
Maturity	ma
Soft to Firm*	SF
Moist	moi
Smooth	sm
Grainy*	gr
Crumbly	cr
Rubbery*	ru
Chewy	chw
Mouth Coating*	MC
After Taste	AT

*Descriptive terms used by McEwan *et al.* (1989)

Procedure

Each cheese was tasted in duplicate over 10 sessions, 6 samples per session. Samples were identified only by a three figure random number. The tastings took place in the isolation of a sensory booth where the lighting, temperature and ventilation could be controlled. The samples were cut into 2 cm cubes except for the sliced cheese, for two main reasons:

Firstly, to provide a uniform sample size. Cardello and Segars (1989) showed that hardness and chewiness tended to increase as a function of sample size irrespective of whether the panelist could see or feel the sample.

Secondly, to restrict consumption of a preferred sample, which might have caused a build up of flavor components within the mouth leading the assessors to give high scores for strength of flavor and aftertaste.

The cubes of cheese were stored in a refrigerator until 15 min before use, when they were removed and allowed to come to room temperature (approximately 20C). This prevented excessive loss of volatile components and ensured that the cheese was not too cold to chill the mouth nor would the texture be affected. The cheese was presented in closed glass containers so that released volatiles would be trapped and the external appearance would not be affected by surface drying.

A glass of deionized water at approximately 20C and a Golden Delicious apple were also supplied to clear the assessors' palate and diminish any carryover between samples. The apple proved to be a popular incentive for panel members to attend tasting sessions.

Moisture Determination

Moisture content was determined by oven-drying in duplicate with the aid of fine-grained sand (Egan *et al.* 1987).

RESULTS

Mean profiles over assessors and duplicates were calculated, omitting the color characteristic "White to Orange". This was done because overall color did not vary with maturation and depended only on the original carotenoid content of the raw milk and the amount of colouring added. The resulting data matrix was analyzed by principal components analysis (PCA; Piggott and Sharman 1986). Using the criteria suggested by Piggott and Sharman, three components, accounting for 83% of total variance, were selected. Loadings of the descriptors are shown in Fig. 1 and 2, after Varimax rotation (Kaiser 1959).

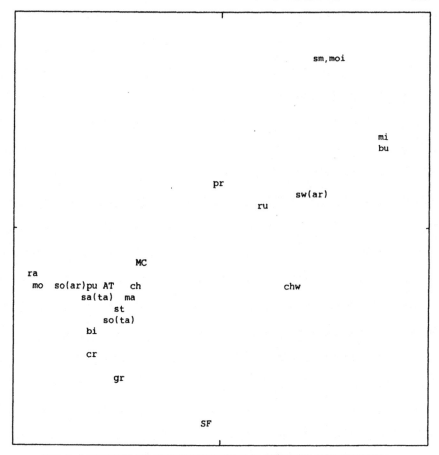

FIG. 1. LOADINGS OF DESCRIPTIVE TERMS ON FIRST (HORIZONTAL) AND SECOND (VERTICAL) ROTATED PRINCIPAL COMPONENTS
Abbreviations are shown in Table 3.

The first component clearly represents flavor, the positive pole being associated with high loadings for a milky, buttery, unmatured flavor, and the negative pole being associated with sharper, more distinct characteristics such as rancid, moldy and pungent along with sour aroma. The second component, however, represents variation in textural properties; smooth and moist had positive loadings while grainy and firm had negative loadings. The third component also explained a textural variation, but to differentiate it from the second it is perhaps more appropriate to describe it as representing mouthfeel. In particular the terms mouthcoating, strength and aftertaste had negative loadings. The sample scores plotted on these three components are shown in Fig. 3 and 4.

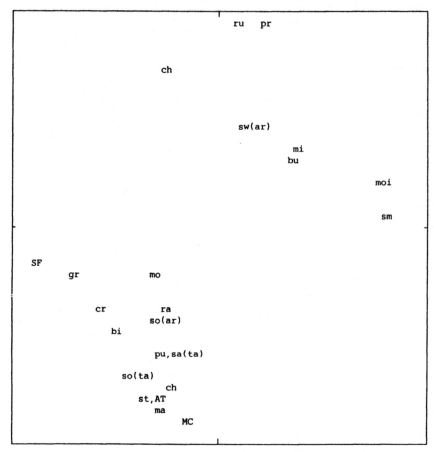

FIG. 2. LOADINGS OF DESCRIPTIVE TERMS ON SECOND (HORIZONTAL) AND THIRD (VERTICAL) ROTATED PRINCIPAL COMPONENTS
Abbreviations are shown in Table 3.

DISCUSSION

Inspection of the sample scores showed that all the cheeses in the mature, extra mature and vintage classifications (including the 9 months old HRI samples) fell within a narrow range of negative scores on the third component (Fig. 4). These cheeses also had negative scores on the first component (Fig. 3). The maturation process therefore results in a Cheddar cheese with a distinct mouthcoating effect, a strong flavor and aftertaste. The flavor of the mature cheeses was more moldy, pungent and rancid than the less mature; although these characteristics are usually unacceptable in a food, they are accepted and well established as characteristics of matured hard cheeses. The compounds

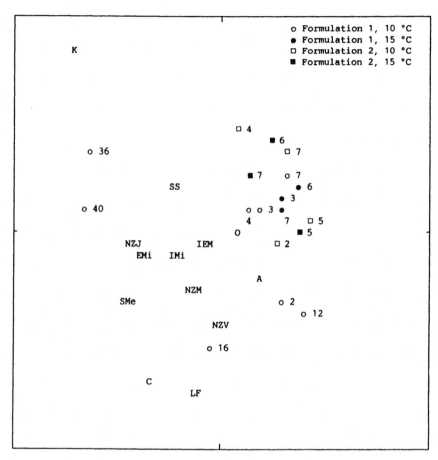

FIG. 3. SCORES OF 31 CHEDDAR CHEESE SAMPLES ON FIRST (HORIZONTAL) AND SECOND (VERTICAL) ROTATED PRINCIPAL COMPONENTS
Commercial sample abbreviations are shown in Table 1. For samples prepared by Hannah Research Institute, formulations are shown in Table 2 and symbols are followed by age in weeks.

which cause these flavors may be: results of proteolysis (Aston *et al*. 1983a, 1983b); sulphur compounds (Manning *et al*. 1976; Kristoffersen 1973); or aliphatic compounds from fatty acid breakdown (Forss and Patton 1966).

The mature cheeses showed a wide range of textural properties (component 2). This indicated that the materials, production method and conditions of maturation have more influence on matured texture than does the length of maturation. The immature HRI cheeses support this hypothesis; they showed relatively little change on component 2 after the initial few weeks of maturation. All these cheeses were produced on the same day using the same batch of milk and production procedures but with different rennet and DVI levels.

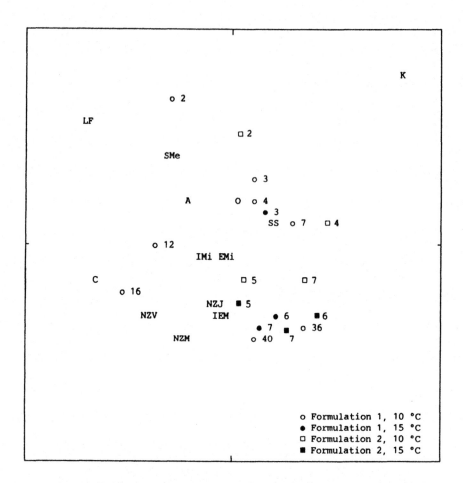

FIG. 4. SCORES OF 31 CHEDDAR CHEESE SAMPLES ON SECOND (HORIZONTAL) AND THIRD (VERTICAL) ROTATED PRINCIPAL COMPONENTS

Commercial sample abbreviations are shown in Table 1. For samples prepared by Hannah Research Institute, formulations are shown in Table 2 and symbols are followed by age in weeks.

During the early stages of maturation studied, the HRI cheeses moved from high scores on component 3 towards the negative scoring mature cheeses with high strength of flavor and strong aftertaste. The movement was greatest for the cheeses matured at elevated temperatures, which could be expected to advance maturation by between 8 and 12 weeks (Law 1984b). Significant flavor development of these cheeses did not occur, as there was little movement towards the negative end of component 1 as would have been expected. There are two possible explanations for this:

Firstly, that the high maturation temperature has led to uncharacteristic flavor development (Law 1984b).

Secondly, McGugan *et al.* (1979) suggested that nonvolatile components entrapped by the protein matrix may play a major role in not only the intensity of background flavor but overall flavor quality generally attributed to the volatile components. The results would support this suggestion in that full flavor development will not occur until there has been sufficient breakdown of the protein matrix to release nonvolatile flavor compounds which leads to a distinct mouthfeel characteristic.

The effect of increasing the rennet and DVI levels can also be seen over the maturation period studied. This is seen most clearly by comparing two cheeses with normal levels (formulation 1) with two cheeses with elevated levels (formulation 2), stored at 15C and examined after 6 and 7 weeks. For both pairs the overall effect was the same; a move towards a characteristic flavor and a firm texture with a movement towards the mature cheeses on component 3 previously discussed. Two points to note are that increasing the rennet and DVI levels amplified the rate of textural and flavor changes, while also causing the cheeses to be positioned more positively on components 1 and 2. Continuing change during maturation is due in part to the continued proteolytic action, after coagulation and pressing, of undamaged rennet. More undamaged rennet would have remained in the cheeses with increased initial levels, and thus would have caused an increased rate of proteolysis and texture and flavor change. Similarly the increase in firmness was due to the increased rate of initial coagulation producing a rubbery curd, affecting the rate of syneresis which in turn resulted in a moist, rubbery, chewy cheese (Davis 1965).

The mild or medium cheeses, as expected, were positioned between the mature cheese and the HRI fresh cheeses, but there were three outlying samples. The Kraft Cheese Slices had high positive scores on components 2 and 3, indicating a very moist, smooth cheese with rubbery, processed and chewy characteristics. Its score on component 1 showed the normally acceptable flavors of sour, moldy, pungent and rancid to be present to the greatest extent of all the samples used.

Another outlier was St. Ivels Low Fat cheese. The degree of flavor development appeared to be satisfactory compared with other mild cheeses, but it had a low score on component 2 and a high score on component 3, indicating a firm but grainy cheese with unusual chewy, rubbery mouthfeel. Banks *et al.* (1989) reported similar findings during experimental production of a low-fat Cheddar type cheese. The product proved to be over-firm and although the flavor intensity appeared to be sufficient it was of poor quality. McEwan *et al.* (1989) reported broadly similar results during sensory work with commercial cheeses.

The mature Canadian cheese fitted into the mature group on components 1 and 3, but it had a low score on component 2, indicating firm and grainy characteristics. Graininess is considered a defect and results from excessive acid production and low moisture retention in the cheese (Bodyfelt et al. 1988), and the Canadian cheese was found to have very low moisture content of 33.4%. Holland et al. (1989) showed Canadian Cheddar to have one of the lowest moisture contents (at 34.3%) of cheeses taken from 6 countries (the average was 36%).

Although the cheeses with similar descriptions e.g., mature, extra mature and vintage or mild and medium showed similarities within the groups, it is not possible to describe a typical pattern of maturation using the data presented here. This is because there are so many factors affecting the path of maturation, in particular concerning component 2.

However, some generalizations can be made from these results:
(1) Cheese changes from a milky/buttery flavor to more distinct sour, rancid and pungent characteristic flavor.
(2) The length of maturation largely dictates the strength, aftertaste and mouthfeel.
(3) Textural changes take place during maturation but a great deal depends on the original production procedure and maturation conditions.

CONCLUSIONS

Descriptive sensory analysis was used to follow the weekly textural, mouthfeel and flavor changes of maturing Cheddar cheese, and identified commercial classifications of cheeses e.g., vintage, mature, medium and mild. The mouthfeel, strength and aftertaste of cheeses used were largely dependent on the maturation length and conditions, as was the flavor conversion from milky/buttery to rancid, sour and pungent. The texture of the cheeses was dependent on the original production procedure and raw materials. The single sample of reduced fat cheese showed a firm, grainy texture and less flavor development.

ACKNOWLEDGMENTS

We are grateful for the advice and assistance of Dr. J.M. Banks and the staff of the Hannah Research Institute, Ayr, for giving up their time to show us some of the secrets of successful cheesemaking and for supplying cheese samples; and to Dr. Alistair Paterson for valuable advice and discussion.

REFERENCES

ARBIGE, M.V., FREUND, P.R., SILVER, S.C. and ZELKO, J.T. 1986. Novel lipase for Cheddar cheese flavor development. Food Technol. *40*(4), 91-98.

ASTON, J.W., DURWARD, I.G. and DULLEY, J.R. 1983a. Proteolysis and flavor development in Cheddar cheese. Austral. J. Dairy Technol. *38*(2), 55-59.

ASTON, J.W., GRIEVE, P.A., DURWARD, I.G. and DULLEY, J.R. 1983b. Proteolysis and flavor development in Cheddar cheeses subjected to accelerated ripening treatments. Austral. J. Dairy Technol. *38*(2), 59-65.

BANKS, J.M., BRECHANY, E.Y. and CHRISTIE, W.W. 1989. The production of low fat Cheddar-type cheese. J. Soc. Dairy Technol. *42*(1), 6-9.

BANKS, J.M. and MUIR, D.D. 1984. A laboratory-scale technique for controlled production of Cheddar cheese. J. Food Technol. *19*, 593-604.

BODYFELT, F.W., TOBIAS, J. and TROUT, G.M. 1988. *The Sensory Evaluation of Dairy Products*, Van Nostrand Reinhold, New York.

BURTON, J. 1989. Towards the digital cheese grader. Dairy Ind. Int. *54*(4), 17-21.

CARDELLO, A.V. and SEGARS, R.A. 1989. Effects of sample size and prior mastication on texture judgements. J. Sensory Studies *4*, 1-18.

COLWILL, J.S. 1989. Sensory sense at Campden. Food Manuf. *64*(4), 4951.

DAVIS, J.G. 1965. *Cheese: Basic Technology*, Churchill Livingstone, London.

EGAN, H., KIRK, R.S. and SAWYER, R. 1987. *Pearson's Chemical Analysis of Foods 8th. Ed.*, Longman Scientific and Technical, London.

FAO 1988. *Production Year Book 42*. Food and Agriculture Organisation, Rome.

FORSS, D.A. and PATTON, S. 1966. Flavor of Cheddar cheese. J. Dairy Sci. *49*, 89-91.

HILL, A.R. and FERRIER, L.K. 1989. Composition and quality of Cheddar cheese. Modern Dairy *68*(2), 58-60.

HOLLAND, B., UNWIN, I.D. and BUSS, D.H. 1989. *Milk Products and Eggs: 4th. Supplement to McCance and Widdowson's The Composition of Foods*, Royal Society of Chemistry, London.

IDF 1986. *The World Market for Cheese, Bulletin No. 203*, International Dairy Federation, Brussels.

KAISER, H.F. 1959. Computer program for Varimax rotation in factor analysis. Educ. Psychol. Measurement *19*, 413-420.

KRISTOFFERSEN, T. 1967. Interrelationships of flavor and chemical changes in cheese. J. Dairy Sci. *50*, 279-284.

KRISTOFFERSEN, T. 1973. Biogenesis of cheese flavor. J. Agric. Food Chem. *214*, 573-575.

LAW, B.A. 1984a. Flavor development in cheeses. In *Advances in the Microbiology and Biochemistry of Cheese and Fermented Milk*, (F.L. Davies and B.A. Law, eds.) pp. 187-208, Elsevier Applied Science, London.

LAW, B.A. 1984b. The accelerated ripening of cheese. In *Advances in the Microbiology and Biochemistry of Cheese and Fermented Milk*, (F.L. Davies and B.A. Law, eds.) pp. 209-228, Elsevier Applied Science, London.

McEWAN, J.A., MOORE, J.D. and COLWILL, J.S. 1989. The sensory characteristics of Cheddar cheese and their relationship with acceptability. J. Soc. Dairy Technol. *42*(4), 112-117.

MGGUGAN, W.A., EMMONS, D.B. and LARMOND, E. 1979. Influence of volatile and nonvolatile fractions on intensity of Cheddar cheese flavor. J. Dairy Sci. *62*, 398-403.

MANNING, D.J., CHAPMAN, H.R. and HOSKING, Z.D. 1976. The production of sulphur compounds in Cheddar cheese and their significance in flavor development. J. Dairy Res. *43*, 313-320.

MANNING, D.J., RIDOUT, E.A. and PRICE, J.C. 1984. Non-sensory methods for cheese flavour assessment. In *Advances in the Microbiology and Biochemistry of Cheese and Fermented Milk*, (F.L. Davies and B.A. Law, eds.) pp. 229-253, Elsevier Applied Science, London.

PIGGOTT, J.R. 1991. Selection of terms for descriptive analysis. In *Sensory Science Theory and Applications in Foods*, (H. Lawless and B. Klein, eds.) pp. 339-351, Dekker, New York.

PIGGOTT, J.R. and SHARMAN, K. 1986. Methods for Multivariate Dimensionality Reduction. In *Statistical Procedures in Food Research*, (J.R. Piggott, ed.) pp. 181-232, Elsevier Applied Science, London.

S.I. 1970. *Statutory Instruments No. 94, The Cheese Regulations 1970*, HMSO, London.

S.I. 1974. *Statutory Instruments No. 1122, The Cheese (Amendment) Regulations 1974*, HMSO, London.

SCOTT, R. 1986. *Cheesemaking Practice 2nd. Ed.*, Elsevier Applied Science, London.

SOOD, V.K. and KOSIKOWSKI, F.V. 1979. Accelerated Cheddar cheese ripening by added microbial enzymes. J. Dairy Sci. *62*, 1865-1872.

WEBSTER, F. and FRYE, B. 1987. Cheese aging and consumer preference. Dairy Field *170*(2), 50-53.

CHAPTER 2.2

SENSORY PROFILING OF DULCE DE LECHE, A DAIRY BASED CONFECTIONARY PRODUCT

GUILLERMO HOUGH, NICHOLAS BRATCHELL[1]
and DOUGLAS B. MACDOUGALL[2]

*At the time of original publication, author Hough was affiliated with
Comision de Investigaciones Cientificas de la Provincia de Buenos Aires
Instituto Superior Experimental de Tecnologia Alimentaria
6500 Nueve de Julio, Buenos Aires, Argentina*

and

*authors Bratchell and Macdougall were affiliated with
AFRC, Institute of Food Research, Reading Laboratory
Shinfield, Berks, RG2 9AT, United Kingdom.*

ABSTRACT

Dulce de Leche is a dairy based confectionary product, popular in most Latin American countries. A 2^4 factorial design was used to study the influence of glucose syrup concentration, type of neutralizer, cooking time and inclusion of vanilla essence on the sensory profile of this product. Fractions of the full design was used to obtain 37 sensory descriptors covering appearance, manual and oral texture, flavor and aftertaste, and a number of objective color and texture measurements. The effects of the design factors on individual descriptors were tested using analysis of variance; aggregate analysis of attributes were carried out using multivariate analysis of variance, principal component analysis and partial least squares regression.

Analysis of the principal components identified several interpretable underlying factors for appearance and texture but not for flavor or aftertaste. Partial least squares regression showed that oral texture is closely related to manual texture, but that aftertaste is quite different from flavor. Instrumental color and texture measurements were found to be highly correlated with the corresponding sensory descriptors.

[1]Present address: Biometrics Department, Pfizer Central Research, Ramsgate Road, Sandwich, Kent CT13 9NJ, UK.
[2]Present address: Department of Food Science and Technology, University of Reading, PO Box 226, Whiteknights, Reading RG6 2AP, UK.

Reprinted with permission of Food & Nutrition Press, Inc., Trumbull, Connecticut.
©Copyright 1992. Originally published in Journal of Sensory Studies 7, 157-178.

INTRODUCTION

"Dulce de Leche" (DL) is a popular Latin American dairy based confection described by Moro and Hough (1985) in a previous paper. It is consumed primarily in Argentina, where per-capita consumption is 2.2 kg/year, and 3% of milk production is used in its manufacture (Anon. 1988). In spite of its wide and popular use, few sensory studies have been carried out on this product. Hough *et al.* (1986) performed a preference study of commercial brands with Argentine consumers. Hough *et al.* (1990) measured the sensory defect of sandiness in this product. Sensory color measurements have been related to objective measurements (Ferreira *et al.* 1989; Buera *et al.* 1991). The aptitude of DL for industrial cake filling machines was measured on a sensory scale by an expert (Pauletti *et al.* 1984).

The formulation and manufacturing variations that could affect a sensory profile of DL are many, but economic considerations and food regulations restrict possibilities. For example, in Argentina (Camera Comercio de Buenos Aires 1988) a minimum percentage of milk solids in DL are required; on the other hand, an increase of milk solids to the detriment of sugar is not economically feasible: changing the milk:sugar ratio is not practical. With this type of limitation, the practical factors most likely to affect the sensory profile are: replacement of sucrose by other sugars, neutralizer, cooking time and added flavoring.

The results of the sensory profiling will be useful to further studies on the consumer acceptability of the product (Hough *et al.* 1992). Any attempts at improving the acceptability require detailed understanding of the sensory impact produced by formulation and processing parameters, accompanied by simplification of the sensory information gained through multivariate statistical tools, such as principal component analysis.

The objectives of this work were: (a) To develop the descriptors and methodology for the sensory profile of DL. (b) To establish the influence of formulation and manufacturing factors on the sensory profile.

MATERIALS AND METHODS

Profile Development: Preparation of Samples

To develop the profile a total of 6 samples were used. Two were the highest selling brands in Argentina: La Serenisima (Mastellone Hnos, Buenos Aires, Argentina) and Sancor (Sancor Coop. Ltda., Santa Fe, Argentina). The other four were made in a 15 L pilot plant cooker (Stephan, Holland), equipped with a surface scrape agitator rotating at 35 rpm, and a central blade agitator rotating at 1500 rpm. The basic formulation for these was: milk (3.5% fat, 11.8% total solids) 100 parts, sugar 20 parts, sodium bicarbonate 0.1 parts, gelling agents (Aglugar from Saporiti, Buenos Aires, Argentina) 0.07 parts, vanilla essence (DL1603 from Saporiti, Buenos Aires, Argentina) 0.30 mL/kg of DL. The gelling agent was added in order to obtain an adequate consistency with the type of cooker used.

The variation in the four samples was generated by a 2^2 factorial design, with one factor being 15 and 30% replacement of sucrose by 40 Dextrose Equivalent glucose syrup (Cargill PLC, Essex, UK), and the second 3 h and 3 h 45 min cooking time. The cooking process was regulated to reach 70 Brix in 3 h. At this point half the DL was removed from the cooker and cooled. Boiling water was added to the remaining half in order to obtain the second cooking time with the same percent solids.

Profile Development: Sensory Analysis

The panel consisted of 12 assessors, each one with a minimum of 240 h of sensory work on a wide variety of foods using Quantitative Descriptive Analysis (Stone and Sidel 1985).

Assessors received 15 g of each sample in 100 mL glass bowls, at room temperature, under artificial daylight illumination, and in individual booths. Between samples they ate half an unsalted cracker and rinsed their mouths with water. A total of 4 h, divided in 3 sessions was used to develop a list of descriptors and definitions by consensus. This was obtained by round table discussion of descriptors produced by each assessor. For flavor and aftertaste, standards of suggested descriptors were presented to help panel uniformity.

To test the panel on the developed descriptors, the six samples were rated using 10 cm unstructured lines anchored at the extremes (Stone and Sidel 1985). The samples were presented as before, numbered with 3 digit codes, and the order of presentation randomized for each assessor. Measurements were done by triplicate, in 3 sessions, each one on a different day.

Formulation and Manufacture Factors

Once the sensory profile had been developed, the effects of four formulation and manufacturing variables were studied; 16 samples were generated using a 2^4 factorial design with each factor at two levels, listed in Table 1. This allows the effect of each factor and their interactions to be tested. The samples were prepared as described above.

TABLE 1.
FACTORS USED IN THE 2^4 FACTORIAL DESIGN TO STUDY THE INFLUENCE OF FORMULATION AND MANUFACTURE ON THE SENSORY PROFILE OF DULCE DE LECHE

FACTOR	Level 1	Level 2
Replacement of sucrose by glucose syrup	10% replacement samples 1 to 8	40% replacement samples 9 to 16
Neutralizer (a)	100% sodium bicarbonate samples 1-2-3-4 9-10-11-12	50% sodium bicarbonate + 50% calcium hydroxide samples 5-6-7-8 13-14-15-16
Cooking time	3 h samples 1-2-5-6 9-10-13-14	3 h 45 min samples 3-4-7-8 11-12-15-16
Addition of vanilla essence	No samples 1-3-5-7 9-11-13-15	Yes samples 2-4-6-8 10-12-14-16

(a) = % expressed in chemical equivalents

Sensory Profiling

DL is a very sweet product. During profile development assessors found, after repeated tasting of 6 samples to quantify all descriptors, that they had reached saturation. Therefore, it was decided to give assessors no more than 4 samples per session.

In order to minimize loss of information on the four factors and their interactions, the 16 samples were divided into groups of four samples following a fractional factorial design (Cochran and Cox 1957) and the four sets allocated to the 12 assessors, three receiving each set. The allocation of samples to assessors in the first session is shown schematically in Table 2 using standard notation: A, B, C and D denotes the design factors; ad denotes the sample with the highest levels of factors A and D and the lowest levels of B and C, (1) denotes the lowest levels of all factors. This also shows that in this session the AD, ABC and BCD interactions were confounded with assessors. The full

TABLE 2.
FRACTIONAL FACTORIAL DESIGN USED IN THE FIRST SESSION OF SENSORY PROFILING OF A 2^4 FACTORIAL WHERE EACH ASSESSOR EVALUATES 4 SAMPLES
Interactions AD, ABC and BCD are confounded with assessors (blocks).

ASSESSORS	Sample 1	Sample 2	Sample 3	Sample 4
1-3	b	c	ad	abcd
4-6	a	abc	bd	cd
7-9	ab	ac	d	bcd
10-12	(1)	bc	abd	acd

experiment used a further three sessions within which different interactions were confounded, resulting in a partially confounded design. This design gives three replicates of each sample in each session and 12 replicates in total. As described above, the order of presentation of samples was randomized for each assessor. This provides a most efficient experimental design strategy, which minimizes assessor fatigue and maximizes the number of samples presented and information gained.

Objective Measurements

The color of each of the 16 samples was measured using a Hunterlab 5100 Color Difference Meter (Hunter Associates Laboratory, USA) using illuminant C, expressing the results in the L, a, b system. Samples were heated to approximately 60C to facilitate pouring into 100 mL plastic culture bottles. The resulting sample thickness was approximately 20 mm, which has been proved to be opaque for DL (Ferreira *et al.* 1989). Measurements were duplicated.

Instrumental texture was measured at room temperature with a Lloyd MSK testing machine (Southampton, UK) equipped with a 5 N load cell. Samples, 200 g each, were placed in 9.5 cm diameter cups, resulting in 3 cm depth. The penetrating body was a cylinder 5.1 cm in diameter and 1.1 cm in height. The cylinder was driven into the sample at a speed of 10 mm/min, to a depth of 10 mm. The force value at the end of penetration was expressed as stress (N/m2). As DL is a thixotropic fluid (Hough *et al.* 1988), the first three measurements on each sample were discarded to ensure structure breakdown. The average of the next three measurements were taken.

Statistical Analysis

Sensory data are necessarily complex, and a number of statistical techniques are needed to analyze them. In this study the following methods were used: univariate and multivariate analysis of variance, Generalized Procrustes analysis, principal components analysis and partial least squares regression. These are outlined below.

The performance of the assessors was tested by Generalized Procrustes analysis (Arnold and Williams 1986). The variation among individual assessments is replaced by a consensus (average) configuration of samples, and assessors are tested for their closeness to the consensus. Assessors close to the consensus are deemed to perform well; those with large residuals perform poorly.

The use of a fractional factorial design and the aim of testing the various processing factors lead naturally to analysis of variance (ANOVA). In this case the design allows the main effects and low-order interactions to be tested after first removing the between-session and between-assessor variation. However, sensory data are multivariate and correlations among the descriptors are of interest. Multivariate analysis of variance (MANOVA) extends the utility of analysis of variance to take account of these correlations and often reveals effects hidden in the simpler univariate analyses.

Principal components analysis (PCA) addresses these correlations directly and exploits duplication of information to provide highly informative low-dimensional plots of the data that complement ANOVA and MANOVA. PCA provides a detailed multidimensional view of the effects of the factors, the relative importances of the factors and the various descriptors and the relationships among these, which is often overlooked in purely univariate analyses. This strategy can be extended to examine the relationships between different categories of descriptors or instrumental measurements by principal components regression or partial least squares regression (PLS) (Wold *et al.* 1983; Martens and Martens 1986). PLS was used in preference to principal components regression, as it seeks to model the relationships between the groups of variables in a more direct way.

The Genstat statistical language (Genstat 1989) was used for all analyses. Most of the procedures used are standard methods which are available in Genstat. The exception is PLS, which was developed by the authors using the Genstat language following an algorithm proposed by Hoskuldsson (1988); it is available from Ian Wakeling, AFRC Institute of Food Research, Reading RG2 9AT, U.K.

RESULTS

Profile Development

The descriptors and definitions developed by the panel are listed in Table 3. Flavor and appearance descriptors are self explanatory; definitions are provided for texture descriptors. The end-anchors for the scales were low-nil to high-extreme.

PCA analysis of the 2 commercial and 4 pilot plant samples, gave the following percentage variances explained by the first two principal components: Flavor 61%, Aftertaste 65%, Manual Texture 93%, Oral Texture 92%, and Appearance 86%. The pilot plant samples were within the same range for these two principal components as the commercial samples.

TABLE 3.
DESCRIPTORS USED FOR SENSORY PROFILING OF DULCE DE LECHE

Flavor	Aftertaste	Appearance
Initial impact	Sweet	Orange
Sweet	Dryness	Yellow
Acid	Bitter	Brown
Salt	Condensed milk	Gloss
Condensed milk	Vanilla	Transluscent edges
Creamy	Creamy	Darkness
Caramel	Cooked	
Vanilla	Duration	
Glucose syrup		
Overall intensity		

Manual Texture		
Sticky	Degree of stickiness to spoon	
Stringy	Length of string when lifting spoon from dish	
Peaks	Length of time peaks hold their shape	
Soft	Degree of softness, as opposed to hardness	
Flow rate	Rate/speed at which the Dulce de Leche flows off the spoon	
Smooth	Degree of smoothness when spreading the sample against the side of the bowl with back of spoon	
Spreadability	Ease required to spread the sample over the bottom of the bowl with back of spoon	

Oral Texture		
Thinness	Ease required to manipulate sample in mouth	
Smooth	Degree of smoothness as opposed to grainy-sandy	
Rate of melting	Speed with which the sample melts in the mouth	
Sticks to mouth	Degree to which the sample sticks to mouth surface during manipulation	
Mouthcoating	Degree of oily mouthcoating after swallowing	
Ease of swallow	Ease with which sample can be completely swallowed	

Generalized Procrustes analysis showed the same discrimination pattern as PCA, which reflects that the assessors were performing well as a whole. Each assessor's performance can be monitored from their residual variance (Sinesio et al. 1990), which is a measure of the assessor's distance from the consensus configuration. Residuals were sufficiently small, therefore no assessor was considered an outlier.

ANOVA showed the majority of descriptors to be significant in discriminating among samples, which supports the validity of the chosen descriptors in developing the sensory profile.

At this stage no reduction in the number of attributes was considered. What might be considered superfluous to describe one set of samples, might become useful for a different set.

Formulation and Manufacturing Factors

ANOVA of Flavor and Aftertaste. Results from ANOVA for each attribute and design factor are in Table 4.

From Table 4, glucose syrup was the factor that influenced most descriptors. Sucrose is sweeter than glucose and the higher sugars that accompany the syrup (Amerine et al. 1965). This was perceived by the panel who found a lower replacement of sucrose sweeter for Flavor and Aftertaste. Glucose syrup is acidic in nature, and this was transmitted to DL flavor with the higher replacement recognized as more acidic. Glucose is a reducing sugar, increasing browning reactions during the cooking process of DL, generating caramel and bitter flavors (Fennema 1982). This explains the higher values for these descriptors for higher replacement values, and also lowering of dairy descriptors such as "condensed milk" and "creamy" for a high glucose level. Values of the specific glucose syrup descriptor indicate that the panel perceived the different levels.

Different concentrations of sodium bicarbonate have a marked influence on browning in DL (Hough et al. 1991), and this would most probably influence flavor; yet a qualitative change in neutralizer, as used in this design, did not affect Flavor or Aftertaste significantly as can be seen in Table 4.

Increase in cooking time enhances the Maillard reaction in DL, producing components that were detected by the panel in the increased dryness with longer cooking and greater creaminess with shorter cooking identified in Aftertaste.

As expected, vanilla essence factor was highly significant for the vanilla descriptor in both Flavor and Aftertaste. Condensed milk and cooked flavor were impaired by the higher level of essence.

TABLE 4.
DESCRIPTIVE PROFILING OF 16 SAMPLES OF DULCE DE LECHE: MEANS FOR EACH DESCRIPTOR AND SIGNIFICANCE LEVELS DERIVED FROM ANOVA ANALYSIS

	DESIGN FACTOR								
	Glucose syrup		Neutralizer		Time		Essence		
DESCRIPTORS	L1(a)	L2	L1	L2	L1	L1	L1	L2	Interactions
Flavor									
Initial impact	63(b)	56	-(c)	-	-	-	58	61	-
Sweet	64	58	-	-	-	-	-	-	-
Acid	17	21	-	-	-	-	-	-	-
Salt	-	-	-	-	-	-	-	-(13)(d)	-
Condensed milk	46	41	-	-	-	-	46	41	-
Creamy	48	44	-	-	-	-	-	-	-
Caramel	45	48	-	-	-	-	-	-	-
Vanilla	-	-	-	-	-	-	35	43	-
Glucose syrup	31	37	-	-	-	-	-	-	-
Overall intensity	-	-	-	-	-	-	-	-(60)	-
Aftertaste									
Sweet	63	56	-	-	-	-	-	-	-
Dryness	-	-	-	-	51	55	-	-	-
Bitter	13	17	-	-	-	-	-	-	-
Condensed milk	39	33	-	-	-	-	39	34	-
Vanilla	-	-	-	-	-	-	30	38	-
Creamy	41	37	-	-	40	37	-	-	-
Cooked	-	-	-	-	-	-	31	24	-
Duration	-	-	-	-	-	-	-	-(59)	-
Manual Texture									
Sticky	38	73	53	57	53	58	-	-	-
Stringy	47	30	-	-	-	-	-	-	-
Peaks	23	76	47	51	46	52	-	-	GxN
Soft	78	37	59	55	62	53	-	-	GxN
Flow rate	71	12	45	38	44	38	-	-	GxN
Smooth	75	65	-	-	74	66	-	-	GxN
Spreadability	80	48	-	-	67	61	-	-	GxN
Oral Texture									
Thinness	49	26	50	45	52	44	-	-	GxN
Smooth	81	72	-	-	78	75	-	-	GxN
Rate of melting	78	48	65	61	65	60	-	-	-
Sticks to mouth	37	57	-	-	-	-	-	-	-
Mouthcoating	44	58	-	-	-	-	-	-	-
Ease of swallow	80	55	70	65	70	64	-	-	-
Appearance									
Orange	41	16	-	-	33	24	-	-	-
Yellow	27	7	15	17	21	12	-	-	GxT
Brown	38	68	54	51	46	60	-	-	GxT
Gloss	62	55	-	-	61	56	-	-	GxT
Translucent	39	7	-	-	-	-	-	-	-
Darkness	34	70	53	51	44	60	-	-	GxT

(a) L1= Level 1, L2= Level 2. See Table 1 for values of levels
(b) Mean values are on a 0-100 unstructured sensory scale
(c) -= non-significant effect (P>.05)
(d) Overall mean for descriptors not affected by any design factor

ANOVA of Manual and Oral Texture. Samples with the lower level of glucose syrup were a lot thinner and softer in texture than the higher level samples. This was due to glucose being more reactive than sucrose during the cooking process. Apart from thin and soft, all other descriptors of Manual and Oral Texture were significant in discriminating between glucose levels.

Replacement of sodium bicarbonate by calcium hydroxide as neutralizer, increased thickness and other descriptors. The exact mechanism for this change in texture in DL is not known, but is related to the salt balance in the system (Webb *et al.* 1980).

Longer cooking time increases the Maillard reaction and protein denaturation, resulting in higher gelation (Webb *et al.* 1980), which was detected by the panel in descriptors such as soft, thin, etc., in both Manual and Oral Texture.

Essence factor had no influence on texture descriptors.

A significant two way magnitude type interaction (Stone and Sidel 1985) for some descriptors was glucose × neutralizer. Glucose syrup produced a higher effect on texture using level 1 neutralizer. Here again, the salt balance would explain this behavior.

ANOVA of Appearance. Higher glucose produced a darker and browner DL, confirming previous studies (Hough *et al.* 1991). Orange and yellow were perceived to be higher in lower level glucose, where browning reactions are less intense.

Different cations affect nonenzymatic browning (Saunders and Jervis 1966; Petriella *et al.* 1988); this was reflected by slight changes in appearance for the neutralizer factor.

As expected, longer cooking times significantly affected Appearance descriptors, due to prolonged nonenzymatic browning. Essence had no effect on Appearance.

A significant two way magnitude type (Stone and Sidel 1985) interaction for some descriptors was glucose × time. For yellow, glucose syrup had a higher effect for shorter cooking time. For a higher cooking time, the same effect was probably present, but the darker color masked it to assessors. This is in agreement with the interaction effect for darkness and brown, where glucose syrup had a higher effect for longer cooking time.

MANOVA Analysis. MANOVA was used to test each attribute as a single entity rather than each attribute independently. In general the results supported those found by univariate analysis of variance, with one exception. With the univariate analyses the effect of time on Flavor was not found to be significant. MANOVA identified an overall significant ($P < 0.05$) time effect for both Flavor and Aftertaste.

PCA Analysis. PCA was used to obtain a graphical overview of the effects identified by analysis of variance, examine the correlations among the descriptors and identify any simplification of the attributes. Table 5 gives the percentage variances accounted for by the first four principal components for each of the attributes. Manual and Oral Texture and Appearance appear to be one-dimensional attributes in which the various descriptors in each category provide essentially similar information. Flavor and Aftertaste are much more complex with information appearing in the first three components. The principal components are plotted in Fig. 1-5.

TABLE 5.
PERCENTAGE VARIANCES EXPLAINED BY THE HIGHER PRINCIPAL
COMPONENTS OF THE DESCRIPTOR CATEGORIES

CATEGORY	Principal component (a)			
	1	2	3	4
Flavor	45	19	13	8
Aftertaste	40	27	15	8
Manual Texture	91	5	-	-
Oral Texture	92	4	-	-
Appearance	84	11	-	-

(a) Only the major principal components are presented.

As noted, both Flavor and Aftertaste appear to be complex attributes. The ANOVA showed that, in each case, some descriptors are affected by the processing factors while others are not; this is reflected in the scores plots (Fig. 1A and 2A) where it is not possible to identify these effects. The correlations plots in Fig. 1B and 2B indicate the complexity of the relationships among the attributes.

This contrasts with the results for Manual and Oral Texture. Manual Texture was a one-dimensional attribute with samples ranging from extreme peaky/sticky properties to smooth/stringy and soft/spreading/flowing properties (Fig. 3B). The scores plot shows a clear effect of sugar composition along this axis, with 40% sucrose replacement leading to sticky/peaky samples. Similar conclusions can be drawn from Fig. 4 for Oral Texture, except that here there was a neutralizer effect at the lower level of sucrose replacement, but not at the higher level.

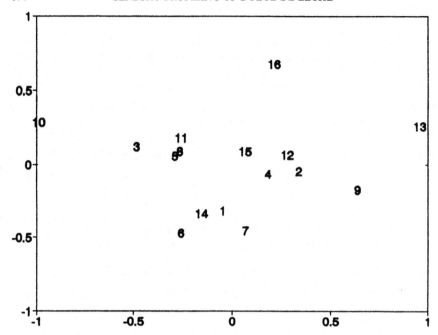

FIG. 1A. PCA FLAVOR SCORES FOR PRINCIPAL COMPONENTS 1 (ABSCISSA) AND 2 (ORDINATE)
Sample codes are in Table 1.

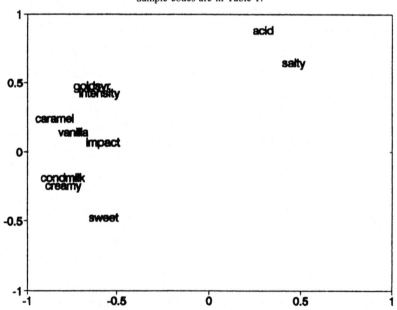

FIG. 1B. PCA FLAVOR DESCRIPTOR CORRELATIONS WITH PRINCIPAL COMPONENTS 1 (ABSCISSA) AND 2 (ORDINATE)

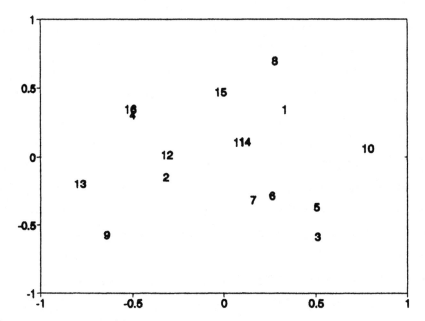

FIG. 2A. PCA AFTERTASTE SCORES FOR PRINCIPAL
COMPONENTS 1 (ABSCISSA) AND 2 (ORDINATE)
Sample codes are in Table 1.

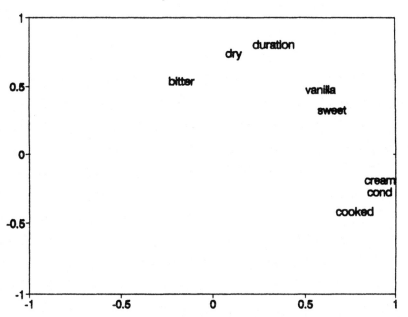

FIG. 2B. PCA AFTERTASTE DESCRIPTOR CORRELATIONS WITH PRINCIPAL
COMPONENTS 1 (ABSCISSA) AND 2 (ORDINATE)

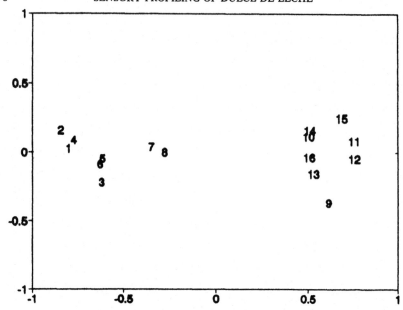

FIG. 3A. PCA MANUAL TEXTURE SCORES FOR PRINCIPAL COMPONENTS
1 (ABSCISSA) AND 2 (ORDINATE)
Sample codes are in Table 1.

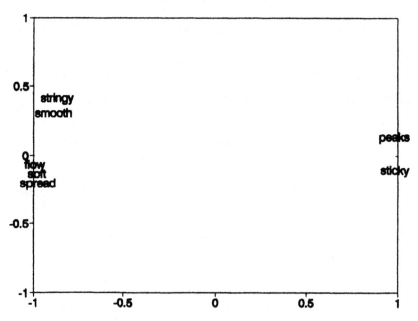

FIG. 3B. PCA MANUAL TEXTURE DESCRIPTOR CORRELATIONS WITH PRINCIPAL
COMPONENTS 1 (ABSCISSA) AND 2 (ORDINATE)
Descriptor definitions are in Table 3.

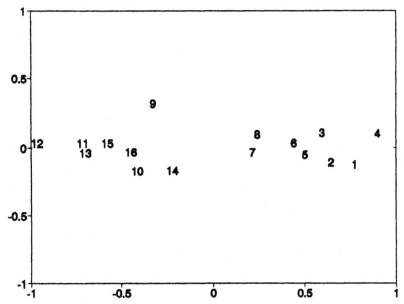

FIG. 4A. PCA ORAL TEXTURE SCORES FOR PRINCIPAL COMPONENTS
1 (ABSCISSA) AND 2 (ORDINATE)
Sample codes are in Table 1.

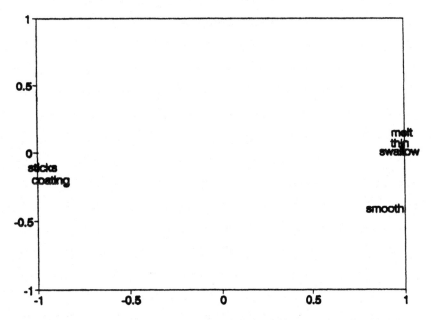

FIG. 4B. PCA ORAL TEXTURE DESCRIPTOR CORRELATIONS WITH PRINCIPAL
COMPONENTS 1 (ABSCISSA) AND 2 (ORDINATE)
Descriptor definitions are in Table 3.

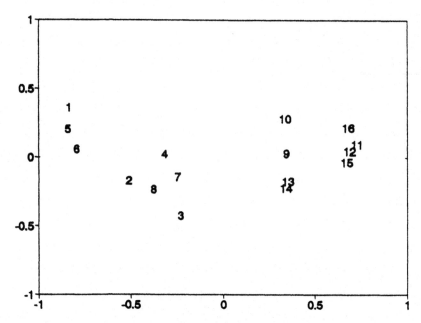

FIG. 5A. PCA APPEARANCE SCORES FOR PRINCIPAL COMPONENTS 1 (ABSCISSA) AND 2 (ORDINATE)
Sample codes are in Table 1.

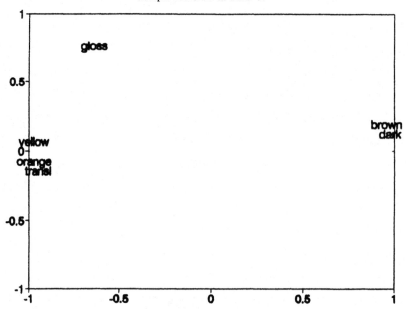

FIG. 5B. PCA APPEARANCE DESCRIPTOR CORRELATIONS WITH PRINCIPAL COMPONENTS 1 (ABSCISSA) AND 2 (ORDINATE)

Appearance was also a one-dimensional attribute with samples separating along a dark/brown to yellow/orange/translucent axis; glossiness is related to yellow/orange/translucent but appears to provide some other information (Fig. 5B). The scores plot in Fig. 5A shows the clear effects of sucrose replacement, with the higher level leading to darker samples, and the subordinate effect of cooking time, with longer cooking leading to relatively darker samples.

Partial Least Squares Analysis of Sensory Profiles. The PCA above revealed close similarities between Oral and Manual Texture and the complex natures of the Flavor and Aftertaste attributes. PLS was used to (a) test the similarity of the two texture attributes and (b) identify a relationship between Flavor and Aftertaste.

Treating Oral Texture as the dependent variables, PLS confirmed the one dimensional nature of the texture attributes and their close similarity. This is emphasized in Fig. 6, which shows the correlations between the descriptors and the first two PLS factors. These factors account for 89% and 5% of the variation in Manual Texture and 86% and 2% in Oral Texture. In Fig. 6, the Manual Texture descriptors peaky and sticky are highly correlated with the Oral Texture descriptors sticky and coating; similarly stringy, swallow, melt, thin, spread, flow and soft are highly positively correlated with each other and highly negatively correlated with the peaky/sticky descriptors. The analysis reveals that Manual and Oral Texture contain the same information and that in routine profiling Oral Texture does not have to be measured.

PLS with Aftertaste as the dependent variable found a poor relationship with Flavor. As found with PCA, several dimensions were needed to account for the variation in both attributes: the first four factors accounted for 86% of variation in Flavor and 64% of variation in Aftertaste. This indicates that Aftertaste cannot be predicted by Flavor.

Objective Color Measurements. PLS was performed to predict sensory Appearance (Y-block) from the instrumental L-a-b values (X-block). The first 3 dimensions explained 93, 6 and 1% of the variation in the instrumental data; and 80, 6 and 2% of the sensory data. The PLS L-a-b scores were correlated with the original L-a-b instrumental values and with sensory Appearance. For the first PLS dimension the correlation coefficients for the instrumental data were: L 0.96, a 0.94 and b 1.00; for the sensory descriptors they were: orange 0.96, yellow 0.91, brown –0.97, gloss 0.63, translucent 0.86 and darkness –0.98. It can be seen that, except for gloss, instrumental L-a-b was a good predictor for sensory appearance values. This confirms conclusions from a previous study (Ferreira *et al.* 1989).

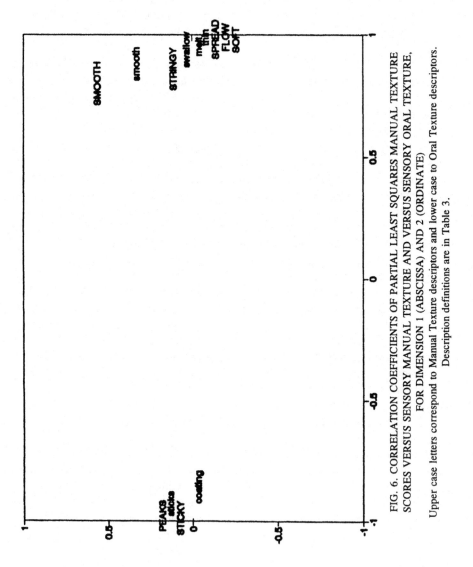

FIG. 6. CORRELATION COEFFICIENTS OF PARTIAL LEAST SQUARES MANUAL TEXTURE SCORES VERSUS SENSORY MANUAL TEXTURE AND VERSUS SENSORY ORAL TEXTURE, FOR DIMENSION 1 (ABSCISSA) AND 2 (ORDINATE)

Upper case letters correspond to Manual Texture descriptors and lower case to Oral Texture descriptors. Description definitions are in Table 3.

Objective Texture Measurement. Manual and Oral Texture scores were modeled by instrumental stress values using an exponential function of the type (Moskowitz 1967):

$$S = kI^n \qquad (1)$$

where S = sensory measurement (scale 0-100)
I = instrumental stress (N/m2),
k, n = constants

Values of k, n and the corresponding correlation coefficients are in Table 6. It can be observed that stress correlated well with most variables, especially those that are dominant in PC 1 (see Fig. 3 and 4), confirming that most of the texture variation in the samples was due to a soft-hard component. Lucisano *et al.* (1989) found similar correlations between stress and sensory hardness for bentonite solutions. The exponent n values of Eq. (1) were also similar to those reported by Moskowitz (1977) for texture attributes, i.e., in the order of 0.40.

TABLE 6.
CORRELATION VALUES OF THE EQUATION LOG
[SENSORY DESCRIPTOR (a)] = LOG (k) + n.LOG [INSTRUMENTAL STRESS (b)],
FOR MANUAL AND ORAL TEXTURE DESCRIPTORS

DESCRIPTOR	k	n	Correlation Coefficient
Manual texture			
Sticky	6.14	0.34	0.97
Stringy	1.42×10^2	-0.21	-0.87
Peaks	0.81	0.63	0.96
Soft	6.99×10^2	-0.41	-0.99
Flow rate	1.25×10^4	-0.97	-0.99
Smooth	1.20×10^2	-0.09	-0.85
Spreadability	3.55×10^2	-0.28	-0.98
Oral texture			
Thinness	1.14×10^3	-0.52	-0.99
Smooth	1.08×10^2	-0.05	-0.90
Melting rate	3.08×10^2	-0.26	-0.98
Sticks to mouth	10.2	0.24	0.93
Mouth coating	20.3	0.15	0.87
Ease of swallow	2.33×10^2	-0.20	-0.97

(a) Sensory descriptor values on a 0-100 unstructured scale
(b) Instrumental stress= N/m2

CONCLUSION

A descriptive sensory profile for Dulce de Leche consisting of Flavor, After-taste, Manual and Oral Texture and Appearance was developed using a set of commercial and experimental samples. Generalized Procrustes analysis confirmed that assessors used the profile well and that the experimental samples were essentially similar to the commercial samples. Subsequently a 2^4 factorial design was devised to test the effects of the main formulation and processing factors: replacement of sucrose by glucose, neutralizer, cooking time and addition of vanilla essence. A confounded fractional design was used to allocate sets of four samples to assessors over several sessions.

ANOVA revealed that Flavor and Aftertaste were significantly affected by replacement of sucrose by glucose and, to a lesser extent, adding vanilla essence, but that there were no significant interactions between the factors. Replacement of sucrose by glucose was found to affect Texture and Appearance significantly, as did both neutralizer and cooking time. For Appearance there was a significant interaction between replacement of sucrose and cooking time and for Texture there was a significant neutralizer by sucrose interaction. PCA and PLS provided further information and some simplification of the attributes. Oral and Manual Texture and Appearance were found to be one-dimensional attributes, and it was possible to predict Oral Texture from Manual Texture. For these attributes the effects of the processing variables was clearly visible in low dimensional plots of the principal components. The Flavor and Aftertaste attributes were more complex, requiring three dimensions to account for the information and showing generally poor correlations among the descriptors. The principal component scores plots did not reveal the effects of the processing variables. PLS revealed that sensory Appearance could be well predicted by instrumental color measurements. Many of the Texture descriptors were highly correlated with the penetration test using an exponential function. The results lead to two broad conclusions. Firstly, the analysis revealed the simplicity of the Texture and Appearance attributes and the ease with which the effects of processing factors can be identified. Secondly, and conversely, they showed the complexity of the Flavor and Aftertaste attributes. Thus the effects of modification of the product on texture and appearance can be studied easily, but similar work with flavor requires more work to devise more informative and less complex flavor profiles.

ACKNOWLEDGMENTS

This work was carried out at the AFRC Institute of Food Research, Reading Laboratory, with financial support from the Consejo Nacional de

Investigaciones Cientificas y Tecnicas de la Republica Argentina, and the Royal Society of the UK.

REFERENCES

AMERINE, M.A., PANGBORN, R.M. and ROESSLER, E.B. 1965. *Principles of Sensory Evaluation of Food*, pp. 95, Academic Press, New York.

Anon. 1988. Tablas de Estadistica. Orientacion Lactea *8*, 44-47.

ARNOLD, G.M. and WILLIAMS, A.A. 1986. The use of Generalized Procrustes techniques in sensory analysis. In *Statistical Procedures in Food Research*, (J.R. Piggott, ed.) pp. 233-253, Elsevier Applied Science Publishers, Essex, UK.

BUERA, M.P., HOUGH, G., MARTINEZ, E. and RESNIK, S. 1991. Colorimetric, spectrophotometric and sensory color measurements of Dulce de Leche. Anal. Assoc. Quim. Argent. (In press).

Camara Comercio de Buenos Aires. 1978. *Codigo Alimentario Argentino*, pp. 99, Imprenta del Congreso, Buenos Aires.

CHATFIELD, C. and COLLINS, A. 1980. *Introduction to Multivariate Analysis*, pp. 148-149, Chapman and Hall, London.

COCHRAN, W.G. and COX, G.M. 1957. Experimental Designs, pp. 279-327, John Wiley & Sons, New York.

FENNEMA, O.R. 1982. *Introduccion a la Ciencia de los Alimentos*, pp. 107-109, Editorial Reverte, Spain.

FERREIRA, V.L.P., HOUGH, G. and YOTSUYAANAGI, K. 1989. Cor de doce de leite pastoso. Colet. ITAL Campinas 19, 134-143.

Genstat. 1989. *User's Manual*, Nag Ltd., Oxford, UK.

HOSKULDSSON, A. 1988. PLS regression methods. J. Chemometrics *2*, 211-228.

HOUGH, G., BRATCHELL, N. and WAKELING, I. 1992. Consumer preference of Dulce de Leche among students in the UK. Private communication.

HOUGH, G., BUERA, M.P., MARTINEX, E. and RESNIK, S. 1991. Effect of composition of nonenzymatic browning rate in Dulce de Leche like systems. Anal. Asoc. Quim. Argint. *79*, 31-40.

HOUGH, G., CONTARINI, A. and MORO, O. 1986. Analysis sensorial de preferencia en dulce de leche. La Aliment. Latinoamer. *20*, 66-71.

HOUGH, G., MARTINEZ, E. and CONTARINI, A. 1990. Sensory and objective measurement of sandiness in Dulce de Leche, a typical Argentine dairy product. J. Dairy Sci. *73*, 604-611.

HOUGH, G., MORO, O., SEGURA, J. and CALVO, N. 1988. Flow properties of Dulce de Leche, a typical Argentine dairy product. J. Dairy Sci 71, 1783-1788.

LUCISANO, M., CASIRAGHI, E. and POMPEI, C. 1989. Optimization of an instrumental method for the evaluation of spreadability. J. Texture Studies 20, 301-315.

MARTENS, M. and MARTENS, H. 1986. Partial least squares regression. In *Statistical Procedures in Food Research*, (J.R. Piggott, ed.) pp. 293-359, Elsevier Applied Science, Essex, UK.

MORO, O. and HOUGH, G. 1985. Total solids and density measurements in Dulce de Leche, a typical Argentine dairy product. J. Dairy Sci. 68, 521-525.

MOSKOWITZ, H.R. 1967. Univariate psychophysical functions. In *Correlating Sensory-Objective Measurements: New Methods for Answering Old Problems*, pp. 36-47, ASTM STP 594, Amer. Soc. for Testing and Materials, Philadelphia, Pa.

MOSKOWITZ, H.R. 1977. Magnitude estimation: Notes on what, how, when, and why to use it. J. Food Quality 3, 195-227.

PANCOAST, H.M. and JUNK, W.R. 1980. *Handbook of Sugars*, pp. 179, Van Nostrand/AVI, New York.

PAULETTI, M.S., CASTELAO, E.L., VENIER, A.M., BERNARDI, C. and SABBAG, N. 1984. Influenza delle variabili di processo sulle caractteristiche del dolce di latte. I: Tessitura. Latte 9, 917-920.

PETRIELLA, C., CHIRIFE, J., RESNIK, S.L. and LOZANO, R.D. 1988. Solute effects of high water activity on non-enzymatic browning of glucose-lysine solutions. J. Food Sci. 53, 987-988.

SAUNDERS, J. and JERVIS, F. 1966. The role of buffer salts in non-enzymatic browning. J. Sci. Food Agric. 17, 245-248.

SINESIO, F., RISVIK, E. and RODBOTTEN, M. 1990. Evaluation of panelist performance in descriptive profiling of rancid sausages: A multivariate study. J. Sensory Studies 5, 33-52.

STONE, H. and SIDEL, J. 1985. *Sensory Evaluation Practices*, pp. 215, Academic Press, Orlando, Fla.

WEBB, B.H., JOHNSON, A.H. and ALFORD, J.A. 1980. *Fundamentals of Dairy Chemistry*, Chap. 11, Van Nostrand/AVI, New York.

WOLD, S. *et al.* 1983. Pattern recognition: Finding and using regularities in multivariate data. In *Food Research and Data Analysis*, (H. Martens and H. Russwurm, Jr., eds.), pp. 147-188, Applied Science Publishers, London.

CHAPTER 2.3

SENSORY PROPERTIES OF FERMENTED MILKS: OBJECTIVE REDUCTION OF AN EXTENSIVE SENSORY VOCABULARY

E. ANTHONY HUNTER and D. DONALD MUIR

At the time of original publication,
author Hunter was affiliated with the
Scottish Agricultural Statistics Service
The University of Edinburgh
James Clerk Maxwell Building, The King's Buildings
Edinburgh, EH9 3JZ, Scotland

and

author Muir was affiliated with
Hannah Research Institute
Ayr, KA6 5HL, Scotland.

ABSTRACT

Sensory laboratories develop vocabularies for new products and even for some old products so that Quantitative Descriptive Profiling can be carried out. It is common to find that the vocabulary developed is too extensive to be used on a routine basis. One method of reduction is to use statistical methods to identify sub-sets of descriptors. This paper describes how these methods have been used in the development of a vocabulary for fermented milks. Sub-sets of the vocabulary were found, using the communalities from the principal component analyses, and these were verified to cover the sensory space by Procrustes Rotation. A confirmatory experiment was carried out using the reduced vocabulary. A Procrustes Rotation of the scores derived from the separate analysis of these experiments verified that they were in excellent agreement.

INTRODUCTION

A coordinated research program has been established at the Hannah Research Institute to define the relation between milk composition and the properties of fermented milks. Sensory evaluation posed a particular problem.

Although a computer-aided literature search located more than 200 relevant papers, few concerned detailed sensory evaluation. Therefore, as prerequisite for routine, quantitative descriptive profiling, it was necessary to develop a sensory vocabulary.

Various strategies are available for development of a sensory vocabulary: all offer some advantages and most an equal number of disadvantages. The extreme strategies, in terms of the freedom they allow judges to use their own descriptors, are the traditional approach on the one hand and free choice profiling on the other. In the traditional approach, judges are exposed to the product range to be evaluated and, in group discussion, a vocabulary is agreed (Stone and Sidel 1985; Piggott 1991). However, the final vocabulary may be disproportionately influenced by the preconceptions of the dominant personality who is often the panel leader. In contrast, free choice profiling allows each judge to develop their own vocabulary. Provided judges use their set of terms consistently, it can offer valuable information on the relative positions of test samples in the perceptual space produced by a Generalised Procrustes Analysis of the data. This technique has now been applied to a range of products with some success (e.g., Oreskovich *et al.* 1991). Nevertheless, there are difficulties associated with the use of free choice profiling. Firstly, in a research environment, judges may only participate in a small number of trials on a product and these may be spread over a long period. Consequently the requirement for judges to consistently use their own vocabulary is particularly demanding. Secondly, interpretation of the principal dimensions of the perceptual space in terms of the attributes used by the judges can be difficult.

We have adopted an evolutionary approach to the development of a vocabulary. Because our judges quickly become fatigued when using a long profile, our aim was to produce a list with a modest number of attributes which was adequate to describe the differences in samples of fermented milk produced experimentally. The number of subjective decisions associated with the traditional method have been reduced by careful design of the studies and by rigorous analysis of the data.

MATERIALS AND METHODS

Sensory analysis was carried out in a newly equipped sensory laboratory which was designed in accordance with the recommendations of Stone and Sidel (1985). Each booth was evenly illuminated with a "daylight" fluorescent light and was ventilated with fresh tempered air. The booths were maintained at a slight positive air pressure to prevent ingress of extraneous odors.

Panel members were drawn from the staff of the institute with previous experience of the sensory evaluation of dairy products. No selection or training took place, and no feedback on how judges compared was permitted. Motivation was maintained by frequent assurance that the sensory work was an important part of the food science program and by the number of reports and papers arising.

Preliminary Study

Six fermented milk products of diverse character were presented to 20 judges who were willing to participate in sensory studies of plain fermented milk products. The initial vocabulary (see Table 1) was suggested by three sensory analysts, skilled in product description, who were members of the scientific team and not of the panel of judges. The judges were asked to score each product (0 = term does not apply or 1 = term applies) for sensory terms describing odor, flavor and texture. Judges were encouraged to add terms that they considered to be relevant. In addition, they were asked to rate each product for aftertaste and to write in a description of any aftertaste they detected.

Main Experiment

Twenty four samples of nine different types of plain, uncolored fermented milks were tested in a sensory profiling experiment. Further details of the samples are given in Muir and Hunter (1992). For eight of the types, samples were obtained from different suppliers or from the same supplier at different times. Twelve samples were evaluated in three sessions in each of two weeks using the same panel of 12 judges. These judges were drawn from the panel who participated in the preliminary experiment. The allocation of samples to session for each judge and the order of presentation within session were determined using a Latin Square design balanced for the residual effects of treatments (Williams 1949).

The sensory vocabulary used was derived from the analysis of the results of the preliminary experiment. In addition the catch-all category "other" was used for odor, flavor and aftertaste. The terms used are given in Table 2. Judges were asked to give suitable terms for the "other" categories.

Judges were invited to score each of the chosen attributes on a 125 mm undifferentiated scale with two anchor points at the end of the scales.

Confirmatory Experiment

Sixteen samples of the nine different types of plain, uncolored fermented milks were tested in a sensory profiling experiment. Of the 16 samples, two were duplicate samples of two of the brands. Also, two were "new" brands of two of the types. Eight samples were evaluated in two sessions in each of two weeks using 12 judges in the first week and 14 in the second week. The judges were drawn from the panel that participated in the preliminary experiment. The allocation of samples to session and the order of presentation for each judge were determined using Latin Square designs balanced for the residual effects of treatments (Williams 1949).

The sensory vocabulary used was that developed from the main experiment and is given in Table 4. Judges used a 125 mm undifferentiated scale with two anchor points at the end of the scale.

ANALYSIS AND DISCUSSION

Preliminary Study

From this study, data were recorded for 13 odor terms, 22 flavor terms and 9 texture terms in total 44 variates (see Table 1). Each variate consists of data values for each, combination of judge and sample. The data can have two values i.e., 0 = term does not apply or 1 = term applies. The data are thus binary and are properly analyzed using the Generalized Linear Model with Binomial variation and a Logit link function (Aitken *et al.* 1989). Parameters were fitted for judge and sample. Terms that were chosen had either high means (i.e, 0.50 or greater) indicating that they were generally applicable, or alternatively they had lower means but discriminated between samples as assessed by the ratio of sample mean deviance to error mean deviance.

For description of odor, the terms meaty, spicy and cowy were seldom used and so were discarded. The terms strength, sour, fruity and creamy were accepted on the basis of their wide applicability. On the grounds of their means, their ability to differentiate between samples and other considerations, buttery and yeasty were accepted while nutty, rancid, cheesy and pungent were rejected. Frequency counts for the extra terms written in suggested that sweet be added to the list of terms for subsequent experiments. The term strength was replaced by intensity.

For flavor, the terms strength (renamed intensity), sour, fruity, buttery, rancid, creamy, bitter, acid, lemon and sweet were accepted. Salty and chemical were also accepted because they discriminated between samples despite having a low mean value.

For texture, the terms gelatinous and grainy were rejected and all others accepted.

Aftertaste proved to be more important than initially thought with a mean of 0.59 i.e., 59% of the data points (samples by judges) resulted in a report of aftertaste. The terms written in to describe the aftertaste were amalgamated into strength (or intensity), bitter and sour/acid.

TABLE 1.
PRELIMINARY STUDY-VOCABULARY FOR ODOR, FLAVOR AND TEXTURE
ASSESSMENT OF SIX FERMENTED MILK PRODUCTS

Odor	Flavor	Texture
\multicolumn{3}{c}{Accepted Terms}		
strength *	strength *	firmness
sour	sour **	creamy
fruity	fruity	thick
buttery	buttery	slimy
yeasty	rancid	curdy
creamy	creamy	mouth-coating
	salty	chalky
	bitter	
	acid **	
	lemon	
	sweet	
	chemical	
\multicolumn{3}{c}{Rejected Terms}		
nutty	nutty	gelatinous
rancid	yeasty	grainy
meaty	meaty	
spicy	cheesy	
cheesy	spicy	
pungent	pungent	
cowy	cowy	
	chalky	
	dry	
	synthetic	

* renamed intensity in subsequent studies

** amalgamated into sour/acid in subsequent studies

Main Experiment

The data from this experiment can be regarded as coming from a resolvable incomplete block design. Judges are replicate blocks and sessions within judges are incomplete blocks within replicate blocks. The design used was balanced for the order of presentation and partly balanced for the residual effects of previous sample. It is important to allow for these effects in the analysis in order to appropriately estimate errors. It is also important to verify that the previous sample is not interfering with sensory evaluation of the current sample. In order to analyze the data appropriately, a sophisticated general procedure was used for the univariate analysis.

Estimates of sample effects were obtained using the Residual Maximum Likelihood (REML) directives of Genstat 5; release 2.2; Copyright 1990; Lawes Agricultural Trust, Rothamsted Experimental Station. For a description of the REML technique, see Payne *et al.* (1992). Fixed effects for sample, previous sample and order of presentation in a session were fitted and random effects for judge and for session within judge.

The sample means were then summarized using Principal Component Analysis. For detailed results of the analysis see Muir and Hunter (1992).

The experimental results allowed the sensory vocabulary to be further refined. Since the 24 samples covered a very wide spectrum of fermented milks, any vocabulary that successfully described these differences would describe the differences produced by experimental treatments.

Nevertheless, this task was approached with caution and, rather than seek to simplify the whole vocabulary in one exercise, the odor, flavor, aftertaste and texture sections were simplified separately. Accordingly, a principal component analysis on each part of the data and the whole was carried out. The mean scores and the variation between samples was different for each attribute, despite the use of the same scale by judges. In order to preserve the information contained in the magnitude of the variation between samples, the principal component analyses were carried out on the variance-covariance matrices.

Before examination of the contribution of individual attributes to each of the principal component analyses the number of components to be retained required to be decided. Jolliffe (1986) discussed methods commonly used to make these decisions. Our method was to select sufficient terms to account for a total of 80-90% of the variation with the aid of bar charts of the latent roots against order, see Fig. 1a-1d. (The total variance explained was large because of the diverse nature of the samples.) This diagnostic aid has become known as a "Scree Diagram" and is used to identify break points in the sequence of latent roots, thus separating signal from noise. Three components were retained for Odor, Flavor and Texture and two for Aftertaste.

Given that most of the variation in each set of variates was summarized by a lower dimensional subspace, the axes of the principal components were redefined by varimax rotation to give more interpretable loadings on the original variables. The original loadings together with the new rotated loadings are given for Odors, Flavors, Aftertastes and Textures in Table 2. Also given are the communalities for each original variate, which were the sum of the squares of the loadings on both the principal and rotated components. The rotation simplified the interpretation of the dimension reduction especially for the flavor variates.

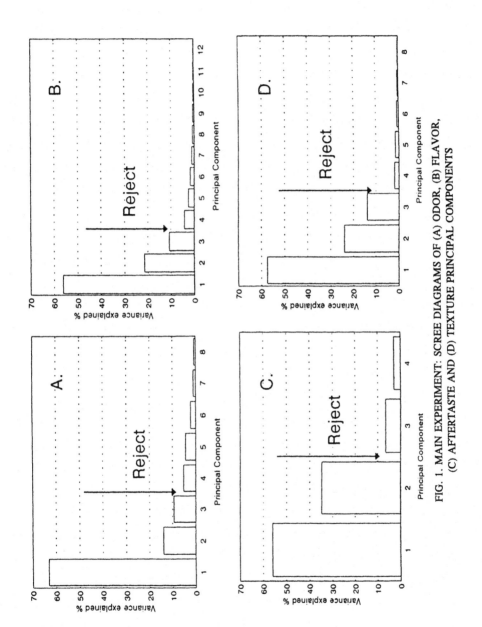

FIG. 1. MAIN EXPERIMENT: SCREE DIAGRAMS OF (A) ODOR, (B) FLAVOR, (C) AFTERTASTE AND (D) TEXTURE PRINCIPAL COMPONENTS

TABLE 2.
MAIN EXPERIMENT-PRINCIPAL COMPONENT ANALYSIS OF ODOR, FLAVOR, AFTERTASTE AND TEXTURE DATA-VARIATION ACCOUNTED FOR PRIOR TO FACTOR ROTATION AND LOADINGS BEFORE AND AFTER ROTATION TOGETHER WITH COMMUNALITIES

Term	1ST.	2ND.	3RD.	1ST.	2ND.	3RD.	Commun-alities
Variation(%) ODOR	63	14 Loadings	9	Loadings after rotation			
intensity	-.14	.24	.51	.12	.09	.56	.33
sour	-.49	.33	.53	-.11	-.08	.78	.63
fruity	.30	.09	.12	.20	.27	-.01	.11
buttery	.38	-.32	.38	.62	.01	-.03	.39
yeasty	-.31	-.37	-.16	-.09	-.48	-.13	.26
creamy	.36	-.26	.50	.66	.05	.10	.45
sweet	.53	.42	-.14	-.00	.66	-.18	.47
other	.08	.58	-.10	-.32	.49	.13	.36
Variation(%) FLAVOR	56	22 Loadings	11	Loadings after rotation			
intensity	.17	-.48	-.06	-.10	-.50	-.09	.27
sour/acid	.57	-.43	.13	.08	-.67	.28	.53
fruity	-.31	-.29	.18	-.45	-.04	-.07	.21
buttery	-.18	-.18	-.42	.02	-.08	-.49	.25
rancid	.17	.07	.11	.09	-.03	.19	.05
creamy	-.25	-.19	-.66	.09	-.07	-.73	.54
salty	.03	-.02	.01	.00	-.04	.02	.00
bitter	.29	.02	.12	.14	-.14	.25	.10
lemon	.22	-.38	-.16	.04	-.44	-.12	.21
sweet	-.52	-.40	.52	-.84	.02	.08	.71
chemical	.11	.20	.04	.16	.10	.14	.06
other	-.04	.27	.01	.11	.25	.06	.08
Variation(%) AFTERTASTE	56	35 Loadings		Loadings after rotation			
intensity	-.33	.61		.19	.66		.48
bitter	-.28	-.19		-.34	.06		.12
sour/acid	-.72	.33		-.29	.74		.63
other	.54	.70		.88	.12		.78
Variation(%) TEXTURE	57	24 Loadings	14	Loadings after rotation			
firmness	.15	-.55	-.43	.10	-.57	-.01	.33
creaminess	-.29	-.46	.07	-.34	-.43	.10	.31
viscosity	.13	-.62	-.02	.06	-.63	.02	.40
sliminess	.16	.14	-.04	.18	.12	-.05	.05
curdiness	.56	-.04	.11	.55	-.09	.11	.33
mouth-coat	-.05	-.27	-.17	-.07	-.27	-.15	.10
chalky	.02	.04	-.97	.02	-.03	-.97	.95
serum sepn.	.73	.02	-.02	.73	-.05	-.02	.53

Principal Component Analysis allows the results to be summarized by a set of coordinates for each sample, which defines its position in a multidimensional sensory space. Not only did we wish to reduce the number of dimensions that describe the existing data set but also to reduce the number of individual attributes that would be required to be measured in the future to obtain the coordinates in the sensory space. This is not a new problem (Jolliffe 1972, 1973; McCabe 1984; Lyon 1987). Recently Krzanowski (1987) has suggested a new way of tackling it using a Procrustes Criterion (Gower 1971): a Principal Component Analysis of each sub-set of variates is carried out and the principal component scores obtained for the same number of components as is judged to be significant in the analysis of the full set of variates. The subset scores are then rotated to match the full-set scores and the proportion of the sum of squares unaccounted for by rotation is obtained. Krzanowski (1987) has shown that this has many advantages over previously described methods.

The reduced sets were suggested by the size of the communalities. The decision was taken to include the catch-all category "other" in the reduced configurations. The subsets were evaluated against the full configuration using the procedure outlined above. These sets were further evaluated by dropping terms in turn and by adding unused terms. Thus, subsets of terms were identified that best preserved the multivariate dimensionality of the four parts of the vocabulary. The Odor subset was sour, creamy, sweet and other and the proportion of variation unaccounted for by rotation was 0.055. The Flavor subset was sour/acid, creamy, sweet and other and the proportion was 0.031. The Aftertaste subset was sour/acid and other and the proportion was 0.051. The Texture subset was viscosity, chalky and serum separation and the proportion was 0.048. Comparing the aggregate of the four subsets with he full vocabulary, left 0.027 of the variation unaccounted for by rotation.

Examination of the "other" responses (see Muir and Hunter 1992) suggested that the term was used to mean "atypical" for Odor and "chalky" for Flavor. For Aftertaste the terms "chalky", "creamy" and "sweet" were used with equal frequency. These results suggested that confusion might be avoided by replacing "other" by unambiguous attributes. For the confirmatory experiment "other" was dropped from Odor and replaced by "chalky" in Aftertaste.

The final check of the adequacy of the derived vocabulary in summarizing the results of the main experiment was to examine the loadings (after rotation) from the principal components analysis using only the 12 terms carried forward. As with the full set of variates five components were of importance, and these had a very similar interpretation to those based on all variates (see Table 3) and Muir and Hunter (1992).

TABLE 3.

MAIN EXPERIMENT, REDUCED VOCABULARY-VECTOR LOADINGS OF THE FIRST FIVE PRINCIPAL COMPONENTS AFTER FACTOR ROTATION

Term		first	second	third	fourth	fifth
Total variance explained (%)		50	17	15	8	4
		Vector Loadings after Rotation				
Sensory Attribute						
Odor	sour	.20	-.16	-.09	.14	-.02
	creamy	.04	.22	.03	-.28	-.01
	sweet	.01	.40	.02	.02	.03
Flavor	sour/acid	.71	-.06	-.02	.03	-.11
	creamy	-.24	-.08	.21	-.46	-.22
	sweet	-.07	.68	-.08	.07	-.07
	other	.05	.11	-.04	-.05	.52
Aftertaste	sour/acid	.60	.09	.09	-.09	.10
	other	.02	.41	.06	-.14	.39
Texture	viscosity	.09	-.14	-.16	-.79	.14
	chalky	-.12	-.29	.04	.16	.69
	serum separation	-.04	.00	-.95	-.01	-.01
Interpretation		sour/acid	sweet	serum/separation	viscosity	chalky

Confirmatory Experiment

The univariate analysis was carried out in a similar manner to that in the main experiment.

Sample effects were then analyzed using Principal Component Analysis on the variance-covariance matrix. Again there was evidence that the results were summarized by five components. The loadings after rotation are given in Table 4. The interpretation of the loadings is very similar to that of the main experiment despite a different mix of products and sources.

A further test of the agreement of the results of the two experiments was provided by taking the first five principal component scores for the 12 types by brand in common to the two experiments (taking averages over repeated samples) and using a Procrustes rotation to align the scores from the confirmatory experiment onto the full vocabulary scores (after factor rotation) from the main experiment. The scores for each experiment were individually

TABLE 4.
CONFIRMATORY EXPERIMENT-VECTOR LOADINGS OF THE FIRST FIVE
PRINCIPAL COMPONENTS AFTER FACTOR ROTATION

Term		first	second	third	fourth	fifth
Total variance explained (%)		61	19	11	4	3
		Vector Loadings after Rotation				
Sensory Attribute						
Odor	sour	.11	.10	-.16	.11	-.12
	creamy	-.02	-.01	.05	.17	.12
	sweet	.07	.01	-.08	.05	.65
Flavor	sour/acid	.77	-.02	.05	.16	.00
	creamy	-.20	-.12	.14	.69	-.09
	sweet	-.10	.01	.16	-.08	.66
	other	.09	-.05	-.30	.32	.32
Aftertaste	sour/acid	.56	-.01	.03	-.18	-.01
	chalky	-.03	-.12	-.78	-.06	-.02
Texture	viscosity	.12	.15	-.09	.56	-.06
	chalky	-.07	.09	-.45	-.05	-.02
serum	separation	-.03	.96	-.02	.01	.01
Interpretation		sour/ acid	serum sep- aration	chalky	visc- ous/ creamy	sweet

centered. Isotropic scaling was allowed. Of the variation in the scores from the confirmatory experiment, 88.1% was explained by rotation. The statistical significance of this result was estimated using a randomization test. For use of this technique in the context of Generalized Procrustes Analysis, see Wakeling et al. (1992) . The order of the scores from the confirmatory experiment was randomized, 1000 times, a rotation performed and the residual sums of squares saved. The results of the randomization are given in Fig. 2. In none of the rotations of the randomized data was the residual sums of squares as low as that in the initial rotation. It was concluded that there was a very high degree of agreement between the main and confirmatory experiments. A more detailed test of agreement of the results was provided by plotting the aligned scores from the confirmatory experiment against those of the main experiment (after factor rotation, see Fig. 3). Consistent with the above results, the plots are linear with correlations of 0.99, 0.91, 0.98, 0.78 and 0.92. The lower correlation of 0.78 for PC4 "creamy texture" is due to one product that was perceived to have a different texture in the confirmatory experiment.

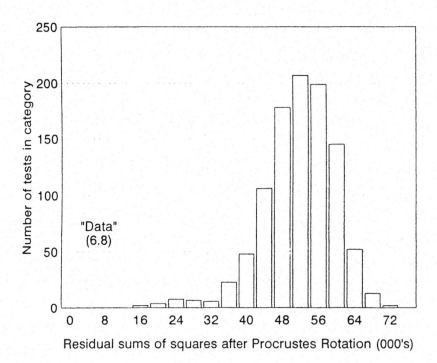

FIG. 2. RANDOMIZATION TEST OF AGREEMENT OF RESULTS OF
MAIN AND CONFIRMATORY EXPERIMENTS

The five significant principal components imply that the multivariate space could be covered by as few as five descriptors. Such a vocabulary would have no redundancy but would also have very little ability to describe new samples. By simplifying the odor, flavor, aftertaste and texture parts of the vocabulary separately, we have preserved a measure of flexibility. The main and confirmatory experiments profiled samples from each of the nine main types of fermented milk available at present in the UK and is, in our opinion, sufficiently flexible to describe differences between experimental samples of every one of these products created by experimental treatments.

The reduced vocabulary captured the same information as the much larger vocabulary of the main experiment and this showed that our sensory methods are capable of consistently measuring sensory properties of fermented milks.

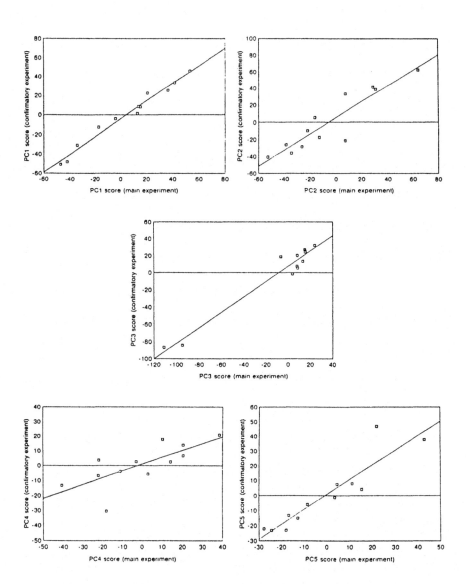

FIG. 3. AFTER A PROCRUSTES ROTATION, GRAPHS OF SCORES FROM THE CONFIRMATORY EXPERIMENT AGAINST SCORES FROM THE MAIN EXPERIMENT

CONCLUSIONS

Using statistical methods, it has been possible to reduce the number of subjective judgments that are required to reduce an extensive sensory vocabulary. Statistical analysis has shown that the reduced vocabulary of 12 terms covers the sensory space, which is defined by 32 terms in the main experiment. The results obtained in a confirmatory experiment are in excellent agreement with those of the main experiment. We recommend the use of these methods.

ACKNOWLEDGMENT

Carol Shankland is thanked for her expert technical assistance. Members of the fermented milk panel are thanked for their enthusiastic participation in this study. This work was funded by The Scottish Office Agriculture and Fisheries Department.

REFERENCES

AITKEN, M., ANDERSON, D., FRANCIS, B. and HINDE, J. 1989. *Statistical Modelling in GLIM*, Oxford University Press, Oxford, UK.

GOWER, J.C. 1971. Statistical Methods of comparing different multivariate analyses of the same data. In *Mathematics in the Archaeological and Historical Sciences*, pp. 138-149, Edinburgh University Press, Edinburgh, Scotland.

JOLLIFFE, I.T. 1972. Discarding variables in a principal component analysis. I. Artificial data. Appl. Statistics *21*, 160-173.

JOLLIFFE, I.T. 1973. Discarding variables in a principal component analysis. II. Real data. Appl. Statistics *22*, 21-31.

JOLLIFFE, I.T. 1986. *Principal Component Analysis*, Springer Verlag, New York.

KRZANOWSKI, W.J. 1987. Selection of variables to preserve multivariate data structure, using principal components. Appl. Statistics *36*, 22-33.

LYON, B.G. 1987. Development of chicken flavour descriptive attribute terms aided by multivariate statistical procedures. J. Sensory Studies 2, 55-67.

McCABE, G.P. 1984. Principal variables. Technometrics *26*, 137-144.

MUIR, D.D. and HUNTER, E.A. 1992. Sensory evaluation of fermented milks: Vocabulary development and the relations between sensory properties and composition and between acceptability and sensory properties. J. Soc. Dairy Technol. *45*, 73-80.

ORESKOVICH, D.C., KLEIN, B.P. and SUTHERLAND, J.W. 1991. Procrustes analysis and its applications to free-choice and other sensory profiling. In *Sensory Science Theory and Applications in Foods*, (H.T. Lawless and B.P. Klein, eds.) pp. 353-393, Marcel Dekker, New York.

PAYNE, R.W. 1992. *Genstat 5 Reference Manual* 2nd Ed., Clarendon Press, Oxford, UK.

PIGGOTT, J.R. 1991. Selection of terms for descriptive analysis. In *Sensory Science: Theory and Applications in Foods*, (H.T. Lawless and B.P. Klein, eds.) pp. 339-351, Marcel Dekker, New York.

STONE, H. and SIDEL, J.L. 1985. *Sensory Evaluation Practices*, pp. 198 et seq., Academic Press, New York.

WAKELING, I.N., RAATS, M.M. and MacFIE, H.J.H. 1992. A new significance test for consensus in generalised Procrustes analysis. J. Sensory Studies 7, 91-96.

WILLIAMS, E.J. 1949. Experimental designs balanced for the estimation of residual effects of treatments. Aust. J. Sci. Res. *A2*, 149-168.

CHAPTER 2.4

MEASURING SOURCES OF ERROR IN SENSORY TEXTURE PROFILING OF ICE CREAM

BONNIE M. KING and PAUL ARENTS

At the time of original publication,
both authors were affiliated with
Quest International Nederland BV
28 Huizerstraatweg
NL-1411 GP Naarden
The Netherlands.

ABSTRACT

This paper presents a general protocol for the sensory texture profiling of ice cream using eleven descriptors. Four ice cream bases, all containing 10% butter fat, were used in the profiling experiments. The composition of the bases was varied to create differences in softness, cold sensation and toughness. In addition, three sources of error were investigated: temperature, manner of serving and order of presentation. The importance of each type of error was determined by ANOVA techniques. Texture variables (descriptors) are influenced not only by the composition of an ice cream base, but also by the temperature of the ice cream during evaluation. When good standard procedures are used to scoop samples and to present them to panelists, sensory evaluations need not suffer from the other two sources of error examined.

INTRODUCTION

Sensory panels are widely used in industrial product development. Sensory profiling is a procedure designed to generate a two-way data matrix of intensity scores (samples × descriptors) for each panelist. In classical profiling experiments a scaling method is agreed upon; panelists are then trained to recognize the sensory property under investigation and to use the scaling method. In the actual experiments, panelists are presented with the samples to be evaluated, one at a time, in as many sittings as required to complete the work.

The purpose of most ice cream evaluations is to determine certain main effects, namely those related to flavors or food ingredients, and the interaction of the two. Unfortunately, it is difficult to find experiments without error in sensory research. Variation is introduced through the use of panelists as human measuring instruments. This variation can be minimized by training panelists until they meet given criteria. The samples themselves are a source of variation, that is why one measures them, but not all the sample variation can be attributed to the main effects under consideration.

Sensory experiments should be designed in such a way that the conclusions drawn from them are valid with respect to the proper partitioning of variance. Each intensity measurement mean has a given variance, the largest portion of which can be attributed (usually) to the panelists. The size of this total variance about an intensity mean is considered when two means are compared. This variance can be increased unwittingly by experimental conditions, sometimes to the point that actual sample differences are masked by experimental error. In this paper, three possible sources of error were considered in texture profiling experiments involving ice cream, and experiments were designed to investigate each source: sample temperature, serving portion, and serving order. The importance of each type of error was determined by ANOVA techniques.

MATERIALS AND METHODS

Samples

Four different types of ice cream bases (A, B, C, D) were used. Base A was a standard recipe for 10% butter fat ice cream. Base B had maltodextrine added to this standard recipe. No emulsifiers were used in base C. Both monosaccharides and glycerol were added to the standard recipe to create base D. The bases were identically flavored.

Sensory Panel

A paid, professional panel consisting of 20 women, 36-55 years old, was used in these experiments. Most of the panelists have worked for Quest's Department of Sensory Research in this capacity for eight years. Each panelist had been involved in ice cream evaluation for at least 20 months at the time of these studies.

Eleven descriptors were agreed upon by the panel to measure texture characteristics of ice cream. These are defined in the protocol given in Table 1. Panelists required a maximum of five minutes to register the eleven measurements on each sample; the average time was normally about three minutes.

TABLE 1.
PROTOCOL FOR MEASURING THE TEXTURE OF ICE CREAM BY THE
AUDIO METHOD USING 11 DESCRIPTORS

A. Cut one heaping spoonful of ice cream and measure

    1. MANUAL FIRMNESS: The more effort it takes to cut with the spoon (hard ice cream), the higher the tone.

    2. MANUAL BRITTLENESS: The more pieces break off, the higher the tone.

B. Put whole spoonful into mouth between tongue and palate to measure

    3. COLD SENSATION: The colder the sample feels on tongue and palate, the higher the tone.

C. Press sample between tongue and palate to measure

    4. ICE CRYSTAL DETECTION: The easier it is to detect ice crystals - large number or large in size - the higher the tone.

    5. SMOOTHNESS: The smoother the sample, the higher the tone.

    6. TOUGHNESS: The more effort required to prepare the sample for swallowing (e.g. by chewing), the higher the tone.

    7. RATE OF MELT: The faster the sample becomes liquid in the mouth, the higher the tone.

    8. VISCOSITY: The thicker the melted sample, the higher the tone.

D. Swallow sample to measure after effects

    9. MOUTH COATING: The more oily coating, the higher the tone.

    10. MOUTH DRYING: The more stiff/powdery feeling, the higher the tone.

    11. LINGERING FEELING: The more tenacious the after effect, the higher the tone.

Data Collection and Treatment

The audio method for measuring intensity by changing the pitch of a tone has been described several years ago (King 1986). The technique was completely automated shortly thereafter so that the tone can be heard through headphones attached to laptop computers in the sensory booths. The audio settings in the control computer make it possible for the panel leader to define certain parameters. For example, the function for the change of pitch with time and the speed of this change may be regulated. It is also possible to vary the lag time between pressing the cursor and activating the tone change, and an initial "start tone" may be defined. The standard default settings were used for all experiments described in this paper.

Panelists change the tone frequency (20-2500 Hz) by pressing the up/down cursors, using the convention that higher intensities are represented by higher tones. Numerical values (Hz) for intensity scores are sent to the control computer when panelists release the cursor key and press ENTER. This action cuts off the tone until panelists reactivate it for the continuation of their measurements. At no point are they given anything but audio feedback. No attempt has been made to coerce panelists into using the audio scale in a similar manner. Although the 500 Hz start tone used in profiling obviously has some centering effect, panelists differ in their level and range usage of the scale.

Panelists must be able to discriminate among samples, and they must be reliable. Discrimination is measured per relevant descriptor by ANOVA F ratios for main effects ($p < 0.05$ unless the samples were found not to differ significantly by difference testing). Panelists are said to be reliable when their ANOVA F ratios for replicated evaluations show $p > 0.05$ (replications are always separated by at least 48 h). Despite the fact that panelists were never asked to develop an absolute scale, their profile data show such a high degree of reliability that statistical techniques can be applied directly to the Hz values. ANOVA for replication effects is done on raw profile data.

In order to quantify the effect of each source of error discussed in this paper, the data were treated as randomized block or split-plot designs using GLM procedures from SAS. In the split-plot designs, the factor of primary interest is the subplot factor. Factors were considered significant when $p < 0.05$. Generalized Procrustes Analysis (GPA) was performed by PROCRUSTES-PC v.2.0. The geometric means of panelists' data were used to create the profiles shown in the figures.

Temperature Effect

It is not unreasonable to assume that some variation in the scores for the eleven texture descriptors is caused by variation in temperature. Even with the normal precautions of good temperature control (freezers as well as evaluation rooms), differences can occur. For this reason extra portions of each sample to be evaluated in a session were prepared, and temperature measurements were conducted on these samples concurrently with the sensory evaluation. All temperature measurements were made by sticking the thermocouple of a microprocessor thermometer (Hanna Instruments, HI 9053) into the sample of ice cream and waiting until the digital display stabilized.

Experiment 1. Vanilla flavored ice cream was prepared on base A. Scooped portions were profiled by the panel at two different temperatures: -11C (samples direct from freezer) and -5C (samples left 30 min at room temperature, 23C). The temperature difference of 6C was chosen as a "worst case" since under normal evaluation conditions the ice cream would not warm up more than two degrees, and freezer temperature differences would seldom be larger than six degrees. Data were analyzed by two-way ANOVA.

Experiments 2-4. One flavor of vanilla ice cream was prepared for each of the other three bases, and scooped samples were profiled on different days after the samples had been equilibrated to different temperatures. The bases and temperatures were as follows:

Experiment 2: base B at -6 and -13C, respectively
Experiment 3: base C at -7 and -14C, respectively
Experiment 4: base D at -3 and -15C, respectively

Data from these experiments were analyzed by two-way ANOVA.

Experiment 5. In order to compare the effects of temperature differences on the perception of these four different bases, a data set consisting of measurements on 12 samples was analyzed. Each base was evaluated under the standard conditions, and then also at a higher and a lower temperature. The range of temperatures was 8 or 9 degrees for each base. (Temperatures for each sample are given in the legend to Fig. 3.) The data were analyzed by ANOVA for a split-plot design and by GPA.

Serving Portions

Experiment 6. When ice cream is produced in individual dixie-cup portions, errors due to scooping are eliminated. It is also easier to equilibrate these portions of ice cream to the desired serving temperature without risking ice crystal formation on the surface. Unfortunately one is quite often asked to evaluate products on the market which are sold only in larger portions, generally ranging from one-half to two liter packages. In these cases one cannot avoid scooping the ice cream into smaller individual portions. Surface properties of the scooped ice cream will differ depending on whether the scoop came from the sides or the middle of the package. It is possible that the texture evaluation will be influenced by the size and form of the scoop. Changes in texture can also be introduced by the thawing/refreezing process.

Ice cream from the same production batch was used to fill half-liter packages and individual dixie-cups. The scooped samples were taken from the half-liter packages according to standard protocol; the individual portions were evaluated as such by the panel. Both kinds of samples contained about the same amount of ice cream served at -12 ± 1C. Data were analyzed by split-plot design ANOVA, 2-way ANOVA per base (panelists, scoop) and comparison of means per base.

Serving Order

Experiment 7. Previous experience has shown that the first evaluation in a sensory tasting session is often made differently from subsequent evaluations (Van der Pers 1988; Fletcher *et al.* 1991). Sequential Sensitivity Analysis considers the effect of stimuli tasted previously on the discriminability of each stimulus in a test protocol. (See Thieme and O'Mahony 1990 and references therein.) In order to investigate the effects of serving order, two separate experiments were carried out using ice cream prepared on bases B, C, and D, all having the same flavor. These bases were chosen because of their distinct texture profiles, as determined in previous experiments.

Panelists profiled samples of bases B, C, and D under standard conditions, using scooped portions. The first sample for each panelist was also her first evaluation for the tasting session. The data analysis used a design variable (FP) to indicate whether or not each sample was the first one to be evaluated. For ANOVA, an unbalanced incomplete block design was used.

TABLE 2.
DISTRIBUTION OF THE 99 EVALUATIONS IN EXPERIMENT 8
TO DETERMINE POSITIONAL EFFECTS

Position	Base			
	B	C	D	Total
1	10	12	11	33
2	12	10	11	33
3	11	11	11	33
Total	33	33	33	99

Experiment 8. The three bases used in experiment 8 are sufficiently different in textural properties that one might expect carry-over effects to be prevalent. Base D, for example, is so much softer than the other two bases that a subsequent sample might be judged firmer by comparison than it would be judged when it was not evaluated directly after Base D. In order to investigate these kinds of effects, three variables (V_B, V_C, V_D) were defined, one for each base. Each of these variables has three levels to classify the preceding sample: none, or one of the other two bases.

Each panelist profiled samples of bases B, C, and D once, and some panelists twice (on separate days, using a different presentation order) until the 99 evaluations given in Table 2 were completed. All samples were served in individual dixie-cup portions at -12 ± 1C. These evaluations were not the first ones to be made at the given tasting session; the panel had previously profiled savory samples.

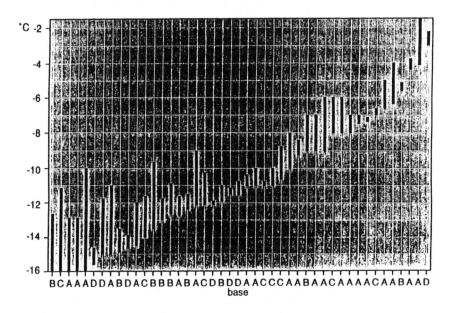

FIG. 1. THE INCREASE IN TEMPERATURE OF 10% BUTTER FAT ICE CREAM DURING ITS EVALUATION AT A ROOM TEMPERATURE OF 23-24C

RESULTS AND DISCUSSION

Temperature Effect

Figure 1 shows the rise in sample temperature measured during the evaluation period for ice cream served at different initial temperatures. Under normal conditions (-12 ± 1C) the temperature of the sample increases by only one or two degrees. Very cold ice cream (at or below -16C) warms up faster than ice cream served at the normal consumption temperature. Moreover, very cold samples cannot be evaluated, so panelists wait until they become warmer. The trend shown for temperature increases appears to be independent of the ice cream base.

Table 3 gives the p-values for panelist ($F_{10,10}$) and temperature effects ($F_{1,10}$) from the two-way ANOVA for the data from experiment 1. The texture profiles are shown in Fig. 2. Four descriptors show significant decreasing scores for warmer ice cream: manual firmness, manual brittleness, cold sensation, and toughness. When p-values for temperature and panelist effects are compared for each of these descriptors, it becomes apparent that the temperature effect is very important. Ice crystal detection, which also decreased, was significant at the 10 % level. The temperature effect for this descriptor was just about as important as the panelist effect. Rate of melt increases significantly with temperature. Here again the temperature effect is much more important than the panelist effect. For viscosity, mouth coating, mouth drying and lingering feeling, only the panelist effects were significant. According to the protocol, these descriptors are evaluated after the ice cream has been in the mouth a while. It would appear logical, therefore, that any initial temperature differences had been equalized.

Ice cream from the same batch was profiled repeatedly in separate experiments under standard temperature conditions. Table 3 gives the p-values for panelist ($F_{17,17}$) and temperature effect ($F_{1,17}$) from these experiments. No descriptors showed significant differences when both temperatures were the same (± 1C). Most of the p-values for the panelist factor, on the other hand, were significant. Clearly the error associated with this experiment can be attributed almost exclusively to panelists, as expected.

Table 4 gives the results from experiments 2-4. Although temperature differences with base C were significant for fewer descriptors than with bases B and D, the trends for all bases are the same. Generally the same effects were found in these experiments as in experiment 1 with base A. These results suggest that the importance of temperature differences is not only independent of the base, but also that it is independent of the *way* that the temperature difference was created (equilibration in freezers versus simply allowing the ice cream to warm up) and the experimental design (repeated measures on the same day or across different days).

TABLE 3.
P-VALUES FROM TWO-WAY ANOVAS FOR ($\Delta T=6°$) AND ($\Delta T=0°$),
EXPERIMENT 1: PANELIST AND TEMPERATURE EFFECTS

Descriptor	$\Delta T=6°$		$\Delta T=0°$	
	p^a	p^b	p^c	p^d
Manual firmness	.8299	.0000	.0001	.3108
Manual brittleness	.6413	.0000	.3524	.2677
Cold sensation	.0311	.0029	.0000	.4614
Ice crystal detection	.0540	.0754	.0001	.6706
Smoothness	.2377	.3191	.0000	.3816
Toughness	.5980	.0000	.0050	.3177
Rate of melt	.8729	.0013	.0587	.5230
Viscosity	.0132	.3921	.0000	.8175
Mouth coating	.0269	.4851	.0003	.3257
Mouth drying	.0034	.1378	.0000	.6174
Lingering feeling	.0018	.8521	.0001	.7871

[a] $P(F_{10,10} > F)$ for Panelist effect
[b] $P(F_{1,10} > F)$ for Temperature effect
[c] $P(F_{17,17} > F)$ for Panelist effect
[d] $P(F_{1,17} > F)$ for Temperature effect

A consensus map of the twelve samples from Experiment 5 (four ice cream bases measured at three different temperatures) is shown in Fig. 3. The first dimension is clearly related to temperature: manual firmness, toughness, and manual brittleness were significantly correlated with the negative (left) direction by all panelists, while rate of melt was correlated with the positive direction. The second dimension is related to smoothness (positive correlation) and cold sensation, ice crystal detection (negative correlation). The consensus explained 83% of the total variance. According to the criterion specified by King and Arents (1991), this was a good consensus: the p-value for this consensus was 0.0032.

Three-way ANOVAs for these data showed significant panelist effects for all descriptors except toughness, and significant base effects for six descriptors: manual firmness, cold sensation, ice crystal detection, smoothness, toughness and rate of melt. The temperature effect was significant for seven descriptors: manual firmness, manual brittleness, cold sensation, toughness, rate of melt, mouth coating and lingering feeling. Temperature effects were more important than base effects for seven of the eleven descriptors; they were more important than panelists effects for three out of eleven descriptors. The base*temperature interaction term was significant for manual brittleness, ice crystal detection, smoothness and toughness.

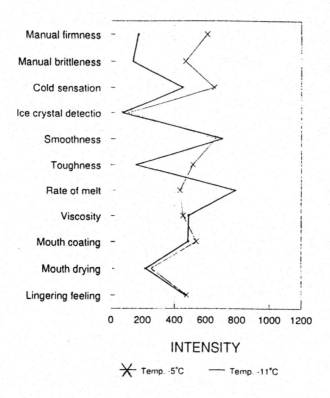

FIG. 2. TEXTURE PROFILES FOR ICE CREAM BASE A, EXPERIMENT 1

A comparison of means (LSD) from one-way ANOVAs for the samples measured under standard conditions showed that bases A and B were not significantly different for the four descriptors showing a significant base effect: manual firmness, manual brittleness, ice crystal detection, and smoothness. (Compare samples 2 and 5 on the map in Fig. 3.) This is not surprising given their composition. On the other hand, one would expect bases C and D to be quite different, and indeed they were significantly different for all four of these descriptors. (Compare samples 8 and 11 on the map.) Bases C and D were also significantly different when means were compared by one-way ANOVAs on each of the first four GPA consensus dimensions. (The first four dimensions account for 89% of the variance explained by the consensus.) These differences all disappear, however, when base D is measured at a lower temperature (sample 12).

TABLE 4.
P-VALUES FROM TWO-WAY ANOVAS PER BASE,
EXPERIMENTS 2-4: PANELIST AND TEMPERATURE EFFECTS

Descriptor	Base B p^a	Base B p^b	Base C p^a	Base C p^b	Base D p^a	Base D p^b
Manual firmness	.8532	.0000	.6044	.0000	.8648	.0000
Manual brittleness	.0136	.0071	.0023	.0902	.0032	.0000
Cold sensation	.0003	.0019	.0016	.5937	.0030	.0005
Ice crystal detection	.0261	.1889	.0247	.7574	.3247	.0650
Smoothness	.1520	.8061	.0001	.2275	.0006	.0021
Toughness	.8998	.0002	.9956	.0053	.2114	.0001
Rate of melt	.4614	.0078	.0950	.0705	.0346	.0289
Viscosity	.4791	.0600	.6767	.0617	.0028	.0017
Mouth coating	.0001	.0191	.2281	.0416	.0045	.2516
Mouth drying	.0277	.0341	.0118	.3411	.0003	.1611
Lingering feeling	.0032	.1654	.0000	.8620	.0046	.9244

^a $P(F_{16,16} > F)$ for Panelist effect
^b $P(F_{1,16} > F)$ for Temperature effect

Serving Portions

Table 6 shows the effects on the eleven texture descriptors when ice cream from the same production batch was evaluated in scooped versus dixie-cup portions (experiment 6). It can be seen that the scoop-effect depended on the ice cream base, although several trends were evident for all bases. Scores for mouth drying were all higher in the scooped portions, but this effect was significant only for base B. All bases showing significant effects for the descriptors manual firmness and ice crystal detection had higher scores for scooped portions. The same could be said for two additional descriptors, with one directional exception each: manual brittleness had lower scores on base C, and toughness had lower scores on base A. On the other hand, the dixie-cup portions had higher scores for smoothness, viscosity, mouth coating, and lingering feeling.

Scoop effects were only comparable in importance to panelist and base effects for three descriptors: manual firmness, viscosity and mouth coating. While the firmness might reasonably be influenced by scooping, it is hard to find a physical explanation for differences in either of the latter two descriptors.

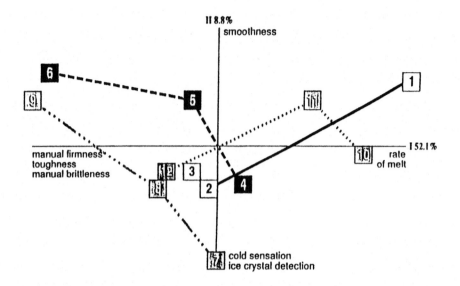

FIG. 3. FOUR ICE CREAM BASES EVALUATED AT THREE TEMPERATURES PLOTTED RELATIVE TO THE FIRST TWO PRINCIPAL COMPONENT DIMENSIONS OF THE GPA CONSENSUS PLOT

The samples are identified by the following points, with temperatures indicated in parentheses: Base A: 1(-5), 2(-11), 3(-14); Base B: 4(-7), 5(-12), 6(-15); Base C: 7(-8), 8(-11), 9(-17); Base D: 10(-9), 11(-11), 12(-17).

Serving Order

Table 7 gives the p-values for the F ratios from the three-way ANOVAs on data from Experiment 7. The first position effect was significant for five of the texture descriptors: manual firmness, manual brittleness, cold sensation, ice crystal detection and smoothness. For each of these descriptors the first-position effect was either more important than the base effect or comparable to it; the effect was always less important than that attributed to panelists, however. When the sample was the first one evaluated in the session, scores were higher for cold sensation and smoothness. These results might be explained by adaption of the mouth to the normal irritation induced during eating. Manual firmness and manual brittleness, on the other hand, were rated lower in the first sample.

The three bases have significantly different textural properties, as can be seen from the Base effect column in Table 7 and the profiles in Fig. 4. Base B is the smoothest and toughest ice cream; it has the slowest rate of melt, and it is the least cold. Base C has the least mouth coating, while base D is the least firm. These properties could be expected from the composition of the bases.

TABLE 5.
P-VALUES FROM THREE-WAY ANOVAS, EXPERIMENT 5: PANELIST,
BASE AND TEMPERATURE EFFECTS IN A SPLIT-PLOT DESIGN

Descriptor	P-effect[a]	B-effect[b]	P*B-effect[c]	T-effect[d]	B*T-effect[e]
Manual firmness	.0227	.0000	.9209	.0000	.0735
Manual brittleness	.0000	.3205	.0491	.0044	.0024
Cold sensation	.0000	.0045	.7915	.0186	.0704
Ice crystal detection	.0000	.0221	.8428	.8560	.0422
Smoothness	.0012	.0217	.8267	.7300	.0071
Toughness	.1014	.0007	.8249	.0001	.0293
Rate of melt	.0191	.0228	.7096	.0020	.0902
Viscosity	.0015	.0764	.0901	.0627	.5792
Mouth coating	.0076	.0686	.5230	.0392	.6264
Mouth drying	.0000	.0618	.6760	.1266	.1211
Lingering feeling	.0000	.1524	.4285	.0174	.6045

[a] $P(F_{4,32} > F)$ for Panelist effect
[b] $P(F_{3,12} > F)$ for Base effect, with Panelist*Base interaction as error term
[c] $P(F_{12,32} > F)$ for Panelist*Base interaction effect
[d] $P(F_{2,32} > F)$ for Temperature effect
[e] $P(F_{6,32} > F)$ for Base*Temperature effect

TABLE 6.
P-VALUES FROM THREE-WAY ANOVAS, EXPERIMENT 6: PANELIST,
BASE AND SCOOP EFFECTS IN A SPLIT-PLOT DESIGN

Descriptor	P-effect[a]	B-effect[b]	P*B-effect[c]	S-effect[d]	B*S-effect[e]
Manual firmness	.0001	.0000	.8765	.0000	.0000
Manual brittleness	.0000	.0000	.2346	.4899	.0011
Cold sensation	.0000	.0002	.1203	.6404	.0257
Ice crystal detection	.0000	.0093	.9211	.0603	.0015
Smoothness	.0000	.0039	.0604	.0093	.0056
Toughness	.0002	.0000	.9287	.0277	.0100
Rate of melt	.0000	.0034	.1571	.1312	.6030
Viscosity	.0000	.0004	.1210	.0000	.3497
Mouth coating	.0000	.0000	.5280	.0000	.4285
Mouth drying	.0000	.0120	.9562	.0088	.3030
Lingering feeling	.0000	.0819	.9989	.0765	.6602

[a] $P(F_{15,60} > F)$ for Panelist effect
[b] $P(F_{3,45} > F)$ for Base effect, with Panelist*Base interaction as error term
[c] $P(F_{45,60} > F)$ for Panelist*Base interaction effect
[d] $P(F_{1,60} > F)$ for Scoop effect
[e] $P(F_{3,60} > F)$ for Base*Scoop effect

TABLE 7.
P-VALUES FROM THREE-WAY ANOVAS, EXPERIMENT 7:
PANELIST, BASE AND SERVING ORDER EFFECTS

Descriptor	P-effect[a]	B-effect[b]	FP-effect[c]	B*FP-effect[d]
Manual firmness	.0004	.0000	.0099	.4373
Manual brittleness	.0000	.8596	.0314	.3700
Cold sensation	.0002	.0190	.0139	.3809
Ice crystal detection	.0052	.3853	.0209	.3532
Smoothness	.0003	.0154	.0181	.7238
Toughness	.0020	.0008	.0849	.1508
Rate of melt	.0089	.0834	.3462	.4396
Viscosity	.2375	.0890	.9345	.0895
Mouth coating	.0114	.0398	.9585	.3002
Mouth drying	.0000	.2373	.9365	.2262
Lingering feeling	.0000	.4145	.8317	.9089

[a] $P(F_{16,29} > F)$ for Panelist effect
[b] $P(F_{2,29} > F)$ for Base effect
[c] $P(F_{1,29} > F)$ for the effect of First Serving Position
[d] $P(F_{2,29} > F)$ for Base*First Serving Position interaction effect

Table 8 gives the p-values for panelist ($F_{16,45}$), base ($F_{3,29}$), positional effects ($F_{1,45}$) and interaction ($F_{2,45}$) from the three-way ANOVAs on data generated according to the design in Table 2 for Experiment 8. The base effect was significant for each descriptor except rate of melt and lingering feeling, but the positional effect and interaction were not significant. No significant effects were found for the variables V_B, V_C, V_D (Table 9.) There were, however, trends in the direction of the expected carry-over effects. For example, both bases C and D were evaluated lower in toughness when preceded by base B, the toughest base. Likewise, both bases B and D were evaluated as less cold when preceded by the coldest base, base C.

The profiles for Experiment 8 are shown in Fig. 5. Most of the trends discussed under Experiment 7 can be seen to advantage in this figure. For example, the fact that base C contains no emulsifier should make it appear colder in the mouth. Base C scored significantly higher in both cold sensation and ice crystal detection in Experiment 8.

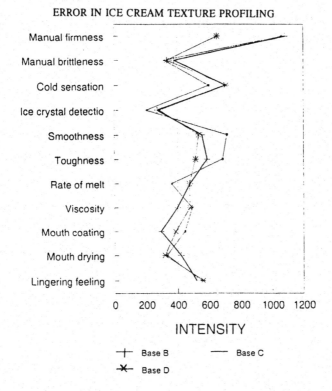

FIG. 4. TEXTURE PROFILES FOR THREE ICE CREAM BASES, EXPERIMENT 7

CONCLUSION

Serving order was shown to contribute the least amount of undesirable variance in these experiments. Given that evaluation sessions can be planned to allow for a warm-up sample, and presentation order can be varied over a panel, texture measurements need not suffer from order or carry-over effects.

The use of individually pre-packaged portions, such as ice cream in dixie-cups, can decrease experimental error due to serving portions. When ice cream must be scooped, however, the use of a well-established standard procedure does not introduce unacceptable levels of error.

Differences in sample temperature contribute most to the experimental error encountered in measurements of ice cream texture. Although this fact is hardly surprising, the importance of the error may not be fully appreciated. It is hoped that the data presented in this paper will serve as a caveat to others studying ice cream texture.

TABLE 8.
P-VALUES FROM THREE-WAY ANOVAS, EXPERIMENT 8: PANELIST, BASE AND SERVING ORDER EFFECTS IN A SPLIT-PLOT DESIGN

Descriptor	P-effect[a]	B-effect[b]	P*B-effect[c]	S-effect[d]	B*S-effect[e]
Manual firmness	.9971	.0000	.9993	.5058	.4575
Manual brittleness	.0078	.0339	.8055	.7463	.8988
Cold sensation	.0000	.0000	.9721	.7507	.9880
Ice crystal detection	.0009	.0002	.7003	.8462	.5937
Smoothness	.0000	.0000	.6452	.4089	.4954
Toughness	.0909	.0000	.9968	.5115	.1502
Rate of melt	.0720	.2724	.3114	.8898	.3519
Viscosity	.0000	.0008	.0117	.3093	.4201
Mouth coating	.0000	.0000	.8068	.2622	.3817
Mouth drying	.0000	.0002	.9783	.4721	.6412
Lingering feeling	.0000	.1025	.9101	.5628	.3330

[a] $P(F_{16,45} > F)$ for Panelist effect
[b] $P(F_{2,29} > F)$ for Base effect, with Panelist*Base interaction as error term
[c] $P(F_{29,45} > F)$ for Panelist*Base interaction effect
[d] $P(F_{1,45} > F)$ for Serving Order effect
[e] $P(F_{2,45} > F)$ for Base*Serving Order effect

TABLE 9.
P-VALUES FROM TWO-WAY ANOVAS PER BASE, EXPERIMENT 8: EFFECT OF PRECEDING SAMPLE ON THE MEASUREMENTS FOR THE BASE INDICATED BY THE SUBSCRIPT

| Descriptor | V_B | | V_C | | V_D | |
	p[a]	p[b]	p[a]	p[b]	p[a]	p[b]
Manual firmness	.9360	.6792	.9468	.4492	.9943	.9178
Manual brittleness	.0448	.8340	.1995	.8006	.1279	.9889
Cold sensation	.1056	.9214	.0240	.9954	.0507	.4406
Ice crystal detection	.0000	.3648	.5147	.8048	.1339	.9187
Smoothness	.0003	.4571	.0408	.2316	.1660	.4126
Toughness	.0782	.3573	.2901	.3679	.8897	.2191
Rate of melt	.2631	.5187	.1307	.4415	.0533	.2786
Viscosity	.0843	.2742	.1167	.4516	.0007	.3367
Mouth coating	.0000	.5734	.0364	.8368	.0126	.2770
Mouth drying	.0000	.6715	.0158	.9474	.0068	.4631
Lingering feeling	.0014	.6442	.0043	.0875	.0109	.4334

[a] $P(F_{16,14} > F)$ for Panelist effect
[b] $P(F_{2,14} > F)$ for Preceding Sample effect

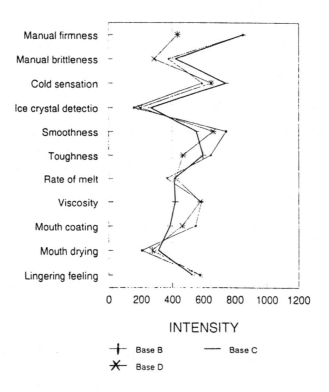

FIG. 5. TEXTURE PROFILES FOR THREE ICE CREAM BASES, EXPERIMENT 8

ACKNOWLEDGMENTS

This work was presented by the first author as an invited paper at the 3èmes Journées Européennes Agro-Industrie et Méthodes Statistiques, 30.11.92 to 1.12.92 at Montpellier, France.

REFERENCES

FLETCHER, L., HEYMANN, H. and ELLERSIECK, M. 1991. Effects of visual masking techniques on the intensity rating of sweetness of gelatins and lemonades. J. Sensory Studies 6, 179-191.

KING, B.M. 1986. Odor Intensity Measured by an Audio Method. J. Food Sci. 51(5), 1340-1344.

KING, B.M. and ARENTS, P. 1991. A statistical test of consensus obtained from Generalized Procrustes Analysis of sensory data. J. Sensory Studies 6, 37-48.

THIEME, U. and O'MAHONY, M. 1990. Modifications to sensory difference test protocols: the warmed up paired comparison, the single standard duo-trio and the A-not A test modified for response bias. J. Sensory Studies 5, 159-176.

VAN DER PERS, B.M. 1988. The effect of sample presentation order in profiling exercises: a methodological point. Internal report number P QN 88 0042.

CHAPTER 2.5

EFFECTS OF STARTER CULTURES ON SENSORY PROPERTIES OF SET-STYLE YOGHURT DETERMINED BY QUANTITATIVE DESCRIPTIVE ANALYSIS

H. ROHM, ALEŠA KOVAC and W. KNEIFEL

*At the time of original publication,
all authors were affiliated with
Department of Dairy Science and Bacteriology
University of Agriculture
A-1180 Vienna (Austria).*

ABSTRACT

A series of yoghurts was produced by using ten commercially available starter cultures under laboratory conditions and subjected to descriptive sensory analysis. The score sheet was developed by a ten member panel and included eight sensory categories. Additionally, a hedonic scale was used for evaluating acceptability impression. Statistical analysis showed good performance of both individual subjects and the whole panel. Furthermore, homogeneity between replicate assessments was observed. Significant differences between products were found in each sensory category except texture (gel firmness). Correlation analysis showed significant interrelations between some sensory continua. By using multiple regression analysis hedonic scores were found to be mainly determined by 'Flavor' and 'Ropiness' showing positive and negative weightings, respectively.

INTRODUCTION

Flavor and texture play an important role in consumer acceptance of plain yoghurt in set-style form, i.e., obtained by fermenting inoculated milk in beakers or glasses thus resulting in firm, gel-type products. Today, yoghurt with tailor made characteristics, e.g., mildly taste, reduced overacidification behavior or increased viscosity, are manufactured by using starter cultures with specific properties. Usually, yoghurt produced with starters of reduced acidifying activity is relatively low-flavored (Kneifel 1992). Moreover, changes in viscosity can affect intensity of taste and flavor perception (Kokini *et al.* 1984; Moskowitz and Arabie 1970). Apart from regional expressions of preference patterns (Kneifel 1992) products largely differing from 'classical' yoghurt obviously influence hedonic responses.

Complete sensory evaluation of yoghurt has been carried out mainly with respect to defects in texture, appearance and flavor (McGill 1983; Richter 1979; Tamime *et al.* 1987). Barnes *et al.* (1991) and Harper *et al.* (1991) used both consumer and descriptive panels for assessment of sensory properties of commercial flavored and plain yoghurt, respectively. Muir and Hunter (1992) developed a vocabulary for descriptive analysis of commercially available fermented milks and related sensory properties to overall acceptability. Regarding both thermophilic and mesophilic starter cultures sensory screening data have been published most recently (Kneifel *et al.* 1992a, 1992b; Ulberth and Kneifel 1992).

Quantitative Descriptive Analysis, or QDA°, was developed for description of sensory characteristics of food materials with precision in mathematical terms (Stone and Sidel 1985; Stone *et al.* 1974, 1980). Essential features of QDA° are the use of judges screened and trained on the specific product to be described and guided by a trained panel leader, the use of descriptive score cards developed by the panel, the use of unstructured line scales and repeated evaluations, and the use of statistical analysis by analysis of variance (ANOVA) and of circular plots for graphically displaying data (Mecredy *et al.* 1974; Stone *et al.* 1974). With respect to these features QDA° allows to describe product differences as well as panel performance and variability between products. In the present study a QDA° technique was applied to evaluation of yoghurt produced by different starters in order to describe main sensory properties and to characterize these starters sensorically.

EXPERIMENTAL

Yoghurt Preparation

UHT milk (2.5% fat) was warmed to 40C and fortified with 2% (w v^{-1}) skim milk powder. Five 1-mixtures were preheated for 20 min at 90C in stainless steel pots using a laboratory steam vessel (Loeblich GmbH, Vienna, Austria) and recooled to incubation temperature. Inoculation with starter cultures containing *Lactobacillus bulgaricus subsp. thermophilus* and *Streptococcus salivarius subsp. thermophilus* was in accordance with specific recommendations provided by culture manufacturers (Table 1). Starters were supplied by Chr. Hansen's Laboratorium, Hørsholm, Denmark; Lacto-Labo, Dange St. Romain, France; Laboratoires Miles-Marschall, Epernon, France; Laboratoire Roger, Duesseldorf, Germany; Laboratorium Wiesby, Niebuell, Germany. Inoculated

TABLE 1.
DESCRIPTION OF YOGHURT STARTER CULTURES AND CORRESPONDING FERMENTATION CONDITIONS

Culture code	Supplier code[a]	Culture type[b]	Yoghurt short characteristics[c]	Inoculation strength (%)	Fermentation conditions min (°C)[d]
A	1	FD	classical	3	240 (40)
B	1	FD	viscous	3	260 (40)
C	2	DF	classical	3	390 (43)
D	3	FD	classical	3	180 (45)
E	4	DF	viscous	3	260 (42)
F	4	FD	classical	2	250 (42)
G	5	FD	classical	2	215 (42)
H	3	FD	highly viscous	4	200 (43)
I[e]	3[e]	DF[e]	n.d.[f]	4	240 (43)
K	4	FD	classical	1	245 (43)

[a]Supplier codes do not follow the listing in the text.
[b]FD, freeze-dried; DF, deep-frozen.
[c]Main characteristics as described by the suppliers.
[d]Duration of fermentation (min) until a pH of 4.6 was obtained.
[e]Yoghurt produced with culture H was deep-frozen for 48 h at -30 C and again used for fermentation under same conditions.
[f]n.d., no description possible (see e).

milk was transferred aseptically into 150 mL polystyrene beakers in a laminar flow chamber. After sealing beakers with aluminum covers these were incubated in W38 water baths (Grant Ltd., Cambridge, UK). The pH was monitored by a 405-60-TT-S7 glass electrode (Ingold GmbH, Steinbach, Germany) connected to a 537 pH meter (WTW GmbH, Weilheim, Germany) and a μr-100 strip-chart recorder (Yokogawa Hokushin Electric, Tokyo, Japan). Fermentation was stopped at a pH of 4.60 by rapidly cooling to 6C. Yoghurt products were subjected to sensory analysis after storage for 2 days at 5C.

Sensory Evaluation

Panel Training and Score Sheet Development. Fourteen panelists familiar with sensory analyses contributed to preliminary panel sessions. In the first training schedule descriptive terms, their definitions and methods of assessments were developed by using commercially available set-style yoghurt.

To obtain good panel participation both quadrant rating technique and rank order technique (Zook and Pearce 1988) were applied. After four sessions of an approximative length of 2 h results of 4 panelists were found to be either inconsistent or deviating largely from those of all other panelists. Therefore, the final group consisted of 4 male and 6 female members aged between 26 and 38 years.

The sensory profile developed by the panelists included 8 sensory attributes and one hedonic scale. On the score sheet these were graded in order of perception. Unstructured scales were used in all attributes. Scales were 120 mm long and had marks 10 mm from either end. Verbal anchors defined and limited each attribute (Fig.1). Guidelines for sensory evaluation were worked out by the panel and, during subsequent sessions, provided on a separate sheet:

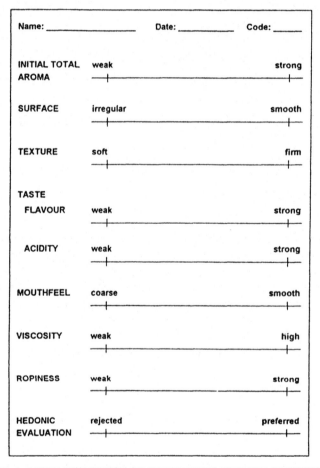

FIG. 1. SCORE SHEET USED IN SENSORY EVALUATION OF YOGHURT

Initial Total Aroma: Intensity of yoghurt-typical aroma (i.e, acetaldehyde) perceived by smelling immediately after opening the yoghurt beaker;

Surface: Perceived by visual inspection of yoghurt surface;

Texture: Gel firmness perceived by penetrating the gel with a teaspoon and removing an appropriate amount of yoghurt without exerting any shearing motion;

Taste, Flavor: Perceived intensity of yoghurt-typical flavor (slightly astringent due to content of acetaldehyde and minor flavor compounds) released within the mouth;

Taste, Acidity: Degree of acidity perceived in the mouth;

Mouthfeel: Degree of smoothness perceived by squeezing yoghurt between tongue and palate;

Viscosity: Perceived resistance against stirring with a teaspoon;

Ropiness: Perceived cohesiveness of the stirred product after pouring it from a teaspoon;

Hedonic Evaluation: Perceived acceptability after abstraction from all assessments made before.

Testing Procedure. Yoghurt beakers were placed in the room 1 h before testing which resulted in a product temperature of about 15C. Assessments were performed in a room of quiet atmosphere. As the testing room was not equipped with individual booths panelists were seated at two tables with their faces towards the wall and a side-by-side distance of about 1.5 m. Within each session 2-3 different samples encoded with 3-digit random numbers were subjected to analysis one after the other. Between evaluation of different samples panelists were requested to rinse their mouth with tap water.

Generally, each panelist had to perform 3 replicate assessments on each sample. Replicates were given in different sessions. Due to some absence not all out of 10 yoghurt samples included in this study were assessed by all of the panelists. The panel leader, however, focused attention on obligate conduction of replicates. The 21-27 individual assessments (no. of panelists × 3) were performed on each sample, and 21-30 assessments were performed by each subject. Therefore, the grand total was 231 assessments.

Statistical Analysis

For statistical evaluation of sensory results the General Linear Model (GLM) procedure of the SAS/Stat™ software package (SAS® Institute 1987) was used. Methods applied were ANOVA and multiple comparisons to describe panel performance, differences between products and correlations between sensory attributes. Simple and multiple linear regressions were performed by using the REG procedure.

RESULTS AND DISCUSSION

Panel Performance

In order to obtain information on individual reliability all replicate data of each assessor for each single descriptive term across all yoghurt products were pooled and tested statistically by two-way ANOVA (Powers 1988) with samples and replicates being treated as main effects. Whereas insignificant F-values for replicates were found in all cases probabilities of F-values for individual subjects caused by samples' effects are given in Table 2. According to Stone et al. (1974) any assessor whose F-value of performance has a probability ≤ 0.50 is contributing to discrimination. Only here and there an assessor had an F-value with a probability of $p > 0.2$ suggesting that the panelists had no problems in yoghurt evaluation. As proposed by Cross et al. (1978) F-values for all scales within each panelist were summed. The distribution of summed F-values did not indicate a clear cut-off level for panelist rejection, and only minor differences were found ($108.9 \leq \Sigma F \leq 142.8$). Regarding the whole panel only in texture assessments a global probability value of $p > 0.01$ was observed.

Estimation of Product Differences

Prior to estimation of product differences ratings of each panelist within each scale and across all yoghurts were standardized to zero mean and unit variance. This data treatment is a commonly used tool of overcoming part of the variance arising by disagreement among assessors about scale values assigned (Powers 1988). Consequently, this standardization is useful to separate errors caused by differences in scaling from those caused by scoring products in different order.

Three-factor ANOVAs (products, assessors, replications) were computed by using standardized sensory scores. Additionally, product-assessor interactions were included in the model. Exemplarily, ANOVA output for the viscosity scale is given in Table 3. By using the error term as denominator the respective F-value points out that yoghurt products differ significantly ($p < 0.001$). Effects of subjects and replications were, however, insignificant. The product × assessor — interaction term showed an F-value of 4.91 ($p < 0.001$) implying that the trend for mean viscosity scores was not followed by each subject (O'Mahony 1985). Effects of such interactions on between-products differences can be considered in calculating F-values by using the interaction mean square instead of the error mean square as denominator. Significantly remaining between-products F-values point to the existence of real product differences, which are not caused by artefacts (Stone et al. 1974, 1980).

TABLE 2.
PROBABILITY LEVELS ASSOCIATED WITH F-VALUES FOR INDIVIDUAL SUBJECTS AND SENSORY CATEGORIES ACROSS ALL YOGHURT PRODUCTS

Sensory category	Assessor										Panel
	1	2	3	4	5	6	7	8	9	10	
Aroma	<0.001	0.026	0.018	<0.001	0.005	<0.001	<0.001	<0.001	0.002	0.064	0.001
Surface	0.485	<0.001	0.121	<0.001	0.050	<0.001	<0.001	<0.001	0.009	0.008	<0.001
Texture	0.284	<0.001	0.004	0.104	0.257	0.021	0.060	<0.001	0.365	0.002	0.027
Flavour	0.001	0.007	0.079	0.024	0.028	0.025	0.007	0.001	<0.001	0.181	<0.001
Acidity	0.010	0.011	0.006	0.075	<0.001	<0.001	0.013	0.059	0.001	0.354	<0.001
Mouthfeel	0.037	0.093	0.123	<0.001	0.008	<0.001	<0.001	0.146	0.492	0.010	<0.001
Viscosity	0.002	<0.001	0.001	<0.001	<0.001	<0.001	0.004	<0.001	<0.001	<0.001	<0.001
Ropiness	<0.001	<0.001	0.003	<0.001	<0.001	<0.001	<0.001	<0.001	<0.001	<0.001	<0.001
Hedonic score	<0.001	<0.001	0.015	0.002	<0.001	<0.001	<0.001	0.001	<0.001	<0.001	<0.001

TABLE 3.
ANOVA RESULTS FOR VISCOSITY

Source	df	Mean square	F-value (error)	p > F	F-value (interaction)	p > F
Total	230					
Products	9	14.19	60.55	<0.001	12.33	<0.001
Assessors	9	0.32	1.35	0.216		
Replications	2	0.44	1.88	0.157		
Products x assessors	58	1.15	4.91	<0.001		
Error	152	0.23				

Despite standardization of sensory data significant between-subjects effects were observed. These were, however, valid only in 'Aroma' and 'Ropiness' scales at a significance level of $0.05 < p < 0.01$ and can mainly be attributed to different product ranking by different subjects (O'Mahony 1985). In all sensory properties evaluated between-products F-values and significance levels are given in Table 4. Testing against error mean square resulted in significant differences between products evident in all scales ($p < 0.001$). Hypothesis testing against interaction, however, decreased F-values and, therefore, increased error probability levels. Except texture (gel firmness) significance of F-values remained, which implies that product differences as such override any noise produced by interaction.

In sensory continua showing significant between-product effects standardized mean scores of yoghurts were related to each other by multiple comparison (Zook and Pearce 1988). Although ANOVA significance levels varied to some extent (cf. Table 4) minimum differences to be significant in the Duncan test were generally computed at $p < 0.05$ for sake of data clarity. In sensory properties showing significant differences yoghurt products were ranked (Table 5). It is evident that a large number of significant differences exist. Especially in yoghurt produced with starters H and I extreme ratings were observed. These will be discussed later in this paper.

Correlations Between Sensory Continua

The correlation matrix based on mean panel estimates for each scale and each product (Table 6) shows several significant relationships between sensory categories. As expected, an interrelationship between 'Aroma' and 'Flavor' scores was observed as both sensations are based on olfactory perceptions.

TABLE 4.
F-VALUES AND SIGNIFICANCE LEVELS OF SENSORY CONTINUA
OF YOGHURT

Sensory category	F-value (vs error)	p > F	F-value (vs interaction)	p > F
Aroma	7.38	<0.001	2.43	0.020
Surface	19.37	<0.001	5.91	<0.001
Texture	4.26	<0.001	1.32	0.245
Flavour	9.15	<0.001	2.28	0.029
Acidity	12.82	<0.001	4.90	<0.001
Mouthfeel	9.95	<0.001	3.44	0.002
Viscosity	60.55	<0.001	12.33	<0.001
Ropiness	237.65	<0.001	33.20	<0.001
Hedonic score	44.33	<0.001	5.91	<0.001

TABLE 5.
MEAN RANKS OF YOGHURT PRODUCTS IN DIFFERENT SENSORY CONTINUA

Sensory category	high score				Product code[a]				low score	
Aroma	I^a	D^{ab}	A^{ab}	B^{abc}	G^{abcd}	K^{bcde}	E^{cde}	C^{de}	F^{ef}	H^f
Surface	H^a	K^a	B^{ab}	I^{ab}	C^b	F^{bc}	G^{cd}	A^{de}	E^e	D^f
Flavour	I^a	C^{ab}	G^{bc}	A^{bcd}	F^{bcd}	E^{cd}	K^{cd}	D^{cd}	B^d	H^e
Acidity	A^a	C^{ab}	F^{abc}	B^{abc}	G^{bcd}	I^{cd}	D^{de}	K^{ef}	H^e	E^f
Mouthfeel	H^a	K^a	I^{ab}	C^{bc}	B^{bc}	F^{bc}	E^{bc}	A^{bc}	G^c	D^d
Viscosity	H^a	E^b	B^b	K^c	A^c	G^d	C^d	D^{de}	F^{de}	I^e
Ropiness	H^a	B^b	E^c	K^d	G^e	A^e	I^e	F^f	D^f	C^f
Hedonic score	I^a	G^b	A^b	C^b	F^b	D^c	E^{cd}	K^{cd}	B^d	H^e

[a] Products within a line showing the same superscript are not significantly different in the Duncan test (P<0.05).

Aroma responses result from volatiles diffusing from the food which are drawn through the nose and detected by olfactory receptor cells whereas, in case of flavor, chewing movements cause tastants to stimulate gustatory and trigeminal receptors throughout the mouth and cause food volatiles to be carried up to olfactory and nasal trigeminal receptors via the retronasal route (Maruniak 1988; Mozell 1988). 'Surface'-'Mouthfeel' interrelations might be explained by the fact that a coarse or grainy structure results in an irregular surface which can be observed visually. The significant interrelationship between 'Viscosity' and 'Ropiness' may be attributed to extracellular polysaccharides produced by starter bacteria (Schellhaass and Morris 1985; Tamime et al. 1984) which increase scores of both sensory categories.

Although acidified to a constant pH of 4.6 yoghurt samples showed largely differing sensorically perceived acidity values. These differences can be explained by starter-specific overacidification (Kneifel et al. 1993) as well as by starter-specific degree of proteolysis (Kneifel et al. 1992b). Positive correlations between 'Acidity' and 'Flavor' scales may be attributed to varying metabolic activity of starter bacteria. Besides flavor components such as diacetyl or acetone acetaldehyde is known to act as a key substance in classical yoghurt products (Tamime and Deeth 1980; Tamime and Robinson 1985). (Over)acidification is accompanied by concomitant enrichment of acetaldehyde. It has been shown that instrumentally evaluated acidity values as well as sensory flavor scores are positively correlated with acetaldehyde content (Ulberth and Kneifel 1992).

TABLE 6.
PEARSON CORRELATION COEFFICIENTS FOR EACH SCALE WITH OTHER SCALES

Sensory category	Aroma	Surface	Texture	Flavour	Acidity	Mouth-feel	Viscosity	Ropiness
Aroma	1.000[a]	-0.248	-0.461	0.646	0.358	-0.360	-0.492	-0.383
Surface		1.000	0.126	-0.321	-0.192	0.754	0.383	0.461
Texture			1.000	-0.043	-0.065	0.586	0.246	0.249
Flavour				1.000	0.690	-0.317	-0.775	-0.724
Acidity					1.000	-0.259	-0.693	-0.638
Mouthfeel						1.000	0.516	0.594
Viscosity							1.000	0.974
Ropiness								1.000

[a] Probability > |R| under H_0 $R(H_0) = 0$ (two-tailed): $|R| \geq 0.632$, $p < 0.05$; $|R| \geq 0.766$, $p < 0.01$; $|R| \geq 0.873$, $p < 0.001$.

Significant inverse correlations exist between 'Acidity' and 'Flavor' on the one hand and 'Viscosity' and 'Ropiness' on the other hand. These findings are in line with some literature reports on reduced intensities of taste and flavor perceptions in foods with increased viscosity (Moskowitz and Arabie 1970; Pangborn and Szczesniak 1973; Pangborn *et al.* 1973) and have been explained by diffusion coefficients of dissolved substances varying with viscosity (Kokini 1987; Cussler *et al.* 1979).

Relation of Hedonic Scoring to Sensory Categories

The final scoring panelists had to make during sensory evaluation of yoghurt regarded general hedonic impression or, more simply, acceptability. Generally, combined sensory evaluation of sensory categories and acceptability by a trained panel is undesirable due to the panelist's high experience, distorted data may result. Therefore, only trends can be read from their results which have to be validated by using consumer panels.

Correlations between descriptive and hedonic scoring were calculated by using the panelist's standardized mean values. In order of decreasing coefficient magnitude significant correlations ($p < 0.01$) in simple models were found in case of 'Flavor' (+), 'Ropiness' (−), 'Viscosity' (−), 'Acidity' (+) and 'Aroma' (+). In multiple 2-parameter models a number of combinations with significant correlation coefficients was found. Best fit was obtained by relating acceptability to flavor and ropiness scores ($r=0.708$, $p < 0.001$). As standardized scores were used in calculation no significant intercept estimate was found. Parameter estimates for 'Flavor' and 'Ropiness' were 0.444 and -0.429 ($p < 0.001$), respectively. By using the 'Extra-sum-of-squares' principle (Pike 1986) inclusion of any additional descriptor did not increase the variance explained by the model significantly. Predicted versus measured acceptability scores are depicted in Fig. 2. It is evident that the predictive equation clearly fits the particular data set.

By using 4 yoghurt samples both acceptability and continua scores are shown in a circular plot (Fig. 3). Considering the comments made by Powers (1988) angles in the spider-web diagram were spaced equal. This diagram visualizes product differences outlined in Table 5 and results from regression analyses between sensory scales as well as between sensory scales and acceptability. Apart from 'Surface' and 'Mouthfeel' which showed no relation to acceptability, increasing hedonic scores are accompanied by increasing 'Aroma' and 'Flavor' scores. Regarding 'Acidity' low scores are in line with poor acceptability. However, product C with significantly higher acidity in sensory evaluation (cf. Table 5) showed medium acceptance. Products with high 'Viscosity' and 'Ropiness' scores were generally rated poor in acceptability.

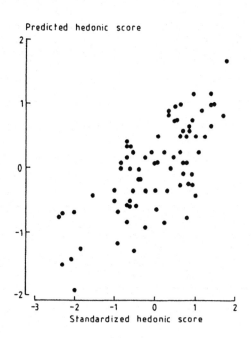

FIG. 2. STANDARDIZED PREDICTED VERSUS STANDARDIZED
ACTUAL SCORES FOR ACCEPTABILITY OF YOGHURT

Especially yoghurt H which was produced by using a highly mucogenic starter was rejected in hedonic analysis.

Originally, starter H was provided in freeze-dried form. It has been observed in another study (Schmid 1992) that yoghurt products lose its typical viscosity and mucogenic character when produced from a deep-frozen and rethawed mother culture instead of using the genuine freeze-dried inoculum. When producing such a 'modified' culture I from starter H we observed that the two resulting yoghurt products showed differing sensory properties. Apart from 'Viscosity' and 'Ropiness' this is also valid in case of 'Aroma', 'Flavor' and 'Acidity'. Surprisingly, this yoghurt has been found to be the most preferred one.

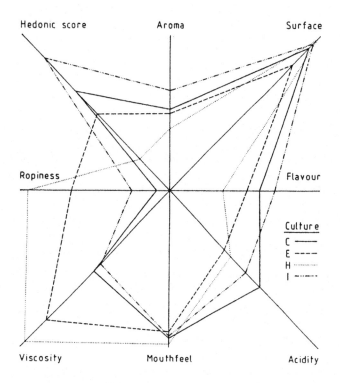

FIG. 3. SPIDER-WEB DIAGRAM SHOWING SENSORY DIFFERENCES
BETWEEN SELECTED YOGHURT PRODUCTS
Distance from center to the end of the lines represents full scale length.

ACKNOWLEDGMENT

The study is part of the research grant P7318 supported by the Austrian Science Foundation, Vienna.

REFERENCES

BARNES, D.L., HARPER, S.J., BODYFELT, F.W. and MCDANIEL, M.R. 1991. Correlation of descriptive and consumer panel flavor ratings for commercial prestirred strawberry and lemon yogurts. J. Dairy Sci. 74, 2089-2099.

CROSS, H.R., MOEN, R. and STANSFIELD, M.S. 1978. Training and testing of judges for sensory analysis of meat quality. Food Technol. *32*, 48-54.

CUSSLER, E.L., KOKINI, J.L., WEINHEIMER, R.L. and MOSKOWITZ, H.R. 1979. Food texture in the mouth. Food Technol. *33*, 89-92.

HARPER, L.J., BARNES, D.L., BODYFELT, F.W. and MCDANIEL, M.R. 1991. Sensory ratings of commercial plain yogurts by consumer and descriptive panels. J.Dairy Sci. *74*, 2927-2935.

KNEIFEL, W. 1992. Starterkulturen fuer Sauermilchprodukte. Ernaehrung/Nutrition *16*, 150-156.

KNEIFEL, W., JAROS, D. and ERHARD, F. 1993. Microflora and acidification properties of yogurt and yogurt-related products fermented with commercially available starter cultures. Int. J. Food Microbiol. *18*, 179-189.

KNEIFEL, W., KAUFMANN, M., FLEISCHER, A. and ULBERTH, F. 1992a. Screening of commercially available mesophilic dairy starter cultures: Biochemical, sensory, and microbiological properties. J. Dairy Sci. *75*, 3158-3166.

KNEIFEL, W., ULBERTH, F., ERHARD, F. and JAROS, D. 1992b. Aroma profiles and sensory properties of yogurt and yogurt-related products. 1. Screening of commercially available starter cultures. Milchwissenschaft *47*, 362-365.

KOKINI, J.L. 1987. The physical basis of liquid food texture and texture-taste interactions. J. Food Eng. *6*, 51-81.

KOKINI, J.L., POOLE, M., MASON, P., MILLER, S. and STIER, F. 1984. Identification of key textural attributes of fluid and semisolid foods using regression analysis. J. Food Sci. *49*, 47-50.

MARUNIAK, J.A. 1988. The sense of smell. In *Sensory Analysis of Foods* (J.R. Piggott, ed.) pp. 25-68, Elsevier, Barking, UK.

MCGILL, A.E.J. 1983. Evaluation and prediction of the consumer acceptability of commercially manufactured yogurt. South African J. Dairy Technol. *15*, 139-140.

MECREDY, J., SONNEMAN, J.C. and LEHMAN, S.J. 1974. Sensory profiling of beer by a modified QDA method. Food Technol. *28*, 36-37, 40-41.

MOSKOWITZ, J.R. and ARABIE, P. 1970. Taste intensity as a function of stimulus concentration and solution viscosity. J. Texture Studies *1*, 501-510.

MOZELL, M.M. 1988. Olfaction. In *Sensory Systems II. Senses Other than Vision* (G. Adelman, ed.) pp. 59-62, Birkhaeuser, Boston.

MUIR, D.D. and HARPER, E.A. 1992. Sensory evaluation of fermented milks: vocabulary development and the relations between sensory properties and composition and between acceptability and sensory properties. J. Soc. Dairy Technol. *45*, 73-80.

O'MAHONY, M. 1985. *Sensory Evaluation of Food*. Marcel Dekker, New York.

PANGBORN, R.M. and SZCZESNIAK, A.S. 1973. Effects of hydrocolloids and viscosity on flavor and odor intensities of aromatic flavor compounds. J. Texture Studies *4*, 467-482.

PANGBORN, R.M., TRABUE, I.M. and SZCZESNIAK, A.S. 1973. Effects of hydrocolloids on oral viscosity and basic taste intensities. J. Texture Studies *4*, 224-241.

PIKE, D.J. 1986. A practical approach to regression. In *Statistical Procedures in Food Research* (J.R. Piggott, ed.) pp. 61-100, Elsevier, Barking, UK.

POWERS, J.J. 1988. Current practices and application of descriptive methods. In *Sensory Analysis of Foods* (J.R. Piggott, ed.) pp. 187-266, Elsevier, Barking, UK.

RICHTER, R.L. 1979. Results of the 1978 American dairy products scoring clinic. Cult. Dairy Prod. J. *13*, 15-17.

SAS® Institute. 1987. *SAS/Stat^{TM} Guide for personal computers*. Version 6 Edition. SAS® Institute, Cary, NC.

SCHELLHAASS, S.M. and MORRIS, H.A. 1985. Rheological and scanning electron microscopic examination of skim milk gels obtained by fermenting with ropy and non-ropy strains of lactic acid bacteria. Food Microstructure *4*, 279-287.

SCHMID, W. 1992. Vergleichende chemisch-physikalische, mikrobiologische und sensorische Charakterisierung von stichfestem Joghurt, erzeugt mit verschiedenen Trockenmasseanreicherungssubstraten. Diploma Thesis, University of Agriculture, Vienna.

STONE, H. and SIDEL, J.L. 1985. *Sensory Evaluation Practices*. Academic Press, Orlando, FL.

STONE, H., SIDEL, J.L. and BLOOMQUIST, J. 1980. Quantitative descriptive analysis. Cereal Foods World *25*, 642-644.

STONE, H., SIDEL, J.L., OLIVER, S., WOOLSEY, A. and SINGLETON, R.C. 1974. Sensory evaluation by quantitative descriptive analysis. Food Technol. *28*, 24-34.

TAMIME, A.Y., DAVIES, G. and HAMILTON, M.P. 1987. The quality of yoghurt on retail sale in Ayrshire. Part II. Organoleptic evaluation. Dairy Ind. Int. *52*, 40-41.

TAMIME, A.Y. and DEETH, H.C. 1980. Yogurt: Technology and biochemistry. J. Food Prot. *43*, 939-977.

TAMIME, A.Y., KALAB, M. and DAVIES, G. 1984. Microstructure of set-style yoghurt manufactured from cow's milk fortified by various methods. Food Microstructure *3*, 83-92.

TAMIME, A.Y. and ROBINSON, R.K. 1985. *Yoghurt. Science and Technology*. Pergamon Press, Oxford.

ULBERTH, F. and KNEIFEL, W. 1992. Aroma profiles and sensory properties of yogurt and yogurt related products. II. Classification of starter cultures by means of cluster analysis. Milchwissenschaft 47, 432-434.

ZOOK, J.L. and PEARCE, J.H. 1988. Quantitative descriptive analysis. In *Applied Sensory Analysis of Foods, Volume I* (H.R. Moskowitz, ed.) pp.43-71, CRC Press, Boca Raton, FL.

CHAPTER 2.6

THE USE OF STANDARDIZED FLAVOR LANGUAGES AND QUANTITATIVE FLAVOR PROFILING TECHNIQUE FOR FLAVORED DAIRY PRODUCTS

CHANTAL R. STAMPANONI

At the time of original publication, the author was affiliated with Givaudan-Roure Flavors Ltd. 8600 Dubendorf, Switzerland.

ABSTRACT

Sensory terminology is very important in descriptive analysis and perceptions are greatly influenced by the language. Definitions can be very useful for specifying and describing a sensation. However, many sensory attributes are not easily defined and physical reference standards can contribute a great deal to smoothing language problems. They can be used to develop the proper descriptive language, to reduce the amount of time required to train the sensory subjects, and to calibrate the panel in the use of the intensity scale, all the while providing documentation of the sensory terminology.

The Quantitative Flavor Profiling (QFP) technique was applied to evaluate flavored cheese analog, yoghurt and sweetened milk samples. Specific standardized flavor languages were developed for each product type and included reference standards for each sensory descriptor. The results of QFP were analyzed by principal component analysis.

INTRODUCTION

Cultural differences influence the language utilized to describe taste, odor and flavor. For example, sweet is "amai" in Japan, "dolce" in Italy, "dulce" in Spain, "süss" in Germany, and "maru" in Tahiti. In other countries, for example Uganda, "kuwoma" means sweet but also salty and pleasant and, in Fiji, "kami-kamidha" means both sweet and succulent. With other taste sensations, it is even more ambiguous: in Tahiti, "avaáva" means both acid and bitter. This very limited taste language generates confusion. Similarly, for odor and flavor description, sensory attributes definition and term selection tend to differ from country to country. Risvik *et al.* (1992) compared results from descriptive analysis of milk chocolate samples, using an English and a Norwegian trained panel. The data indicated differences in the terminology used to profile the products.

Civille and Lawless (1986) underlined the importance of the terminology in sensory evaluation and reported that perceptions are greatly influenced by the language used. In descriptive analysis, sensory dimensions should convey information to show differences and similarities among samples. As sensory attributes should be abstracted, i.e., conceptualized, analytical judges should have their sensory concepts aligned (Ishii and O'Mahony 1990). Each sensory subject must understand and be able to use the language, whose descriptors should be defined and anchored in a similar manner. If this is not the case, sensory concepts will be idiosyncratic (Miller and Johnson-Laird 1976).

Except for the "umami" taste concept, for which Ishii and O'Mahony (1990) found that it was defined very similarly for Japanese and American subjects, using either verbal descriptions or reference standards, physical references can contribute a great deal to smoothing language problems. Rainey (1986) emphasized the importance of using standards for developing the proper descriptive language and to reduce the amount of time to train the sensory subjects, all the while providing documentation of the sensory terminology. She also suggested that references can be used to establish intensity ranges and to show the effect of a specific ingredient. Stampanoni (1993a,b) indicated that references are helpful to decrease judge variability, to counterbalance cultural differences in interlaboratory studies, and to allow calibration of the panel in the use of the intensity scale.

Physical references or standards are chemicals, spices and extracts, ingredients or finished products, and are used to specify a selected characteristic of a food or nonfood item (Rainey 1986; Stampanoni 1993a,b). References should be simple, reproducible, diluted without changing character (Rainey 1986), very "clear" to the subjects, and very specific, i.e. illustrate only a single sensory descriptor. Not all attributes are so easily described by an ideal reference; sometimes a single standard is not enough for proper concept alignment. O'Mahony (1991) and Ishii and O'Mahony (1987, 1990) showed that the use of several physical standards was even superior to using only one.

In spite of its importance in descriptive analysis, the choice of descriptors and the establishment of standardized sensory languages have received too little attention. The majority of the authors do not discuss or specify the references used to train the panel and to anchor the sensory dimensions. In this paper, the Quantitative Flavor Profiling method (Stampanoni 1993 a,b,c) was applied to show three concrete examples in which three flavor languages were created, standardized using physical references and tested on finished flavored products. This technique employs nonambiguous standardized flavor languages to try to minimize language problems and cultural differences in descriptive language.

MATERIALS AND METHODS

Sample Preparation

Strawberry Flavored Yoghurt. Fifteen strawberry flavors (Givaudan-Roure Flavors Ltd., Switzerland) were applied in sweetened (11 g sucrose/100 g yoghurt) full fat (3.8% fat) yoghurts. For 100 g yoghurt, skim milk powder (2 g) and gelatine 240 Bloom (0.3 g) were added to the mix, which after pasteurization was incubated (3-4 h at 42C, final pH 4.5) with a culture of *Lactobacillus bulgaricus* and *Streptococcus thermophilus* (Laboratorium Wiesby GMBH & Co. KG, Germany). The flavors chosen were felt to be different from one another and to be representative enough for the development of a quite comprehensive language covering all major flavor dimensions (Civille 1987).

The samples were tested at 8C and were served in randomized order in 3-digit coded, hermetically closed, 40 mL plastic portion cups (Pakoba AG Switzerland).

Caramel Flavored Milk Drink. As explained in the section on strawberry flavors, 15 caramel flavors (Givaudan-Roure Flavors Ltd., Switzerland) were applied to pasteurized milk drink (3.8% fat) and were sweetened with sucrose (50 g/1000 mL milk). For 1000 mL milk, a blend of plant hydrocolloids (0.2 g, Meyprogen GN, Meyhall Chemical AG, Switzerland) was added for the stabilization of milk drinks. The samples were tested at 15C and were served in randomized order in 3-digit coded 100 mL plastic cups (Pakoba AG, Switzerland).

Flavored Cheese Analog and Flavored Fresh Cheese. Eight cheese flavors (type Mozzarella, Camembert, Blue, Cheddar, Emmental, and Parmesan by Givaudan-Roure Flavors Ltd., Switzerland) were applied in a cheese analog base and tested at 15C. A 100 g portion of cheese analog contained the following ingredients: 22.6 g margarine (83% fat, Margo AG, Switzerland), 25 g rennet casein (Alaren 779, New Zealand Dairy Board, New Zealand), 5 g skim milk powder, 3.5 g stabilizer (Kasomel 2394, Rhone-Poulenc, France), 1 g NaCl and 0.5 g lactic acid. For the preparation of the samples, water was added to the melted margarine, stirred to a premix at 40C and heated to 75C. Casein, salt, stabilizer and flavor were then dissolved in the premix. After pasteurization [83C, 4 min, while stirring (1500 rpm)], lactic acid was added into the mix, which was then stirred under vacuum (3000 rpm, 2 min, 0.4 bar). The samples were filled into plastic portion cups (Pakoba AG, Switzerland), cooled to room temperature, sealed and stored at 6C for 3 days. They were served at 10C in randomized order in 3-digit coded 40 mL plastic portion cups (Pakoba AG, Switzerland).

Additionally, goat and sheep cheese flavors were applied in Philadelphia Cream Cheese (28% fat, Kraft General Foods) and tested at 10C in 3-digit coded 40 mL plastic portion cups (Pakoba AG, Switzerland).

Sensory Evaluation

Quantitative Flavor Profiling of the Strawberry Yoghurts. The Quantitative Flavor Profiling (QFP) method (Stampanoni 1993a,b), which is a modification of the QDA technique (Stone *et al.* 1974, 1980) and focuses on the quantification of flavor notes, was used to evaluate the samples. QFP uses a limited number of trained subjects and a standardized flavor language that is developed in several iterative round-table sessions by 6-8 experienced subjects (generally flavorists) who will not participate in the formal testing but receive all samples to be profiled by the trained judges. In this way, the trained sensory subjects do not have any preconceived ideas about the product characteristics since they do not participate in the language development phase. In addition, the panel will not develop erroneous terms, because of lack of formal instruction, technical background or direction, and, as indicated by Civille and Lawless (1986), experienced subjects may use specific terms that are ignored by the average person.

The standardized flavor language consists of a list of words that are defined and linked to standards illustrating their quality. References anchor the 100 mm unstructured graphic scales (ranging from "none" to "high"), mostly at about 75% intensity.

Fifteen Givaudan-Roure trained judges (Stampanoni 1993a) (19-55 years of age) attended 7 further training sessions in which they became familiar with the sensory methodology, the samples and the descriptive language by smelling selected standards and by participating in 8 data-collection sessions in individual booths. The samples were tested in a balanced, completely randomized experimental design, in 2 replications. In the design, 4 randomly selected samples were served to each judge on the first day, another 4 on a second and third day and the last 3 on the fourth day. Each subject had a different set of samples. For the second replication, the judges received 4 more subsets in the following 4 days.

Testing was conducted in separate booths at about 20C, under normal fluorescent illumination. The subjects were allowed to swallow the samples and rinsed with tap water (15C) between samples.

Quantitative Flavor Profiling of the Caramel Milk Drinks. In this experiment, 12 trained subjects (Stampanoni 1993a) attended 6 further training and 6 data-collection sessions. The QFP technique (Stampanoni 1993a,b) was used to describe and evaluate caramel flavor notes.

Testing was conducted as explained in the section on strawberry flavors and used a completely randomized design in which 5 randomly selected samples were tested per session. Each judge had a different set of samples. For the second replication, the subjects received 3 more subsets in the following 3 days.

Quantitative Flavor Profiling of the Cheese Analogs and the Fresh Cheeses. The cheese flavored samples were evaluated using the QFP technique (Stampanoni 1993a,b) by 11 trained subjects (Stampanoni 1993a), who had previously attended 6 further training and 6 data-collection sessions.

Testing was conducted as explained in the section on strawberry flavors, and utilized a completely randomized design in which fresh and soft cheese flavored analogs were tested in a first session. In a second session, hard cheese flavored analogs were evaluated and noncow milk flavored cream cheese samples were tested in a third session. The experiment was replicated twice and samples were served in a balanced random order.

Data Analysis

The sensory responses were digitized using a Summagraphic Summa-Sketch II Digitizer (Tragon Corporation, Redwood City, CA) and quantified by measuring the distance in mm from the left end of the graphic scale to the mark placed by the judges. Data analysis was performed using the Statistical Analysis System (SAS 1988). Correlations and principal component analyses (PCA) were performed using Proc Factor (SAS 1988) on the correlation matrix of the sample means from Quantitative Flavor Profiling, with no rotation.

RESULTS AND DISCUSSION

In the area of sensory evaluation, there has been a considerable amount of research in descriptive analysis methods to study flavor characteristics (Cairncross and Sjöström 1950; Caul 1957; Meilgaard 1987; Stone *et al.* 1974). Johnsen and Civille (1986) proposed a standard lexicon for warmed-over meat flavor and Johnsen *et al.* (1988) developed a lexicon for peanut flavor. However, standard references were mostly used to align concepts in the training phase and most of the attributes were specified only by means of written definitions.

Only quite recently, efforts were concentrated on the development of precise vocabulary for flavor descriptive analysis systematically employing physical standard stimuli to anchor descriptive terms. For example, Meilgaard *et al.* (1982) created a standardized language for beer flavor, whereas written definitions and selected references were employed to describe the flavor of pond-raised catfish (Johnsen *et al.* 1987), sherbets and noncarbonated beverages (Stampanoni 1993c). The same approach was used by Lyon (1987) for chicken flavor. Very comprehensive work with physical references for each descriptor was performed by Noble *et al.* (1987), who proposed a wine aroma wheel to standardize wine aroma terminology, whereas McDaniel *et al.* (1987) concentrated on physical standards for descriptive analysis of pinot noir, and Noble and Shannon (1987) profiled Zinfandel wines. Heisserer and Chambers (1993) employed a highly trained descriptive panel to identify, define and reference 30 flavor attributes for cheese.

Language Development

In each experiment, descriptors were freely generated in iterative sessions (max. 3 h per language) with the objective of creating a flavor language allowing the most complete, however not redundant, description of differences and similarities among the products under study. As explained by Stampanoni (1993a,b), 8 flavorists were served with representative sets of coded samples that were described by recording descriptors, then compiled into a master list. This individual preliminary evaluation of the samples was complemented by moderated round-table discussions, whereby new attributes could be added and physical references were discussed. The three final standardized flavor languages are summarized in Tables 1-3. In each example, the 5 "basic tastes" and the trigeminal sensations were not included in the final descriptive terminology because they were found irrelevant by the sensory subjects. Twelve attributes were necessary to show differences and similarities among the strawberry flavored yoghurts (Table 1). The 15 terms employed to describe the caramel flavors were derived from a much larger initial list of 31 attributes. It included terms such as woody, phenolic, coffee, chocolate, nutty, almond, malty, cotton candy, spicy-balsamic, custard and egg-yolk, which were eliminated by agreement because nonpertinent or redundant (caramel, malty, brown sugar were all related to burnt sugar, which was used in the final language, Table 2). The cheese language generated for the 8 cheese flavors included 27 flavor descriptors (Table 3), which were quite similar to those reported by Heisserer and Chambers (1993), even if the present study was completed prior to its publication. Butyric acid was used to describe a "rancid" note, which was called butyric acid and "baby vomit" by Heisserer and Chambers (1993). The "spicy-pungent" note was not related to trigeminal effects similar to those caused by horseradish, but it was defined as the "typical pungent note present in ripe Appenzell or Tilsit cheese".

TABLE 1.
STANDARDIZED FLAVOR LANGUAGE FOR THE QUANTITATIVE DESCRIPTION
OF STRAWBERRY FLAVORS USING THE QFP TECHNIQUE:
TERMINOLOGY, REFERENCES AND THEIR INTENSITIES
(TA = Triacetin, PG = Propylene glycol, VO = Vegetable oil, CH = Switzerland).

FRUITY-GREEN	TRANS-2-HEXENAL (0.5%TA)
GRASS-GREEN	CIS-3-HEXENOL (1%PG)
RIPE-FRUITY	ETHYL-BUTYRATE (1%PG)
JAMMY-COOKED	STRAWBERRY JUICE CONCENTRATE (PURE)
	STRAWBERRY JAM 42% FRUIT (MIGROS-BRAND, CH)
FLORAL	LINALOOL (1%PG) + HEDIONE (0.1%PG), 1:1
ESTERY-CANDY	AMYL-ISO-ACETATE (1%PG)
RAISIN	DAVANA OIL (1%TA)
SWEET-COTTON CANDY	ETHYLMALTOL (PURE)
VANILLA	ETHYL-VANILLIN (PURE)
BUTTERY	DIACETYL (1%PG)
CREAMY	UHT CREAM 35% FAT (VALFLORA-BRAND, CH)
MUSTY	COLA-INFUSION IN ETHANOL (PURE)
METHYL-ANTRANILATE	METHYLANTRANYLATE (1%PG)
JUICY	ACETALDEHYDE (10PPM H2O) (BY MOUTH)
HAY-LIKE	BLACK TEA (LIPTON,YELLOW LABEL QUALITY No.1), 3 MONTHS OLD
	HAY-OIL (10%VO)
LACTONY	GAMMA-DECALACTONE (1%PG)
JAMMY-COOKED	Weak Strong --+--------------========---------+--
OTHER NOTES	--+-------------------------========+--

The selection of terms is a difficult problem in sensory analysis (Meiselman 1993). Although the sensory subjects were encouraged to use precise, primary and orthogonal terms, more related to the perception rather than the stimulus (Civille 1987), the words employed to describe flavors were related to a stimulus (e.g., phenolic, lactony or methyl-anthranilate), a sensation (e.g., spicy-balsamic or pungent), a treatment (e.g., roasted or jammy-cooked), an abstract concept (e.g., green, blue or floral) and a concrete concept (e.g., honey, walnut or cottage cheese). Johnsen et al. (1987) and Civille (1987), suggested using descriptive terms illustrating specific and common flavors rather than a presumable causative agent. Whereas, Civille (1987) supported the utilization of "specific" terms (e.g., citral or methyl-anthranilate) rather than "general" ones (e.g., candy-like lemon, wild-note). In addition, Civille and Lawless (1986) indicated that "primary" attributes, also called "elemental" terms, reduce confusion among judges. "Integrated" or complex descriptors might be difficult because each subject might weight the underlying "primary" terms differently. As demonstrated by Gestalt psychologists, subjects tend not to be analytical in their perception (Garrett 1930). Thus, complex words are, in general, used by naive observers who tend to represent stimuli holistically

(Sternberg 1986). For example, green is an integrated word and needs to be separated into more primary attributes such as green-grass, green-fruity, green-melon, green-cucumber. Juicy or fresh are also considered integrated words. However, it does not necessarily matter what a sensation is called (be geosmin, earthy, forest or 57) as long as the subjects feel comfortable with it, agree on the type of sensation to associate with the label, and conceptualize the sensation labelled equivalently.

As shown in Tables 1-3, each attribute was related to a specific physical reference which illustrated the attribute quality and a specific intensity range on the graphic scale. Thus, as indicated by Johnsen *et al.* (1987) and Civille (1987), the subjects had precise standards with which to compare the samples under study. This step helps the calibration of the sensory scale so that sensory judgments are less influenced by cultural differences (Stampanoni 1993a,b). In fact, Rainey (1986) sustained that, without reference standards, "intensities are subjective" because "how strong is strong?"

TABLE 2.
STANDARDIZED FLAVOR LANGUAGE FOR THE QUANTITATIVE DESCRIPTION OF CARAMEL FLAVORS USING THE QFP TECHNIQUE:
TERMINOLOGY, REFERENCES AND THEIR INTENSITIES
(PG = Propylene glycol, CH = Switzerland).

CREAMY	UHT CREAM 35% FAT (VALFLORA-BRAND)
BUTTERY	DIACETYL (1%PG)
CONDENSED MILK	CONDENSED MILK (SWEETENED, MIGROS-BRAND)
LACTONY	DELTA-DECALACTONE (1%PG)
COCONUT	GAMMA-NONALACTONE (1%PG)
WALNUT	FRESH WALNUTS (SUNRAY-BRAND, US)
	NORKNORRLI by GIVAUDAN-ROURE (0.1%PG)
POWDERY	HELIOTROPIN (10%PG, WARM UP TO SOLVE)
VANILLA	ETHYL-VANILLIN (PURE)
COUMARIN	DIHYDRO-COUMARIN (1%PG)
BURNT SUGAR (SUGARY)	HOMOFURONOL (20%PG)
BURNT-SMOKY	5-METHYL FURFURAL (5%PG)
ROASTED	2-ETHYL-3,5or6-DIMETHYL PYRAZINE (0.1%PG)
CITRUS	LEMON OIL ITALY (PURE) +
	ORANGE OIL BRAZIL (PURE), 1:1
RUM/ALCOHOLIC	RUM ETHER-ALCOHOLIC, PURE
HONEY	78845-33 HONEY FLAVOR by GIVAUDAN-ROURE (5%PG)
	METHYL-PHENYL-ACETATE (1%PG)
CREAMY & CONDENSED MILK	Weak Strong --+---------------========---------+--
OTHER NOTES	--+----------------------========+--

TABLE 3.
STANDARDIZED FLAVOR LANGUAGE FOR THE QUANTITATIVE DESCRIPTION OF CHEESE FLAVORS USING THE QFP TECHNIQUE: TERMINOLOGY, REFERENCES AND THEIR INTENSITIES
(PG = Propylene glycol, CH = Switzerland)

CHEESE GENERAL	
MILKY	PASTEURIZED MILK 3.6%FAT (VALFLORA-BRAND)
COOKED MILK	UHT MILK 3.6%FAT (MIGROS-BRAND), COOKED (10')
FATTY	PALM KERNEL FAT (FLORIN AG-BRAND, CH)
BUTTERY	PASTEURIZED COOKING BUTTER (VALFLORA-BRAND) DIACETYL (1%PG)
CREAMY	UHT CREAM 35% FAT (VALFLORA-BRAND)
NUTTY	ROASTED PEANUTS (VACUUM-SEALED, SUNQUEEN) + GROUND HAZELNUTS (SUNQUEEN, CH) + GROUND ALMONDS, (SUNQUEEN), 1:1:1 10005-73 NUT BASE by GIVAUDAN-ROURE, 10%PG
FRESH CHEESE	
BUTTER MILK	PASTEURIZED BUTTER MILK (MIGROS-BRAND)
YOGHURT	YOGHURT 3.2% FAT (MIGROS-BRAND)
COTTAGE CHEESE	COTTAGE CHEESE 25%FAT (MIGROS-BRAND)
CASEINATE	SODIUM-CASEINATE-POWDER (EMMI-BRAND, CH)
WHEY	WHEY POWDER (BIOREX, SERAMO-PRODUKTE, D)
SOAPY	LAURIC ACID (PURE)
SOFT CHEESE	
FERMENTED	FERMENTED MILK 12%FAT (M-DESSERT,MIGROS-BRAND)
MUSHROOM	1-OCTEN-3-OL (1%PG)
EARTHY	GEOSMIN (0.001%PG)
MUSTY	COLA-INFUSION IN ETHANOL (PURE)
SPICY-PUNGENT	VALERIAN ACID (1%PG)
BLUE	OCTAN-2-ONE (1%PG)
AMMONIA	AMMONIA SOLUTION (0.25%H20)
GREEN-GRASS	CIS-3-HEXENOL (1%PG)
HARD CHEESE	
CHEESE RIND	CHEESE RIND (TILSIT MILD, PASTEURIZED FULL FAT, MIGROS-BRAND)
PROPIONIC ACID	PROPIONIC ACID (1%PG)
CAPRIC ACID	CAPRIC ACID (PURE)
BUTYRIC ACID	BUTYRIC ACID (1%PG)
SOAPY	LAURIC ACID (PURE)
FRUITY	CANNED FRUIT SALAD (IN SYRUP, M-QUEEN, MIGROS)
GOAT/SHEEP	
FERMENTED	FERMENTED MILK 12%FAT (M-DESSERT,MIGROS-BRAND)
CHEESE RIND	CHEESE RIND (TILSIT MILD, PASTEURIZED FULL FAT, MIGROS-BRAND)
MUSTY	COLA-INFUSION IN ETHANOL (PURE)
SWEATY	BUTYRIC-ISO-ACID (5%PG)
CAPRIC ACID	CAPRIC ACID (PURE)
BUTYRIC ACID	BUTYRIC ACID (1%PG)
ANIMAL	4-METHYL-OCTANOIC ACID (2%PG)
SOAPY	LAURIC ACID (PURE)
OTHER REFERENCES: MOLECULES:	Weak Strong --+--------------========-------+-- --+-----------------------========+--

Written definitions were found to be redundant and were only used for some difficult attributes (e.g., methyl-anthranilate: "the typical note present in wild strawberries and not in cultivated ones" or lactony: "fruity-aromatic character generated by specific lactons") in the training phase. The descriptive panel was familiarized with the flavor terms by smelling (or tasting, for market samples) the references repeatedly, as the most efficient method to describe flavor to a person is to have the subject taste or smell it. O'Mahony *et al.* (1990) indicated that a color cannot be described to a blind man. In fact, a color can only be communicated by showing objects of that color and then labelling them. The judges were trained with reference standards to achieve alignment of their concepts. In addition to reference sets, the subjects had the possibility to discuss the references. In research on conceptualization, Laughlin and Dogherty (1967) showed that group discussions facilitated learning.

References should be very clear and specific, i.e. illustrate only a single specific sensory descriptor, otherwise they are confusing. For example, in an experiment (Stampanoni *et al.* 1993), 39 male and 28 female subjects performed odor recognition tests with 28 different flavor components. The untrained subjects were instructed to smell the odor, then to freely describe what they smelled. The results showed that the green-grass cis-3-hexenol [1% in propylene glycol (PG)] was described as grassy or leafy by a significant percent of the subjects (65.7%, $p<5$%). Thus, cis-3-hexenol, in that concentration, was a very specific and "singular" reference for green-leafy/grass. Whereas, trans-2-hexenal (1% PG), mostly described as green-fruity or green-apple by flavorists, was found to be green-fruity or green-apple by 39.5% of the subjects and almond or marzipan-like by 16.7% of the subjects. Obviously, trans-2-hexenal (1% PG) was not necessarily the ideal reference for the green-fruity term because it was not "singular". However, in the same experiment, Stampanoni *et al.* (1993) showed that upon training (especially when feedback was provided after each test) the flavor language could be modified and learned (Civille and Lawless 1986; Homa and Cultice 1984). After two testing sessions, a significant portion of the subjects (67.2%, $p < 5$ %) described trans-2-hexenal as "green-fruity" or "apple-green". Trans-2-hexenal was used for the green-fruity note in strawberry flavored yoghurts.

As discussed by O'Mahony *et al.* (1990), the formation of a concept is a two-step process including abstraction and generalization. Analytical sensory subjects must have their concepts aligned: the panel must unequivocally agree on which range of sensations is included in or excluded from the concept (Ishii and O'Mahony 1990). In fact, understanding the boundaries of a concept is very important for the correct utilization of a descriptor and the evaluation of its intensity in a sample (Civille and Lawless 1986). In flavor terminology, there is even a higher degree of overlapping of descriptors and blurring of concept boundaries (Civille and Lawless 1986). Thus, refinement of concept alignment may be obtained by employing multiple physical standards. For example, the jammy-cooked note of strawberry flavor was defined by a strawberry jam and by a strawberry juice concentrate, whereas a strawberry distillate (40% ethanol, pure) was given as a reference falling out of the concept. More than one reference was also used for other attributes (e.g., hay-note, nutty, and buttery). In another study, multiple standards were used to illustrate the "roasted hazelnut-note concept" (Stampanoni 1993a,b); two types of hazelnuts at different roasting degrees were utilized to demonstrate roasted notes falling into the concept, whereas raw hazelnuts were falling outside the concept.

Most of the references used in profiling caramel and strawberry flavors were diluted chemicals or extracts, whereas for the description of cheese flavors, several market samples were also used. Physical standards can in fact be chemicals, spices or extracts, ingredients or finished products (Rainey 1986; Stampanoni 1993a,b). Reference chemicals can be the same ones as employed in the sample formula (e.g. butyric acid or whey). However, they might not necessarily be present: the roasted note in caramel flavors was referenced using pyrazines, which were not used in the flavors tested. As suggested by Rainey (1986), the less complex the standard is, the less difficult the conceptualization of a specific characteristic will be. Spices and extracts (e.g., cinnamon bark oil) are sensorially more complex than single chemicals (e.g., cinnamic aldehyde) (Stampanoni *et al.* 1993). Finished products are also quite complex standards and might convey many sensations. They may be adapted to local taste, be discontinued, or change with time; they are hence not always suitable. Fresh products are even more variable and might not be available throughout the experiment.

Quantitative Flavor Profiling Results

The results of QFP showed that the Mozzarella flavored cheese analog was characterized by milky (e.g. fresh and cooked milk), creamy, buttery and fatty notes. Camembert flavored analog (Fig. 1) showed creamy, fatty and slight milky notes counterbalancing a quite strong mushroom-like note; slight musty and earthy notes complemented the flavor. On the other hand, the blue cheese flavored analog was a balanced mixture of creamy, slightly fatty and mushroom notes, which were combined with the characteristic intense blue cheese note of Penicillium roqueforti. The Parmesan flavored analog showed the sharp notes of cheese rind and capric acid (typical in aged cheese), smoothened by creamy, fatty, and fresh milk notes. In the quite mild and creamy Emmental flavored analog, capric acid was balanced by fatty, milky and slight buttery notes. In the cheddar cheese flavored analog, quite intense buttery, fatty, milky and creamy notes were combined with cheese rind and fermented notes. Quite strong creamy, milky, buttery and fermented notes were blended with cheese rind, butyric acid and animal/sweaty notes typically found in goat milk cheese. In flavored sheep cheese, animal, butyric and capric acid, cheese rind and fermented notes were counterbalanced by creamy, slightly milky and buttery ones.

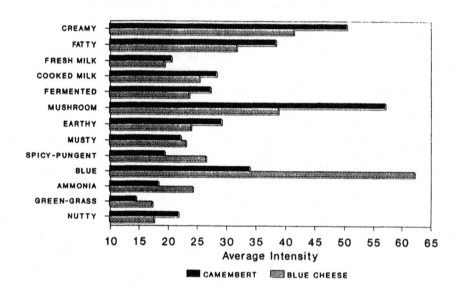

FIG. 1. QUANTITATIVE FLAVOR PROFILES OF CAMEMBERT
AND BLUE CHEESE FLAVORED ANALOGS
(n = 11 × 2 replications)

The results from QFP could be further analyzed using PCA to determine the placement of the samples tested in a flavor space, which also illustrates similarities and differences among the samples and the relationship between flavor attributes. The strawberry flavored samples covered the whole multi-dimensional flavor space, which was illustrated by methyl-anthranilate-floral juicy, green-ripe, lactony-estery, sweetjammy, musty-hay-like, and creamy- vanilla flavor characteristics. Discontinuity test on the scree plot of the eigen-values showed that a model including the first four factors was satisfactory and explained 79.7% of the total variance. Typical notes of fresh strawberries such as fruity-green, grass-green, ripe-fruity, floral and methyl-antranilate notes were positively loaded on factor 1 (loadings >0.56), whereas jammy, sweet, vanilla, creamy and buttery notes were negatively loaded (loadings > -0.65). Factor 2 could be described as a combination of lactony, raisin and musty notes (factor loadings > 0.61), whereas factor 3 was a fruity-green and estery-candy one and factor 4 contraposed raisin and methyl-anthranilate notes to hay-like. The number of attributes used in flavor profiling can be reduced using PCA (Hunter and Muir 1993; Lyon 1987). However, as reported by Heisserer and Chambers (1993), one should distinguish between attributes that are tracking together in a factor and redundancy of attributes that are highly loaded on the same factor. Individual descriptors may load on the same factor, but their intensity might vary. For a more adequate description of the strawberry yoghurts all attributes were necessary. As depicted in Fig. 2, all flavor sensory characteristics tested could be grouped into 6 major flavor dimensions (wild-juicy, green-ripe, aromatic, sweet-jammy, musty-hay and creamy-vanilla), which corresponded to 6 flavor families. 75696-33 and 77696-33 belonged to the wild strawberry family, characterized mainly by methyl-anthranilate, floral and juicy notes. 10394-33 and 10361-33 were typical representative of the green-fruity-ripe family, with 10394-33 being riper and 10361-33 being greener than its counterpart. 10363-33, 77880-33, 77850-33 and 59178-DO were all aromatic (lactony and estery) strawberries, additionally the last two flavors were also more candy-like. The sweet-jammy family included 10359-33 and 78835-33, which were also more aromatic than 78729-33, 78018-33 and 10365-336. A creamy-vanilla character was quite important in 10369-33, mainly creamy, and in 10367-33, also very vanilla-like.

The results from caramel flavor profiling were also analyzed by PCA. The caramel milk drinks also covered the whole space, which was depicted by milky-buttery-creamy, coconut-lactony, burnt-roasted-nutty, rum, coumarin, powdery-vanilla and honey notes. Discontinuity test on the scree plot of the eigenvalues showed that a model including the first four factors explained 79.6% of the total variance. Sweet-spicy (honey, vanilla, condensed milk) and creamy (buttery and creamy) notes were positively loaded (loadings > 0.74) on principal component 1 and were contraposed to nutty, roasted and burnt sugar ones (factor loadings > -0.75). On factor 2, were lactony (0.81), coconut (0.45), citrus (0.90) and rum (0.56), whereas coconut was contraposed to coumarin (-0.68) on factor 3. Principal component 4 was a combination of burnt (also loaded on factor 5), coumarin and powdery notes. As shown in Fig. 3, all flavor characteristics tested could be structured into five major flavor dimensions (powdery-vanilla, milky-creamy, brown-nutty and coumarin-like), although all attributes were important to describe the samples tested. Caramel flavors 10603-33, 10602-33, 10604-33 and to a lesser extent 54474-DO were characterized by brown sugar, roasted, nutty notes. 10608-33, 10443-37, 10606-33 and to a lesser degree 10607-33 were mainly distinguished from the others by quite intense condensed milk, buttery, and creamy notes. Flavor 10601-33 was described by coumarin, and 10600-33 was clearly butterscotch with a vanilla-powdery character.

78787-33 was both coumarin and powdery- vanilla. The toffee flavor 10609-33 showed powdery-vanilla and creamy-milky dimensions, whereas 10610-33 was milky-creamy and brown-nutty, lacking in coumarin notes.

The standardized flavor languages developed in the three experiments were able to clearly show similarities and differences among the samples evaluated. However, they are by no means complete; the main shortcoming is the lack of off-notes (e.g. "decaying" for cheese flavor) as they were not present in the flavors tested.

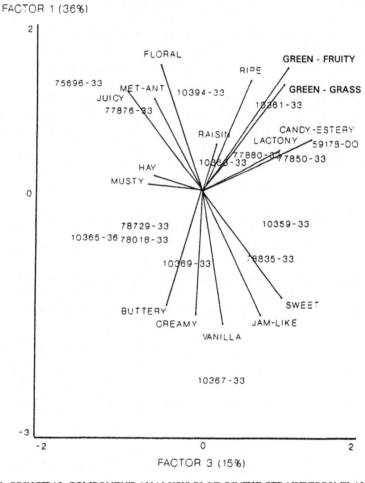

FIG. 2. PRINCIPAL COMPONENT ANALYSIS PLOT OF THE STRAWBERRY FLAVORS
Factor loadings (vectors) and sample scores for principal components 1 and 3. The codes represent all Givaudan-Roure flavors tested and "met-ant" means methyl-anthranilate.

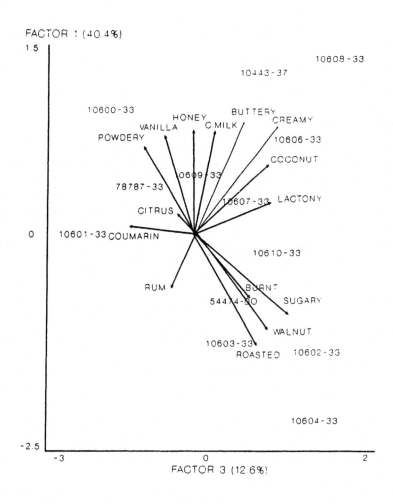

FIG. 3. PRINCIPAL COMPONENT ANALYSIS PLOT OF THE CARAMEL FLAVORS
Factor loadings (vectors) and sample scores for principal components 1 and 3.
The codes represent all Givaudan-Roure flavors tested.

CONCLUSIONS

"When I use a word," Humpty Dumpty said, "it means just what I choose it to mean — neither more nor less." "The question is," said Alice, "whether you can make words mean so many different things." "The question is," said Humpty Dumpty, "which is to be master, that's all." (Carroll 1967).

In sensory science, the meaning of a sensory descriptor has to be mastered. Definitions can be very useful to specify and describe a sensation. However, many sensory attributes are not easily defined; hence, the utilization of reference standards is essential. In the best-case scenario, each descriptor should be defined by means of a written definition (if necessary) and the utilization of one reference standard or, better still, several standards for proper concept alignment. In addition to specifying the quality of a sensory characteristic, references should represent a specific intensity. This is very important in the training phase and allows calibration of the panel in the use of the sensory scale. Together with the standardization of the sensory procedures and evaluation protocols, the use of reference standards will provide another tool to help ensure comparable results across countries. As most international trading companies "think globally and act locally", they require universal and unambiguous communication for fast and well-targeted product development. The use of standardized sensory languages and objective sensory procedures such as the QFP technique will facilitate and expedite this process.

ACKNOWLEDGEMENTS

The author wishes to thank all Givaudan-Roure employees in this project, especially Ms. Susanne Sigrist and Ms. Patrizia Piccinali for coordinating and conducting all sensory tests. The help of Mr. Ralph Bigger, Mr. Silvio Zahnd and Ms. Claudia Schnapp in developing all sample preparation procedures and in preparing the samples was greatly appreciated. Special thanks to all sensory subjects for their kind participation in the tests, and to the flavorists for their stimulating contributions to the development of the standardized sensory languages.

REFERENCES

CAIRNCROSS, S.E. and SJÖSTRÖM, L.B. 1950. Flavor profiles: A new approach to flavor. Food Technol. *4*, 308-311.
CARROLL, L. 1992. *Through the Looking Glass*, Everyman's Library, Childrens Classic, D. Campbell Publ. Ltd., London.
CAUL, J.F. 1957. The profile method of flavor analysis. Adv. Food Res. 7, 1-40.
CIVILLE, G.V. 1987. Development of vocabulary for flavor descriptive analysis. In *Flavor Science and Technology*, (M. Martens, G.A. Dalen and H. Russwurm Jr., eds.) pp. 357-368, John Wiley & Sons, New York.

CIVILLE, G.V. and LAWLESS, H.T. 1986. The importance of language in describing perception. J. Sensory Studies *1*, 203-215.

GARRET, H. 1930. *Great Experiments in Psychology*, The Century Co., New York.

HEISSERER, D.M. and CHAMBERS, IV, E. 1993. Determination of the sensory flavor attributes of aged natural cheese. J. Sensory Studies *8*, 121-132.

HOMA, D. and CULTICE, J. 1984. Role of feedback, category size, and stimulus distortion on the acquisition and utilization of ill-defined categories. J. Exp. Psychol.: Learn. Memory Cognition *10*, 83-90.

HUNTER, E.A. and MUIR, D.D. 1993. Objective reduction of an extensive sensory vocabulary. J. Sensory Studies *8*, 213-227.

ISHII, R. and O'MAHONY, M. 1987. Defining a taste by a single standard: aspects of salty and umami tastes. J. Food Sci. *52*, 1405-1409.

ISHII, R. and O'MAHONY, M. 1990. Group taste concept measurement: verbal and physical definition of the umami taste concept for Japanese and Americans. J. Sensory Studies *4*, 215-227.

JOHNSEN, P.B. and CIVILLE, G.V. 1986. A standardized lexicon of meat WOF descriptors. J. Sensory Studies *1*, 99-104.

JOHNSEN, P.B., CIVILLE, G.V. and VERCELLOTTI, J.R. 1987. A lexicon for pond-raised catfish flavor descriptors. J. Sensory Studies *2*, 85-91.

JOHNSEN, P.B., CIVILLE, G.V., VERCELLOTTI, J.R., SANDERS, T.H. and DUS, C.A. 1988. Development of a lexicon for the description of peanut flavor. J. Sensory Studies *3*, 9-17.

LAUGHLIN, P. and DOGHERTY, M.A. 1967. Discussion vs. memory in cooperative group concept attainment. J. Ed. Psychol. *58*, 123-130.

LYON, B.G. 1987. Development of chicken flavor descriptive attribute terms aided by multivariate statistical procedures. J. Sensory Studies *2*, 55-67.

McDANIEL, M., HENDERSON, L.A., WATSON, JR., B.T. and HEATHERBELL, D. 1987. Sensory panel training and screening for descriptive analysis of the aroma of Pinot noir wine fermented by several strains of malolactic bacteria. J. Sensory Studies *2*, 149-167.

MEILGAARD, M.C., CIVILLE, G.V. and CARR, B.T. 1987. Sensory Evaluation Techniques, Vol. 2, pp. 353, C.R.C. Press, Boca Raton, FL.

MEILGAARD, M.C., REID, D.S. and WYBORSKI, K.A. 1982. Reference Standards for Beer Flavor Terminology System, pp. 119-128, Am. Soc. Brewing Chemistry, St. Paul, MN.

MEISELMAN, H.L. 1993. Critical evaluation of sensory techniques. Food Qual. Pref. *4*, 33-40.

MILLER, G.A. and JOHNSON-LAIRD, P.N. 1976. *Language and Perception*, Cambridge Univ. Press, London.

NOBLE, A.C., ARNOLD, R.A., BUECHSENSTEIN, J., LEACH, E.J., SCHMIDT, J.O. and STERN, P.M. 1987. Modification of a standardized system of wine aroma terminology. Am. J. Enol. Vitic. *38*, 143-149.

NOBLE, A.C. and SHANNON, B. 1987. Profiling Zinfandel wines by sensory and chemical analysis. Am. J. Enol. Vitic. *38*, 1-5.

O'MAHONY, M. 1991. Descriptive analysis and concept alignment, in sensory science. In *Theory and Application in Foods*, (H.T. Lawless and B.P. Klein, eds.) pp. 223-269, IFT Basic Symposium Ser., Marcel Dekker Publishing, New York.

O'MAHONY, M., ROTHMAN, L., ELLISON, T., SHAW, D. and BUTEAU, L. 1990. Taste descriptive analysis: concept formation, alignment and appropriateness. J. Sensory Studies *5*, 71-103.

RAINEY, B.A. 1986. Importance of reference standards in training panelists. J. Sensory Studies *1*, 149-154.

RISVIK, E., COLWILL, J.S., McEWAN, J.A. and LYON, D.H. 1992. Multivariate analysis of conventional profiling data: a comparison of a British and a Norwegian trained panel. J. Sensory Studies *7*, 97-118.

SAS. 1988. SAS/STAT User's Guide, Ver. 6.03, SAS Institute, Cary, NC.

STAMPANONI, Ch.R. 1993a. The "Quantitative Flavor Profiling" technique. Perf. Flavorist 18, Nov./Dec., 19-24.

STAMPANONI, Ch.R. 1993b. Quantitative Flavor Profiling. An effective tool in flavor perception. Int. Food Market. Technol. *2*, 4-6,8.

STAMPANONI, Ch.R. 1993c. Influence of acid and sugar content on sweetness, sourness and the flavor profile of beverages and sherbets. Food Qual. Pref. *4*, 168- 176.

STAMPANONI, Ch.R., PICCINALI, P. and SIGRIST, S. 1993. Unpublished results of the Sensory Givaudan-Roure Flavors Ltd., Dubendorf.

STERNBERG, R.J. 1986. Inside intelligence. Amer. Sci. *74*, 137-142.

STONE, H., SIDEL, J., OLIVER, S., WOOLSEY, A. and SINGLETON, R.C. 1974. Sensory evaluation by descriptive analysis. Food Technol. Nov., 24-34.

STONE, H., SIDEL, J.L. and BLOOMQUIST, J. 1980. Quantitative Descriptive Analysis. Cereal Foods World *25*, 642-644.

CHAPTER 3.0

MEATS

Off-flavor development is a major concern in meats and meat by-products, especially during refrigeration storage. The ability to identify, describe, and quantify any type of perceived odor is useful in the continuing research on the biochemical causes of off-flavor development. Hence, the role of descriptive analysis is increasingly very important. A new development in the use of descriptive analysis is given in Chapters 3.2 and 3.4.

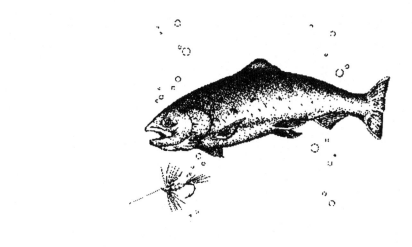

CHAPTER 3.1

DEVELOPMENT OF A TEXTURE PROFILE PANEL FOR EVALUATING RESTRUCTURED BEEF STEAKS VARYING IN MEAT PARTICLE SIZE

B.W. BERRY and G.V. CIVILLE

*At the time of the original publication,
author Berry was affiliated with
Meat Science Research Laboratory
Agricultural Research Service
U.S. Department of Agriculture
Beltsville, MD 20705*

and

*author Civille was affiliated with
Sensory Spectrum, Inc.
44 Brentwood Drive
East Hanover, NJ 07936.*

ABSTRACT

A texture profile panel was developed for measuring textural properties of restructured beef steaks differing in meat particle size. For steaks of different particle sizes, considerable differences existed in the type of sample breakdown and shape of chewed pieces after just two chews. Panelists also found restructured steaks made from large meat particle sizes to be visually more distorted and to contain more gristle than steaks made from small meat particle sizes. Several characteristics (chunkiness after two chews, coarseness of chewed mass at 15 chews) were dropped from the profile over time, while several characteristics (type of sample breakdown and shape of chewed pieces at two chews, size of chewed pieces at 10 chews) not used initially, were added. The texture profile panel approach appears suitable for discerning the textural differences in restructured steaks that can arise from using different meat particle sizes during processing.

INTRODUCTION

The development and expanded acceptance of intermediate value meat products such as restructured beef steaks has been identified as a top priority goal by the U.S. beef industry (Breidenstein 1982). Texture is an important attribute of any meat product. As a result of a variety of meat sources and a multitude of processing systems, texture in restructured steaks following cooking can range from that simulating ground beef to something approximating intact muscle. Several studies (Durland *et al.* 1982; Seideman *et al.* 1982ab; Cardello *et al.* 1983) have shown that differences in meat particle size arising from various processing systems exert considerable influence on textural properties. Obviously, control of texture to a desired end-point would appear to be important.

The use of a texture profile panel would seem to have merit as a means of determining texture and measuring the impact of various processing and cooking procedures. Several studies (Cohen *et al.* 1982; Cardello *et al.* 1983) have reported the use of a texture profile panel for measuring the effects of processing procedures on texture of restructured meats. This particular paper deals with the identification and development of procedures necessary for a texture profile panel to evaluate restructured beef steaks differing in meat particle size.

EXPERIMENTAL

Manufacture of Restructured Steaks

Generally, a common procedure was followed in the processing of the restructured steaks. Frozen meat was flake-cut with an Urschel Comitrol Model 3600. Steaks classified as having large particle or flake pieces were flake-cut with the Comitrol 2-001600-10 head, steaks with intermediate flake pieces were flake-cut with the Comitrol 2J-030750 head, while steaks processing with small flake pieces were flake-cut with the Comitrol 2K-020060 head. The flaked meat was mixed with 0.5% NaCl and 0.25% $Na_5P_3O_{10}$. The meat was stuffed into polyethylene casings using a vacuum stuffer. Meat logs (usually 3.5 kg) were frozen to $-18C$ and then tempered to $-1C$. Logs were pressed in a Bettcher press (Model 70) at 500 psi. Steaks (170 g, 1.70 cm thick) were sliced with a Bettcher cleaver (Model 39). Steaks were frozen to $-18C$ before packaging.

Training of the Texture Profile Panel

For all studies reported in this paper, frozen steaks were cooked to 70C on Farberware (Model 350A) electric broiling units. Steaks were turned frequently to maintain even browning. Iron constantan thermocouples were inserted into the geometric center of each steak to monitor temperature.

Originally, 135 individuals responded to newspaper requests to apply for training to become a member of a texture profile panel. Telephone interviews reduced this number to 35 individuals. Personal interviews which concentrated on an individual's: (1) interest in food, (2) concept of texture, (3) availability, and (4) teeth structure, etc. reduced this number to 15. These individuals then were subjected to the texture profile training outlined by Civille and Szczesniak (1973) and Civille and Liska (1975). The restructured steaks provided the panelists during their initial training and development of a texture profile differed mainly in the size of flaked meat particles. The training process reduced the panel of 15 to a final number of 10. The total training process lasted four months with the group meeting approximately three times weekly for an average duration of three hour/session. The panel trainer provided suggestions as to possible texture characteristics worthy of measuring in the early phases of training. However, later during the training process, the panel determined the appropriate textural characteristics worthy of consideration and developed the terminology and procedures for measuring the characteristics. When the panelists were repeatable and consistent, individually and collectively, in their texture evaluations, they were considered ready to function as a panel.

Visual characteristics of intact steaks and steaks cut lengthwise were evaluated immediately following cooking. Samples (2.54 cm^2) for partial compression, first bite and mastication procedures were placed in a foil pouch (with small holes to allow steam to escape) and then placed in a covered roaster maintained at 51C. Panelists were seated at a conference table and went to the roaster to obtain samples as necessary. Scores were marked on 15 cm long unstructured lines with a low level or degree of a characteristic marked on the left side of the line (low numerical value). Rulers were used to determine the numerical location of the marks and thus the values reported are based on 15-point scales.

Data were subjected to analyses of variance procedures (Snedecor and Cochran 1979). When significant ($P < 0.05$) differences were detected in analyses of variance, Duncan's multiple range test (Duncan 1955) was used to test differences between means.

RESULTS AND DISCUSSION

Data pertaining to changes in texture profile procedures over time are given in Tables 1-4. Texture profile panel characteristics, procedures and definitions for restructured beef steaks that are currently being used are presented in Table 5. Data illustrated in Tables 1-4 highlight the differences in textural properties attributable to meat particle size used in processing. Higher values represent higher degrees of a particular characteristic.

Changes between early studies and later studies for evaluation of visual characteristics were mainly for terminology rather than procedural purposes (Table 1). In general, what was classified as distortion in early studies was changed to macro distortion in later studies, while the term smoothness was changed to micro distortion in later studies. Distortion (macro) has been identified as a major problem in restructured beef steaks following cooking (Field 1983). In our studies, processing of flaked and formed steaks from large size flaked meat pieces produced more macro and micro distortion than what was produced from intermediate and small size pieces.

TABLE 1.
PROCEDURES FOR VISUAL EVALUATION OF DISTORTION
IN RESTRUCTURED BEEF STEAKS

I. Procedures in early studies

Intact cooked steak is visually evaluated for:

A. **Distortion:** The degree to which the overall sample has warped.
B. **Smoothness:** The degree to which the surface appears even.

II. Present procedures

Intact cooked steak is visually evaluated for:

A. **Macro distortion:** The degree to which the overall sample is uneven or warped.
B. **Micro distortion:** The degree to which small sections of the cooked surfaces look uneven or rough.

III. Recent Data Example

Meat particle size used in processing restructured steaks	Macro Distortion	Micro Distortion
Large	5.0 ± 1.3^a	5.5 ± 1.5^a
Intermediate	3.7 ± 1.7^b	3.6 ± 1.0^b
Small	3.8 ± 2.1^b	2.8 ± 1.4^b

[a,b] Means in the same column with different superscripts are significantly different ($P < 0.05$).

Very early in the training process, it was observed that differences were apparent to panelists in the breakdown and shape of chewed pieces even after minimal chewing. In fact, it was felt that these differences were probably more apparent following minimal rather than extensive chewing. The panel first attempted to measure these configurational and breakdown differences using a term identified as chunkiness (Table 2). Chunkiness was defined as the degree to which samples broke into chunks. However it was soon determined that some of the processing procedures used for restructuring steaks yielded breakdowns and shapes that had to be classified in terminology other than chunkiness. What then evolved was a classification system (shape of all pieces) following two bites with the molars. This classification system provided information concerning the type of shearing action, presence of crumbliness, and compacting of the samples before shearing.

Data relating to the use of this classification system with restructured steaks processed from different meat particle sizes are given in Table 2. The values are percentages of evaluations within the meat particle size treatment that were classified into the various type of sample breakdown categories. For steaks manufactured with large flaked meat particle sizes a high proportion of the evaluations consisted of chunky separations with threads (connective tissue) or crust (steak surface) holding the chunks together. These forms or types of chewed pieces were almost never found for restructured steaks processed from intermediate or small meat particle sizes. Most of the sample breakdown and resultant pieces for intermediate particle size steaks consisted of incomplete shearing resulting from pieces attached by the threads or crust, while clean and complete shearing typified the breakdown of samples from restructured steaks processed from small flaked meat particles. The texture profile panel approach reported by Cardello *et al.* (1983) did not include an evaluation of sample breakdown and shape of resultant pieces following minimal chewing. However, our data would indicate that if restructured steaks are processed from different size meat particles, substantial differences exist in breakdown and shape of the chewed pieces after only two chews. While many of texture profile procedures we have developed for restructured steaks require considerable mastication, this particular procedures does not and thus may closely relate to texture differences that consumers might perceive.

Coarseness of the mass following 15 chews was explored as a possible textural characteristic of restructured steaks that might differ as related to raw material and processing variation (Table 3). However, after extensive application of this procedure, it was eventually eliminated due to the minimal differences detected between various experimental formulations. It might be anticipated that variations in meat particle sizes during formulation could result in different amounts of irregular shaped pieces following a certain period of chewing. However, such was not found to be the case (Table 3).

TABLE 2.
PROCEDURES FOR DETERMINING THE TYPE OF SAMPLE BREAKDOWN AND SHAPE OF CHEWED PIECES IN RESTRUCTURED BEEF STEAKS FOLLOWING TWO CHEWS

I. Procedures in early studies

Place one warm 2.54 cm^2 piece in the mouth and using the molars against the cooked surfaces make incision for first bite. Then turn the two pieces 90° and take a second bite. Measure:

   A. Chunkiness: The degree to which the sample breaks into chunks.

II. Present procedures

Same as in early studies; however data are now collected on:

   A. Type of sample breakdown during first two chews and shape of all pieces following two chews.
The following shapes are checked if they are present:

      ___ Shears cleanly and complete
      ___ Incomplete shearing, threads present
      ___ Incomplete shearing, crust present
      ___ Crumbly and complete separation
      ___ Incomplete crumbly separation with threads
      ___ Incomplete crumbly separation with crust
      ___ Compacts along shear line
      ___ Chunky and complete separation
      ___ Incomplete chunky separation with threads
      ___ Incomplete chunky separation with crust
      ___ Layered separation
      ___ Other

III. Recent data example[a]

Type of sample breakdown during two chews and shape of resultant pieces	Meat particle size used in processing restructured steaks		
	Large	Intermediate	Small
Shears cleanly and complete	0.0	9.1	75.0
Incomplete shearing, threads present	15.6	45.4	8.3
Incomplete shearing, crust present	6.2	33.3	16.7
Crumbly and complete separation	0.0	0.0	0.0
Incomplete crumbly separation with threads	0.0	0.0	0.0
Incomplete crumbly separation with crust	0.0	0.0	0.0
Compacts along shear line	6.2	0.0	0.0
Chunky and complete separation	0.0	3.0	0.0
Incomplete chunky separation with threads	37.5	9.1	0.0
Incomplete chunky separation with crust	18.8	0.0	0.0
Layered separation	3.1	0.0	0.0
Other	0.0	0.0	0.0

[a]Values are percentages of sample evaluations within meat particle sizes that were classified into the type of sample breakdown categories.

TABLE 3.
PROCEDURES FOR DETERMINING COARSENESS OF THE CHEWED MASS AND SIZE OF CHEWED PIECES IN RESTRUCTURED BEEF STEAKS

Coarseness of chewed mass

I. Procedure in early studies

Place a warm 2.54 cm^2 sample in mouth and chew with molars. At 15 chews determine the degree to which irregular shaped particles can be detected whether free-floating or connected.

II. Present procedures

This term and procedure were deleted.

III. Data example

	Meat particle size used in processing restructured steaks		
	Large	Intermediate	Small
Coarseness of chewed mass	8.1 ± 1.8	8.5 ± 2.0	7.5 ± 2.0

Size of chewed pieces

I. Procedure in early studies

Not a part of early studies

II. Present procedure

Place a warm 2.54 cm^2 sample in mouth and chew with molars. After 10 chews, move the bulk of chewed mass to the center of the mouth and using the tongue as a feeler, evaluate for the perceived size of clearly separate pieces or pieces held together only by connective tissue web.

III. Data example

Meat particle size used in processing restructured steaks		
Large	Intermediate	Small
10.7 ± 0.8[a]	10.0 ± 1.2[ab]	9.4 ± 0.6[b]

[a,b]Means on the same line with different superscripts are significantly different ($P < 0.05$).

However, it was felt that size rather than shape of chewed meat particles following a set number of chews might be more closely rated to processing and/or raw material variations. Ten chews was found to be a proper degree of chewing that would provide consistent chewed meat size differences between steaks of different particle sizes. Steaks processed from large size flaked meat pieces were found to possess larger size pieces following 10 chews than steaks made from smaller size pieces, with steaks made from intermediate size flaked meat particles not different from either large or small flake steaks in size of chewed pieces. Cardello *et al.* (1983) reported a similar finding for size of chewed pieces as related to variations in flaked meat particle size. They defined size of chewed pieces as the perceived volume of individual particles, however, no mention was given as to which point in the mastication process this characteristic was measured.

Procedures and data pertaining to the detection of gristle are presented in Table 4. In the early stages of texture profile panel development, the appraisal for gristle occurred after a sample had already undergone a substantial amount of chewing. This number of chews (30-60) was selected based on previous experience with descriptive attribute panels evaluating steak and ground beef. However, it was felt that perhaps this amount of chews could conceivably create some webbed tissue from the gristle which could thus be erroneously categorized into that form of connective tissue rather than as gristle.

TABLE 4.
PROCEDURE FOR DETERMINING GRISTLE IN RESTRUCTURED BEEF STEAKS

I. Procedure in early studies

The amount of rubbery pieces are determined following 25 chews.

II. Present procedure

The amount of rubbery pieces are determined following 10 chews and again when the mastcation phase is complete and the sample is ready to be swallowed.

III. Data example

	Meat particle size used in processing restructured streaks		
	Large	Intermediate	Small
Gristle-25 chews-early studies	6.1 ± 1.8^a	3.1 ± 3.0^b	0.3 ± 0.7^c
Gristle-10 chews-present studies	4.7 ± 1.6^a	1.7 ± 1.0^b	0.1 ± 0.2^c
Gristle-following mastication-present studies	5.5 ± 1.9^a	2.0 ± 1.2^b	0.1 ± 0.2^c

[a,b,c]Means on the same line with different superscripts are significantly different ($P < 0.05$).

Next, it was decided to consider evaluation for gristle following 10 chews, since it appeared that sufficient sample breakdown was occurring in this period of time to detect gristle without the gristle being further masticated or shredded into webbed tissue. It was also decided at this time to establish a determination for gristle after all mastication was complete and the sample was ready for swallowing. This procedure would thus be similar to those used by some of our descriptive attribute panels. However, it should be pointed out that while the final evaluation is not given until mastication is complete, appraisal is continuously being done during mastication. Data resulting from evaluation at the three stages of mastication for steaks processed from different meat particle sizes are given in Table 4. Regardless of the procedures involved for gristle determination, steaks processed from large meat particles were rated as having more gristle than steaks made from intermediate and small size meat particles and steaks processed from intermediate size meat particles had more gristle than those made from small size meat particles. Since the raw materials were the same for all formulations, it would appear that the small size Urschel Comitrol flaking heads (2K-02060) are capable of producing a restructured steak where gristle is almost undetectable. Levels of gristle detection appear to be highest following 25 chews and lowest after 10 chews with gristle detection at the point of complete mastication being intermediate. It would appear that between 25 chews and the end of mastication (usually another 20-25 chews) the amount of gristle capable of being detected might decrease. As mentioned before, perhaps gristle is being masticated into webbed tissue and thus it's detection as gristle is reduced. With the highest detection being at 25 chews we now feel that this is perhaps the most accurate of the three procedures for determining gristle and thus in the future we will be evaluating gristle at that endpoint.

The complete texture profile system used at present for restructured beef steaks is presented in Table 5. The major differences between the texture profile system for restructured steak that we use versus those of Cardello *et al.* (1983) are for visual appraisal and sample breakdown at two chews. In our visual appraisal, distortion and fibrousness are determined, while the Cardello *et al.* (1983) profile does not include these characteristics, but includes coarseness which we do not evaluate. The Cardello *et al.* (1983) system does not include sample breakdown at two chews.

CONCLUSIONS

Restructured beef steaks can be made from a variety of raw muscles and ingredients and processed by numerous systems. These different materials and systems can influence textural properties and necessitate other sensory approaches to measuring texture than descriptive attribute scalar measurements.

TABLE 5.
PRESENT TEXTURE PROFILE PANEL CHARACTERISTICS, PROCEDURES AND DEFINITIONS FOR RESTRUCTURED BEEF STEAKS

I. Visual

    A. Distortion--Steak is visually evaluated for the degree that the steak has warped or changed in configuration from its orginal raw-frozen shape. Macro distortion is degree overall steak has distorted. Micro distortion is the degree to which cooked surfaces look uneven or rough.

    B. Fibrousness--Steak is cut in half and the cross section is visually evaluated for the degree that the sample resembles steak or has no disruption of components.

II. Partial Compression

    A. Springiness--Place a warm 2.54 cm^2 piece in the mouth and using the molars against the cooked surfaces, press lightly five times. Wait two sec between each press. Springiness is the perceived degree and speed with which the sample returns to original height and thickness.

III. First Bite

    Take a warm 2.54 cm^2 piece and place it in the mouth in the same manner as for partial compression and evaluate for:

    A. Hardness--Amount of force required to bite through sample.

    B. Cohesiveness--The degree to which the sample deforms before shearing.

    C. Moisture Release--Amount of juiciness perceived during the first bite.

    D. Uniformity--The degree to which the force needed to shear the sample is the same across the bite area.

IV. Mastication

    Take one warm 2.54 cm^2 sample, make the first incision as for first bite. Then turn the two pieces 90° and take a second bite. Evaluate for:

    A. Sample Breakdown at Two Chews - Check the appropriate breakdown category(ies). These are identified in Table 2.

        - Continue chewing and evaluate for -

    B. Juiciness--The amount of juice released following seven chews.

    C. Size of Chewed Pieces--The perceived size of clearly separate pieces or pieces held together only be connective tissue web. Evaluated following 10 chews.

    D. Gristle - The amount of rubbery particles present following 10 chews.

    E. Cohesiveness of Mass--The degree to which particles stick together. This is evaluated at its maximum degree between 10 and 35 chews.

    F. Uniformity of Mass--Degree to which components of the mass are the same. Evaluated following 25 chews.

    G. Webbed Connective Tissue--Amount of connective tissue present just before swallowing.

    H. Number of chews--Total number of chews to accurately determine the amount of webbed connective tissue.

    I. Overall Gristle--Overall impression of the amount of rubbery particles throughout mastication.

TABLE 5. *(Continued)*

> J. Overall Webbed Connective Tissue--Amount of firm thread-like connective tissue present throughout mastication.
>
> V. After-swallow
>
> A. Tooth Pack--Amount of sample remaining in between teeth after swallowing.
>
> B. Mouthcoating--Amount of film residue left on mouth surface following swallowing.

A texture profile system can provide considerable information. We have found that the system should always be under revision, since changes in processing, cooking etc. can result in the need for a new texture characteristic or a totally new profiling approach.

ACKNOWLEDGMENTS

This project was a contributing project to Western Regional Research Project W-145. Mention of trade names does not imply endorsement by the U.S. Government. The authors gratefully acknowledge the assistance of J. Secrist and J. Smith of the U.S. Army Natick Research and Development Laboratories for providing steaks for this study and Mrs. S. Douglass and Miss M. Stanfield for cooking the steaks used in this study.

REFERENCES

BREIDENSTEIN, B.C. 1982. *Intermediate value beef products.* National Livestock and Meat Board, Chicago, IL.

CARDELLO, A.V., SEGARS, R.A., SECRIST, J., SMITH, J., COHEN, S.H. and ROSENKRANS, R. 1983. Sensory and instrumental texture properties of flaked and formed beef. Food Microstruc. 2, 119–133.

CIVILLE, G.V. and LISKA, I.H. 1975. Modifications and applications to foods of the General Foods sensory texture profile techniques. J. Texture Studies 6, 19–31.

CIVILLE, G.V. and SZCZESNIAK, A.S. 1973. Guidelines to training a texture profile panel. J. Texture Studies 4, 204–223.

COHEN, S.H., SEGARS, R.A., CARDELLO, A., SMITH, J. and ROBBINS, F.M. 1982. Instrumental and sensory analysis of the action of catheptic enzymes on flaked and formed beef. Food Microstruc. *1*, 99–105.

DUNCAN, D.B. 1955. New multiple range and multiple F tests. Biometrics *11*, 1–42.

DURLAND, P.R., SEIDEMAN, S.C., COSTELLO, W.J. and QUENZER, N.M. 1982. Physical and sensory properties of restructured beef steaks formulated with various flake sizes and mixing times. J. Food Prot. *45*, 127–131.

FIELD, R.A. 1983. New restructured meat products--food service and retail. In *International Symposium of Meat Science and Technology*, (K.R. Franklin and H.R. Cross, eds.). National Live Stock and Meat Board, Chicago, IL.

SEIDEMAN, S.C., DURLAND, P.R., QUENZER, N.M. and MICHELS, J.D. 1982a. Precooking and flake size effects on spent fowl restructured steaks. J. Food Prot. *45*, 38–40.

SEIDEMAN, S.C., QUENZER, N.M., DURLAND, P.R. and COSTELLO, W.J. 1982b. Effects of hot-boning and particle thickness on restructured beef steaks. J. Food Sci. *47*, 1008–1009.

SNEDECOR, G.W. and COCHRAN, W.G. 1979. *Statistical Methods*. The Iowa State University Press, Ames.

CHAPTER 3.2

A STANDARDIZED LEXICON OF MEAT WOF DESCRIPTORS[1]

PETER B. JOHNSEN and GAIL VANCE CIVILLE

*At the time of original publication,
author Johnson was affiliated with
Monell Chemical Senses Center
Philadelphia, PA 19104*

and

*author Civille was affiliated with
Sensory Spectrum, Inc.
East Hanover, NJ 07936.*

ABSTRACT

A new descriptive language for evaluating the taste of warmed-over meat flavor has been developed. These terms should help researchers elucidate the causes of the phenomena of off-flavor in meats. The meat industry should also be able to apply these terms in new meat product development, quality control, and shelf-life stability.

INTRODUCTION

Participants at a recent industry-scientific research meeting concluded that warmed-over flavor (WOF) in meats is a major stumbling block to the introduction of new meat products to the market place and, therefore is considered to be one of the important problems facing the meat and food industries. WOF is a sensory phenomenon and any analytical chemical work to determine causes and find solutions to this problem must be conducted in combination with sensory analysis of WOF. To this end, a program was initiated by the USDA-Southern Regional Research Center to develop a standardized lexicon of flavor descriptors derived by agreement through the collective experience of meat flavor experts from Industry and Government.

[1]Supported by Contract No. 53-7B31-5-003 from USDA-ARS, SRRC, New Orleans, LA 70179 to the Monell Chemical Senses Center.

*Reprinted with permission of Food & Nutrition Press, Inc., Trumbull, Connecticut.
©Copyright 1986. Originally Published in* Journal of Sensory Studies **1**, 99–104.

Because the mechanism(s) of WOF formation is not understood, precise descriptors representing a variety of flavors were needed. These descriptors should be well defined and should allow flavor panelists accurately to track the development of WOF with time. The selection of actual descriptors as not necessarily made to represent presumed biochemical changes giving rise to WOF but rather to describe adequately the changes in flavor over time as perceived by a trained flavor panel. In this paper we present a flavor descriptor vocabulary as well as a standardized protocol of meat sample preparation that can provide a common model for the study of WOF in meats.

MATERIALS AND METHODS

The expert panel was convened in Philadelphia under the aegis of the Monell Chemical Senses Center at the Eastern Regional Research Center of the USDA. Sensory consultant Gail Vance Civille (Sensory Spectrum Incorporated) led a panel of seven experts in the development of the flavor descriptors. Participants included Sharon Hargett (Oscar Mayer Co.), Brenda Lyon (USDA), George LoPresti (Campbell Soup Co.), Sharon Payton (USDA), Linda Beck (Beatrice Refrigerated Foods), Susan Mayer (Gagliardi Brothers Inc.) and Richard Whiting (USDA). Following the meeting, a draft of the flavor descriptor lexicon was presented to participants, and the final lexicon reflects their comments and suggestions.

Following a review of the literature on WOF in meat, a variety of meats, cooking procedures, storage times, and reheating procedures which would manifest the flavor or combination of flavors representing WOF were selected. The samples used in our panel work were meat patties of chicken, turkey, pork, or beef as well as preformed roasts of beef. There were some 1200 patty samples tested by the panelists.

Except for roast slices, samples were formed into 85 g patties after grinding the meat through a 0.95 cm plate followed by a 0.32 cm plate. Beef samples were made from fresh and frozen (tempered) 85, 80, 75, and 50 chemical lean trimmings formulated to achieve a 25% fat level. The turkey and chicken samples were made using half light and half dark meat. Pork samples were from pork loins trimmed of visible fat. Frozen, preformed, uncooked, beef roasts (5 kg) were obtained from Dr. Russell Cross of Texas A & M University. The roasts were defrosted in the refrigerator and then cooked in a preheated convection oven at 175–185C for 3.5 h (50–54C internal temperature). The roasts were cooled for 20 min and then cut in 3 mm slices. Samples were packaged 3 slices per heat-sealable polyethylene boiling bag and frozen.

For the initial cooking, meat patty samples were either grilled or stream cooked. Patties were grilled on a griddle at 160C until the internal temperature of 55C was reached for beef and 77C for pork, turkey and chicken. Following cooking, samples were refrigerated at 4C for up to seven days (1,2,4,5,6,7 days) and then frozen at −10C, separated by patty paper (James River) in a covered container. Uncooked control samples were also made and frozen without cooking to be used as benchmarks in panel evaluations. Samples of pork, turkey, and chicken were steam-cooked in sealable polyethylene bags to an internal meat temperature of 77C. Beef samples were similarly steam-cooked to an internal temperature of 55C.

Three methods were used to reheat the cooked meat patties: baking, broiling and boiling. To bake, a convection oven was preheated for 30 min at 150C. The frozen, precooked patties were heated at 150C for 13 min when internal temperature reached 65C. Raw, frozen control patties were heated at 150C. Cooked patties were also reheated in the broiler (preheated for 10 min) for 5 min on each side. Steam-cooked samples, in sealed polyethylene pouches, were boiled for 10 min in two quart pots to rewarm.

RESULTS AND DISCUSSION

The first panel session was designed to determine if the sensory perception of WOF was similar across beef, pork, chicken and turkey patties, and across beef patties and beef roasts. Panelists found that WOF was equally identifiable in meat from different species or different treatments within a species, although the samples varied in intensity. This conclusion allowed us to proceed with beef variations alone assuming that, although all the fresh beef descriptors would not apply across the different meats, the WOF character would.

During the second session, the panel began to characterize the flavors detected in beef across cooking procedures (grilled, baked and boiled), storage times, and forms (patties and roasts). The descriptors developed by agreement during the second session and clarified during the third are listed with definitions in Table 1.

Panelists noted that the intensity of fresh cooked beef notes (cooked beef lean, cooked beef fat, browned, serum/bloody, grainy/cowy) was reduced in the stored samples. The cardboard note generally appeared before the oxidized note, at about the same time the reduction in fresh beef characteristics was observed (1-3 days). The oxidized/rancid/painty character was noted in samples that had been stored for 3-7 days. The use of the terms full (meaning a large group of the fresh character notes) and flat (a general decrease in the intensity of fresh notes) was discussed. However, the group decided by majority that by using the descriptors in conjunction with an intensity scale to indicate the strength of each note, the generalized full/flat characterizations could be eliminated.

During the fourth session, the panel evaluated several samples to "exercise" or determine the applicability of the descriptors on blind coded samples.

TABLE 1.
BEEF FLAVOR DESCRIPTIONS

AROMATICS	DEFINITIONS
Cooked Beef Lean:	the aromatic associated with cooked beef muscle meat.
Cooked Beef Fat:	the aromatic associated with cooked beef fat.
Browned:	the aromatic associated with the outside of grilled or broiled beef (seared but not blackened/burnt).
Serum/Bloody:	the aromatic associated with raw beef lean.
Grainy/Cowy:	the aromatic associated with cow meat and/or beef in which grain/feed character is detectable.
Cardboard:	the aromatic associated with slightly stale beef (refrigerated for a few days only) and associated with wet cardboard and stale oils and fats.
Oxidized/Rancid/Painty	the aromatic associated with rancid oil and fat (distinctly like linseed oil).
Fishy	the aromatic associated with some rancid fats and oils (similar to old fish).
TASTES	
Sweet:	taste on the tongue associated with sugars.
Sour:	taste on the tongue associated with acids.
Salty:	taste on the tongue associated with sodium ions.
Bitter:	taste on the tongue associated with bitter agents such as caffeine, quinine, etc.

Panelists evaluated independently, then reported their scores verbally to the panel coordinator. Panelists generally agreed on the intensity score that should be assigned to a sample to indicate the degree to which a specified flavor was present. Agreement was good in the disappearance of the fresh flavors; the appearance, then disappearance, of the cardboard flavor; and the final dominance of the other flavors by the oxidized/rancid flavor. There was a lack of agreement for some of the flavors, especially cowy. Fishy flavor notes were detected only in an occasional sample. The results demonstrated that the group could use the selected descriptors and a scaling (intensity) system to define product differences. Intensity scores for representative samples are shown in Tables 2 and 3. The scores in these tables represent the impression of the majority of the panelists for each flavor note intensity and are derived from individual panelists' scores.

TABLE 2.
COMPARISON OF FRESH CONTROL BAKED BEEF PATTY
WITH STORED STEAMED AND BAKED PATTY

	Control*	Reheated**
Cooked Beef Lean	6	4
Cooked Beef Fat (fresh)	4	1
Browned	4	1
Serum/Bloody	3	2
Grainy/Cowy	2	1
Cardboard	1	4
Oxid/Rancid/Painty	0	3
Fishy	0	0
Sweet	3	2
Sour	1	3
Salty	2	2
Bitter	0	0

NOTE: Lower fresh beef notes and higher cardboard and oxidized/rancid/painty in the reheated patty than the control (fresh). Also note the lack of browned character in the reheated patty since it was initially steamed to cook. Scale used is 10 pts (0 = none, 10 = extremely strong).
*Fresh, frozen beef patty, cooked by baking.
**Steamed beef patty with 5 day storage and reheated by baking.

TABLE 3.
COMPARISON OF FRESH CONTROL BAKED BEEF PATTY
WITH STORED GRILLED PATTY

	Control*	Reheated**
Cooked Beef Lean	6	3
Cooked Beef Fat (fresh)	5	2
Browned	4	3
Serum/Bloody	3	1
Grainy/Cowy	2	2
Cardboard	1	5
Oxid/Rancid/Painty	1	4
Fishy	0	0
Sweet	3	3
Sour	1	2
Salty	2	2
Bitter	0	0

NOTE: Lower fresh beef notes and higher cardboard and oxidized/rancid/painty in aged grilled beef patty than the control (fresh). Scale used is 10 pts (0 = none, 10 = extremely strong).
*Fresh, frozen beef patty, cooked by baking.
**Grilled beef patty with 7 day storage and reheated by baking.

A new descriptive language for evaluating the taste of warmed-over meat flavor has been developed. These terms should help researchers elucidate the causes of the phenomena of off-flavor in meats. The meat industry should also be able to apply these terms in new meat product development, quality control, and shelf life stability.

ACKNOWLEDGMENTS

The authors wish to acknowledge Dr. John R. Vercellotti for his valuable contribution throughout this project. We also wish to thank Drs. John Cherry and James Craig of the Eastern Regional Research Center U.S.D.A.-A.R.S., Philadelphia, PA, for their cooperation in this project; Dr. Russell Cross, Texas A & M University for supplying the meat and assisting in the sample preparations; Pat Panzer for assisting ably in the organization of the meeting; and Florence Tally for helping in the preparation of the samples.

CHAPTER 3.3

DEVELOPMENT OF CHICKEN FLAVOR DESCRIPTIVE ATTRIBUTE TERMS AIDED BY MULTIVARIATE STATISTICAL PROCEDURES[1]

B.G. LYON

At the time of original publication,
author Lyon was affiliated with
U.S. Department of Agriculture
Poultry Meat Quality and Safety Research Unit
Russell Research Center
Athens, Georgia 30613.

ABSTRACT

Chicken patties (50% light/50% dark meat) were used as training and testing samples by a 10-member panel to develop descriptive terms appropriate for profiling the taste/aroma character notes of fresh and reheated chicken meat. A free word association list of 45 terms was developed in initial training sessions. These terms were used to obtain frequency of use and intensity data for statistical analyses to determine appropriateness of the terms for profiling and discriminating among fresh and reheat treatments of the chicken patties. The initial list was reduced to 31 terms by a frequency-use delimiter of 40% over four replications and by MANOVA probability values (P > .05) for sample differences. FACTOR (principal components) analysis applied to the reduced list of terms indicated six logical associative groupings of terms that explained 77% variation in the data. A final list of 12 descriptors was developed by omitting redundant terms in a factor grouping.

INTRODUCTION

Poultry meat is highly susceptible to flavor changes during short-term refrigerated storage. The off-flavor character is generally referred to as 'warmed-over' flavor (WOF). Raw poultry meat held for long periods of time

---

[1]Mention of trade name does not imply endorsement by the author or by the U.S. Dept. of Agriculture to the exclusion of others not named.

Originally published in Journal of Sensory Studies 2(1987) 55–67. *This article was prepared by a U.S. Government employee as part of her official duties. The official information materials are in the public domain and legally cannot be copyrighted.*

in frozen storage is susceptible to another off-flavor development referred to as 'rancid'. Both types of off-flavor are considered to be related to autoxidation of the lipid fractions, although different lipid fractions (i.e. triglycerides, phospholipids or individual fatty acids) of the muscle may be involved in each type of off-flavor (Igene et al. 1980). The mechanisms involved in the chemical changes that result in off-flavors are of growing interest to the poultry industry as precooked, nonfrozen products are introduced to the consumer.

Because WOF and rancid flavors are sensory phenomena, studies involving analytical chemical work to examine mechanism(s) and causes of these flavor problems must be done in conjunction with sensory evaluation (Johnson and Civille 1986). However, meat flavor is very complex. Poste et al. (1986) noted that a lack of standardization in sensory methods and terminology has made assessment of off-flavor problems in cooked meat more difficult.

Descriptive sensory analysis is a method in which a trained panel develops language to describe perceived sensations. The result of a developed profile is a verbal blueprint of a product type which allows tracking of flavor changes due to process and procedure variations. Johnson and Civille (1986) proposed a set of terms appropriate for sensory evaluation of off flavors in beef. Their protocol was used as a model in this study to develop appropriate descriptors of flavor change in cooked chicken patties. A variety of samples and sample treatments (involving formulation, cooking, storage, and reheating methods) was presented throughout the language development process to provide panelists with a range of taste sensations. A primary goal of this work was to combine the greatest possible freedom to panelists during initial term selection with later application of multivariate statistical procedures (MVA) (Powers 1984) to aid in final selection of the terms to define and use. A vocabulary developed in this manner should aid standardization of sensory assessment of flavor change in cooked poultry and should be useful to researchers and to industry for quality control.

MATERIALS AND METHODS

Panel

A 10-member panel was recruited from the staff of the Russell Research Center. Qualifications for selection were availability and willingness to cooperate and commit the time to the project. Panelists were screened for ability to discriminate odors and tastes and for ability to express themselves verbally. Initial training sessions focused on basic taste and odor recognition and on threshold sensitivity. These initial sessions were designed to gain an appreciation for the complexities of human beings as sensitive research tools.

Test Product

Chicken patties made of 50% white meat and 50% dark meat with natural proportions of fat were chosen as the test product. Individually, white and dark muscle may have different taste characteristics. By combining the products, the range of tastes likely to be encountered with either muscle types would be exhibited. The pattie product was chosen because the individual muscles may also differ in texture. Elimination of textural differences allowed the panel to focus on sensory perceptions by mouth and nose. Winger and Pope (1981) also reported advantages of using minced or ground muscle instead of intact muscle to train panelists for evaluation of off-flavors in lamb.

All patties were made by grinding each muscle type twice, then mixing equal proportions of white and dark meat in a ribbon mixer (Keebler Mfg. Co., Chicago, IL). Ninety grams of the mixture were weighed and molded into petri dishes (9.5 cm diam × 1.4 cm high) to form individual patties. Patties were tempered about 1 h at 2C to firm slightly. A set of raw patties was frozen at −34C to firm slightly. A set of raw patties was frozen at −34C to use as a newly cooked control. The remaining patties were batch cooked in a Despatch rotary reel oven set at 177C to an internal temperature of 75 to 80C, measured by a handheld thermometer with probe (Doric Scientific, Model 450-ET, San Diego, CA). Cooked patties were cooled in trays, loosely covered with foil, in a 2C walk-in cooler for a designated number of days prior to bagging (5 patties per bag) and freezing at −34C until evaluated. In Test Set #1, storage times were 1, 3, or 5 days. In Test Set #2, storage times were 0 through 5 days.

To prepare product to use for training and terminology development sessions, thawed raw samples were cooked in a 177C oven to an internal temperature of 75 to 80C. Precooked stored samples were thawed at room temperature for 1 h, wrapped in aluminum foil, and reheated in a 163C conventional oven for 23 to 25 min. Cores (2 cm diam) were taken from patties in a set and presented to panelists. Throughout training and terminology development phases, other cooking and reheating methods and patty compositions were presented to exhibit as many flavor variables as possible. When quantitative data were collected on patties from Test Set #1, panelists evaluated intensity of the attributes on a 0 to 9 category scale, with 0 meaning 'not detectable' and 9 meaning 'very strong'. Each product was evaluated by each panelist four times. For Test Set #2, the scoring system was changed to semi-structured 10-cm line scales for each attribute, anchored on the left side by the term, 'weak', and on the right side by the term, 'strong' (Stone *et al.* 1974). Each product in Test Set #2 was evaluated by each panelist three times. The experimenter recorded panelists' responses by measuring the distance in mm (1 to 100) from the left side of the scale. By this time, panelists had been in training for over 8 weeks and made the transition to the semi-structured scales easily.

Statistical Analysis

At various phases of the terminology development, quantitative data were analyzed to determine panelists' performance, appropriateness of terms to discriminate product differences, redundancy of terms, and product differences (Power 1984, 1986). Statistical programs of SAS (1985) were used. They included ANOVA, MANOVA, FACTOR ANALYSIS, PRINCIPAL COMPONENT ANALYSIS, DISCRIMINANT ANALYSIS, VARCLUS and others. In one-way ANOVA to analyze individual panelist's performance, a probability of $<.30$ was used as the criterion for acceptable discriminating ability (Stone et al. 1974; Powers 1984).

RESULTS AND DISCUSSION

A free word association list was developed over several sessions by allowing each panelist to list any term that he/she felt would describe the individual character notes perceived. Panelists were encouraged to use associative terms (eg., like paper) or cognitive terms (eg., bloody). Qualitative or affective terms, such as good, bad, mild aroma, or don't like, were not allowed. The initial working list (Table 1) included 45 terms. Among the descriptive terms describing meat flavor were the basic taste terms (sweet, salt, sour, and bitter) and the feeling factors (chemical, astringent, metallic). Although many of the terms in the list were very similar, no words were eliminated on the basis of assumed synonymity. The objective at this point was to list as many potential descriptors as possible.

The 45-term list was comprehensive and included flavor terms noted by other authors (Table 1). Only three of the eight studies cited in Table 1 involved chicken. Although some of the terms listed were derived by the sensory panels used in those studies, MacLeod and Coppock (1978) and Persson et al. (1973) were the only researchers focusing on the terminology and panel use of the terms rather than citing the terms as the means of focusing on treatment of process variations.

The initial 45 terms were used to collect quantitative intensity data on patties from Test Set #1. Newly cooked and 1-, 3-, and 5-day-stored reheated samples were prepared, coded, and served under 40-watt green incandescent lights in individual panel stations. The newly cooked sample was presented at each session along with two of the stored samples, necessitating an incomplete balanced design. Order of the test samples was randomized among the panelists at each of six sessions. Each of the precooked reheated samples received four replicate evaluations by each panelist, and the newly cooked sample received six evaluations. Nine of the ten panelists attended all six sessions.

TABLE 1.
THE INITIAL LIST OF 45 TERMS DESCRIBING THE RANGE OF SENSATIONS
PERCEIVED IN A VARIETY OF TEST CHICKEN PRODUCTS AND
USED TO EVALUATE TEST SET #1

1. Chickeny[d]	16. Bouillion-like[a]	31. Reheated
2. Meaty[b,c,d]	17. Serum/bloody[a,f,g]	32. Greasy
3. Meaty, chickeny	18. Liver/organy[a,c]	33. Fatty[d,f,g]
4. Meaty, cooked[f,g]	19. Bone-marrow	34. Oily[d,f,g]
5. Meaty, raw[f]	20. Earthy[f,g]	35. Oxidized
6. Gamey, fowl-like	21. Grassy[a,g]	36. Rancid[b,d,f,g,h]
7. Roasted[f]	22. Feedy[g]	37. Painty[f]
8. Boiled	23. Vegetable (cooked)[f,g]	38. Fishy[g]
9. Canned[g]	24. Musty[a,c,f,g]	39. Salt
10. Browned[a]	25. Moldy[a,f,g]	40. Sweet[b,c,f,g]
11. Burned[f,g]	26. Papery	41. Sour[b,g]
12. Toasted[f]	27. Nutty[a]	42. Bitter[g]
13. Scorched	28. Cardboard	43. Metallic[a,f,g]
14. Heated protein	29. Stale[b,c,h]	44. Chemical[a,h]
15. Brothy[b,h]	30. Warmed-over[d]	45. Astringent[a,h]

a-h
Referenced literature citing similar use of terms:
a. Berry, et al., 1980. Beef.
b. Cipra and Bowers, 1971. Turkey.
c. Jacobson and Koehler, 1970. Chicken.
d. Landes. 1972. Chicken.
e. Lynch, et al., 1986. Beef.
f. MacLeod and Coppock, 1978. Beef.
g. Persson, et al., 1973. Beef.
h. Van de Riet and Hard, 1979. Beef.

A multivariate analysis of variance (MANOVA) was applied to the data of each scale using sample, session, and sample × session interaction as sources of variation in the model. The MANOVA gives a universal F value and associated probability level for significant differences over all the data. Sample and session differences were statistically different ($P < .05$), but the sample × session interaction was not ($P = .99$). Session differences would be expected due to the balanced incomplete block design used. It was concluded that differences in the test samples did exist in the overall data using all 45 terms.

To determine how individual terms were used by the panelists and whether or not the terms could discriminate differences among the samples, individual ANOVA's by term and one-way ANOVA by panelist by term were conducted. Additionally, the frequency of individual term use by the panelists was calculated.

In the first evaluation step, a probability value of $P > .05$ for F ratios from ANOVA by term for sample as source of variation was selected as an elimination criterion. If sample differences were not statistically significant at the $P = .05$ level for an individual term, then the term made no contribution to discrimination of sample differences. The terms from Table 1 that were eliminated in this step were meaty, cooked; meaty, raw; roasted, boiled, heated protein, serum/bloody, grassy, feedy, nutty, greasy, fatty, oily, fishy, and salt.

A second step in the evaluation of terms involved conducting Factor Analysis on the 31 remaining terms, consideration of the frequency-use delimiter of 40%, and panel discussions. The primary purpose of Factor Analysis is to detect redundancy (Bieber and Smith 1986; Powers 1984). In this analysis, terms are grouped together according to their correlation to newly created Factors (SAS 1985). In interpreting the analysis, some logical associations can be made as to the meaning of the underlying dimension of the new Factor variable. Sometimes the terms in the Factors are, in fact, redundant.

In the Factor Analysis of the 31 terms, eight factors were extracted that explained 77% of variation in the data (Table 2). Logical associations of terms began to emerge, although interpretation of some groupings was not clear at this point. Some terms were targeted for elimination because they were used less than 40% of the time by the panelists. Additionally, most terms used less than 40% did not have high loadings on any Factors.

In addition, evaluation of individual panelist performance by term indicated that few of the panelists had probability values of $P < .20$ for the terms that were used infrequently and had low loadings on Factors. The probability level of $P < .20$ for greater F-values was selected as the panelist performance criterion based on considerations cited by Stone *et al.* (1974), Stone and Sidel (1985) and ASTM (1981). Panelists may not have shown greater discriminating ability on these attributes because of inconsistent use, or because differences between intensity scores assigned to the samples were not great enough to show statistical differences. Panel discussions and occasional retesting revealed that some of these terms were meaningful to only a few panelists. In some instances, panelists had considered some terms to be too similar to other terms on the list and did not consider using more than one term to score for the sensation they perceived. For example, bouillon-like was similar to brothy and toasted was similar to browned. The terms eliminated by panel consensus, by information from the Factor Analysis, and by the 40% use criterion, were chemical, astringent, bone-marrow, toasted, moldy, cooked vegetable, earthy, bouillon-like and sour.

TABLE 2.
FACTOR ANALYZED GROUPINGS OF 31 TERMS
REMAINING AFTER FIRST ELIMINATION STEP[a]

FACTORS	TERMS
1	Gamey/fowl-like, canned, metallic, chemical*, astringent*
2	Burned, scorched, musty, stale, warmed-over, reheated, oxidized, rancid
3	Liver/organy, bone-marrow*, bitter
4	Browned, toasted*, moldy*
5	Chickeny, meaty, meaty chickeny, vegetable (cooked)*
6	Earthy*, papery, cardboard, painty
7	Brothy, bouillon-like*, sweet
8	Sour*

a - The proportion of variation in the data explained in the analysis was 77.3%.
*- Terms marked with an asterisk had low factor loadings and frequency of use was <40%.

A principal component analysis was performed for the 22 terms which remained to examine panelist agreement and to detect outliers. Figure 1 reveals that Panelist 2 was an outlier. Panelist 4, 5, and 8 were also outside the cluster of other panelists, who were in reasonable agreement. Examination of individual panelist data revealed that some of the lack of agreement was due to the range of the scale used. For example, Panelist 2 tended to have marked responses on all 22 scales and used the extreme right hand side (higher intensity scores). Panelist 4 tended to mark fewer scales and preferred the mid portion of the scale, avoiding the extremes. Panelist 8 tended to use the left side of the scale (weaker intensities). Panelist 5 tended to show large differences among the stored samples, while most panelists showed large differences between the fresh cooked samples and the stored sample, but scored the stored samples much closer together.

A canonical discriminant analysis was applied to evaluate classification of the samples by panelists based on canonical variables of the multivariate set of 22 terms (Fig. 2). Each plotted value represents a sample canonical mean score for each panelist. Based on the 22-term data set and new canonical variables constructed for the classification criteria, only two points are questionable.

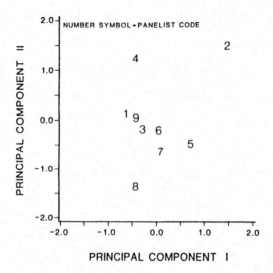

FIG. 1. PRINCIPAL COMPONENT ANALYSIS TO DETERMINE PANELIST AGREEMENT ON EVALUATION OF NEWLY-COOKED AND COOKED, STORED AND REHEATED PATTIES USING THE 22 DESCRIPTIVE ATTRIBUTES
Plotted numbers are panelist code.

These two points represent sample 1 (newly-cooked) and sample 2 (cooked, 1 day stored) for Panelist 6. This panelist did not show much differentiation between these two samples, whereas all other panelists clearly separated the newly cooked sample from the precooked stored samples. Among the stored samples, cooked, 1-day stored was separated from 3- and 5-day stored samples based on canonical variable #2. Mahalanobis distances were significant ($P < .05$) for differences between fresh and all other stored samples and between 1-day storage and 3- and 5-day storage. Three and 5-day storage were not significantly different according to this analysis.

A Factor Analysis applied to the 22 terms extracted 6 factors explaining 85.8% of the variation in the data (Table 3). Panel discussions and retesting were conducted to form a consensus of the terms to eliminate, if any. Reference material to better illustrate each term was presented to panelists. For example, cardboard was sniffed. Because it was described as "papery" and "stale," the panel agreed that "cardboard" was more specific. "Cardboard" was retained with the terms, "stale" and "papery," forming part of the definition. The panel agreed that "meaty, chickeny" was redundant for the two separate terms of "meaty" and "chickeny". "Scorched" was deleted in preference for the term "burned".

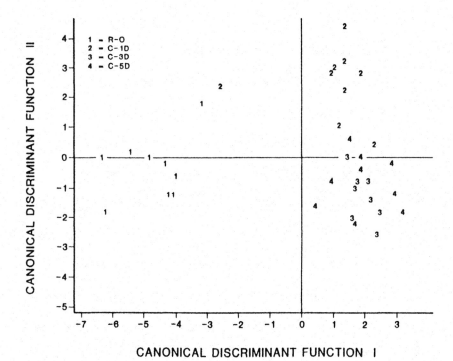

FIG. 2. SPATIAL RELATIONSHIPS OF NEWLY-COOKED (R-O) AND COOKED, STORED, REHEATED PATTIES BY CANONICAL DISCRIMINATE ANALYSIS BASED ON THE USE OF 22 DESCRIPTIVE ATTRIBUTES

Sample code 1 = Raw-0, newly-cooked; 2 = cooked, stored 1 day;
3 = cooked, stored 3 days; 4 = cooked, stored 5 days.

TABLE 3.
FACTOR ANALYZED GROUPINGS OF 22 TERMS
REMAINING AFTER SECOND ELIMINATION STEP[a]

FACTORS	TERMS
1	Scorched*, rancid, burned, bitter, stale*, browned, painty
2	Canned*, gamey/fowl-like, metallic, liver/organy
3	Chickeny, meaty chickeny*, meaty
4	Warmed-over, oxidized*, reheated*, musty
5	Cardboard, papery*
6	Sweet, brothy

a - The proportion of variation in the data explained in the analysis was 85.8%.

* Terms marked with an asterisk had low loadings on factors and were deleted by panel consensus.

"Canned" was evaluated a number of times to resolve whether this represented "retorted" flavor or "metallic" feeling factor. It could not be resolved and was deleted. "Gamey, fowl-like" received a great deal of attention because of confusion between (old fowl) and undomesticated fowl such as dove. This attribute was separated to give two terms, "gamey" and "fowl-like," each describing a distinct character note. "Oxidized" and "reheated" were deleted.

With 16 terms remaining on the list and tentative definitions given for each term, the panel evaluated patties from Test set #2. This test set included newly cooked samples, and cooked patties held under refrigerator storage for 0 through 5 days. Panelists evaluated each storage sample three times. Both univariate and multivariate analyses of variance were applied to the data. "Gamey" and "fowl-like" did not show discrimination of samples, and four panelists did not discriminate using these terms. Although there was strong feeling among some panelists that these terms were necessary, they were omitted due to lack of panel agreement. Also through panel discussion, decisions were made to combine "musty" with "cardboard" and to combine "painty" with "rancid".

Factor Analysis, Variable Analysis, and Stepwise Discriminant Analysis all verified deletion or combination of these terms. Table 4 gives results of the Variable Cluster analysis of the 16 descriptive terms. The proportion of variation in the data explained by the analysis was 85.4%. Eight groups of terms were formed. Groups containing terms with similar meaning are obvious. However, the panel did not agreed to select just one term in the cluster because they felt some terms represented distinct character notes. One term that did not fit the

TABLE 4.
GROUPING OF 16 DESCRIPTIVE TERMS BY VARIABLE CLUSTER ANALYSIS
(Varclus, centroid option)[a]

GROUPS	TERMS
1	Burned, rancid, painty
2	Chickeny, brothy
3	Gamey, fowl-like, liver/organy
4	Bitter, metallic
5	Sweet
6	Browned
7	Meaty
8	Musty, cardboard, warmed-over

a - The proportion of variation in the data explained in the analysis was 85.4%.

criteria of specificity, association, or cognition was "warmed-over." However, the panel insisted on its inclusion in the list of terms. A compromise was reached to allow retention of "warmed-over" in the list of terms with subsequent monitoring of its use in related studies. The final twelve terms that were selected and defined are given in Table 5.

TABLE 5.
TWELVE SENSORY DESCRIPTIVE TERMS WITH DEFINITIONS
DEVELOPED FOR EVALUATION OF CHICKEN FLAVOR

TERM	DEFINITION
	AROMATIC/TASTE SENSATION ASSOCIATED WITH:
Chickeny	– cooked white chicken muscle
Meaty	– cooked dark chicken muscle
Brothy	– chicken stock
Liver/organy	– liver, serum or blood vessels
Browned	– roasted, grilled or broiled chicken patties (not seared, blackened, or burned)
Burned	– excessive heating or browning (scorched, (seared, charred)
Cardboard/Musty	– cardboard, paper, mold, or mildew; described as nutty, stale
Warmed-over	– reheated meat; not newly-cooked nor rancid/painty
Rancid/Painty	– oxidized fat and linseed oil
	PRIMARY TASTE ASSOCIATED WITH:
Sweet	– sucrose, sugar
Bitter	– quinine or caffeine
	FEELING FACTOR ON TONGUE ASSOCIATED WITH:
Metallic	– iron/copper ions

Descriptive analysis is a powerful sensory tool. By incorporating multivariate statistical techniques (MVA) as aids in evaluating data collected during the training and language development process, panelists were better able to comprehend the meaning of terms, and how terms and scales were used by others. Adjustments by the individual panelists to be more in line with the panel group as a whole were more easily made by presenting results of the MVA statistical procedures.

In summary, a 45 word list to describe flavor changes in cooked and reheated chicken patties was reduced to 12 terms using panel input and MVA statistical procedures. We noted that flavor change due to short term refrigerated storage after cooking was detectable within 24 h, and description of the change is more complex than simply being off-flavored. The twelve descriptors derived through the panel process should be appropriate to assess this sensory change methodically.

ACKNOWLEDGMENTS

The technical assistance of Elizabeth Savage and Fredda Gillen in preparing products and conducting panel sessions, and the secretarial and administrative assistance of Frankee Simpson, are gratefully acknowledged.

The author also gratefully acknowledges the following Center employees who served as panelists: Carole Bassett, Carol Walter, Sherry Turner, Debbie Posey, Herb Morrison, Andra Dickens, Carl Davis, Gene Lyon, Roy Forbus, and Don Shackelford.

REFERENCES

ASTM. 1981. Guidelines for the selection and training of sensory panel members. STP 758. 35 pp, American Society for Testing and Materials: Philadelphia, PA.

BERRY, B.W., MAGA, J.A., CALKINS, C.R., WELLS, L.H., CARPENTER, A.L. and CROSS, H.R. 1980. Flavor profile analysis of cooked beef loin steaks. J. Food Sci. *45*, 1113–1115, 1121.

BIEBER, S.L. and SMITH, D.V. 1986. Multivariate analysis of sensory data: a comparison of methods. Chemical Senses *11*, 19–47.

CIPRA, J.S. and BOWERS, J.A. 1971. Flavor of microwave- and conventionally-reheated turkey. Poultry Sci. *50*, 703–706.

IGENE, J.O., PEARSON, A.M., DUGAN, L.R. and PRICE, J.F. 1980. Role of triglycerides and phospholipids on development of rancidity in model meat systems during frozen storage. Food Chem. *5*, 263–276.

JACOBSON, M. and KOEHLER, H.H. 1970. Development of rancidity during short-term storage of cooked poultry meat. J. Agr. Food Chem. *18*(6), 1069-1072.
JOHNSEN, P.B. and CIVILLE, G.V. 1986. A standardized lexicon of meat WOF descriptors. J. Sensory Studies *1*, 99-104.
LANDES, D.R. 1972. The effects of polyphosphates on several organoleptic, physical, and chemical properties of stored precooked frozen chickens. Poultry Sci. *51*, 641-646.
LYNCH, N.M., KASTNER, C.L., CAUL, J.F. and KROPF, D.H. 1986. Flavor profiles of vacuum packaged and polyvinyl chloride packaged ground beef: A comparison of cooked flavor changes occurring during product display. J. Food Sci. *51*, 258-276.
MACLEOD, G. and COPPOCK, B.M. 1978. Sensory properties of aroma of beef cooked conventionally and by microwave radiation. J. Food Sci. *43*, 145-161.
PERSSON, T., VON SYDOW, E. and AKESSON, C. 1973. Aroma of canned beef: sensory properties. J. Food Sci. *38*, 386-392.
POSTE, L.M., WILLEMOT, C., BUTLER, G. and PATTERSON, C. 1986. Sensory aroma scores and TBA values as indices of warmed-over flavor in pork. J. Food Sci. *51*, 886-888.
POWERS, J.J. 1981. Perception and analysis: a perspective view of attempts to find causal relations between sensory and objective data sets. In *Flavour '81*. (P. Schrier, ed.) pp. 103-131, Walter de Gruyter, Berlin.
POWERS, J.J. 1984. Using general statistical programs to evaluate sensory data. Food Technol. *38*(6), 74-84.
POWERS, J.J. 1986. Applying multivariate statistical methods to food industry problems. Special Report, No. 59: Sensory Evaluation. pp. 15-22, Twentieth Annual Symposium, Cornell Univ., Ithaca, NY.
SAS User's Guide. 1985. Statistics, Version 5 Edition. 956 pp, SAS Institute, Cary, NC.
STONE, H. and SIDEL, J.L. 1985. *Sensory Evaluation Practices*. 311 pp, Academic Press, Orlando, FL.
STONE, H., SIDEL, J., OLIVER, S., WOOLSEY, A. and SINGLETON, R.C. 1974. Sensory evaluation by quantitative descriptive analysis. Food Technol. *28*, 24-34.
VAN DE RIET, J.J. and HARD, M.M. 1979. Flavor quality of antioxidant-treated, cooked, ground beef patties. J. Am. Diet. Assn. *75*, 556-559.
WINGER, R.J. and POPE, C.G. 1981. Selection and training of panelists for sensory evaluation of meat flavors. J. Food Technol. *16*, 661-669.

CHAPTER 3.4

A TECHNIQUE FOR THE QUANTITATIVE SENSORY EVALUATION OF FARM-RAISED CATFISH

PETER B. JOHNSEN and CAROL A. KELLY

At the time of original publication,
all authors were affiliated with
USDA Agricultural Research Service
Southern Regional Research Center
P.O. Box 19687
New Orleans, LA 70179.

ABSTRACT

A descriptive analysis panel was trained to use a refined lexicon of flavor descriptors to evaluate farm-raised catfish flavor using a referenced intensity rating scale. To reduce variance due to within sample variability, a procedure using Blended Individual Fish Samples (BIFS) was developed. The reproducibility of the panel performance, utility of the BIFS and the sensitivity of the method to discriminate differences in fish flavor was demonstrated in three experiments.

INTRODUCTION

American per capita consumption of edible fishery products has risen from an estimated 17.1 lb in 1980 to 20.2 lb in 1987 (Dicks and Harvey 1988). To meet this growing demand, aquaculture output has increased over 20% annually since 1980 making it the fastest growing sector within American agriculture. Farm-raised catfish (*Ictalurus punctatus*) dominate fin-fish aquaculture in the United States with 295.1 million pounds processed in 1988 (Anon 1989). The growing demand for this product is based on its perceived high quality.

The first step of quality control in the processing of farm-raised catfish is a flavor evaluation to check for the absence of off-flavors and the presence of desirable flavors. If off-flavors are detected in test fish, the harvest of the source pond is delayed until off-flavors are eliminated. If harvested fish arrive at a processing plant and are determined to be off-flavor, they are returned to the pond. Only fish which are judged to possess suitable flavor quality are then processed.

Reprinted with permission of Food & Nutrition Press, Inc., Trumbull, Connecticut.
©Copyright 1990. Originally published in Journal of Sensory Studies **4**, 189-199.

Because flavor quality is essential to the marketability of catfish and is one of the most attractive features of the product, considerable attention is now being focused on factors which might impact flavor quality. The role of genetics, feed, production practices, processing and storage are all being investigated. An essential element to all of the ongoing research is sensory evaluation of the fish to determine the impact of the factor under study.

Previously, typical taste testing has involved cooking several fish or fillets either by baking or broiling (Boggess et al. 1971; Lovell 1983; Silva and Ammerman 1984). Screened, experienced and occasionally semitrained tasters evaluated several pieces of fish. Most often intensity of hedonic attributes are rated on an anchored but unreferenced scale. These procedures have proved to be adequate for many studies but as the complexity of research increases, the need for more rigorous evaluation methods becomes apparent.

In most studies no attempt to determine the precision and reliability of the evaluation is reported. Concern over the performance capabilities of individual panelists and the panel as a whole, as well as, the material being evaluated prompted us to develop a trained Descriptive Analysis (Meilgaard et al. 1987) panel and a unique method of sample preparation and presentation. Through this process we hoped to maximize the precision and reliability of the sensory evaluation of farm-raised catfish.

MATERIALS AND METHODS

Panel Selection and Training

Prospective panel members were recruited from the neighboring community. Candidates underwent a screening process which included tests to determine their ability to recognize and name basic tastes and common aromatics. Triangle tests at near threshold intensity for some common off-flavors were used to ensure that panel members were not insensitive to compounds of interest. In addition, a questionnaire to determine eating, drinking and smoking patterns as well as general awareness to flavors and the ability to communicate chemosensory experiences was administered. While panelists were to be financially compensated for participation, the one year time commitment required that an interest in the work be the principal motivation for becoming a member of the panel.

The selected panel of 16 members ranged in age from 19 to 74 years, contained no smokers and included four members with partial dentures. Because all members were trained to rate intensities on the basis of a reference scale, possible sensitivity differences arising from age were eliminated.

Panel training covered a period of 5 months (75 total hours). During training, panelists were instructed in general sensory descriptive analysis (Civille 1979; Meilgaard *et al.* 1987) and were led through the development of a lexicon of descriptors for fish in general and then farm-raised catfish in particular. Panelists individually listed terms that described the flavor notes they perceived in the samples. Through guided discussions, redundant terms were eliminated and agreement was reached on terms for the sensory ballot.

Panel performance during training was evaluated by simple techniques, such as examining the range of scores for each attribute. Individual mean scores were compared to the panel mean scores and those members using restricted ranges of the intensity scale or showing a large standard deviation in their assessment of repeated samples were coached to improve performance.

Fish Sample Preparation

Experimental fish are processed using a modification of commercial processing practices. Fish are decapitated, eviscerated and placed into an ice bath to chill. Skinning is accomplished by a Jaccard model A35-P membrane skinner adjusted to remove skin and fascia. Shank fillets are prepared by hand as are most commercially prepared fillets. The experimental method differs from commercial in the equipment used for decapitation and evisceration. The critical skinning procedure which leaves the subcutaneous fat layer intact is the same.

Often a major shortcoming of sensory evaluation of fish is the flavor variation found both between fish as well as within a fish. Our *ad hoc* fish panel often noted that anterior and posterior portions of fish had different intensities of flavors and the flavor of different fish taken from the same pond could have quite disparate flavor profiles. These variations often led to large standard deviations in sensory evaluation scores. To reduce this we sought to create a sample which would be more homogenous and thus better representative of the population.

Samples for sensory analysis are made into Blended Individual Fish Samples (BIFS). These individual portions are prepared by combining fillets from all fish of an experimental sample. The fillets are shredded without excessive cell disruption by a food processor run for four seconds. Following thorough mixing, ten gram samples are then placed in seal-a-meal bags (7 × 7 cm). Excess air is expelled from the bags which are then heat sealed. Samples are frozen at −20C until presentation to the panel. This pooled sample technique assures that replicates are true and that subsamples taken for chemical analysis are properly representative of the source sample.

Sensory Panel Protocols

BIFS are placed in boiling water and cooked for five minutes after the water returns to a boil. The BIFS are presented under red light to the panelists who open the bags with scissors, smell the aromatics and place approximately half of the contents into the mouth for flavor by mouth assessment. Intensity of aromatics, tastes and feeling factors are recorded. Texture is not assessed. The second half of the sample is then evaluated and initial scores are corroborated or are adjusted to represent the integrated sample. Specific intervals between sample presentations are chosen depending on the design of the experiment. Unsalted crackers and distilled, deionized water are used to rinse the mouth between samples. A 0.03% citric acid rinse is sometimes used as an additional rinse when working with strong off-flavor samples. Typically three replicates for each sample are evaluated by the panelists.

Descriptive analysis profiles are prepared using a lexicon modified from Johnsen et al. (1987). Terms and definitions are presented in Table 1. While similar to the descriptors presented previously, the revised descriptor list included the new terms "eggy/sulfury", "fishy", "bitter" and "peppery". The "grainy/corn/green vegetable" complex of Johnsen et al. (1987) was further resolved into two different terms "green vegetable/grassy" and "corn". Similarly, "boiled chicken/buttery" was differentiated to "chickeny" and "fat complex". These changes primarily reflect our interest in enhanced resolution in the evaluation of desirable flavor attributes as opposed to simple identification of major off-flavors.

Intensities are based on an open-ended scale established in reference to flavor intensities that are assigned to specific characteristics apparent in several commercially available food products (Table 2). Farm-raised catfish flavor intensities are less than 10 on this scale.

Sensory evaluations are recorded via a computer system using a light pen to indicate the descriptor intensity on 16 cm lines (61 pixels). Values are anchored by the numbers one through ten above the line. Data are captured by Compusense Inc. (Guelph, Ontario) Computerized Sensory Analysis — Descriptive Analysis (CSADA ver. 1.02A) software operated by a Novell Inc. (Provo, Utah) STF Netware 286 (ver. 1.02) network of 12 PC-XT computer stations. Sample identification, individual panelist responses, panel means and standard deviations are processed and returned to panel members at the conclusion of each session for discussion or coaching. For further analysis, data are transposed to SAS files (SAS Institute Inc., Cary NC) and subjected to specific statistical tests.

Panel sessions are initiated with members tasting and reviewing the intensity reference standards. A commercially obtained "standard" fish sample chosen for the duration of the experiment is then presented and evaluated. The

TABLE 1.
CATFISH FLAVOR DESCRIPTORS

DESCRIPTOR	DESCRIPTION
AROMATICS	
Nutty	The aromatic associated with fresh pecans and other hardshell nuts.
Chickeny	The aromatic associated with sweet cooked chicken meat.
Fat Complex	The aromatic associated with dairy lipid products, melted vegetable shortening, and cooked chicken skin.
Corn	The aromatic associated with cooked corn kernels.
Green Vegetable/Grassy	The aromatic associated with fresh grassy vegetation and green vegetables.
Eggy/Sulfury	The aromatic associated with boiled old-egg proteins.
Geosmin/Dry musty	The aromatic associated with old books. Geosmin is the reference.
MIB/Wet musty	The aromatic associated with mud. 2-methylisoborneol is the reference.
Decaying vegetation	The aromatic associated with decaying vegetation particularly pondweed, decaying wood, and swamp grass.
Cardboardy	The aromatic associated with slightly oxidized fats and oils, and reminiscent of wet cardboard or a brown paper bag in the mouth.
Fishy	The aromatic associated with the reference trimethylamine.
TASTES	
Sweet	The taste on the tongue associated with sugars.
Salty	The taste on the tongue associated with sodium ions.
Sour	The taste on the tongue associated with citric acid.
Bitter	The taste on the tongue associated with caffeine.
FEELING FACTORS	
Astringent	The sensation on the tongue, described as puckering/dry, and associated with strong tea.
Metallic	The sensation on the tongue described as flat, and associated with iron and copper.
Peppery	The sensation on the tongue described as tingly, and associated with pepper.

panel means for individual attributes are calculated and discussed. Consensus values are then agreed upon. This exercise serves to assist individuals to establish their daily calibration and as a warm-up (O'Mahony *et al.* 1988). The unknown samples are then presented in a random order previously determined for

TABLE 2.
INTENSITY REFERENCE SCALE[1]

Intensity	Descriptor	Description of Product
1	cooked wheat	Wheat Thins (Nabisco Brands, Inc.)
2	oil	Lays Potato Chips (Frito-Lay, Inc.)
3	buttery	Land-O-Lakes Margarine (Land O'Lakes ,Inc.)
4	grape	Grape Kool-Aid (General Foods Corp.)
5	apple	Mott's Natural Apple Sauce (Mott's USA, Cadbury Schweppes, Inc.)
6		
7	orange	Minute Maid Frozen Concentrated Orange Juice (Coca-cola Foods)
8		
9		
10	grape	Welch's Grape Juice (Welch's)

[1] Adopted from Meilgaard et al., 1987

the experiment. Individual samples are identified by three digit random codes written on the BIFS bags with an indelible black marker. Typically, six experimental samples are evaluated in two hour panel sessions meeting twice a week.

RESULTS AND DISCUSSION

Evaluation of Panel Performance

BIFS *versus* Fillets. Concerned that between fish and even within fish differences could mask minor effects from experimental treatments, BIFS were developed as a sample presentation protocol. To determine if this concern was valid, an experiment was conducted to examine within sample variation. The variance of the sensory scores for BIFS was compared to the variance of scores for whole fillets. Using samples created by purging off-flavor fish for various times, three treatments levels (low or no off-flavor, 14 day purge; slight off-flavor, 2 day purge; strong off-flavor, no purge) were obtained. BIFS were prepared from one side of the fish while selected fillets from the opposite side were packaged in boiling bags. As with BIFS, panelists received 10 g samples from the fillet samples. Following flavor evaluations, scores for off-flavor attributes were submitted to the Bartlett test for homogeneity of variances (Snedecor and Cochran 1967). Because it is uncertain whether the variable being tested for homogeneity of variances meets the normality assumption, we used $\alpha = 0.001$ to reject the null hypothesis as suggested by Anderson and McLean (1974). This analysis indicates that there were significant differences for the variances of off-flavor attributes (geosmin, MIB, decaying vegetation, green vegetable/grassy) between BIFS and fillets for the 0 and 2 day purge samples but no differences in the 14 day purge samples. This might be expected for the 14 day samples because both BIFS and fillets had no off-flavor. Thus, the advantage of using BIFS to present a more homogeneous sample to the panel and thereby reducing experimental variance was confirmed.

Panel Reproducibility. Two experiments were conducted to evaluate panel performance. In Experiment 1, four groups of fish were evaluated (3 farm-raised ponds and 1 wild river fish). Likewise, in Experiment 2, four groups (small farm-raised; large farm-raised; wild lake fish; experimental pond fish) were evaluated. Fish were subjected to six replicate evaluations. To examine the reproducibility of repeated evaluations, panel means of individual attributes for each fish of the experiments were subjected to SAS Inc. (1985) General Linear Models (GLM) of Analysis of Variance (ANOVA). Using replicates as the test factor, significant differences between repeated sensory evaluations for the different attributes of individual fish in Experiment 1 were observed (Table 3). Following further training and practice, Experiment 2 was conducted. This time no differences for replicate evaluations were noted indicating that the panel was adequately trained (Table 4). This conclusion was further supported in subsequent experiments to be reported elsewhere.

Panel Sensitivity. An experiment was conducted to determine if the panel was functioning as a discriminating instrument and could differentiate among groups of fish with presumably minor flavor variations. The four groups of fish from Experiment 2 were evaluated. Seven attributes were always observed and are included in this analysis. Again, SAS (1985) GLM ANOVA was used to examine the data. Duncan's Multiple Range Test (SAS 1985) indicated the different fish could be discriminated using the indicated attributes (Table 5).

The distinctive Nutty flavor was significantly higher in intensity in small and large farm-raised catfish compared to wild and experimental pond fish. At the same time, large catfish exhibited the most intense Chickeny flavor with wild and experimental pond fish significantly lower than both farm-raised groups.

TABLE 3.
EXPERIMENT 1: SUM OF SQUARES AND MEANS FROM ANALYSIS OF VARIANCE

SOURCE	df	NTY	CHY	FCX	CRN	GRV	SWT	STY
Rep	5	27.0**	35.51**	30.02**	10.93**	13.55**	9.10**	4.48*
Fish	3	13.37**	32.55**	10.78**	1.63	2.91	18.13**	3.05*
Rep X Fish	15	9.11	13.81	10.65	12.16	8.85	10.92*	5.84
Grand Means	-	2.08	2.59	1.38	1.34	0.76	1.14	1.30

NTY = nutty; CHY = chickeny; FCX = fat complex; CRN = corn; GRV = green vegetable/grassy
SWT = sweet; STY = salty

* = P < 0.05; ** = P < 0.01

TABLE 4.
EXPERIMENT 2: SUM OF SQUARES AND MEANS FROM ANALYSIS OF VARIANCE

SOURCE	df	NTY	CHY	FCX	CRN	GRV	SWT	STY
Rep	5	1.19	4.17	8.13	1.69	8.04	1.40	0.99
Fish	3	49.62**	85.86**	18.96**	18.30**	10.28**	28.77**	6.60**
Rep X Fish	15	14.54	16.44	7.71	9.54	9.83	7.35	1.68
Grand Means	-	1.81	2.50	1.57	0.89	1.25	0.89	1.41

NTY = nutty; CHY = chickeny; FCX = fat complex; CRN = corn; GRV = green vegetable/grassy
SWT = sweet; STY = salty

* = P < 0.05; ** = P < 0.01

Both Corn and Sweet were lower in the wild and experimental fish. These four sensory attributes represent the desirable flavor compliment for catfish. The two farm-raised catfish groups fed the high grain based diets had the highest scores for these flavors. The experimental pond fish were fed a trout chow which is higher in fish meal than vegetable source protein. Stomach analysis indicates that invertebrates are the principal diet items in wild catfish.

The farm-raised and experimental pond fish were grown in a nutrient enriched environment which results in significant blooms of algae. Likewise, Lake Des Allemands is a eutrophic environment with significant algae populations (Malone and Burden 1985). While these microorganisms are known to produce many compounds which might be absorbed by the fish and contribute to flavor composition (Juttner 1983), the intensity of Nutty, Chickeny, Corn and Sweet appears to be more related to diet than environment.

A flavor attribute which does seem to be related to the environment is the Green vegetable/grassy. The large farm-raised catfish were lower in intensity for this attribute than the other three groups. This might be explained by the fact that brood fish ponds are generally less eutrophic than production ponds due to lower feeding rates. Feed added to the pond increases available nutrients which in turn promotes algae growth. Brown and Boyd (1982) have correlated the severity of off-flavor problems in catfish production ponds with the amount of feed and thereby nutrients added to the system.

Determinations of fat content in terms of per cent fillet weight were made by the perchloric-acetic acids method (Koniecko 1985) and indicate that there was a range from 7.5% to 11.25% (Table 5). The experimental pond fish fat composition values were the highest as were the Fat Complex sensory scores. However the relationship between fat composition and this sensory attribute does not follow for the other three groups. Fat Complex sensory scores for wild fish were the lowest while the fat content values were second highest. This incongruity, however, may be due to the type of lipids found in farmed fish versus wild fish but remains to be confirmed by an appropriate experiment.

TABLE 5.
EXPERIMENT 2: FAT CONTENT AND MEAN SENSORY SCORES FOR FOUR CATFISH SAMPLES

FISH SAMPLE	%FAT	NTY	CHY	FCX	CRN	GRV	SWT	STY
Small farm-raised	7.25	2.32 a	3.00 b	1.54 b	1.15 a	1.49 a	1.14 b	1.45 a
Large farm-raised	8.25	2.40 a	3.42 a	1.50 b	1.15 a	0.73 b	1.43 a	1.61 a
Wild lake fish	9.25	1.39 b	1.94 c	1.05 c	0.53 b	1.24 a	0.38 c	1.15 b
Experimental pond	11.25	1.08 b	1.57 c	2.26 a	0.69 b	1.55 a	0.60 c	1.42 a

NTY=nutty; CHY=chickeny; FCX=fat complex; CRN=corn; GRV=green vegetable/grassy SWT=sweet; STY=salty

Means with same letter in a column are not significantly different.

CONCLUSIONS

Descriptive Analysis panels can be trained to evaluate fish samples with a high degree of reproducibility and precision. It is possible to discriminate among samples with presumably small flavor differences on the basis of specific flavor attributes. The procedure makes use of Blended Individual Fish Samples (BIFS) which were developed to reduce experimental variation due to differences between and within samples. Experimental support for this procedure was presented. An additional advantage of the procedure is that true subsamples may be drawn from this pooled sample for further chemical analysis. This tends to minimize possible sources of variance between sensory and chemical analyses.

The demonstrated sensitivity of this sensory evaluation technique will allow analyses based on specific attributes. The technique permits more detailed evaluation of experimental treatments than do hedonic preferences and/or difference tests. In this way, individual flavors can be evaluated and related to the source or cause of particular flavor notes. In addition, off-flavors such as geosmin, 2-methylisoborneol or lipid oxidation products (Cardboardy and Painty) can be quantitated and abatement measures adequately evaluated.

The described method should be useful for flavor evaluation of all fish species after appropriate descriptor lexicons have been developed. While the BIFS preparation protocol does eliminate texture cues and long-term frozen storage samples most probably oxidize faster than intact fish, the method does have advantages over many previously used methods.

ACKNOWLEDGMENTS

The authors wish to thank Dr. Karen L. Crippen for her contributions to panel training, statistical advice and management of the sensory laboratory at SRRC. Additionally, the dedicated efforts of all panel members and sensory laboratory staff are greatly appreciated. Use of a product name by the department does not imply approval or recommendation of the product to the exclusion of others which may also be suitable.

REFERENCES

ANDERSON, V.L. and McLEAN, R.A. 1974. *Design of Experiments — A Realistic Approach*, 418 pp. Marcel Dekker, New York.

ANON. 1989. Catfish Processing Report, January 1989. USDA National Agricultural Statistics Service.

BOGGESS, T.S., HEATON, E.K. and SHEWFELT, A.L. 1971. Storage stability of commercially prepared and frozen pond-raised channel catfish (*Ictalurus punctatus*, Rafinisque). J. Food Sci. *36*, 969-973.

BROWN, S.W. and BOYD, C.E. 1982. Off-flavor in channel catfish (*Ictalurus punctatus*) from commercial ponds. Trans. Am. Fish. Soc. *111*, 379-383.

CIVILLE, G.V. 1979. Descriptive analysis-flavor profile, texture profile, and qualitative descriptive analysis. In *Sensory Evaluation Methods for the Practicing Food Technologist* (M.R. Johnson, ed.) Institute of Food Technologists, Chicago, IL.

DICKS, M. and HARVEY, D. 1988. Aquaculture: Situation and outlook report. USDA Economic Research Service. AQUA 1, October 1988. 39 pp.

JOHNSEN, P.B., CIVILLE, G.V. and VERCELLOTTI, J.R. 1987. A lexicon of pond-raised catfish flavor descriptors. J. Sensory Studies *2*, 85-91.

KONIECKO, E.S. 1979. *Handbook of Meat Analysis*, 289 pp. Avery Publishing Group, Wayne, N.J.

LOVELL, R.T. 1983. New off-flavors in pond-cultured channel catfish. Aquaculture *30*, 329-334.

MALONE, R.F. and BURDEN, D.G. 1985. A condition index system for Louisiana lakes and reservoirs: Final report, 156 pp. Louisiana Dept. Environmental Quality, Baton Rouge.

MEILGAARD, M., CIVILLE, G.V. and CARR, B.T. 1987. *Sensory Evaluation Techniques*, Vol. 11, 159 pp. CRC Press, Boca Raton, FL.

O'MAHONY, M., THIEME, U. and GOLDSTEIN, L.R. 1988. The warm-up effect as a means of increasing the discriminability of sensory difference tests. J. Food Sci. *53*, 848-850.

SAS. 1985. *SAS/STAT Guide for Personal Computers*, Version 6, 378 pp. SAS Institute, Cary, NC.

SILVA, J.L. and AMMERMAN, G.R. 1984. Effect of the size of channel catfish (*Ictalurus punctatus*) on its storage stability and on its quality attributes. Catfish Processors Workshop *13*, 18-20. Miss. State University.

SNEDECOR, G.W. and COCHRAN, W.G. 1967. *Statistical Methods*, 6th ed. 593 pp. Iowa State University Press, Ames, Iowa.

CHAPTER 4.0

ALCOHOLIC BEVERAGES

Amerine and Roessler (1976) pioneered the use of sensory evaluation in wines by publishing this important book. Since then, several works have been published profiling alcoholic beverages by descriptive analysis. However, a healthy debate continues between wine quality experts and descriptive analysis panel advocates. In this chapter, research findings on the versatility of descriptive analysis for profiling alcoholic beverages are demonstrated. Chapter 4.6 discusses the use of quality experts in descriptive analysis.

REFERENCE

AMERINE, M.A. and ROESSLER, E.B. 1976. Wines, Their Sensory Evaluation. W.H. Freeman and Co., San Francisco.

CHAPTER 4.1

SENSORY PROFILING OF BEER BY A MODIFIED QDA METHOD

JAMES M. MECREDY, JOHN C. SONNEMANN
and SUSAN J. LEHMANN

*At the time of original publication,
all authors were affiliated with
the Consumer Product Research Dept.
Jos. Schlitz Brewing Co.
1610 N. 2nd St., Milwaukee, Wis. 53212.*

Sensory evaluation has been a routine, everyday operation at the Jos. Schlitz Brewing Co. for years. The brewmasters meet daily in their taste test laboratory to evaluate current production and to assess competitive trends in beer quality. As a result of years of experience, they have developed an acuity to recognize not only the subtle differences influenced by variables in the brewing process, but also — to a surprising degree — changes that competitors make in their brewing processes.

To provide the brewmasters with information on the degree to which consumers can recognize variations in beer flavors, we decided to develop an additional taste test laboratory, as one arm of a consumer product research facility. The role of the laboratory would be to quantitatively describe the flavor differences that exist among competitive beers.

TASTE PANEL ESTABLISHED

Work was started to select and train a beer taste panel using the basic principles of classic descriptive analysis (Amerine *et al.* 1965), modified to fit the needs and experiences of the brewing department. Prospective panel members were recruited from salaried personnel at the corporate headquarters in Milwaukee. Approximately 200 prospective panel members were screened, using standard gustatory techniques of rank order testing with sweet, sour, bitter, and salt water samples (Harrison and Collins 1968; Harrison 1970). The 50 most sensitive people were further screened, using triangle taste tests with beer. Those who discriminated at the 95% confidence level were trained to

Reprinted with permission of the Institute of Food Technologists, Chicago, Illinois.
©Copyright 1974. Originally published in Journal of Food Technology (Nov.) 36-37, 40-41.

recognize the various beer characteristics shown in Table 1 by giving them beers that had been modified by the addition of appropriate chemicals to demonstrate each attribute. Once panelists were familiar with the beer attributes, work was started on profiling samples of competitive beers.

Panel members attended taste test sessions several times a week to evaluate five beers at each session. They evaluated beers one at a time, recording their responses on the score sheet shown in Fig. 1. Each beer was rated for the intensity of each of the attributes, using the 9-point rating scale which ranged from "not noticeable" to "very strong." Note that the score sheet does not contain an "overall quality" or "preference" rating. The reason is that "true" preference measures are obtained directly from consumers by the Field Research arm of the Consumer Product Research Department.

TABLE 1.
GLOSSARY OF BEER CHARACTERISTICS USED IN ORIGINAL TESTING

Attribute	General description
Fruity/floral	Perfumy fruit flavor, resembling ripe fruit or flowers. For example, apple
Oxidized	Old beer, stale
Yeasty	Baker's yeast
Hoppy	Typical odor of the essential oil of hops
Bitter	Bitter taste—as in tonic water or quinine sulfate
Grainy/worty	A cereal grain-type flavor
Tart	Sour taste. For example, vinegar or lemon
Musty	Dank cellar, moldy
Sulfur dioxide	Odor of burnt matches
Malty	Malt syrup or malt powder used in malted milk. Heavy body
Diacetyl	Typical butter-like odor
Skunky	Characteristic odor of a skunk
Phenolic	Medicinal
Onion	Taste of onions, disagreeable, mercaptan
Aftertaste	Astringent. A hanging in the throat as in drinking whiskey neat

Beer Evaluation

Taster ______________________ (   ) 14-17   Sample No. ______________

32-39 (z

		-18 Fruity floral	-19 Oxidized	-20 Yeasty	-21 Hoppy	-22 Bitter	-23 Grainy worty	-24 Tart	-25 Musty	-26 Sulfur dioxide	-27 Malty	-28 Diacetyl	-29 Skunky	-30 Phenolic	-31 Onion	-40 After-taste	-41 Other
Not noticeable	0																
Trace, not sure	1																
Faint	2																
Slight	3																
Mild	4																
Moderate	5																
Definite	6																
Strong	7																
Very strong	8																

Date ______________________ 42-47

Form 832

80(T

FIG. 1. SCORE SHEET USED IN ORIGINAL TESTING

IMPROVEMENTS NEEDED

Between 1971 and 1972, using an average of 25 panel members per sample, profiles were obtained with 607 beer samples across 83 different brands-over 10,000 total observations. An analysis of variance was carried out on the entire file of data. Results showed that variations due to individual panel member scoring were considerably larger than variations due to differences among brands of beer while differences among brands of beer were statistically significant, subsequent analysis showed that approximately 10 samples of a particular beer (250 observations) were required to draw reliable conclusions about beer differences.

The analysis indicated that the efficiency and statistical reliability of the sensory evaluation program would have to be substantially increased for the laboratory to be of major value. An intensive review of the collected data and laboratory procedures led to a number of areas where improvements could be made:

· **Panel Members** needed retraining in weak areas.

· **A Larger Pool** of panel members was needed; panel members tended to become less discriminating over a long period of time, and the indication was that they discriminated better if given 3- to 4-week rest periods followed by refresher training.

· **The List of Attributes** needed revising to eliminate those on which beers differ very little and to add attributes that panel members were requesting.

· **Revisions in Serving Techniques** would reduce chances for errors in serving.

· **Efficiency of Screening** and training could be increased by starting with triangle testing using beer instead of water samples, which were less predictable of subsequent panel performance.

· **Data Handling Procedures** were needed to allow statistical analysis on a routine basis.

· **Statistical Analysis** could be strengthened by modifying the 9-point intensity scale to reduce skewing of responses to the lower end of the scale.

QDA METHOD EVALUATED

A joint experiment with Stanford Research Institute was undertaken to evaluate their Quantitative Descriptive Analysis (QDA) method (Stone *et al.* 1974) as a possible way to increase the efficiency and statistical stability of our sensory evaluation program.

In this experiment, four different brands of beer were rated by panel members using our original methodology. Another panel was trained to use the QDA method. Results showed that a definite improvement could be gained by using the QDA method. The panel members trained by the QDA method discriminated differences among beers on 13 of the 15 scales at the 99% level of statistical significance. The panel using our original method discriminated differences on 6 of the 15 scales but only at the 95% level of statistical significance.

Two important principles in the QDA method were recognized as a means of obtaining some of the desired improvements. These principles are:

· **The Use of Experimental Procedures** that yield quantitative data to which statistical analysis can be applied.

· **An Emphasis on Rigorous Control** of the testing environment and data collection to minimize the effect of extraneous influences on the results.

TABLE 2.
MODIFIED GLOSSARY OF BEER CHARACTERISTICS USED AFTER ADOPTION OF THE QDA METHOD

Attribute	General description
Aroma[a]	The total intensity of the odor of a product
Fruity	Perfumy fruit flavor, resembling fruit or flowers
Hoppy	Odor of the essential oil of hops
Sweet[a]	The taste of sugar, sensed primarily at the tip of the tongue
Diacetyl	Typical butter-like flavor
Tart	Taste sensation caused by acids, sensed primarily along the sides of the tongue. For example, vinegar or lemon
Bitter	The bitter taste from hop isohumulones. It is sensed primarily at the back of the tongue. Tonic water and caffeine produce this sensation
Flavor strength[a]	The intensity of overall taste sensation of a product. For example, weak/light or strong/heavy, full-bodied
Oxidized	Aged beer. May resemble a papery or cardboardy taste
Malty	Malt syrup
Aftertaste	The duration of a taste sensation perceived after swallowing
Sulfur compounds:	
Skunky	Characteristic odor of a skunk
Sulfur dioxide	Taste or odor of burnt matches
Hydrogen sulfide	Odor of rotten eggs
Onion	Flavor of onions

[a] Not on original list

QDA METHOD MODIFIED

Modifications to the general QDA methodology were made in three areas:
· **Brewmasters' Terminology,** which had been used in prior work, was retained for continuity with the brew-masters' historical perspective. Since the results are used primarily by brewmasters, credibility demands that the characteristics be technically definable insofar as possible. On the other hand, understanding what the consumer experiences requires inclusion of those characteristics which are generally recognized to exist even though they cannot be precisely defined technically. Therefore, group discussions were held with the panel members, and additional characteristics were added at their request. The attributes adopted from this review (shown in Table 2) included most of the original terms plus "aroma," "flavor strength," and "sweet."
· **Panel Members Were Retrained** using modified beers but with more emphasis on the identification of characteristics in several brands of beer. We had empirically found that modifiers were either masked or accentuated by inherent characteristics in a particular brand of beer, so it was necessary to use more than one brand to train panel members.
· **"Overall Quality"** or "preference" measures were excluded from our adaptation of the QDA method for the same reasons that they were excluded from our original work.

HOW THE METHOD WORKS

Today we have a pool of 50 people trained to evaluate sensory characteristics of beer. For a particular experiment, 10-14 judges are chosen at random from the pool. Daily sessions are conducted so that experiments may be completed in a relatively short period of time. During each session, panelists evaluate five beers, one at a time, recording their responses on the score sheet shown in Fig. 2. The panelist marks each continuous scale with a single vertical mark, and the results are later transformed into numbers, using an overlay. The data are then statistically analyzed by analysis of variance to determine judges' performance as well as product differences. Post hoc comparisons are made to determine significant attribute differences among the beers.

For presentation, attribute means are plotted on polar coordinate scales (Fig. 3) which show the flavor dimensions of the beers. The arrangement of flavor dimensions around the center point is for convenience only, with the taste attributes on the bottom and the aroma attributes on the top. The center point of the axis represents the minimum perceptible intensity, with the intensity of each characteristic increasing with distance from the center.

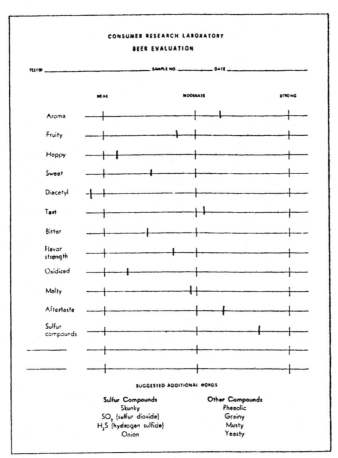

FIG. 2. MODIFIED SCORE SHEET USED AFTER ADOPTION OF THE QDA METHOD

For illustrative purposes, the attribute mean values are connected to provide an individual beer profile, such as that for a typical American beer shown in Fig. 4. The width of the band connecting the dimensions is approximately equal to the 95% confidence limit for each sample.

For comparative purposes, profiles may be superimposed on each other. An open space between bands represents a statistically significant difference between samples. Figure 5 shows a typical imported beer compared with the American beer. It is apparent that the imported beer is more "hoppy," "bitter," and "malty" and less "tart" than the American beer.

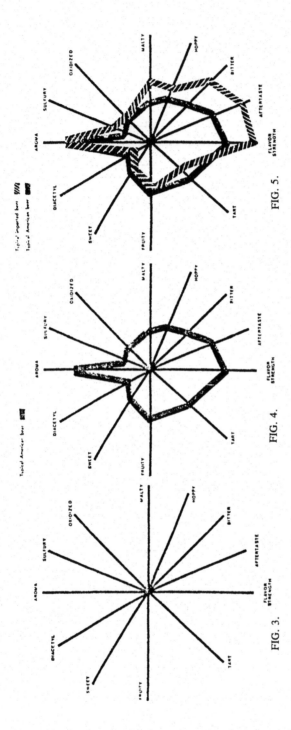

FIG. 3. POLAR COORDINATE SCALE USED FOR PLOTTING THE ATTRIBUTE MEANS;
Arrangement of the attributes around the center point is for convenience only.
FIG. 4. PROFILE OF A TYPICAL AMERICAN BEER; THE WIDTH OF THE BAND
Connecting the attributes is approximately equal to the 95% confidence limit.
FIG. 5. COMPARISON OF A TYPICAL IMPORTED BEER AND A TYPICAL AMERICAN BEER;
Open space between the bands represents a statistically significant difference.

DESCRIPTIVE SENSORY ANALYSIS IN PRACTICE 311

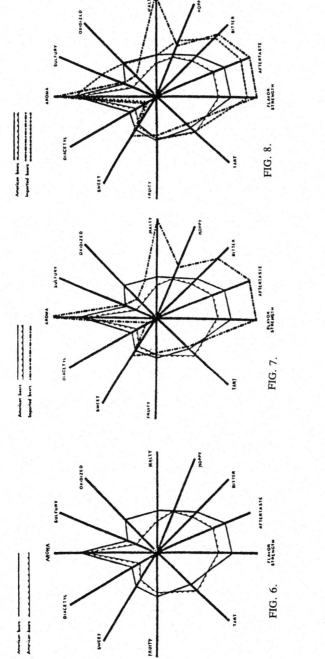

FIGS. 6, 7, AND 8. PROFILES FOR VARIOUS BEERS CAN BE SUPERIMPOSED ON EACH OTHER FOR COMPARISON
(note that in order to differentiate the various beer samples when the figures are reduced for publication, the profiles are drawn with narrow lines connecting the attribute mean values; in actual use, the lines would be equal in width to the 95% confidence limits)

The utility of this presentation technique can be illustrated with the results of a recent study on some American and imported beers, shown in Fig. 6-8. Note that in order to differentiate the various beer samples when Fig. 6-8 are reduced for publication, the profiles are drawn with narrow lines connecting the attribute mean values; in actual use, the lines would be equal in width to the 95% confidence limits for each sample.

Figure 6 shows a comparison of two American beers which differ in "aroma," "diacetyl," "flavor strength," "oxidized," and "after-taste." An imported beer (Fig. 7) shows dramatic differences in "aroma," "malty," and "bitter." A second imported beer (Fig. 8) is closer to American beers in "aroma" and "malty."

The beers used in this particular experiment were all purchased at retail, so the characteristics of each are representative of what the consumer actually gets.

A VALUABLE TOOL

Thus, the adoption of the Quantitative Descriptive Analysis principles of experimental design — using quantitative data to which statistical procedures are applied and emphasizing rigorous control of the testing environment — has provided a valuable tool for understanding sensory differences among beers.

REFERENCES

AMERINE, M.A., PANGBORN, R. and ROESSLER, E.B. 1965. *Principles of Sensory Evaluation of Food*, p. 377. Academic Press, New York.

HARRISON, G.A.F. and COLLINS, E. 1968. Determination of taste thresholds for a wide range of volatile and nonvolatile compounds in beer. Am. Soc. Brew. Chem. Proceedings, St. Paul, Minn.

HARRISON, G.A.F. 1970. The flavour of beer — A review. J. Inst. Brewing 76, 486.

STONE, H., SIDEL, J., OLIVER, S., WOOLSEY, A. and SINGLETON, R.C. 1974. Sensory evaluation by Quantitative Descriptive Analysis. Food Technol. 28 (11).

Based on a paper presented at the Sensory Evaluation Division Symposium on Recent Developments in Descriptive Testing, at the 34th Annual Meeting of the Institute of Food Technologists New Orleans, La., May 12-15, 1974.

CHAPTER 4.2

FACTOR ANALYSIS APPLIED TO WINE DESCRIPTORS

LOUISE S. WU, R.E. BARGMANN and JOHN J. POWERS

*At the time of original publication,
all authors were affiliated with
University of Georgia
Athens, GA 30602.*

ABSTRACT

Eighty-six descriptors commonly used to characterize wines were evaluated by 86 respondents. The list was reduced to 33 descriptors which had the greatest meaning. Panels of 33 and 18 judges, respectively, evaluated 14 red and 12 white table wines for acceptability. For each descriptor, each wine was compared against a red or white reference wine. The wines had originally been purchased based on price and general repute as to quality. The panels did not rank the most expensive wines as the most acceptable. For the red wines, 27 descriptors were used for factor analysis. Eight factors were extracted and 14 well defined, oblique simple-structure planes were identified. For the white wines, six factors were extracted and 11 planes identified.

INTRODUCTION

To assess quality, foods are frequently scored for such broad categories as acceptability, appearance, color, flavor, or mouthfeel. The last four are, of course, components of the first. In turn, a component such as flavor is itself composed of numerous taste and odor notes. If these sub-components could be categorized and weight attached to their importance, flavor quality might be evaluated more precisely. The same would apply to the other components of acceptability.

One of the problems in dealing with various flavor or mouthfeel nuances is to describe them. Considerable effort usually has to be expended to articulate a list of descriptors because the response of each individual is so subjective. Among the means of classifying the various descriptors with some degree of objectivity is factor analysis. A long-term goal of this laboratory is the development of better methods to relate particular descriptors to particular

*Reprinted with permission of the Institute of Food Technologists, Chicago, Illinois.
Copyright 1977. Originally published in Journal of Food Science **42**, 944-952.

chemical or mechanical measurements. The immediate objectives of this study were (1) to apply factor analysis to wine descriptors to ascertain whether panelists found the descriptors useful in discriminating among wines and (2) whether the descriptors as subcomponents of flavor appeared to be important in determining flavor acceptability. Wine was selected for this study because, unlike most other foods, there already is an abundance of descriptive terms; thus, a word list did not need to be originated.

Vuataz et al. (1974) applied multivariate analysis of variance (MANOVA) and T^2 tests to 18 milk chocolates evaluated for 12 attributes by 7-11 panelists. Stepwise discriminant analysis, principal components analysis, canonical analysis, cluster analysis, and multidimensional scaling were applied to the data. Bargmann et al. (1976) evaluated descriptors for canned blueberries by component analysis. Mecredy et al. (1974) profiled beer through a series of descriptive terms. Palmer (1974) used MANOVA for flavor terms applied to tea, and Williams (1975) studied a vocabulary and profile assessment of aroma constituents of cider and perry. McDaniel and Sawyer (1975) used descriptive terms to evaluate whiskey sours. Lehrer (1975, 1976) carried on two extensive linguistic studies of terms used to characterize wines.

There are many papers dealing with factor analysis or related methods. Among them are those of Harper (1956), Woskow (1964), Moskowitz and Gerbers (1972), Yoshida (1969, 1972), Harries (1973), Duchamp et al. (1974) and Schiffman (1975).

EXPERIMENTAL

Survey of Terms

The study was divided into three phases. First a list of wine descriptors was assembled from books, articles, and personal knowledge. A list of 86 words was then distributed to 115 people; 86 individuals returned completed questionnaires. Each person was asked to check off whether the word had no, a little, or a lot of meaning with reference to wine qualities. The respondents were also asked to cross check words as to degrees of similarity. The respondents were asked whether they drank wine frequently, occasionally, or rarely. The ages of the respondents were from 18-65. The respondents were about evenly divided as to sex. Approximately 50% of the respondents were connected with the Food Science Department or had previously served on sensory panels of the Department. Students and faculty members in journalism and English were included among the respondents so as to obtain judgments from individuals who should have special competency with regard to the meaning of words.

From the responses, the list of descriptors was reduced to 33 for use in the wine trials themselves. Words were eliminated because they were said to have little or no meaning with reference to wine or because there was so much similarity between words that the use of both words would have been superfluous.

The purpose of the survey was to determine beforehand descriptors which average consumers *think* have meaning and, from the wine evaluation trials themselves, to determine the extent to which the terms *actually* are used in distinguishing among wines.

Wines Used

Fourteen red table wines and 12 white wines were evaluated by separate panels. For comparison among the red wines, Gallo's Hearty Burgundy was selected as a reference wine. For the white wine trials, Sebastiani's chablis was used as the reference wine. The wines were bought by price and general repute to give a reasonable range of differences for wines on the retail market. The intent was that the reference wine would be intermediate in acceptability and in the various quality attributes. For both sets of wine, the appropriate reference wine was inserted as a coded wine among the other specimens being evaluated to provide a check on the panelists. The wines are listed in Tables 1 and 2. The two red and two white experimental wines listed were pure varietal wines available from another project dealing with trace minerals of wine.

Red Wine Trials

The second stage consisted of having 37 panelists compare the 14 red wines against the reference wine using the 33 terms selected The words are listed in column 2 of Table 3. The procedure was the same as used by Vuataz *et al.* (1974) except we dropped any judge who missed an evaluation session. Vuataz *et al.* pointed out that some of their panelists did not evaluate every sample because the judge could not be available throughout the trial. We too had that trouble except we had 37 judges originally so we could afford to eliminate four judges who missed some sessions.

Each panelist was given 15 mL of each of four coded wines and 30 mL of the reference wine at each session. The panelists were instructed to sample the reference wine and then check off for each descriptor whether the coded wines were weaker, the same, or stronger than the reference wine. The scale ranged from -4 to +4. The panelists did not know the identity of the reference wine and they did not know that the reference wine was being included among the test samples as a coded sample. They were merely told that the coded samples were to be rated for degrees of difference from the reference wine.

TABLE 1.
RED WINES

Number	Wine	Price/500 ml
1	Inglenook Cabernet Sauvignon '69	$3.96
2	Calvet Medoc '72	3.30
3	Sebastiani Barbera	2.50
4	Jaboulet-Vercherre Beaujolais '72	2.38
5	Charles Krug Claret '72	2.35
6	Inglenook Vintage Zinfandel '72	2.15
7	Almaden Chianti	2.08
8	Taylor Lake Country Red	1.42
9	Taylor Burgundy	1.42
10	Mogen David Concord	1.32
11	Gallo Hearty Burgundy	1.12
12	Roma Burgundy	0.87
13	Cabernet Sauvignon (experimental)	—
14	Zinfandel (experimental)	—

TABLE 2.
WHITE WINES

Number	Wine	Price/500 ml
1	Christian Brother's Pinot Chardonney	$3.26
2	Charles Krug Gerwurz Traminer	3.26
3	Christian Brother's Johannisberg Riesling	2.74
4	Charles Krug Dry Sauterne	2.35
5	Meier's Catawba	1.62
6	Almaden Rhine	1.44
7	Sebastiani Chablis	1.05
8	Italian Swiss Colony Rhineskeller Moselle	1.03
9	Gallo Rhinegarten	1.01
10[a]	Italian Swiss Colony Rhineskeller	0.96
11	Sauvignon Blanc (experimental)	—
12	Chenin Blanc (experimental)	—

[a] Wine purchased 1 yr prior to use.

DESCRIPTIVE SENSORY ANALYSIS IN PRACTICE 317

TABLE 3.
DESCRIPTORS USED, NUMBER OF PANELIST SIGNIFICANT FOR EACH DESCRIPTOR, NUMBER USING THAT DESCRIPTOR, AND F VALUE FOR EACH DESCRIPTOR

	Descriptors		Red wine			White wine	
No.	Term	No. panelists significant	No. panelists using term	F value	No. panelists significant	No. panelists using term	F value
1	tart	10	27	16.1	12	18	33.12
2	dry	16	27	37.6	15	18	47.77
3	sweet	27	27	72.3	18	18	121.50
4	bitter	12	27	20.1	10	17	19.49
5	salty	4	18	3.03	4	15	6.42
6	metallic	3	20	5.28	5	17	4.37
7	biting	11	24	17.2	10	17	26.81
8	astringent (puckery)	13	27	19.8	12	18	28.13
9	smooth	10	27	11.8	10	18	20.38
10	coarse	7	25	6.73	11	18	15.36
11	syrupy	21	27	19.4	13	18	50.16
12	watery	5	26	4.1	5	17	8.03
13	aromatic	8	27	9.23	8	18	4.32
14	hearty	5	25	1.57	7	18	1.06
15	delicate (soft, light)	11	27	6.94	10	10	14.74
16	insipid (flat)	1	22	1.81	5	16	1.32
17	medicinal	7	25	3.01	5	18	3.30
18	sharp	8	27	12.1	9	18	19.98
19	winey	11	27	5.97	9	16	2.91
20	fruity	19	25	33.1	10	16	25.98
21	grapey	16	24	30.2	7	16	12.70
22	woody	6	19	6.26	3	14	4.47
23	musty (earthy, moldy)	6	21	5.25	2	14	4.49
24	musk-like	2	10	2.71	1	12	1.59
25	yeasty	2	16	0.91	3	12	1.33
26	burnt-smokey	3	10	3.59	4	14	6.73
27	spicy	2	19	0.96	3	12	0.84
28	vinegary	4	23	0.47	5	10	10.08
29	sulfurous	1	12	1.34	2	15	1.07
30	fresh	7	21	5.62	4	15	4.89
31	mature	8	22	2.40	4	17	1.73
32	balanced (round)	8	25	5.75	9	18	12.40
33	desirable aftertaste	7	25	8.30	13	18	15.62

Independently, the panelists were also asked to check each wine for acceptability on a hedonic basis which was later transformed to a 7-point scale.

The panelists were instructed to taste the reference wine first, then to evaluate the coded specimens from left to right. The order and the particular wines to be evaluated on a given day had been predetermined before the trials actually began, using a table of random numbers to make the session and position assignments. Eventually every panelist sampled the 14 red wines twice.

During the session, the panelists could refresh their memories by returning to the reference wine whenever they wished. They had their choice of crackers or bread between samples. They were instructed they could swallow the wine or expectorate it, but whatever they did, they had to do consistently throughout. Most chose to swallow the wine because "pleasant aftertaste" was one of the descriptors but swallowing was not really essential to making that judgment.

Under results, the red wine panel will be treated as one panel. Actually, we had two panels. Nineteen of the panelists evaluated the wine in the sensory room of the Food Science Building between 3:00 and 4:30 p.m. Fourteen individuals evaluated the wine at home between 5:30 and 6:30 p.m. The second panel was used to secure responses from a more diverse group than was available in the Food Science Department chiefly in terms of age, exposure to a wide range of ordinary and fine wines or frequent use of wine with meals or at social gatherings. The wine was brought to these individuals in small vials. They were given small shot glasses and they were told the order in which the various coded wines were to be evaluated. They were instructed to judge the wine in the same manner as those who evaluated it as an assembled panel. Hereafter, the results of the assembled and nonassembled panels will be treated as one, because the task being almost exclusively one of discrimination, there was no evidence that the assembled judges performed any better or differently than the nonassembled ones.

White Wine Trials

The third phase consisted of evaluating 12 white wines, replicated four times. Ten of the judges assembled as a panel in the Food Science Building; eight had the wines delivered to them at home. The 18 judges were drawn from the red wine panel, using those judges who were most discriminating. All of the judges for the red and white wine panels had been cautioned that the task, except for the acceptability phase, was to determine differences, not their preference for an individual quality.

A tabulation of the panelists by age, sex, and frequency of wine use is given in Table 4. None of the panelists were expert wine tasters; a few might be called wine connoisseurs because of their general knowledge, interest and experience with wines.

TABLE 4.
PANEL COMPOSITION

	Red wines	White wines
Total	33	18
Female	16	10
Male	17	8
Age 20—29	15	8
30—30	8	4
40—49	6	5
Over 50	4	1
Frequency of wine use		
Frequently	12	10
Occasionally	16	4
Rarely	5	4

Statistical Analysis

The first thing that was done once the sensory trials were over was to subject each panelist's score to analysis of variance (using wines as the independent variable) to ascertain standard errors and ability to discriminate among wines. This was done for each panelist and each descriptor, using the MANOVA program of the SAS User's Guide (Barr and Goodnight 1972). Of the 33 red-wine judges who had been retained because they had completed the series of tasks, six were eliminated since their F values among treatments (wines) was non-significant. Using the same criterion as above, all the white wine judges were retained.

As a part of the evaluation of judges, a matrix for judge/judge correlations was printed out. These correlations were used to ascertain whether the acceptability results should be partitioned into different sub-groups because of judges having different preferences among the wines (Powers and Quinlan 1974; Powers 1976).

For the red wine, seven descriptors were discarded either because there was not a significant difference in score among wines or because the term had been used by very few judges. Five descriptors were discarded after the white wine trials.

Having eliminated nondiscriminating panelists and descriptors, factor analysis was then applied to the results, separate analyses being performed for the red and white wines. Factors were extracted by stepwise maximum likelihood solution and rotated (oblique rotation) by Thurstone's Analytical Method of Rotation (Harman 1968).

The data, edited to remove nondiscriminating panelists and descriptors, was also subjected to contingency analysis with and without transformation of the variables to normalize them (PREPRO program, Kundert and Bargmann, 1972). Contingency analysis permits examination of the scores to ascertain whether assignment seems to be truly influenced by treatment or is more nearly random (Quinlan *et al.* 1974). Through transformation of the score levels, an indication can be obtained of the degree to which the panelists are holding constant the interval between score levels (Powers and Quinlan 1974; Powers 1976).

RESULTS & DISCUSSION

Factor Analysis

For the red wine, correlations between the 17 descriptors used are given in Table 5 and the communalities for eight factors in Table 6. Of 14 oblique, well-determined, simple-structure planes, eight were distinct and interpretable (Table 7). The others were either aliases or doubletons.

For the white wine, 25 descriptors and their communalities for six factors are given in Tables 8 and 9. Of 11 well-defined, oblique, simple-structure planes, six were distinct and interpretable. The six rotated factors are given in Table 10.

The use of a survey to determine the meaning of words as applied to wines was effective. Most of the words that the respondents claimed had meaning to them were later verified by the actual wine trials. Only a few additional terms were discarded. For the red wines, the descriptors, salty, hearty, insipid, musk-like, yeasty, burnt-smoky, spicy, and sulfurous were either nonsignificant as descriptors or else used by very few panelists. Some of the significant descriptors with the lowest F values were dropped, along with the nonsignificant descriptors, because the computer program had a limitation of 30 factors for the correlation matrix. Terms which were used by only a few of the judges could mean that only those judges were capable of recognizing that particular flavor note, if it existed; some of the other terms which were eliminated would not be likely to apply anyway to red wines of good commercial quality.

For white wines, hearty, insipid, musk-like, yeasty, spicy, sulfuruous, and mature had nonsignificant F values or F values not much beyond the 0.05 probability level. As was true for the descriptors discarded for the red wines, these words apparently had little application in characterizing the flavor notes of the particular wines used. Table 3 shows the extent to which various descriptors were used, the number of panelists who were significant for that descriptor, and the F value for each descriptor.

The factor analysis showed that some words normally assumed to be opposites were not necessarily opposite in the minds of our panelists. Examples are sweet-dry and fresh-mature. In red wine, dryness did not mean solely lack of sweetness, but it also had connotations with bitter, sharp, astringent and other words of that factor. Fresh and mature did not come out as opposites for either the white or red wines. Some words used on the labels of the wines, such as spicy, apparently had no meaning at all to the panelists because they did not use that word to distinguish among the wines.

About the same time we were conducting our study, Lehrer (1975) was also conducting a similar study from a linguist's point of view. She used a panel of 22 subjects. Her findings and ours agree in several respects. She commented that sweet, for example, had two antonyms: sour and bitter. Robinson (1970) reported that untrained tasters often confuse sour and bitter tastes. Lehrer stated that the use of words, either for descriptive or evaluative purposes, varied according to whether the subject liked or disliked the wine. She carried on an experiment where descriptions for three wines were prepared by part of the judges and then the remainder of the judges took the descriptions and endeavored to match the descriptions to the wines. She commented that "The results are not an impressive record of successful communication. If this rate of precision was maintained between airplane pilots and control towers, most of us would probably give up flying." This is one of the reasons, of course, for trying to partition out those terms which do have meaning for others. Robinson (1970) pointed out: "Surprisingly enough, educated subjects exercising a sense modality which ought to be important to them on solutions of relatively high concentration often fail in identification even though they can discriminate and give the tastes different names." Winton *et al.* (1975) had 8- or 9-member panels of wine experts attempt to identify the variety of grape from which they believed the wine had been made. The percentages of correct identification were fairly high for some varieties but very low for others. Palmer (1974) commented that different descriptions were used to discriminate between the teas he studied than to describe the teas. He pointed out that this could be a sensible result, for a minor component in flavor could still be the best criterion for discrimination. Panelists who had previous experience at profiling foods, but not teas, were nearly as good as experts in discriminating among teas whereas those with no sensory experience were considerably less discriminating.

In our trials, some of the panelists who had no formal training in sensory evaluation were among the best at the task of discrimination. Table 3 lists the number of panelists who were significant in using each term and the total number

TABLE 5.
CORRELATION VALUES (X 100) BETWEEN DESCRIPTORS APPLIED TO RED TABLE WINES

Descriptors	2	3	4	6	7	8	9	10	11	12	13	14	15
1	56	56	57	66	66	64	35	24	31	-2	-11	6	-26
2		72	49	26	53	57	-34	29	-49	7	-10	4	-12
3			-56	-26	-53	-53	39	-30	51	-15	19	3	17
4				42	68	61	-48	33	-24	1	-19	-5	-33
6					41	31	-32	23	-15	5	-13	-13	-18
7						76	-51	40	-28	-4	-20	0	-38
8							-50	39	-31	-5	-15	6	-34
9								-62	16	-3	28	21	49
10									-20	-1	-19	-7	-28
11										-27	1	0	-10
12											-14	-16	13
13												25	17
14													14
15													
16													
17													
18													
19													
20													
21													
23													
28													
30													
31													
32													
33													

DESCRIPTIVE SENSORY ANALYSIS IN PRACTICE

	16	17	18	19	20	21	23	28	30	31	32	33	Accp.
1	3	14	56	-7	-29	-30	21	43	-16	-7	-11	-16	12
2	1	15	49	-11	-43	-42	14	34	-17	10	-1	-10	-3
3	-6	-14	-44	20	53	51	-20	-40	22	-3	-12	21	10
4	-14	32	49	-20	-36	-34	37	47	-20	-14	-25	-31	-23
6	18	47	24	-16	-13	-15	24	40	-10	-22	-28	-24	-18
7	7	32	66	-12	-34	-33	26	54	-25	-12	-24	-30	-21
8	2	25	61	-4	-33	-34	20	45	-26	-4	-18	-25	-13
9	-12	-36	-37	23	30	29	-30	-40	26	28	50	56	34
10	6	21	36	-10	-16	-18	15	28	-12	-12	-27	-36	-26
11	1	3	-33	15	42	43	3	-19	7	-11	-7	-6	-5
12	26	6	-6	-13	-16	-13	6	7	-6	-5	-9	-4	2
13	-18	-21	-11	18	26	29	-17	-23	22	21	26	26	17
14	-10	-16	-5	25	8	9	4	-11	13	47	44	36	21
15	5	-25	-30	14	17	17	-28	-31	27	38	50	49	28
16		33	-2	-15	-2	-1	27	11	1	-9	-16	-14	-5
17			21	-8	-5	-4	27	31	-17	-22	-34	-32	-25
18				-7	-31	-32	5	41	-19	-7	-18	-20	-15
19					44	42	-10	-14	10	27	28	24	19
20						86	-14	-26	27	3	12	20	9
21							-13	-27	31	2	13	20	8
23								30	-5	-15	-25	-30	-15
28									-23	-25	-34	-34	-17
30										4	19	19	15
31											67	51	26
32												72	42
33													51

[a] The descriptor's number matches those shown in Table 3.

of panelists who used that term. Another measure of the panelist's effectiveness is the degree of closeness or departure from the base line in evaluating the coded reference wines. No mean score departed more than 0.5 unit from zero for the red wine nor more than 0.2 unit for the white wine. Though unknown to themselves, the panelists were thus "recognizing" the coded reference wines because their scores clustered around zero.

TABLE 6.
RED WINE, COMMUNALITIES FOR EIGHT FACTORS

Descriptor[a]	Value	Descriptor[a]	Value
1	0.606	16	0.317
2	0.689	17	0.577
3	0.732	18	0.579
4	0.632	19	0.331
6	0.412	20	0.861
7	0.770	21	0.853
8	0.703	23	0.381
9	0.723	28	0.426
10	0.576	30	0.168
11	0.528	31	0.578
12	0.280	32	0.786
13	0.209	33	0.693
14	0.382	Accp.	0.274
15	0.513		

[a] Descriptor's numbers matches those in Table 3.

Acceptability

One of the reasons for carrying on factor analysis is to search for the main variables reflecting differences. In our study, a second objective was to ascertain the relation of the different discriminators to flavor acceptability. The correlation coefficients for the different discriminators with flavor acceptability may be seen in Tables 5 and 8.

Because our panels were small for acceptability studies, only a few comments will be made relative to the acceptability results. The results were subjected to analysis of variance and Duncan's multiple range test. In general, our panelists scored the slightly-sweet to semi-sweet wines highest in acceptability. As we have pointed out before, judge/judge correlations should be examined because naturally the preferences of all judges are not alike (Powers

TABLE 7.
ROTATED FACTORS, RED WINE

		Factor			
1.	Pungency		2.	Overal Quality[a]	
	Tart	0.55		Mature	0.61
	Biting	0.52		Balanced	0.58
	Astringent	0.46		Hearty	0.50
	Sharp	0.46		Desirable Aftertaste	0.39
	Bitter	0.42		Winey	0.21
	Dry	0.35		Acceptability	0.20
	Vinegary	0.35			
	Sweet	−0.33			
	Coarse	−0.26			
	Remaining loadings	18			21
	Less than 0.1 (absolute value)	15			17
	Between 0.1 + 0.2	2			4
3.	Smoothness vs coarseness		4.	Sweetness vs dryness	
	Coarse	−0.43		Sweet	0.42
	Smooth	0.37		Syrupy	0.36
	Medicinal	0.32		Dry	−0.35
	Metallic	0.27		Medicinal	0.29
	Desirable Aftertaste	0.25		Watery	−0.25
	Delicate	0.20		Tart	−0.22
	Remaining loadings	21			21
	< 0.1	18			18
	0.1 < x < 0.2	3			3
5.	Flavor		6.	Disagreeableness (1)	
	Watery	0.37		Insipid	0.29
	Insipid	0.26		Musty	0.28
	Hearty	−0.24		Bitter	0.26
	Delicate	0.24			
	Syrupy	−0.22			
	Remaining loadings	22			24
	< 0.1	16			17
	0.1 < x < 0.2	6			7
7.	Fruitiness		8.	Disagreeableness (2)	
	Grapey	0.59		Medicinal	0.58
	Fruity	0.58		Insipid	0.46
	Winey	0.36		Metallic	0.45
	Remaining loadings	22			24
	0.1	19			17
	0.1 x 0.2	3			6
	0.2 x 0.25	2			1

[a] The last two descriptors would not have been included, usually, but were reported here because acceptability was one of them.

and Quinlan 1974; Powers 1976). Partitioning of the panelists into sets of judges having similar preferences revealed that there were different patterns of acceptability among the judges. The fact that different panelists have different preferences for different reasons is, of course, one of the strongest arguments for attempting to find out which components truly affect acceptability.

TABLE 8.
CORRELATION VALUES (× 100) BETWEEN DESCRIPTORS APPLIED TO WHITE TABLE WINES

Descriptors	2	3	4	6	7	8	9	10	11	12	13	14
1	54	-57	48	22	69	65	-35	-34	-45	-14	-15	10
2		-71	36	19	42	50	-25	25	-49	8	-8	8
3			-48	-23	-52	-53	38	-33	59	-22	27	4
4				40	57	46	-37	33	-31	15	-16	3
6					26	20	-18	16	-8	7	-13	-2
7						75	-45	40	-38	10	-22	11
8							-48	46	-37	9	-17	11
9								-77	14	-13	36	-1
10									-11	5	-38	5
11										-43	10	2
12											-8	8
13												11
14												
15												
16												
17												
18												
19												
20												
21												
30												
31												
32												
33												

	15	16	17	18	19	20	21	30	31	32	33	Accp.
1	-22	6	12	64	6	-42	-33	-8	10	-12	-18	-15
2	-12	-2	5	35	-1	-39	-30	-11	19	3	-10	-4
3	21	-7	-10	-46	8	52	35	17	-10	11	21	16
4	-30	3	27	48	-10	-41	-33	-5	-4	-24	-31	-30
5	-14	6	28	20	-14	22	-18	-4	4	-16	-18	-19
6	-30	0	20	72	-2	-44	-35	9	3	-21	-27	-21
7	-32	-6	14	62	7	-37	-31	-11	6	-22	-28	-20
8	63	-11	-29	-43	15	35	20	17	13	57	59	45
9	-53	14	29	37	-11	-31	-15	-21	-7	-51	-51	-38
10	-3	-4	2	-33	3	44	40	8	-19	-19	-8	-9
11	-1	15	0	13	-14	-21	-19	-5	-4	-6	-7	-10
12	20	-27	-7	-14	20	31	15	26	16	30	27	24
13	-10	-17	-11	13	22	3	1	15	27	22	9	16
14		6	-26	-35	3	22	8	8	17	55	58	43
15			-7	-5	-20	-9	-2	-14	-20	-11	-10	-11
16				17	-13	-14	-12	-12	5	-28	-36	-30
17					7	-40	-34	-10	4	-20	-31	-20
18						46	36	2	22	22	21	18
19							70	21	-9	16	25	23
20								8	-9	5	16	12
21									14	10	4	18
30										43	31	16
31											75	54
32												55

[a] The descriptor's number matches those shown in Table 3.

TABLE 9.
WHITE WINES, COMMUNALITIES FOR 6 FACTORS

Descriptor[a]	Value	Descriptor[a]	Value
1	0.641	15	0.552
2	0.648	16	0.221
3	0.779	17	0.168
4	0.453	18	0.617
6	0.140	19	0.442
7	0.734	20	0.747
8	0.667	21	0.623
9	0.776	30	0.111
10	0.744	31	0.347
11	0.544	32	0.768
12	0.144	33	0.721
13	0.329	Accp.	0.410
14	0.285		

[a] Descriptor's numbers matches numbers in Table 3.

Our panel did not rate the dry and generally more mature varietals highest in acceptability, but in both trials wines rated as somewhat sweet and with desirable aftertaste in the discrimination trials were the wines rated toward the top in acceptability. Lehrer (1975) also reported that her panelists liked wines with from 1 to 2% sugar although beforehand they had said they prefer dry wines. Filipello et al. (1958) reported that their panelists preferred rose wine with 3% sugar to that with 1% and the preference increased with age. Our acceptability scores were examined to learn if there were differences according to age, sex, or whether the panelists sampled the wine in the laboratory or at home. Only age seemed to have a bearing on acceptability. Panelists under 30 yr of age liked the highly-sweet, Concord-type red wine the least while panelists over 40 scored it toward the top in acceptability. There was also a tendency for the most-sweet white wine to be scored in the same fashion. The sweetest red and white wines were scored as having the least desirable balance and aftertaste. Figure 1 illustrates the differences between the most acceptable and least acceptable red and white wines.

Our results as to acceptability naturally relate only to our panel and the particular wines used. It might be argued that our wine tasters were naive subjects not familiar with good wine qualities. This was probably not so. Even among the young, a high percentage serve wine at home, as members of the armed forces they have traveled widely in Europe, and among the older

TABLE 10.
ROTATED FACTORS, WHITE WINE

Factor			
1. Overall quality		2. Flavor	
Smooth	0.64	Insipid	−0.38
Coarse	−0.64	Hearty	0.34
Delicate	0.58	Aromatic	0.32
Desirable Aftertaste	0.51	Mature	0.25
Balanced	0.49	Winey	0.22
Acceptability	0.34	Watery	−0.22
Syrupy	−0.22	Delicate	0.22
Medicinal	−0.22		
Remaining loadings	17		18
Less than 0.1 (absolute value)	12		14
Between 0.1 + 0.2	5		4
3. Pungency		4. Smoothness vs coarseness	
Biting	0.59	Aromatic	0.37
Sharp	0.50	Coarse	−0.37
Tart	0.49	Insipid	−0.34
Astringent	0.48	Smooth	0.29
Bitter	0.38	Fresh	0.22
Remaining loadings	20		20
< 0.1	15		15
0.1 < x < 0.2	5		5
5. Sweetness vs dryness		6. Fruitiness	
Syrupy	−0.60	Fruity	0.63
Dry	0.56	Grapey	0.62
Sweet	−0.36	Winey	0.57
Tart			
Remaining loadings	21		22
< 0.1	16		13
0.1 < x < 0.2	5		6
			3 between 0.2 + 0.25

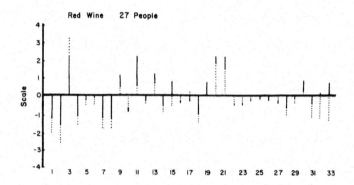

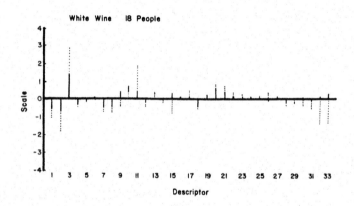

FIG. 1. DIFFERENCES IN FLAVOR COMPONENTS FOR WINES RECEIVING
HIGHEST AND LOWEST ACCEPTABILITY SCORES IN FLAVOR
The solid line shows the score assigned to the most acceptable wine;
the dotted line shows the score assigned to the least acceptable wine.

members a fairly higher percentage had their origins in countries or were members of ethnic groups where wine is used from childhood. The simple fact is, stripped of their labels and any mystique attached to certain types of wines, the panelists liked the characteristics of wines which were not completely dry. Some of the bulk-produced, blended wines thus came out ahead of some of the dry and more elegant varietals in acceptability. Singleton and Ough (1962) pointed out that blending of wines made them more complex in character and caused acceptability scores to be higher.

Aside from any possible parochialism of the particular panel used, two of the reasons for the acceptability scores coming out as they did goes back to the taste-testing situation itself. Sampling of wines analytically possibly produced a different result than might have been attained had the wines been drunk in their normal setting. The characteristics that make them complement other foods perhaps did not serve them well when they were sampled alone and under the stress of analytical though instead of emotional appeal. Powers et al. (1976) compared acceptability, flavor, mouthfeel, appearance, and color scores for green beans scored prior to panelists having to check off a series of descriptive terms and after having done so. While each of the corresponding scores was correlated, the scores were different when analytical though preceded scoring for general quality factors.

A second consideration which must be kept in mind relative to the acceptability scores is the matter of contrast. It was recognized before the trials started that if a sweet wine came ahead of a dry or tart wine this would be contrary to the usual practice in evaluating wines. However, to have always placed the dry wines first would have caused even more serious biases because the judges would have then known what to expect and subconsciously they might have begun to score according to their expectations. The fact that the panelists ate bread or crackers between samples and they could re-sample the reference wine tended to lessen the contrast between the sweet and dry (or tart) wines when sweet came first. Refreshing with the reference wine served to purge the lingering sensation of sweetness. Furthermore, if there was an effect from sweet wine coming before dry wine one time and after dry wine another time, this would have increase variance and in turn made it more difficult to detect differences between samples. The fact that significance was obtained in spite of possible increased variance suggests that the differences in acceptability were sufficiently robust to override positional disadvantages. Purposeful assignment of the order of sampling probably would have introduced more error than the randomized order of testing did. The fact that the two panels discriminated among the wines statistically both as to descriptors and acceptability indicates that the taste-testing procedure employed was effective. The procedure we followed, which was adopted from Vuataz et al. (1974), was simple to use in that panelists with only a nominal amount of training performed well.

Factors analysis should perhaps be used more widely to develop lists of discriminators. The objectivity it brings into play is desirable. Descriptors which are ineffective or redundant can be eliminated and insight is obtained as to the meaning terms really have for panelists such as the fact that our panelists did not always use certain words as antonyms though they are normally assumed to be opposite in meaning. Lehrer (1975) reached the same conclusion. Factor analysis is but one step in devising better means of assessing the quality of foods. Once a pertinent list of descriptors has been selected, then other methods of multivariate analysis should be used to relate components of the major sense modalities to each modality itself or to compositional differences.

REFERENCES

BARGMANN, R.E., WU, L. and POWERS, J.J. 1976. Search for the determiners of food quality ratings-description of methodology with application to blueberries. In *Correlating Sensory/Objective Measurements-New Methods for Answering Old Problems*, Ed., Powers, J.J. and Moskowitz, H.R. Am. Soc. for Testing & Materials, Philadelphia, PA, Publ. STP 594, 56-71.

BARR, A.J. and GOODNIGHT, J.H. 1972. A user's guide to the statistical analysis system (SAS). Student Supply Store, North Carolina State University, Raleigh, NC 27606.

DUCHAMP, A., REVIALK, M.F., HOLLEY, A. and MacLEOD, P. 1974. Odor discrimination by frog olfactory receptors. Chemical Senses and Flavor 1: 213.

FILIPELLO, F., BERG, H.W. and WEBB, A.D. 1958. A sampling method for household surveys. 2. Panel characteristics and their relation to usage of wine. Food Technol. 12: 508.

HARMAN, H.H. 1968. *Modern Factor Analysis*, 2nd ed., University of Chicago Press.

HARPER, R. 1956. Factor analysis as a technique for examining complex data on foodstuffs. Applied Statistics 5: 32.

HARRIES, J.M. 1973. Complex sensory assessment. J. Sci. Fd. Agr. 24: 1571.

KUNDERT, D.R. and BARGMANN, R.E. 1972. Tools of analysis for pattern recognition, Themis Rept. No. 22, Dept. of Statistics & Computer Science, Univ. of Georgia, Athens, GA 30602.

LEHRER, A. 1975. Talking about wine. Language 51: 901.

LEHRER, A. 1976. We drank wine and we talked about it. Manuscript submitted for publication.

MECREDY, J.M., SONNEMANN, J.C. and LEHMANN, S.J. 1974. Sensory profiling of beer by a modified QDA method. Food Technol. 28: 36.

McDANIEL, M.R. and SAWYER, F.M. 1975. Magnitude estimation versus category scaling in food preference testing and food quality rating. Paper presented at the annual meeting, Canadian Institute of Food Science and Technology, Halifax, Nova Scotia, June 1975 (Abstract in Newsletter of Sensory Evaluation Div., IFT).

MOSKOWITZ, H.R. and GERBERS, C. 1972. Dimensional salience of odors In *Odor: Evaluation, Utilization, and Control*, Annals of the N.Y. Acad. of Sci. 237: 3.

PALMER, D.H. 1974. Multivariate analysis of flavor terms used by experts and non-experts for describing teas. J. Sci. Fd. Agr. 25: 153.

POWERS, J.J. 1974. Validation of sensory and instrumental analyses for flavor. Proceedings IV International Congress of Food Science & Technol. 11: 173.

POWERS, J.J. 1976. Experiences with subjective/objective correlation. In *Correlating Sensory/Objective Measurements-New Methods for Answering Old Problems*, Ed. Powers, J.J. and Moskowitz, H.R. Am. Soc. for Testing & Materials, Publ. STP 594, 111-122.

POWERS, J.J., BARGMANN, R.E. and GODWIN, D. 1976. Cluster and component analyses as objective-subjective tools. Paper presented at the 172nd National Meeting, American Chemical Society, San Francisco, Aug. 29-Sept. 3.

POWERS, J.J. and QUINLAN, M.C. 1974. Refining of methods for subjective-objective evaluation of flavor. J. Agr. Fd. Chem. 22: 744.

QUINLAN, M.D., BARGMANN, R.E., EL-GALALLI, Y.M. and POWERS, J.J. 1974. Correlations between subjective and objective measurements applied to grape jelly. J. Food Sci. 39: 794.

ROBINSON, J.O. 1970. The misuse of taste names by untrained observers. British J. Psychol. 61: 375.

SCHIFFMAN, S.S. 1974. Physicochemical correlates of olfactory quality. Science 185: 112.

SINGLETON, V.L. and OUGH, C.S. 1962. Complexity of flavor and blending of wines. J. Food Sci. 27: 189.

VUATAZ, L., SOTEK, J. and RAHIM, H.M. 1974. Profile analysis and classification. Proceedings Fourth Intnl. Congress of Food Sci. & Technol., Madrid, Spain, Sept. 1974 (In Press).

WILLIAMS, A.A. 1975. The development of a vocabulary and profile assessment method for evaluation of flavour contribution of cider and perry aroma constituents. J. Sci. Fd. Agric. 26: 567.

WINTON, W., OUGH, C.S. and SINGLETON, V.L. 1975. Relative distinctiveness of varietal wines estimated by the ability of trained panelists to name the grape variety correctly. Am. J. Enol. Viticult. 26: 5.

WOSKOW, H.M. 1964. Multidimensional scaling of odors. Ph.D. dissertation, Univ. of Calif., Los Angeles.

YOSHIDA, M. 1972. Studies in psychometric classification of odors (6). Japanese Psychological Res. 14(2): 70.

YOSHIDA, M. 1969. Multidimensional scaling of emotion, taste of amino acids, odors, and tactual impressions. Bulletin republishing research papers for Japanese Psychological Research, Japanese Psychological Association, Tokyo.

ACKNOWLEDGMENTS

Presented as an undergraduate research project by Mrs. Louise Wu at the Student Division II Undergraduate Research Paper Forum, 35th Annual Meeting, Institute of Food Technologists, Chicago, IL, June 8-12, 1975.

Appreciation is expressed to Professor Dominique Reymond, Nestle Products Technical Assistance Co., Ltd., La Tour de Peilz, Switzerland who provided a copy of "Le Vin, caracteres organoleptiques" by A. Vedel, G. Charle, P. Charney, and J. Tourmeau and published by the Institut National des appellations d'origine des vins et eaux-de-vie, Macon, France, and to Mrs. Bernadette Allard for discussion concerning shades of meaning in translating the above from French into English.

CHAPTER 4.3

DESCRIPTIVE ANALYSIS AND QUALITY RATINGS OF 1976 WINES FROM FOUR BORDEAUX COMMUNES

ANN C. NOBLE, ANTHONY A. WILLIAMS and STEPHEN P. LANGRON

*At the time of original publication,
author Noble was affiliated with
Department of Viticulture and Enology
University of California, Davis
Davis, California 95616, USA*

and

*authors Williams and Langron
were affiliated with
Long Ashton Research Station
University of Bristol
Long Ashton, Bristol BS189AF.*

Twenty-four wines from four communes in Bordeaux were evaluated by descriptive analysis by trained assessors. The same 1976 wines were assigned quality ratings by Masters of Wine (MW). The major aroma difference between the wines was attributed to variation in the intensity of the 'green bean/green olive' character by canonical variates analysis (CVA) of the aroma descriptor ratings across wines. The CVA of the flavour by mouth ratings showed the wines to be discriminated primarily on the basis of astringency and bitterness. By multivariate analysis of variance across regions, and by examination of the configurations derived from the CVA across wines, it was shown that the wines did not vary significantly between communes. No significant difference between the wines in quality ratings of the MWs was found.

1. INTRODUCTION

Many far reaching claims about the uniqueness of wines from different Bordeaux communes have been made by wine writers.[1-4] However, despite the extensive literature on the chemistry of Bordeaux wines, no sensory analytical evaluations have been reported.

To profile flavour analytically, the technique of descriptive analysis has been applied to a variety of beverages, including wines,[5-7] cider,[8] beer,[9,10] and whisky.[11] Success in the use of this technique depends on the selection of appropriate, well defined and consistently used flavour attributes. To define the terms rated in beer, Mecredy et al.[9] prepared standards in several different brands of beer, so that specific flavour notes were recognisable against different backgrounds. Similarly, panelists were trained in the consistent use of descriptors in wine descriptive analysis by Schmidt[7] using reference standards prepared by adulteration of a base wine. To train panelists for evaluating flavours of cider and perry, Williams[8] defined flavour terms by addition of specific compounds or essences to paraffin wax.

In this paper the descriptive analysis of wines from four Bordeaux communes by trained assessors is presented and compared with their quality ratings by experts.

2. EXPERIMENTAL

2.1 Wines

Five wines, selected from each of four Bordeaux communes, varied in quality designation from unclassed to second growths. Four additional regional Bordeaux wines were included for comparison. Details of the 24 1976 wines are provided in Table 1. The wine used for preparation of all reference standards was a Carignane made from French grapes at Long Ashton Research Station.

2.2. Descriptive analysis panel training

A 17 member panel was selected from available personnel at Long Ashton Research Station (nine men and eight women, aged between 22 and 50 years). The majority of the judges had participated in previous descriptive sensory tests, but only a few had extensive wine tasting experience.

Following two orientation sessions in which terms were generated by the assessors from individual assessment of two wines, six sessions were held, in each of which a different set of three of the 1976 Bordeaux wines were presented with 15 to 20 reference standards. The aroma reference standards were prepared by adding to the base red wine food products, flavours or chemicals to define terms resulting from discussions in the session. No standards were provided for the flavour by mouth terms. At each session assessors smelt the reference wines, and then rated the intensities of the terms in each wine. The appropriateness of each term for the wines was then discussed, both to select

TABLE 1.
COMMUNE OR DISTRICT OF ORIGIN, CHÂTEAU AND GROWTH
DESIGNATION OF 1976 BORDEAUX WINES[a]

Code	Commune or district of origin	Château	Growth
1	St Estèphe	Houissant	
2		Montrose	2nd
3		Calon Segur	3rd
4		de Pez	
5		Haut Marbuzet	
6	St Julien	Lagrange	3rd
7		Ducru Beaucaillou	2nd
8		Gloria	
9		Talbot	4th
10		Léoville Lascases	2nd
11	Margaux	du Tertre	5th
12		Brane Cantenac	2nd
13		Malescot St Exupéry	3rd
14		Giscours	3rd
15		Rauzan Gassies	2nd
16	St Emilion	Roudier	
17		Grand-Corbin-Despagne	Grand Cru Classé
18		Fombrauge	
19		l'Angelus	Grand Cru Classé
20		Canon la Gaffelière	Grand Cru Classé
21	Haut Médoc	Cissac	
22	Haut Médoc	la Tour St Joseph	
23	Médoc	la Cardonne	
24	Bordeaux	du Pradeau	

[a] According to 1855 Médoc and 1954 St Emilion classification.[12]

important descriptors and achieve a consensus as to the meaning of each term. From the discussions in the training sessions, a final set of ten aroma and five flavour by mouth terms was selected for use in the formal descriptive analysis. The compositions of these 10 aroma standards are listed in Table 2.

Because of the lack of intensity of most attributes in the wines, seven additional training sessions were held prior to the formal testing, using the same protocol, wines and conditions as those used in the formal sessions.

TABLE 2.
AROMA TERMS SELECTED FOR DESCRIPTIVE EVALUATION AND COMPOSITION OF THE REFERENCE STANDARDS CREATED TO DEFINE THEM

Term	Composition of reference standard[a]
1. Berry (blackberry/raspberry)	10 ml liquor from canned blackberries 2–3 (thawed) frozen raspberries 5–6 g strawberry jam 5–6 g raspberry jam
2. Black currant (canned/'Ribena')	7–10 ml liquor from canned black currants 2–4 ml Sainsbury's black currant drink
3. Synthetic fruit	5 ml cherryade (Corona drinks) 1 pear drop
4. Green bean/green olive	4–5 ml liquor from canned green olives 8–10 ml liquor from canned green beans
5. Black pepper	4 particles black pepper (fine ground)
6. Raisin	10 raisins
7. Soy/'Marmite'	0.5 ml soy sauce 0.5–1 g 'Marmite' yeast extract
8. Vanilla	0.25 ml vanilla flavouring essence
9. Phenolic spicy	5–10 μl 4-ethyl guaiacol
10. Ethanolic	5 ml (950 ml litre^{-1}) ethanol

[a] In 30 ml Carignane wine.

2.3. Descriptive analysis protocol and design

Using a partially balanced incomplete block design to permit duplicate evaluation of each of the 24 wines, three wines were presented at each of 16 sessions. No wine was presented in combination with any other more than once. At each session, assessors smelt the 10 aroma reference standards before evaluating the test wines. The intensity of aroma attributes was scored on each of the wines in the randomised order in which they were presented, after which the intensities of the flavour by mouth terms were rated. The intensity of each attribute was scored on a 10 point scale, where 0 = not present, 1 = low, and 9 = high intensity.

All wines were presented in coded, standard, tulip-shaped, clear 215 mL wine glasses[13] and evaluated at 16-22C in isolated booths illuminated by fluorescent lighting. A 25 mL sample of wine was poured into each glass and covered with a watch glass at least 30 min prior to testing. Distilled water was provided for cleaning the palate between wines, and all samples were spat out.

2.4. Quality rating by wine experts

Ten MWs evaluated the overall quality of the wines in one session. In an incomplete block design, each MW rated 16 of the 24 wines, in blocks of four, providing 6-8 replicate assessments of each wine. The MWs rated overall quality of the wines on a nine point scale, where one was defined as an unacceptable, defective 1976 Bordeaux wine, five was a standard 1976 Bordeaux with no defects and nine was an excellent 1976 Bordeaux. The wines were rated on two criteria: 'if immediately consumed' and 'when ready to drink'. In addition, for each wine the MWs were requested to describe briefly the attributes of the wines which influenced their assignment of these overall quality ratings.

Wines were presented in the same coded, covered glasses as used for descriptive analysis. Judges were seated at individual tables in a room (20C) illuminated with northwest light.

2.5. Data analysis

Individual analyses of variance (ANOVA) were run on each term rated by the trained assessors and on the two overall quality ratings of the experts. The first nine aroma terms and the five flavour by mouth attributes, respectively, were then analysed by multivariate analysis of variance (MANOVA) and canonical variate analysis (CVA) using a Genstat program;[14] (the ethanol term was not included in the data analysis since many of the panelists did not rate it).

Whereas ANOVA, coupled with the F test enables one to test for significant sample differences over one attribute. MANOVA, coupled with the Bartlet or approximate F test using the Wilks Λ statistic,[15] enables the inspection of the data as a whole and determination of the number of dimensions over which sample differences lie. CVA enables the construction of these dimensions from the original variations, once the MANOVA have ascertained the number of significant dimensions. The canonical variate scores are defined as $a_1^T X$, a linear compound of the original response variable X. The canonical variate loadings a_1^T are chosen in such a way that between-sample or treatment differences are maximised and within-sample or treatment differences minimised by $a_1^T X$. The best discriminating combination $a_1^T X$ is known as the first canonical variate, the next best discriminating combination independent of $a_1^T X$ is the second canonical variate $(a_2^T X)$ and so on.

MANOVA were performed to test whether wines were significantly different across all aroma (or flavour by mouth) attributes, and then to test if the communes of origin (regions) were significantly different over either set of attributes. Where significant differences were found between wines, for aroma and for flavour by mouth terms, canonical variates were calculated. To illustrate the differences among the wines, the canonical variate scores and their

3. RESULTS

3.1. Descriptive analysis

In Tables 3 and 4 the correlation coefficients among the descriptors are shown. Although there were several significant correlations, the only highly significant coefficient was between 'green' and 'synthetic fruity' (-0.74; d.f.=22, $P<0.001$). This corresponds to an explained variation of only 54% of one in terms of the other. Thus, the terms were considered to have been used to describe different characteristics. Any existing correlations between attributes are, nevertheless, taken into account by the CVA.

Analysis of the aroma data across the 24 wines for all 17 assessors showed highly significant wine × assessor interaction for the MANOVA and the ANOVA of the individual terms reflecting the inconsistent use of several of the terms by the judges. By visual inspection of the data, seven assessors were removed and the analyses repeated: Examination of the seven assessors eliminated from consideration revealed that they did not form a separate consistent population, but were randomly using the terms. Using the data set for the remaining ten assessors, the summaries of the ANOVA for each aroma term are presented in Table 5. In Table 6 the summary of the MANOVA of the nine aroma attributes across wines is presented.

With the exception of very highly significant variation which was due to the assessors ($P<0.001$), no significant sources of variation were found for the MANOVA across regions.

TABLE 3.
CORRELATION MATRIX AMONG THE AROMA DESCRIPTIVE TERMS (d.f.=22)

				Term					
Term	1	2	3	4	5	6	7	8	9
1. Berry	1.00								
2. Black currant	0.12	1.00							
3. Synthetic fruit	0.54**	−0.02	1.00						
4. Green bean/olive	−0.62**	−0.13	−0.74***	1.00					
5. Black pepper	0.32	0.19	0.27	−0.16	1.00				
6. Soy/'Marmite'	−0.30	−0.16	−0.37	0.56**	−0.09	1.00			
7. Phenolic/spicy	−0.20	0.16	0.03	−0.06	0.22	0.12	1.00		
8. Vanilla	0.24	0.04	0.30	−0.41*	0.47*	−0.04	0.20	1.0	
9. Raisin	0.23	0.08	0.32	−0.19	0.16	−0.15	−0.05	0.04	1.0

*, **, *** $P<0.05$, $P<0.01$, and $P<0.001$, respectively.

TABLE 4.
CORRELATION MATRIX AMONG THE FLAVOUR BY MOUTH
DESCRIPTORS (d.f.=22)

	Term				
	1	2	3	4	5
1. Sour	1.00				
2. Bitter	0.00	1.00			
3. Astringent	0.21	0.61**	1.00		
4. Berry	−0.31	−0.10	−0.26	1.00	
5. Black pepper	−0.02	0.17	0.03	0.04	1.00

** $P<0.01$.

Assessors were a very highly significant source of variation ($P<0.001$) in all cases. Significant differences in intensities across wines were found in the ANOVA of 'black currant' ($P<0.05$) and 'green bean/green olive' ($P<0.001$) aromas. Although a significant wine × assessor interaction occurred in the 'soy/"Marmite"' ANOVA, this was not a significant source of variation in the MANOVA across wines. Although the intensity ratings of the aroma terms taken together varied significantly across wines ($P<0.001$), they did not differ between regions. Hence, CVA was performed only on the data across wines. The first three canonical variates constructed from the aroma data account for 30.2, 17.7, and 16.0% of the variance, respectively. The first canonical variate is highly significant ($\chi^2=280.6$, d.f.=207, $P<0.001$). The added variation explained by the second is only significant at the 90% level of confidence ($\chi^2=202$, d.f.=176), whereas the third was nonsignificant ($\chi^2=152.4$, d.f.=147).

In Figure 1 the wine means are plotted on the first two canonical variates; the contribution each descriptor makes to the axes is given in the legend. Around each wine mean 95% confidence limits are drawn as a rough indication of their relative uncertainty. Fewer confidence limits overlap on variate 1 than on variate 2, tying in with the results of the MANOVA which attached less significance to this variate when determining differences between the wines.

The largest difference among the aromas of the wines is due to variation along the first dimension in the 'green bean/green olive' attribute, although differences in the 'vanilla' and 'soy/"Marmite"' characters also contribute. The second dimension is primarily a contrast between the 'black currant' and 'phenolic/spicy' notes. Wines 1 and 12, highest in the 'green bean/green olive' aroma and lowest in the 'vanilla' character, were from St Estèphe and Margaux respectively. In contrast, those lowest in 'vegetative' character and highest in

TABLE 5.
ANALYSES OF VARIANCE OF AROMA ATTRIBUTE RATINGS (10 ASSESSORS): DEGREES OF FREEDOM (d.f.) AND MEAN SQUARES

Source of variation	d.f.	Berry	Black currant	Synthetic fruit	Green bean/ olive	Black pepper	Soy/ 'Marmite'	Phenolic/ spicy	Vanilla	Raisin
Assessor (A)	9	114.66***	148.53***	100.18**	59.77***	12.13***	19.08***	65.45***	35.16***	76.73**
Wine (W)	23	1.35	5.09*	1.14	6.95***	0.64	1.25	1.55	1.42	0.69
Reps (R)	1	0.13	0.00	0.13	0.83	0.03	3.68*	0.92	0.25	0.00
A × W	207	1.77	2.41	1.18	2.51	0.52	1.26**	1.55	1.08	0.60
A × R	9	3.49	2.72	2.08	1.17	1.30*	0.69	1.93	0.78	0.48
W × R	23	2.06	1.49	0.76	1.60	0.56	0.99	1.08	1.20	0.49
Error	207	2.07	2.65	1.09	2.10	0.56	0.88	1.23	0.92	0.53

*, **, *** $P < 0.05$, $P < 0.01$, $P < 0.001$ respectively.

'vanilla' were from St Julien (10) and the Médoc (23). A St Emilion wine (17) was higher in the 'phenolic/spicy' character and lower in 'black currant' than all but five wines. Examination of Figure 1, further confirms the lack of significance found across regions by MANOVA. No apparent clustering of the wines by commune occurs, nor is there any partitioning based on growth designation.

TABLE 6.
MULTIVARIATE ANALYSIS OF VARIANCE OF NINE AROMA ATTRIBUTES ACROSS WINES (10 ASSESSORS)
Degrees of freedom (d.f.) and approximate F ratios

Source of variation	d.f.$_1$[a]	d.f.$_2$[a]	Approximate F
Assessors (A)	81	1295	41.53***
Wines (W)	207	1694	1.38***
Reps (R)	9	199	0.64
A x W	1863	1817	1.07
A x R	81	1295	1.24
W x R	207	1694	0.94

*** $P < 0.001$.
[a] Functions based on number of variates and degrees of freedom of hypothesis and residual.[16]

Using the data for all 17 assessors, the ANOVA for the five flavour by mouth terms are summarised in Table 7, with the MANOVA across wines shown in Table 8. All the terms were used consistently and reproducibly by the entire panel. However, only bitterness ($P < 0.05$) and astringency ($P < 0.001$) varied significantly across the wines. Similarly to the results found for the aromas of the wines, significant differences were found across the wines by MANOVA of the flavour by mouth terms ($P < 0.001$), whereas no significant differences were found between regions. In the MANOVA across regions, only assessors varied significantly ($P < 0.001$).

The first three canonical variates from the analysis of the flavour by mouth terms across wines accounted for 49.3, 19.9 and 16.0% of the variance respectively. However, only the first canonical variate is significant ($2 = 173.3$, d.f. = 115, $P < 0.001$). In Figure 2 the wine means, with their 95% confidence intervals, are plotted for the first two canonical variates. Attribute loadings for the two axes are given in the legend. The significant dimension in the flavour by mouth data represents a variation in astringency and bitterness among the wines. The most astringent (and bitter) wines, 2, 14 and 21, were from St

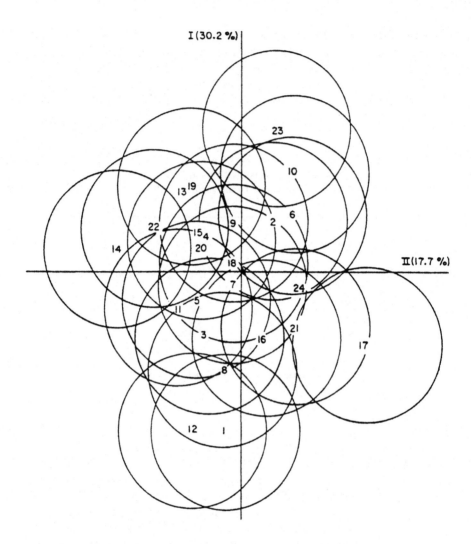

FIG. 1. CANONICAL VARIATES ANALYSIS OF NINE AROMA ATTRIBUTES ACROSS 24 BORDEAUX WINES.
Canonical variate scores and their 95% confidence intervals.

	Variate loadings × 10³								
	Berry	Black currant	Synthetic fruit	Green bean/ green olive	Black pepper	Raisin	Soy/ 'Marmite'	Vanilla	Phenolic/ spicy
Canonical variate 1	−4.8	−1.5	+14.7	−40.2	+2.7	+4.9	−21.9	+20.4	−6.5
Canonical variate 2	−5.9	−42.4	+17.8	+0.5	−9.9	+12.0	−9.3	+13.0	+29.2

Estèphe, Margaux and Haut Médoc, respectively. The least astringent (and bitter) wine (23) was from the Médoc. As shown by nonsignificance in the MANOVA across regions and by the widely scattered wine means in Figure 2, the wine flavours (by mouth) were not significantly different between regions.

TABLE 7.
ANALYSES OF VARIANCE OF FLAVOUR BY MOUTH ATTRIBUTE RATINGS
(17 ASSESSORS): DEGREES OF FREEDOM (d.f.) AND MEAN SQUARES

Source of variation	d.f.	Mean squares				
		Sour	Bitter	Astringent	Berry by mouth	Black pepper by mouth
Assessor (A)	16	86.21***	114.79***	86.99***	77.79***	144.67***
Wine (W)	23	1.36	3.19*	5.68***	1.05	2.79
Rep (R)	1	4.27	5.67	0.71	2.94	0.00
A × W	368	1.69	1.98	1.95	1.65	1.61
A × R	16	2.53	2.25	2.24	1.53	2.28
W × R	23	1.78	1.43	1.00	2.32	1.08
Error	368	2.05	1.86	1.76	1.49	1.79

*, *** $P < 0.05$, $P < 0.001$ respectively

TABLE 8.
MULTIVARIATE ANALYSIS OF VARIANCE OF FIVE FLAVOUR BY MOUTH
ATTRIBUTES ACROSS WINES (17 ASSESSORS)
Degrees of freedom (d.f.), approximate F ratios.

Source of variation	d.f.$_1$[a]	d.f.$_2$[a]	Approximate F
Assessors (A)	80	1757	37.36***
Wines (W)	115	1791	1.53***
Reps (R)	5	364	1.02
A × W	1840	1826	1.03
A × R	80	1757	1.19
W × R	115	1791	0.92

*** $P < 0.001$.
[a] See footnote, Table 6.

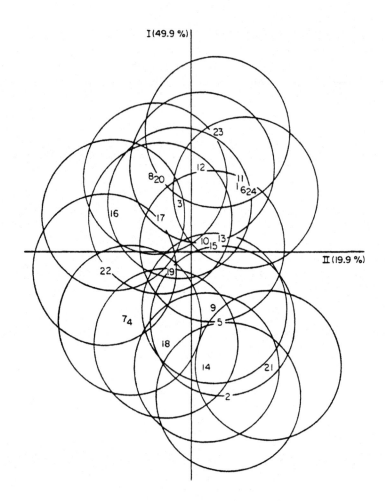

FIG. 2. CANONICAL VARIATES ANALYSIS OF FIVE FLAVOUR BY MOUTH ATTRIBUTES ACROSS 24 BORDEAUX WINES
Canonical variate scores and their 95% confidence intervals.

	Variate loadings × 10^3				
	Sour	Bitter	Astringent	Fruity/berry	Black pepper
Canonical variate 1	+11.6	−14.2	−36.5	−2.1	+12.3
Canonical variate 2	+3.1	−7.8	+1.5	+6.3	−37.3

3.2. Masters of Wine: quality ratings

The overall quality of the wines was rated significantly higher when 'ready to drink' than 'if consumed immediately' ($t=13.38$, d.f. $=23$, $P<0.001$). There was no significant difference among the wines for either quality rating, as shown by the summaries of the ANOVA of the quality ratings in Table 9. This is due to differences existing among the assessors assigned quality ratings over and above systematic differences accounted for by ANOVA. The various sensory aspects of the wines were obviously weighted differently by the assessors in arriving at an overall quality rating. As an example of these differences in interpretation of quality among the MWs, comments on one wine ranged from 'rich and concentrated fruitiness', 'pleasant, but cheap fruit taste', and 'not much depth', to 'totally defective, nose appalling'.

In contrast to their different methods of integrating the wines' attributes and estimating quality, the MWs were often consistent in their specific descriptions of the wines. In descriptive analysis of the wines by the trained assessors, wines 1 and 10 were rated as having very intense 'green bean/green olive' and 'berry' characters, respectively. The MWs described wine 1 as having the 'smell of rotting vegetation' and 'pronounced vegetal (*sic*) aroma, bell pepper'. Six of the eight assessors evaluating wine 10 described it as 'fruity', with the more specific descriptions, 'plummy', 'raisins', and 'black currant', also being used.

TABLE 9.
ANALYSES OF VARIANCE OF OVERALL QUALITY 'IF CONSUMED IMMEDIATELY' (NOW) AND 'WHEN READY TO DRINK' (LATER) DEGREES OF FREEDOM (d.f.), MEANS SQUARES (MS) AND F RATIOS

Source of variation	d.f.	Quality 'now'		Quality 'later'	
		MS	F	MS	F
Assessors ignoring wines	9	17.82	—	9.94	—
Wines adjusted for assessors	23	2.43	1.12 NS	3.05	1.34 NS
Residual	125	2.16	—	2.28	—
Wines ignoring assessors	23	3.04	—	3.35	—
Assessors adjusted for wines	9	16.25	7.52***	8.74	3.83***
Residual	125	2.16	—	2.28	—

NS = not significant.
*** $P<0.005$.

4. DISCUSSION

For this set of 1976 Bordeaux wines, the differences in flavour were primarily due to variations in the intensities of the vegetative (green bean/green olive) and fruit (black currant) aromas and astringency and bitterness. Greater differences were observed among wines than among communes or growth designations. In contrast to the analytical profiles of the wine flavours provided by the descriptive analysis, the overall quality ratings made by the MWs yield little information. Quality is a composite response to the sensory properties of a wine based on the assessor's expectations and hence is an individual response based on preferences and experiences. Accordingly, although the wines were shown to have different flavours, no significant differences in quality were found, because of the differences in preferences of the expert judges. These results are consistent with those of others.[17-19] In contrast to the reproducible and consistent performance of trained assessors in descriptive analysis of systems varying in colour, viscosity or sweetness, Trant *et al.*[17] found considerable variation in hedonic or preference responses to the system made by the same assessors. Further, many of the individuals changed their preference over time.

Although they were not rated as being significantly different, wines 22 and 23 were rated the highest and lowest respectively for both quality 'now' and 'when ready to be consumed'. The sensory characteristics of these wines can be defined from the mean intensity ratings for the attributes in Table 10. Wine

TABLE 10.
MEAN INTENSITY RATINGS FOR WINES 22 AND 23 AND CORRESPONDING
LEAST SIGNIFICANT DIFFERENCES FOR DESCRIPTIVE ANALYSES

	Mean ratings[c]		
	Wine 22	Wine 23	L.s.d. (0.05)
Aroma			
Berry	3.60	3.85	
Black currant	5.15	3.65	1.008
Synthetic fruit	1.50	1.85	
Green bean/green olive	1.80	1.15	0.897
Black pepper	1.15	0.70	
Soy/'Marmite'	0.80	0.50	
Phenolic/spicy	1.45	0.75	
Vanilla	1.05	0.95	
Raisin	1.05	0.95	
Flavour by mouth			
Sour	3.41	3.35	
Bitter	3.44	2.68	0.648
Astringent	3.88	3.06	0.631
Berry by mouth	3.24	3.59	
Black pepper by mouth	2.32	1.77	

[c] For aroma terms, $n=20$ (10 assessors × 2 reps); for flavour by mouth terms, $n=34$ (17 assessors × 2 reps)

22 was rated significantly higher in the intensities of black currant aroma, bitterness and astringency. Although not significantly so, wine 23 was lower in 'green bean/green olive', 'phenolic/spicy', and 'black pepper' and higher in synthetic fruit aroma. Examining Figures 1 and 2, it is apparent that neither axis corresponds to a 'hedonic' dimension, significant to the MWs in separating the two wines. Wine 23 was one of the lowest in the vegetative character and astringency, but 22 had only mid-range intensity values for these attributes.

Ideally, by descriptive analysis, the sensory characteristics of wines can be defined and used to interpret the patterns in quality ratings by experts or in preference of consumers. Only by combining the two evaluations can the reasons for quality assessments be interpreted in terms of variation in specific sensory attributes.

ACKNOWLEDGMENTS

The authors thank John Harvey & Sons Ltd, Grants of St James's Ltd, International Distillers and Vintners Ltd, Saccone & Speed Ltd and Stowells of Chelsea Ltd for supplying wines and for financial assistance for this work.

REFERENCES

1. JOHNSON, H. *World Atlas of Wine* Mitchell Beazley Ltd. London, 1971, pp. 74-92.
2. BROADBENT, M. *Wine Tasting/Enjoying/Understanding* Christie's Publications, London, 1977, pp. 43-44.
3. DURAC, J. *Wines and the Art of Tasting* E.P. Dutton & Co Inc, New York, 1974, pp. 131-147.
4. GROSSMAN, H.J. *Grossman's Guide to Wines, Beers and Spirits* Charles Scribner's Sons, New York, 1977, p. 42.
5. NOBLE, A.C. Sensory and instrumental evaluation of wine aroma. In *Analysis of Foods and Beverages* (Charalambous, G., Ed.), Academic Press, New York, 1978, pp. 203-228.
6. WILLIAMS, A.A., BAINS, C.R., ARNOLD, G.M. Towards the objective assessment of sensory quality in inexpensive red wines. *Centennial Symposium of Department of Viticulture and Enology,* University of California, Davis, California, USA, June 1980.
7. SCHMIDT, J.O. *Comparison of methods of rating scales used for sensory evaluation of wines* M.Sc. Thesis, University of California, Davis, California, USA, June 1981.

8. WILLIAMS, A.A. Development of a vocabulary and profile assessment method for evaluating the flavour contribution of cider and perry aroma constituents. J. Sci. Food Agric. 1975, 26, 567-582.
9. MECREDY, J.M., SONNEMAN, J.C., LEHMANN, S.J. Sensory profiling of beer by a modified QDA method. Food Technol. 1084. 29 (11), 36-39.
10. CLAPPERTON, J.F., PIGGOTT, J.R. Flavour characterization by trained and untrained assessors. J. Inst. Brew. 1978, 84, 275-277.
11. PIGGOTT, J.R., JARDINE, S.P. Descriptive sensory analysis of whiskey flavour. J. Inst. Brew. 1979, 85, 82-85.
12. COCKS, C., FERET, C. *Bordeaux et ses vins*. 12th edn, Peret et Fils, Bordeaux, 1969, pp. 90-127.
13. Anon. Sensory Analysis Apparatus, Part 1. *Specification for wine tasting glass* BS55556, part 1, 1978. ISO 3591 1977. British Standards Institute.
14. *Genstat: A General Statistical Program Statistics Department*, Rothamsted Experimental Station, Harpenden, Herfordshire, 1977.
15. CHATFIELD, C., COLLINS, A. *Introduction to Multivariate Statistics*. Chapman & Hall, London, 1980, pp. 148-154.
16. RAO, C.F. An asymptomic expansion of the distribution of Wilks' Criterion. Bull. Int. Stat. Inst. 1941, 32, 177-180.
17. TRANT, A.S., PANGBORN, R.M., LITTLE, A.C. Potential fallacy of correlating hedonic responses with physical and chemical measurements. J. Food Sci. 1981, 46, 583-588.
18. LUNDGREN, B., PANGBORN, R.M., SONTAG, A.M., PIKIELNA, N.B., PIETRZAK, E., GARRUT, R., CHAIB, M.A., YOSHIDA, M. Taste discrimination vs hedonic response to sucrose in coffee beverage. An inter-laboratory study. Chemical Senses and Flavor, 1978, 3, 249-265.
19. GIOVANNI, M.E., PANGBORN, R.M. Measurement of taste intensity and degree of liking of beverages by graphic scales and magnitude estimation. J. Food Sci. 1983.

CHAPTER 4.4

SENSORY PANEL TRAINING AND SCREENING FOR DESCRIPTIVE ANALYSIS OF THE AROMA OF PINOT NOIR WINE FERMENTED BY SEVERAL STRAINS OF MALOLACTIC BACTERIA

MINA MCDANIEL, LEE ANN HENDERSON, BARNEY T. WATSON, JR.,
and DAVID HEATHERBELL

*At the time of original publication,
all authors were affiliated with
Department of Food Science and Technology
Oregon State University
Corvallis, OR 97331-6602.*

ABSTRACT

A sensory panel was selected and trained to describe the aroma of Pinot noir wines whose only difference was that they were fermented by different strains of malolactic bacteria. The panel developed a ballot consisting of 33 aroma descriptors and rated the wines using a balanced incomplete block statistical design. A final screening of the individual panelist's data was accomplished by use of correlation analysis. Significant differences were found in twenty of the aroma descriptors, showing that the strain of malolactic bacteria selected for use can effect aroma perception.

INTRODUCTION

Malolactic fermentation is encouraged in many wines to lower the acidity by converting malic to lactic acid, to increase aroma and flavor complexity, and to increase biological stability (Amerine *et al.* 1982; Davis *et al.* 1985; Kunkee 1974). Many winemakers disagree, however, on exactly how the malolactic fermentation affects the aroma of the final wine.

Davis *et al.* (1985) reviewed studies of the effect of malolactic fermentation on sensory characteristics of wine. They found that sensory procedures differed widely as did the results. Most studies utilized preference or difference tests to determine the effect of malolactic fermentation as compared to a wine which had not undergone the secondary fermentation. Few of the studies employed any panelist training so that precise differences in flavor attributes could not be quantified. None of the studies conducted descriptive analysis of different strains of malolactic bacteria.

In 1981, a trial was established to evaluate the performance of several strains of commercially available malolactic bacteria with two new strains, Ey2d and Erla, isolated from Oregon wines. Pinot noir wine was inoculated with several strains of malolactic bacteria and their performance was reported by Henick-Kling (1983) and Watson et al. (1984). Trained descriptive analysis (Cairncross et al. 1950) panels have been successfully used to describe and to show sensory differences among commercial wines (Heymann and Noble 1987; Noble and Shannon 1987). However, this current project involved evaluation of the same starting wine where only one secondary processing step varied. Only subtle differences in aroma were expected to be perceivable. Because subtle differences were of interest, the panel could not be limited to just general descriptive terms. Therefore, part of the purpose of this study was to evaluate the use by the panel of descriptors used to describe subtle differences. A vocabulary of descriptive terms for wine aroma in the format of an aroma wheel developed by the American Society of Enology and Viticulture (Noble et al. 1984) was used as a guide for panel training. In descriptive analysis it is advisable to evaluate each judge's performance for each attribute (Malek et al. 1986). Even though each judge's training is identical, individual judges have difficulty agreeing on common descriptors for certain attributes. Therefore, a major part of the project was to develop and test a panelist screening procedure to be utilized on an individual attribute basis. Because panelists were required to rate only seven of the 33 available terms, incidence of voluntary descriptor usage was also of interest.

This current study of Pinot Noir fermented by different strains of malolactic bacteria was undertaken with the following objectives in mind: (1) To elucidate descriptive terms from a panel. (2) To test the use of those terms by observing frequency of use. (3) To select the most sensitive panelists for each attribute of the wines tested. (4) To develop a descriptive profile for each wine. (5) To evaluate subtle aroma differences created by fermentation of Pinot Noir with different strains of malolactic bacteria.

MATERIALS AND METHODS

Panel Selection

Twenty-one people with an interest in sensory evaluation of wine volunteered to participate in training for a special wine evaluation panel. All subjects were put through a series of screening tests. The tests involved ranking intensities of various concentrations of solutions. Solutions used in testing were of sucrose, tartaric acid, caffeine, ethanol and mixtures of two or three in both spring water and white wine. Panelists also practiced describing aroma characteristics of experimental and commercial Pinot noir wines. After a training period of about three months, twelve final panelists were selected based on consistent performance.

Panel Training

In order to become familiar with the aroma of Pinot noir and to practice describing it, the panel spent three sessions evaluating a selection of commercially available Pinot noir wines. In order to refine the vocabulary of descriptive terms for Pinot noir, the Wine Aroma Wheel (Noble *et al.* 1984) was used. The wine aroma wheel is divided into groups so that terms which describe similar aroma characters are grouped together. The main groups (first tier terms) are further divided into specific characters (second and third tier terms). For example: Fruity, a first tier term, is broken down into citrus, berry, tree, tropical, estery and dried, second tier terms. Specific or third tier terms would then be lemon, raspberry, prunes, etc. Working with a section of group of sections at a time, standards (Table 1) were presented to the panel along with a commercial sample of Pinot Noir. The panel evaluated the wine sample by comparing its aroma to the aromas of the standards. By limiting the standards available, the panelists were able to concentrate on specific characters in the wine samples and familiarize themselves with those sensations.

For the following nine sessions, panelists were presented actual tests samples (see materials section). During these sessions, panelist generated attributes were rated on a nine-point intensity scale (1 = none, 9 = extreme). The experimenters used these results to develop a final set of descriptors. It was not necessary for all panelists to agree that each of the aroma descriptors was useful. For example, several of the panel members may have felt the berry note present was raspberry, while others felt it was blackberry. Both raspberry and blackberry were then included on the final ballot. At this point it was decided to concentrate only an aroma differences for evaluation of these samples as no taste differences were apparent. The final ballot was tested over several sessions to give the panelist practice in using it, and to further refine the descriptive terms. The final ballot consisted of 33 aroma descriptors (Fig. 1).

FACILITIES

During training, the panel was seated around a table for discussion. For testing, panelists were seated in individual testing booths with red lighting to mask color differences.

TABLE 1.
TRAINING STANDARDS DEVELOPED FOR PINOT NOIR AROMA DESCRIPTION TRAINING

TERM	STANDARD*
cinnamon	ground cinnamon - 1/2 tsp.
cloves	whole cloves - 3 cloves
licorice, anise	pure anise extract - 1/2 tsp.
mint	peppermint oil, spearmint oil -drop on glass wool
grapefruit	fresh grapefruit segments - 1/4 cup
orange	fresh orange segments -1/4 cup
blackberry	canned blackberries in heavy syrup - 1/4 cup
	frozen blackberries - 1/4 cup
	blackberry jam - 1 Tbs.
raspberries	frozen raspberries - 1/4 cup
	raspberry jam - 1 Tbs.
strawberry	frozen strawberries - 1/4 cup
	canned strawberries in heavy syrup - 1/4 cup
	strawberry jam - 1 Tbs.
cherry	canned dark sweet cherries in heavy syrup - 1/4 cup
	canned sour pie cherries in water - 1/4 cup
	frozen cherries - 1/4 cup
apricot	canned apricots in heavy syrup - 1/4 cup
peach	canned peaches in heavy syrup - 1/4 cup
pear	fresh, sliced - 1/4 cup
	canned in heavy syrup - 1/4 cup
apple	fresh, sliced - 1/4 cup
	canned, juice - 1/4 cup
pineapple	canned tidbits in its own juice - 1/4 cup
melon	cantalope, fresh - 1/4 cup pieces
	honeydew, fresh - 1/4 cup pieces
banana	fresh, slices - 1/4 cup pieces
estery	Hi-C fruit drink - 1/4 cup
	artificial strawberry - several drops on glass wool
	artificial raspberry - several drops on glass wool
	artificial blackberry - several drops on glass wool
	artificial blueberry - several drops on glass wool
strawberry jam	strawberry jam (Smuckers) - 1 Tbs.
raisins	Sun Maid raisins - 1/4 cup
prune	dried prunes - 2 prunes
	stewed prunes - 2 prunes
fig	Mission, dried figs - 2 figs
	Mission, stewed figs - 2 figs
	Kalamara, dried figs - 2 figs
	Kalamara, stewed figs - 2 figs
labrusca	Welch's grape juice - 2 Tbs.
stemmy	Thompson grape stems - 3 in. crushed

TABLE 1. *(continued)*

TERM	STANDARD
grass/cut green	grass
bell pepper	fresh slices - 1/4" chopped
hay straw	straw
tea	black tea - 1 Tbs.
green beans	frozen green beans - 2 Tbs.
	canned green beans - 2 Tbs.
asparagus	frozen asparagus - 1/4 cup
	canned asparagus - 1/4 cup
green olives	green olives, packed in glass - 1/4 cup
black olives	canned black olives - 1/4 cup
artichoke	canned artichoke - 1/4 cup
mushroom	fresh mushrooms - 1/4 cup
concrete	dry concrete - 1/4 cup
earthy	soil - 1/4 cup
rubbery	2 rubber bands
onion	fresh onion - 1 tsp.
garlic	fresh garlic - 1 tsp.
cabbage	fresh cabbage - 1/4 cup
burnt match	2 burnt matches
wet dog	dog hair, wetted - 1/4 cup
filter pad	1 wetted filter paper - 3" diameter
wet cardboard	wet cardboard - 1" square
wet paper	wet paper - 1" square
ethyl acetate	ethyl acetate -- 150 ppm - 1 Tbs.
acetic acid	acetic acid -- 1% - 1 Tbs.
ethanol	ethanol -- 10% - 1 Tbs.
sorbate	sorbate -- 200 ppm - 1 Tbs.
acetaldehyde	acetaldehye -- 50 ppm - 1 Tbs.
vanilla	vanilla extract - 1/2 tsp.
cedar	cedar chips - 1 Tbs.
pine	pine chips - 1 Tbs.
fir	fir chips - 1 Tbs.
honey	clover honey - 1 Tbs.
butterscotch	2 candies
soy sauce	soy sauce - 1 Tbs.
chocolate	cocoa powder - 1 Tbs.
molasses	molasses - 1 Tbs.
burnt toast	burnt toast - 1/4 slice
flor yeast	baking yeast -- wetted - 1 Tbs.
sauerkraut	sauerkraut - 2 Tbs.
lactic acid	lactic acid -- 10% - 1 Tbs.
linalool	linalool - several drops on glass wool
geranium	geraniol - several drops on glass wool

*Standards served as described, undiluted, in 8½ ounce wine glasses covered with watch glasses.

```
                            NAME_____
    TRAINED PANEL BALLOT    DATE_____SAMPLE #____

    USING THE 9-POINT INTENSITY SCALE SHOWN BELOW, RATE EACH SAMPLE FOR
    ALL ATTRIBUTES LISTED.

    1 - NONE
    2 - THRESHOLD           FOR AROMA ONLY
    3 - SLIGHT
    4 - SLIGHT TO MODERATE
    5 - MODERATE            OVERALL INTENSITY_____
    6 - MODERATE TO LARGE
    7 - LARGE
    8 - LARGE TO EXTREME
    9 - EXTREME
```

1ST TIER	2ND TIER	3RD TIER
FRUITY____	CITRUS____	GRAPEFRUIT____
	BERRY____	BLACKBERRY____
		STRAWBERRY____
		RASPBERRY____
	TREE FRUIT____	CHERRY ____
	DRIED FRUIT____	STRAWBERRY JAM____
		RAISIN____
		FIG____
		PRUNE____
SPICY____	SPICY____	BLACK PEPPER____
		CLOVES____
VEGETABLE____	CANNED/COOKED____	
EARTHY____		
CARAMELIZED____	CARAMELIZED____	HONEY ____
		BUTTERY ____
		BUTTERSCOTCH ____
CHEMICAL____	PUNGENT____	ETHANOL____
	SULFUR____	
MICROBIOLOGICAL____		LACTIC____

FIG. 1. THREE TIERED SENSORY BALLOT DEVELOPED FOR
AROMA EVALUATION OF PINOT NOIR

Serving Procedure

Standards were reviewed by all panelists prior to sample evaluation. The 60 mL samples were served at ambient temperature (20C) in 8½ oz, clear, tulip-shaped wine glasses coded with three digit random codes and covered with a watch glass. Two samples were evaluated each session. Their presentation order was randomized and they were presented monadically so that no cross-sample comparison was allowed.

Ballot

In using the ballot, panelists were asked to first evaluate the sample for overall intensity. Then the sample was rated for its more general characteristics progressing to the more specific terms. Since the training relied on the Wine Aroma Wheel, general terms such as fruity and chemical were referred to as first tier terms, the more specific groups of berry and pungent were referred to as second tier terms, and the precise descriptors such as raspberry and ethanol were referred to as third tier terms. In evaluating a sample, if the panelist detected a fruity character (first tier), overall fruity intensity was rated. Upon further investigation if the panelist determined the fruitiness to be a raspberry note, both raspberry intensity (third tier) and berry (second tier) were rated. The panelist always rated the first tier fruity character equal to or higher than the more specific berry (second tier) and raspberry (third tier) characters. Panelists were required to rate all first tier terms using the 9 point intensity scale (1 = none, 9 = extreme). For the more specific (second and third tier) terms, the panelists only rated the characteristics that were detected in the samples. Any terms not rated were assigned a value of one (no intensity) in the analysis.

EXPERIMENTAL DESIGN AND STATISTICAL ANALYSIS

A balanced incomplete block (BIB) experimental design was carried out by each of twelve panelists for 33 descriptors of six samples of wine with different malolactic fermentation treatments. Data were analyzed as presented by Gacula and Singh (1984) for each panelist and the panel as a whole. A correlation analysis of estimated descriptor mean scores for each panelist against the panel as a whole was used as a selection criterion to remove panelists from further analysis. The data from the remaining panelists was used to test for no significant difference among the malolactic treatments with respect to each descriptive term.

The BIB design compared, over fifteen sessions, all possible combinations of the six treatments presented two at a time. The design required five replications of each treatment. Occasional missed sessions by panel members were made up at a later date by single sample presentations. A few missing values (five) over all scores were estimated by averaging across replications of that particular treatment for that panelist.

Figures 2 and 3 give the steps followed in the analysis of these data and the selection criteria used to select panelists, respectively. As described by Gacula and Singh (1984), there are two analyses of BIB designs: (1) intrablock analysis and (2) interblock analysis. For this situation "blocks" and "sessions" were synonymous. The distinction was based on treating within session random varia-

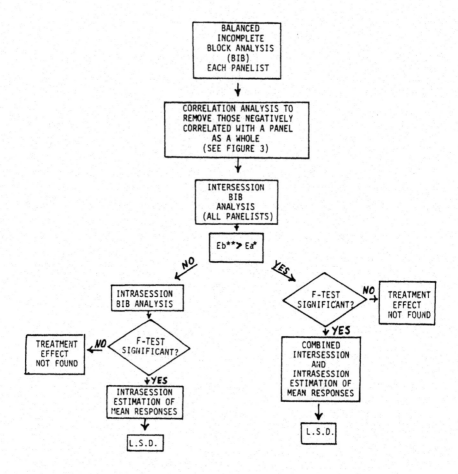

FIG. 2. DATA ANALYSIS FOLLOWED FOR EACH INDIVIDUAL ATTRIBUTE
*Ea = within session random variation
**Eb = between session random variation

tion (Ea) or between session random variation adjusted for sample differences (Eb) as the appropriate error term. When the session to session random variation was no larger than the within session random variation (Ea), a simple F-test of mean square treatments (unadjusted for blocks) against Eb was appropriate. When this was not the case, the appropriate F-test was the mean square treatments (adjusted for blocks) against Ea. Estimated mean responses were calculated directly from the intrablock analysis in the later case; or in the former case, a combined interblock and intrablock estimation of mean responses as suggested by Yates (1940).

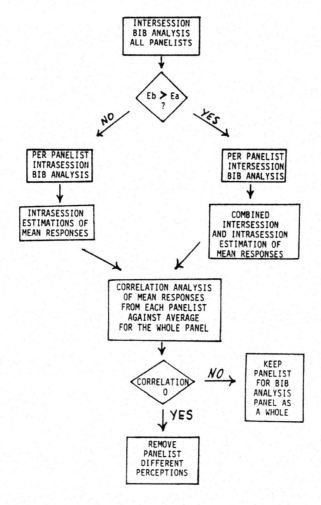

FIG. 3. CORRELATION ANALYSIS TO SCREEN PANELISTS

There were two ways in which a panelist's data could be removed from the analysis. One was for the judge to be classed as a nonperceiver of that particular attribute in the set of wines tested (assigning all samples the category scale value of "none". Being classed as a nonperceiver of cherry, for example, did not mean that the person could not perceive cherry. It did mean that for this particular set of wines, no cherry character was perceived. The other was by correlation analysis which judged how each panelist's values agreed with the panel as a whole. The correlation analysis (Fig. 3) involved calculation of Eb and Ea from all panelist scores, then either using the combined or intrablock estimated mean scores for each panelist. These estimates of mean scores were then correlated against the estimated mean scores for the panel as a whole. Panelists with positive correlations were selected for further analysis.

Those panelists not removed based on the correlation analysis are used to test for a significant treatment effect. When significant effects are found, then pairwise comparisons are made among those estimated mean scores using Fisher's protected least significant difference (L.S.D.). Each of the 33 descriptors were analyzed separately.

A correlation analysis was conducted on all 33 attributes in order to see how the attributes interrelated.

Wine Production

Pinot Noir vintage 1981 was crushed, destemmed, and 50 mg/mL of sulfur dioxide was added. The must was inoculated 1% by volume with UCD #595 Pasteur Champagne yeast and fermented at 18C for six days with frequent punching down of the cap. The new wine was pressed from the skins at 0° Brix, settled for 48 h and racked. The new wine contained 12.1% alcohol by volume, 5.8 g/L titratable acidity as tartaric, 1.74 g/L malic acid (enzymatic procedure), 0.05 g/100 mL residual sugar, 13 mg/L free SO_2 and 23 mg/L total SO_2 (Ripper procedure), 0.013 g/100 mL volatile acidity as acetic, and a pH of 3.46. The wine was divided into duplicate lots of 3 gallons each and inoculated 1% by volume with actively fermenting malolactic bacterial cultures. Plate counts to determine the number of viable cells were done on MR-V8 agar with 70 mg/L cycloheximide (Sigma Chemical Co.) to inhibit yeast growth. The number of viable cells in the wine prior to inoculation was < 100 cfu/mL and on the order of 10^6 cfu/mL after inoculation with each strain. The bacteria cultures were prepared in grape juice media at pH 3.5 (10 days, 27C) from inocula grown in MR-V8 broth at pH 4.6 (4 days, 30C) (Henick-Kling 1983; Noble et al. (1984). The wines were maintained between 16C and 18C until completion of the fermentations. Malolactic fermentations were considered complete when the concentration of malic acid was 0.1 g/L or less, and the evolution of CO_2 had ceased.

The malolactic bacteria strains used included Erla and Ey2d (Oregon State University isolates), ML-34 (U-C-Davis), PSU-I (Penn. State) and MLT (Microlife Technics, Florida). Upon completion of the fermentations the wines contained less than 0.05 g/L malic acid and had an average titratable acidity of 4.5 g/L as tartaric (range 4.5-4.6 g/L), an average pH of 3.58 (range 3.56-3.59) and an average volatile acidity of 0.029 g/100 mL as acetic (range .027-.030 g/100 mL). The replicated lots were pooled, filtered (Scott SG pads), bottled and stored at 12C for three years before sensory evaluation by the trained panel.

RESULTS AND DISCUSSION

Final Panelist Screening

Although 33 descriptors appeared on the ballot for aroma evaluation, not all the panelists used all the descriptors available in evaluating the six wines. For example, panelist number five (Table 2) did not perceive any blackberry notes among the six wines. The judge was then classed as a nonperceiver (NP) and his or her values were not included in the analysis for that attribute. The remaining panelists for each attribute were further screened to determine whether or not each judge's individual ratings across samples correlated with the panel as a whole. If the correlation was less than zero, that judge was classed as a noncorrelator (NC) and his or her values were excluded from the analysis for that attribute. For example, judge number eight was a noncorrelator only once, for overall intensity, while judge number twelve was a noncorrelator a total of twelve times.

It was expected that each judge would be a noncorrelator and a nonperceiver several times. In fact, incidents of noncorrelation ranged from only one, for judge number eight, to a high of twelve for judge number nine with an average of slightly over seven. Incidents of nonperceivers ranged from only once for judge one to a high of thirteen for judge number seven for an average of slightly over five. Judge number eight appeared to be the best overall judge being a noncorrelator only once and a nonperceiver only twice. However, even judge number twelve, who was a noncorrelator and nonperceiver a total of 22 times, still contributed valuable data on eleven other attributes.

Examination of noncorrelators and nonperceivers on an individual attribute basis shows which attributes were difficult for the panel as a whole to rate. The average number of noncorrelators across attributes was 2.64/attribute. The attributes giving the panel the most difficulty were vegetative, canned, cooked, microbiological and overall intensity, each having four noncorrelators, and chemical and pungent with seven and eight noncorrelators, respectively. Microbiological, chemical and pungent were not significant descriptors of the six wines tested (Table 3). The average number of nonperceivers was 1.94/ attribute. Prune and sulfur each had four nonperceivers, honey and butterscotch each had five nonperceivers, while fig and buttery each had six nonperceivers. Of those six attributes, only butterscotch was a significant descriptor of the six wines tested.

Evaluation of Attributes

Of the 33 attributes used to describe the aroma of the six wines, thirteen showed no significant differences (Table 3). The number of panelists data re-

TABLE 2.
PANELISTS ANALYZED AS NONPERCEIVERS (NP) AND NONCORRELATORS (NC)

	Panelist Number													
	1	2	3	4	5	6	7	8	9	10	11	12	NC	NP
OVERALL		NC			NC			NC	NC				4	--
FRUITY									NC	NC			2	
Citrus	--	NC					NP					NC	2	1
grapefruit							NP		NC			NC	2	1
Berry	--	NP					NC		NC	NC			3	1
blackberry		NC		NC	NP		NP			NC		NP	3	3
strawberry							NC	NP	NP		NP		1	3
raspberry	NC				NC		NP					NP	2	2
Tree Fruit				NC		NC	NP					NP	2	2
cherry			NC	NC		NC	NP						3	1
Dried Fruit		NC		NC		NC							3	--
strawberry jam				NC		NC	NP		NC				3	1
raisin		NP	NC						NP	NC		NP	3	2
fig	NP	NC	NP	NP	NP	NP						NP	1	6
prune				NP	NP		NP		NP		NC		1	4
SPICY							NC	NP				NC	2	1
Spicy					NC		NP		NC			NC	3	1
black pepper					NC		NP		NC				2	1
cloves		NP	NC				NP			NP		NC	2	3
VEGETATIVE	NC					NC	NC		NC				4	--
Canned/Cooked					NP		NC	NC			NC	NC	4	1
EARTHY					NP	NP					NP	NC	1	3
CARAMELIZED		NP	NC	NC							NP	NP	2	3
Caramelized		NP	NC	NC					NC		NP	NP	3	3
honey	NC	NP							NP	NP	NP	NP	1	5
buttery	NC	NP	NP		NC			NP		NP	NP	NP	2	6
butterscotch		NP	NP						NC	NP	NP	NP	1	5
CHEMICAL	NC	NC		NC	NC				NC	NC		NC	7	--
Pungent	NC	NC		NC	NC	NC	NC		NC			NC	8	--
ethanol	NC			NC								NC	3	--
sulfur				NC	NP		NC			NP	NP	NP	2	4
MICROBIOLOGICAL	NC		NC	NC								NC	4	--
Lactic				NC	NP								1	1
											TOTAL		87	64
											X̄		2.64	1.94
TOTALS														
NC	8	8	7	11	8	7	6	1	12	5	2	11		
NP	1	8	4	3	4	2	13	2	4	5	8	11		
GRAND TOTAL	9	16	11	14	12	9	19	3	16	10	10	22		

maining in the analysis for each attribute ranged from a low of four for cherry, buttery and pungent to a high of ten for fruity and lactic. Cherry was a statistically significant descriptor ($p \leq 0.01$) even though so few judges used it effectively. Lactic, on the other hand, was a nonsignificant descriptor, even though a large portion of the panel used it and their results correlated well.

Overall aroma intensity was the first attribute rated and a significant difference was found. All of the remaining attributes were evaluated in tiers, working from left to right across the ballot within a general category.

The first first-tier term, fruity, was highly significant ($p \leq 0.001$). Citrus, a second tier term, was not significant. However grapefruit, a more specific term, was significant at the $p \leq 0.05$ level. This may be explained by the fact that the final panelists included in the data analysis were not identical for both attributes. Panelist number two was a noncorrelator for citrus while panelist number nine was a noncorrelator for grapefruit. It is difficult to explain why a judge could discriminate using the specific term grapefruit, when he or she were not able to discriminate using the general term, citrus, unless the more specific term allowed better concentration of better focusing.

Berry and all of the more specific descriptors under it were significant descriptors. Both tree fruit and cherry were significant descriptors. The general descriptor dried fruit was significant as was strawberry jam, but raisin, fig and prune were not.

Spicy was rated twice, as a first tier term by all panelists, and again as a second tier term by those panelists wishing to specify which spice. At the first tier level spicy was not significant, however, it was at the second tier level. Black pepper was also significant. Panelist number twelve was a noncorrelator for spicy, but was able to correlate with the panel on black pepper, again showing how focusing on a more specific attribute can be helpful. Both vegetative and canned/cooked were significant as was earthy.

Caramelized was not significant with seven judges as a first tier term, but was highly significant ($p \leq 0.01$) with six judges as a second tier term. Buttery and honey were not significant descriptors for the six wines, however, butterscotch was. In general, the number of nonperceivers in this category was larger than in other categories.

The general term chemical was highly significant, but the only significant term under it was ethanol ($p \leq 0.05$). Again, pungent, the more general term, was not as good as ethanol, the specific term, in describing the wine aroma differences. Panelists simply did not correlate with each other on pungency ratings, yet performed better on ethanol ratings.

The general term microbiological was rated by the entire panel but was not a significant attribute with these samples. Nor was lactic a significant term.

Results of a correlation analysis of the 33 descriptive terms rated by the panel are found in Fig. 4. In many cases, as expected first tier terms correlate with descriptors in their respective second and third tiers, such as berry correlating with blackberry, and tree fruit correlating with cherry. Overall intensity ratings, tied to no specific descriptor, correlated with cloves, vegetative and canned/cooked. The general term, fruity, correlated with only one specific fruity term, raisin. However, it also correlated with earthy, caramelized (first

TABLE 3.
ADJUSTED MEAN SCORES[a] FOR INTENSITY OF PINOT NOIR AROMA ATTRIBUTES FOR DIFFERENT MALOLACTIC BACTERIA AS JUDGED BY A TRAINED SENSORY PANEL

	Uninoculated Control	MLT	Er-1a	ML-34	Ey-2d	PSU-1	F	P	No. of Panelists
OVERALL INTENSITY	4.75^a	4.95^a	5.60^b	5.10^a	5.10^a	4.60^a	4.76	0.001	10
FRUITY	4.24^a	4.74^{bc}	4.88^c	4.40^{ab}	3.98^a	4.40^{ab}	4.63	0.001	10
Citrus	1.38^a	1.58^a	1.31^a	1.40^a	1.16^a	1.20^a	2.04	NS	9
grapefruit	1.58^{ab}	1.40^{ab}	1.64^b	1.60^b	1.42^{ab}	1.33^a	2.43	0.05	9
Berry									
blackberry	2.15^a	3.18^{bc}	3.15^{bc}	3.33^c	2.33^{ab}	2.70^{ab}	5.04	0.001	8
strawberry	1.00^a	1.67^c	1.60^{bc}	1.60^{bc}	1.37^{abc}	1.23^{ab}	3.22	0.01	6
raspberry	1.53^{ab}	1.60^{ab}	1.55^{ab}	1.73^b	1.33^a	1.45^{ab}	3.06	0.05	8
	2.03^{ab}	1.98^{ab}	2.30^{bc}	2.83^c	1.68^a	1.95^{ab}	3.91	0.01	8
Tree Fruit	2.13^b	2.15^b	2.20^b	1.38^a	2.08^b	1.80^{ab}	2.99	0.05	8
cherry	2.23^b	2.13^b	2.28^b	1.33^a	2.18^b	1.83^b	3.86	0.01	8
Dried Fruit	2.16^{ab}	2.36^{abc}	2.69^c	2.07^a	1.96^a	2.64^{bc}	2.94	0.05	9
strawberry jam	1.53^a	1.75^a	2.33^a	1.63^a	1.68^a	1.90^{ab}	2.65	0.05	8
raisin	1.43^a	1.54^a	1.69^a	1.34^a	1.34^a	1.31^a	1.24	NS	7
fig	1.16^a	1.76^a	1.52^a	1.36^a	1.64^a	1.96^a	2.23	NS	5
prune	1.57^a	1.69^a	1.37^a	1.51^a	1.20^a	1.51^a	1.11	NS	7
SPICY	2.89^a	2.49^a	2.78^a	2.51^a	3.11^a	2.31^a	1.95	NS	9
Spicy	2.80^c	2.08^{ab}	2.40^{bc}	2.30^{ab}	2.68^c	1.88^a	2.43	0.05	8
black pepper	1.96^{bc}	1.44^a	1.69^{abc}	1.53^a	2.02^c	1.60^{ab}	2.36	0.05	9
cloves	1.46^a	1.63^a	1.94^a	1.71^a	1.97^a	1.31^a	1.67	NS	7
VEGETATIVE	1.98^{bc}	2.18^c	1.28^a	1.65^{ab}	1.93^{bc}	2.13^c	4.71	0.01	8
Canned/Cooked	1.74^{abc}	2.00^{bc}	1.40^a	1.63^{ab}	1.83^{abc}	2.23^c	2.40	0.05	7
EARTHY	1.49^a	1.40^a	1.29^a	1.49^a	2.00^b	1.34^a	2.88	0.05	8

TABLE 3. (continued)

	Uninoculated Control	MLT	Er-1a	ML-34	Ey-2d	PSU-1	F	p	No. of Panelists
CARAMELIZED	1.89^a	2.51^a	2.57^a	2.14^a	2.09^a	2.23^a	1.58	NS	7
Caramelized	1.80^a	2.47^{bc}	2.60^c	2.27^{abc}	2.10^{abc}	2.07^{ab}	3.22	0.01	6
honey	1.57^a	1.87^a	1.90^a	1.67^a	1.77^a	1.83^a	1.46	NS	6
buttery	1.30^a	1.10^a	1.25^a	1.40^a	1.05^a	1.05^a	1.27	NS	4
butterscotch	1.20^a	1.73^b	1.73^b	1.33^{ab}	1.40^{ab}	1.53^{ab}	2.45	0.05	6
CHEMICAL	5.76^b	4.76^a	4.44^a	4.76^a	5.44^b	4.72^a	3.83	0.01	5
Pungent	5.20^a	4.15^a	4.15^a	4.50^a	5.00^a	4.70^a	2.30	NS	4
ethanol	4.38^c	3.56^a	3.73^{ab}	3.93^b	3.89^{ab}	3.78^{ab}	2.91	0.05	9
Sulfur	2.07^a	1.80^a	1.83^a	1.70^a	2.47^a	2.17^a	1.21	NS	6
MICROBIOLOGICAL	2.43^a	2.25^a	2.43^a	2.45^a	2.98^a	2.75^a	1.172	NS	8
Lactic	2.18^a	2.16^a	2.26^a	2.28^a	2.52^a	2.48^a	1.05	NS	10

^{a}Scale ranged from 1 = not present to 9 = extreme.
NOTE: Same letter superscripts in the same row indicate no significant difference at p ≤0.05 level.

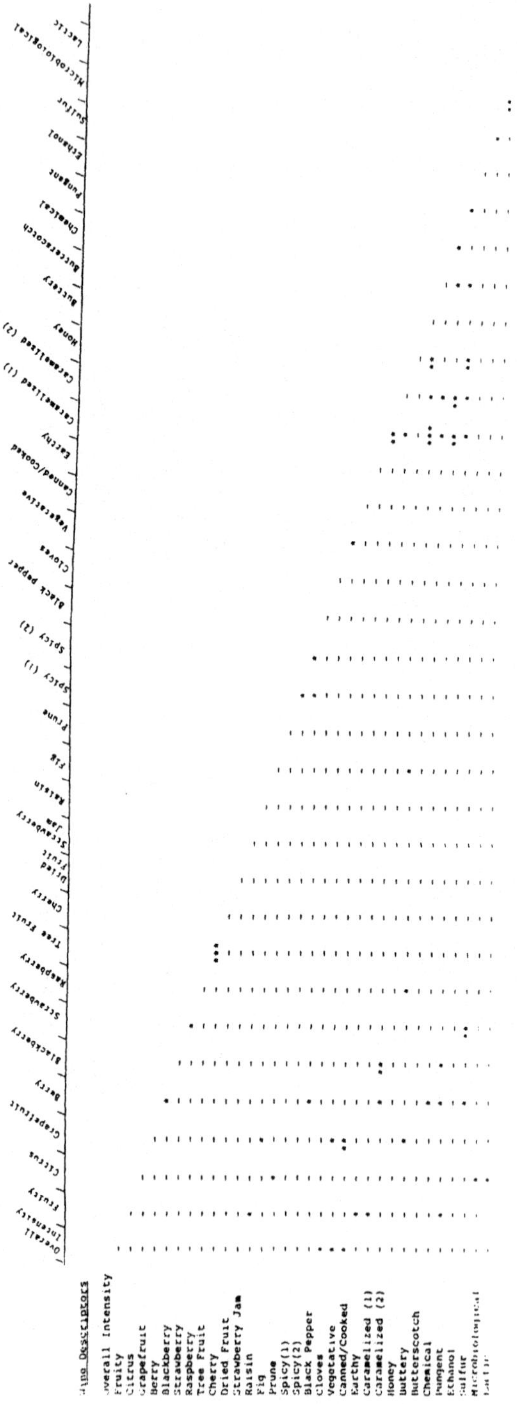

FIG. 4. SIGNIFICANT LEVELS (* = $p < 0.05$, ** = $p < 0.01$) FROM CORRELATION ANALYSIS OF DESCRIPTOR INTENSITY RATINGS

tier) and pungent. Pungent correlated seven times with other descriptors. The terms berry and caramelized (first tier) both had six incidences of correlation, while eleven descriptors were never correlated with any other descriptor.

Comparison Among Wines

Of major importance to the Oregon industry is the comparison of the two new Oregon strains, Erla and Ey2d, and other potentially useful strains, PSU-1 and MLT, to the primary industry standard ML-34. Figure 5 contains a comparison of each of the other strains to ML-34 on the basis of significant attribute differences from this study.

Overall, ML-34 is high in berry notes, but often low in other aroma characteristics. MLT was rated highest in blackberry, vegetable and butterscotch (tied with Erla). Erla rated highest in overall intensity. It also rated highest in intensity for fruity, grapefruit, tree fruit, cherry, dried fruit, strawberry jam, raisin, caramelized, honey and butterscotch (tied with MLT). Ey2d was rated the highest in spicy (Tier 1), black pepper, cloves, earthy, sulfur, microbiological, and lactic. PSU-1 was the highest in canned/cooked (vegetative) intensity. The uninnoculated control, which went through malolactic fermentation "on its own," was rated highest in spiciness (Tier 2), chemical and ethanol intensity.

CONCLUSIONS

Screening of the individual panelists by use of correlation analysis provided a simple statistical approach by which judges could be eliminated on an individual attribute basis. Also, if a large number of judges are eliminated for any one attribute, that attribute's importance as a descriptor should be questioned. The attribute may simply be perceived differently from one panelist to another or more thorough training may be required.

The balanced incomplete design where only two samples were evaluated at any one time, but where five replications of each sample were completed, allowed for small aroma differences present to be perceived. Previous studies may have found few differences simply due to panelist fatigue and/or adaptation.

Also of interest in this study was to determine if fermentation of Pinot noir by different strains of malolactic bacteria created differences in the wines' aroma. There is no question that the strain of malolactic bacteria selected can effect aroma perception. The judgment of whether or not one strain produces a better quality wine, however, is highly subjective and must be left to the winemaker to decide.

FIG. 5. COMPARISON OF SIGNIFICANTLY DIFFERENT AROMA DESCRIPTORS FOR EACH MALOLACTIC STRAIN COMPARED TO ML-34, THE OREGON INDUSTRY STANDARD

ML-34	vs.	MLT
greater ethanol		greater tree fruit
		greater cherry
		greater vegetative

ML-34	vs.	Er-1a
		greater fruity
		greater tree fruit
		greater cherry
		greater dried fruit
		greater strawberry jam
		greater overall intensity

ML-34	vs.	Ey-2d
greater berry		greater tree fruit
greater strawberry		greater cherry
		greater spicy
		greater black pepper
		greater earthy
		greater chemical

ML-34	vs.	PSU-1
greater grapefruit		greater cherry
greater berry		greater dried fruit
greater spicy		greater vegetative
		greater canned/cooked

ML-34	vs.	Uninoculated Control
greater berry		greater tree fruit
greater blackberry		greater cherry
		greater spicy
		greater black pepper
		greater chemical
		greater ethanol

The authors gratefully acknowledge the partial financial support of this work by the Oregon Wine Advisory Board. This work is published as Technical Paper No. 8081 of the Oregon Agricultural Experiment Station.

REFERENCES

AMERINE, M.A., BERG, H.W. and CRUESS, W.V. 1982. *Technology of Wine Making*, Fourth Edition. Chapter 16, Bacteria in Wine, pp. 557-581. Van Nostrand Reinhold/AVI, New York.

CAIRNCROSS, S.E. and SJÖSTRÖM, L.B. 1950. Flavor Profile — A new approach to flavor problem. Food Technol. *4*, 308-311.

DAVIS, C.R., WIBOWO, D., ESCHENBRUCH, R., LEE, T.H. and FLEET, G.M. 1985. Practical implications of malolactic fermentation: A Review. Am. J. Enol. Vitic. *36*(4), 290.

GACULA, M.C. and SINGH, J. 1984. *Statistical Methods in Food and Consumer Research*. Chapter 5. Incomplete Block Experimental Designs. pp. 141-175. Academic Press, San Diego, CA.

HENICK-KLING, T. 1983. Comparison of Oregon Derived Malolactic Bacteria in Pilot Scale Wine Production. M.S. Thesis, Dept. of Microbiology, Oregon State University.

HEYMANN, H. and NOBLE, A.C. 1987. Descriptive analysis of commercial Cabernet Sauvignon Wines from California. Am. J. Enol. Vitic. *38*(1), pp. 41-44.

KUNKEE, R.E. 1974. Malo-Lactic Fermentation and Winemaking. Chemistry of Winemaking, Adv. Chem. Ser. 137, pp. 151-170. Amer. Chem. Soc., Washington, D.C.

MALEK, D.M., MUNROE, J.M., SCHMITT, D.J. and KORTH, B. 1986. Statistical evaluation of sensory judges. J. Am. Soc. Brew. Chem. *44*(1), 23.

NOBLE, A.C., ARNOLD, R.A., MASUDA, B.M., PECORE, S.D., SCHMIDT, J.O. and STERN, P.M. 1984. Progress towards a standardized system of wine aroma terminology. Am. J. of Enol. and Vitic. *35*(2), 107-109.

NOBLE, A.C. and SHANNON, M. 1987. Profiling Zinfandel Wines by sensory and chemical analyses. Am. J. Enol. Vitic. *38*(1), 1-5.

WATSON, B.T., Jr., MICHAELS, N.J., HEATHERBELL, D.A., HENICK-KLING, T. and SANDINE, W.E. 1984. Commercial Evaluation of Two New Oregon Strains of Malolactic Bacteria. Proceedings of the International Symposium on Cool Climate Viticulture and Enology, June 25-28, Eugene, OR.

YATES, F. 1940. The recovery of inter-block information in balanced incomplete block designs. Ann. Eugen. *10*, 317-325.

CHAPTER 4.5

DESCRIPTIVE ANALYSIS OF PINOT NOIR WINES FROM CARNEROS, NAPA, AND SONOMA

JEAN-XAVIER GUINARD and MARGARET CLIFF

*At the time of original publication,
all authors were affiliated with
the University of California
Davis, CA 95616-5270.*

Twenty-eight Pinot noir wines from Carneros, Napa, and Sonoma were evaluated by descriptive analysis by seven trained judges. Wines differed significantly for all 14 sensory attributes except prune aroma. Principal component analysis of the mean ratings showed that Carneros wines differed from Napa and Sonoma wines and that they had unique sensory attributes. Carneros Pinot noir wines were characterized by intense fresh berry, berry jam, cherry, and spicy aromas.

Carneros is a viticultural area which extends across the southern part of Napa and Sonoma counties. It received official viticultural appellation status in 1983 due to its distinct climate and soil conditions. Carneros is classified as a climate region I, and nearby San Pablo Bay causes warmer winter and cooler summer temperatures. Soils are shallow, rich in nutrients, with slow drainage. These viticultural conditions are particularly suited for growth of Chardonnay and Pinot noir varieties, which represent 48% and 37%, respectively, of the total production. Numerous observations suggest that wines made from Carneros Pinot noir grapes are unique. However, no sensory analytical evaluations have confirmed these observations.

The technique of descriptive analysis, combined with univariate and/or multivariate statistics, has been used to profile wine flavor (7,9,10,17). This technique allows for the quantitative characterization of perceived sensory attributes (16). Judges are trained in the use of specific descriptive terms or attributes (8), the intensities of which are subsequently rated across the wines.

In this study, the technique of descriptive analysis was applied to define the sensory properties of Carneros Pinot noir wines by comparison with Pinot noir wines from the Napa and Sonoma areas (excluding Carneros).

*Reprinted with permission of the American Society for Enology and Viticulture.
©Copyright 1987. Originally published in* Am. J. Enol. Vitic. **38**, 211-215.

MATERIALS AND METHODS

Wines: Twenty-eight Pinot noir wines were selected from the Carneros (10), Sonoma (9), and Napa (9) areas. The wines were chosen from two vintages, 1981 and 1983, in order to account for year to year variation. Details of the wines are provided in Table 1. Titratable acidity, pH, volatile acidity, ethanol, reducing sugars, and free and bound SO_2 were determined using the methods given by Amerine and Ough (1). Total phenols were determined using the method given by Slinkard and Singleton (14). A commercial red wine (Mountain Castle Burgundy) and a Pinot noir blend were used, respectively, for the preparation of the aroma and flavor by mouth standards. The wines used for the selection of descriptive terms and for panel training were taken from the 28 wines selected for the study.

Panel: Fourteen judges volunteered to participate in the study. Twelve judges, seven men and five women ranging in age from 25 to 50 years, remained after the selection of descriptive terms and training. All of the judges had extensive wine tasting experience, but only a few had previously been on descriptive analysis panels.

Selection of descriptive terms and preparation of the corresponding standards: In a first session, descriptive terms were generated by the judges from individual evaluation of three wines. The same procedure was repeated in a second session in which the same three wines were presented with 50 reference standards. The standards, corresponding to the terms selected by the judges, were prepared by adding food products or chemicals to the base red wine. In a third session, a different set of five wines was assessed by the judges with 30 reference standards provided. Upon discussion of the wines and standards, a consensus was reached on a final set of 11 aroma and three flavor by mouth terms. The list of terms and the composition of the corresponding standards are presented in Table 2.

Experimental design and procedure: A randomized complete block design was used for each training period and for the formal sessions.

The intensity of each attribute was evaluated across the wines on an unstructured, 10-cm scale in the order in which the attributes are listed in Table 2. All wines were served in coded, tulip-shaped, clear wine glasses at 16C to 22C. A 30-mL sample was poured into each glass and covered with a plastic lid at least 15 min prior to evaluation. The tests were conducted in isolated booths illuminated with daylight and incandescent lighting. At each training or formal session, judges smelled and/or tasted the standards prior to entering the booths and, if necessary, before evaluating each attribute. Judges rinsed between samples with drinking water (Feather River, Sacramento, CA), and all samples were expectorated.

TABLE 1.
IDENTIFICATION AND CHARACTERISTICS OF THE 28 PINOT NOIR WINES

Code	Origin	Winery	Vintage
1	Carneros	A	1981
2			1983
3		B	1981
4			1983
5		C	1981
6			1983
9		D	1981
10			1983
22		E	1981
23			1983
11	Sonoma	F	1981
12		G	1983
13		H	1981
14		I	1983
15		J	1983
16		K	1983
17			1981
21		L	1981
28		M	1983
7	Napa	N	1981
8			1983
18		O	1981
19			1983
20		P	1981
24		Q	1981
25		R	1983
26		S	1983
27		T	1983

In a first training period consisting of five sessions held on different days, five wines were evaluated in triplicate at a rate of three samples per session. Analyses of variance were performed on the scores generated for each attribute to assess the performance of the panel. Significant F-ratios for the replications and for the judge × wine interaction were taken as indications of poor reproducibility and inconsistent use of the terms by the judges. respectively. Seven terms gave a significant judge × wine interaction, and two terms gave significantly different replications.

In a second training period consisting of two sessions held on different days, three wines were evaluated in duplicate at a rate of three samples per session. Upon analysis of variance of the data, five terms gave a significant judge × wine interaction, and one term gave significantly different replications. At the end of training, two judges were dropped because of absenteeism and inconsistency.

TABLE 2.
AROMA AND FLAVOR BY MOUTH TERMS SELECTED FOR DESCRIPTIVE ANALYSIS AND COMPOSITION OF THE CORRESPONDING REFERENCE STANDARDS

Term	Composition of reference standard*
1. Fresh berry (strawberry, raspberry, black currant)	5 mL red berry fruit drink (Capri Sun™) + 3 mL black currant syrup (Vedrenne™)
2. Berry jam (strawberry, raspberry, blackberry)	6.5 g each of strawberry, raspberry and blackberry jam (Empress™)
3. Cherry	5 mL cherry drink (Hi-C™)
4. Prune	10 mL prune juice (Town House™)
5. Spicy (black pepper, cloves)	pinch of black pepper + 2 cloves
6. Mint/eucalyptus	2 cm² green mint + 4 cm² eucalyptus leaf
7. Earthy (potato, mushroom)	7.5 mL canned potato liquor + 10 mL canned mushroom liquor (Town House™)
8. Leather	4 cm² leather
9. Vegetal (green bean, green tea)	10 mL canned green bean liquor (Town House™) + 1.2 g green tea (Dynasty™)
10. Smoke/tar	0.01 mL liquid hickory smoke (Wright's™) + 1 g tar
11. Berry by mouth	1 mL I.F.F.™ strawberry extract in 150 mL Pinot noir
12. Bitterness	0.2 g caffeine in 150 mL water/0.4 g caffeine in 150 mL Pinot noir
13. Astringency	0.6 g aluminum sulfate in 150 mL water/1.2 g aluminum sulfate in 150 mL Pinot noir

* In 30 mL Mountain Castle Burgundy™ (unless otherwise specified)

Fourteen formal sessions were held on different days to evaluate the 28 wines in duplicate. Four wines were presented at each session. No wine was presented in combination with any other wine more than once.

Data analysis: All statistical analyses were performed using Statistical Analysis Systems (SAS) (15). Individual analyses of variance (AOV) were run on each attribute rated by the judges. The mean ratings of the 28 wines for 11 of the 14 attributes rated by the judges were then analyzed by principal component analysis (PCA). A PCA was also performed on the chemical data across the 28 wines.

RESULTS

Analyses of variance of most attribute ratings across the 28 wines for all 12 judges showed a significant judge × wine interaction reflecting the inconsistent use of several of the terms by some judges. Five judges were removed from subsequent analysis after examination of the data. Separate evaluation of the data from these five judges revealed that they did not form a distinct, consistent population and were using the terms improperly.

The results of the AOV of the attribute ratings across the 28 wines for seven judges are summarized in Table 3. Judges were a very highly significant source of variation in all cases ($p < 0.001$). Wines were significantly different for all attributes except prune. Replications were not a significant source of variation except for bitterness ($p < 0.05$). A highly significant judge × wine interaction was found in the analysis of mint/eucalyptus, earthy, and berry by mouth ratings, indicating that despite the reduction of the panel size to seven judges, these terms were still not used in the same way by all judges. Consequently, these terms were not taken into account in the PCA of the mean ratings.

The correlation coefficients among the attributes are shown in Table 4. Very highly significant correlations ($p < 0.001$) were found between fresh berry and berry jam ($r = 0.60$), between fresh berry and cherry ($r = 0.71$), and between leather and smoke/tar ($r = 0.70$). In descriptive analysis, a significant correlation between two terms suggests that they may have been used to describe the same attribute. In this case, however, these terms were not used redundantly but merely described related aromas. For example, the fact that the terms related to fruitiness (fresh berry, berry jam, cherry, and prune) were not all significantly correlated with one another indicates that these terms were not used to describe the same fruity character. Bitterness and astringency ratings were also very highly correlated ($r = 0.61$, $p < 0.001$). These attributes traditionally are highly correlated in wine because they are elicited by the same compounds (13).

The correlation matrix generated from the mean ratings of each wine across the 11 attributes retained after the AOVs was analyzed by PCA with no rotation. The first three principal components (PC) accounted for 33%, 25%, and 13% of the variance, respectively. In the scree test (3), these PCs were the most important ones as the scree plot shows a break at the third eigenvalue. Also, only the first three PCs had eigenvalues above one (4). These two criteria indicate that only the first three PCs are 'significant' ones. Consequently, interpretation of the data was limited to these PCs. In Figure 1, the 28 wines and the 11 attributes are plotted on the first two PCs. Wines are separated along the first PC according to the intensity of their fresh berry, berry jam, and cherry aromas. The intensity of their spicy, leather, smoke/tar, and vegetal aromas also contributes to the separation of the wines, but to a lesser extent, as indicated by the larger angle between these attribute vectors and the first PC. The position

TABLE 3.
ANALYSES OF VARIANCE OF ATTRIBUTE RATINGS (7 JUDGES): DEGREES OF FREEDOM (df), F-RATIOS, AND ERROR MEAN SQUARES (MSE)

	Judges (J)	Reps (R)	Wines (W)	J × R	J × W	R × W	MSE
Red color	30.74***	0.02	61.17***	0.90	1.12	1.97**	94.87
Fresh berry	37.30***	0.08	2.61***	0.59	1.17	1.15	198.73
Berry jam	70.34***	2.20	2.90***	0.28	1.32*	0.85	150.17
Cherry	71.91***	0.35	2.71***	0.39	1.42*	0.83	130.65
Prune	72.46***	0.23	1.39	0.66	1.21	1.38	147.57
Spicy	130.64***	1.17	1.78*	1.82	1.28	0.99	89.32
Mint/eucal.	121.27***	0.90	2.32***	0.55	2.10***	1.40	105.02
Earthy	121.47***	1.33	5.28***	1.03	1.64***	1.93***	125.67
Leather	56.59***	0.63	2.39***	0.98	1.17	0.85	181.54
Vegetal	103.28***	0.01	4.74***	2.64*	1.38*	0.77	130.74
Smoke/tar	110.38***	0.07	4.02***	2.81*	1.19	1.22	129.90
Berry by mouth	36.05***	3.67	2.21**	1.79	1.55**	0.92	171.75
Bitterness	111.21***	5.72*	3.83***	2.22*	1.19	1.06	134.94
Astringency	128.78***	0.29	5.05***	1.45	1.24	1.65	146.91
df	6	1	27	6	162	27	162

*, **, ***, significant at $p < 0.05$, $p < 0.01$, and $p < 0.001$, respectively.

TABLE 4.
CORRELATION MATRIX AMONG THE DESCRIPTIVE TERMS (df = 26)

Term	1	2	3	4	5	6	7	8	9	10	11
1. Red color	1.00										
2. Fresh berry	-0.05	1.00									
3. Berry jam	0.01	0.60***	1.00								
4. Cherry	0.13	0.71***	0.42*	1.00							
5. Prune	-0.22	-0.22	0.04	-0.19	1.00						
6. Spicy	0.32	0.29	0.27	0.47*	-0.25	1.00					
7. Leather	0.18	-0.50**	-0.45*	-0.28	0.14	0.16	1.00				
8. Vegetal	-0.46*	-0.57**	-0.27	-0.61***	0.51**	-0.54**	0.11	1.00			
9. Smoke/tar	0.46*	-0.66***	-0.41*	-0.35	0.04	0.20	0.70***	0.15	1.00		
10. Bitterness	0.09	-0.14	-0.14	-0.08	0.35	0.04	0.35	0.10	0.27	1.00	
11. Astringency	0.57**	0.00	0.00	-0.02	-0.04	0.23	0.23	-0.27	0.36	0.61***	1.00

*, **, ***, significant at p < 0.05, p < 0.01, and p < 0.001, respectively.

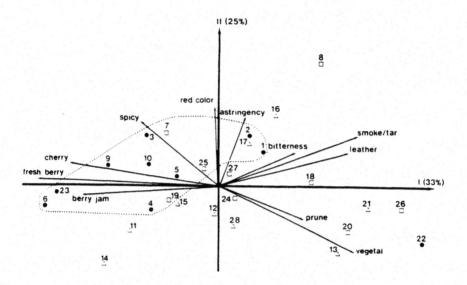

FIG. 1. PRINCIPAL COMPONENT ANALYSIS OF THE MEAN RATINGS
OF THE 28 WINES FOR 11 ATTRIBUTES
The first two principal components are shown.
The attribute vectors and the 28 wines with their codes (See Table 1)
and origin, *i.e.*, Carneros (●), Napa (□), or Sonoma (▲), are plotted.

of the wines on the second PC is determined by the intensity of their red color and astringency attributes.

In the plane formed by the first two PCs (Fig. 1), all Carneros wines except wine 22 are clustered. Furthermore, this cluster is separated from all but four Napa and Sonoma wines. On the other hand, no apparent clustering of the Napa or Sonoma wines occurs. Examination of the coordinates of Carneros wines on the third PC shows that the clustering of Carneros wines observed on the first two PCs is conserved along the third one.

Plotting the attribute vectors and the wines in the same plane allows for the characterization of the wines. Carneros Pinot noir wines generally are high in fresh berry, berry jam, cherry, and spicy aromas and low in vegetal, leather, and smoke/tar aromas.

Principal component analysis of the basic chemical data (pH, titratable acidity, volatile acidity, reducing sugars, ethanol, and total phenols) showed no clustering of Carneros wines on the first three PCs. Nonetheless when the variables free SO_2 and bound SO_2 were added in the analysis, Carneros wines were separated from Napa and Sonoma wines. This separation was achieved along the free SO_2, bound SO_2, and volatile acidity vectors. More specifically, Carneros Pinot noir wines were lower in these parameters (Table 5).

TABLE 5.
AVERAGE FREE SO_2, BOUND SO_2, AND VOLATILE ACIDITY FOR
CARNEROS, NAPA, AND SONOMA PINOT NOIR WINES

Origin	Free SO_2 (mg/L)	Bound SO_2 (mg/L)	Volatile acidity (g/100 mL as acetic acid)
Carneros (n = 10)	7.5	44.9	0.052
Napa (n = 9)	9.3	64.2	0.066
Sonoma (n = 9)	14.7	68.3	0.061

DISCUSSION

The chemical complexity of the flavor of Pinot noir wines is widely recognized (6,11,12), but from a sensory standpoint, little research has been conducted on the topic. Pinot noir wines are commonly described as pepperminty, berry-like, strawberry-like, or blackberry-like (2). This study confirmed the fruity character of California Pinot noir wines with the choice of descriptors such as fresh berry (strawberry, raspberry, black currant), berry jam (strawberry, raspberry, blackberry), cherry, and prune, and emphasized their unique spicy (black pepper, cloves), leather, and smoke/tar characters. Large variations in the sensory attributes of Carneros, Napa, and Sonoma Pinot noir wines were nevertheless evidenced. Indeed, significant differences across wines were found for all descriptive terms except prune (Table 3). This emphasizes the discriminative power of descriptive analysis combined with parametric statistics.

Examination of the judges' comments collected on the individual scorecards revealed that wine 22 was defective and that the defect was probably sulfur-related. This explains the position of wine 22 on the first two PCs (Fig. 1) and the high ratings it received for the vegetal, leather, and smoke/tar attributes. Therefore, it can be considered as an outlier. To include wines 1 and 2, the cluster of Carneros wines had to be expanded (Fig. 1). Both wines were produced from the same vineyard by the same winery. Higher fermentation temperature (90-100F) and a more extensive use of new barrels than is normal in Carneros (Larry Brooks, personal communication, 1986) probably account for the position of wines 1 and 2 at the center of Figure 1. Indeed, a higher fermentation temperature causes a reduction in fruity esters (5), and the use of new barrels causes an increase in bitterness and astringency (13).

The cluster formed by the Carneros wines in the PCA of the sensory data indicates that the sensory attributes of Carneros Pinot noir wines are unique. It also justifies the appellation status of Carneros. Pinot noir wines from Carneros are characterized by intense fresh berry (strawberry, raspberry, black currant), berry jam (strawberry, raspberry, blackberry), cherry, and spicy (black pepper, cloves) characters. On the other hand, they are low in vegetal (green bean/green tea), leather, and smoke/tar aromas, which are found at moderate to high levels in most of the Napa and Sonoma Pinot noir wines which were studied.

Distinction between Carneros wines and Napa or Sonoma wines was also achieved by PCA of the chemical data. However, the three variables which caused clustering of Carneros wines, *i.e.*, free SO_2, bound SO_2, and volatile acidity, are strictly under the control of the winemaker. Therefore, these variables should not be used to discriminate among wines from different regions because they are likely to vary from winery to winery, independent of the variety studied.

Now that the uniqueness of Carneros Pinot noir wines within California has been demonstrated, these wines should be characterized further by comparison with Oregon and Burgundy Pinot noir wines using the same descriptive analysis techniques.

LITERATURE CITED

1. AMERINE, M.A. and C.S. OUGH. *Methods for Analysis of Musts and Wines*. 341 pp. John Wiley and Sons, New York (1983).
2. AMERINE, M.A. and E.B. ROESSLER. Wines. Their Sensory Evaluation. 432 pp. Freeman and Co., New York (1983).
3. CATTELL, R.B. The scree test for the number of factors. Multivar. Behav. Res. 1:245-76 (1966).
4. KAISER, H.F. The application of electronic computers in factor analysis. Educ. Psychol. Measurement 20:141-55 (1960).
5. KILLIAN, E. and C.S. OUGH. Fermentation esters - Formation and retention as affected by fermentation temperature. Am. J. Enol. Vitic. 30:301-5 (1979).
6. MEUNIER, J.M. and E.W. BOTT. Das verhalten verschiedener aromastoffe in Burgunderweinen im verlauf des biologischen saureabbaues. Chem. Mikrobiol. Technol. Lebensm. 6:92-5 (1979).
7. NOBLE, A.C. Sensory and instrumental evaluation of wine aroma. In *Analysis of Foods and Beverages. Headspace Techniques*. G. Charalambous (Ed.). pp 203-28. Academic Press, New York (1978).
8. NOBLE, A.C., R.A. ARNOLD, B.M. MASUDA, S.D. PECORE, J.O. SCHMIDT and P.M. STERN. Progress towards a standardized system of wine aroma terminology. Am. J. Enol. Vitic. 35: 107-9 (1984).
9. NOBLE, A.C., A.A. WILLIAMS and S.P. LANGRON. Descriptive analysis and quality ratings of 1976 wines from four Bordeaux communes. J. Sci. Food Agric. 35:88-98 (1984).

10. SCHMIDT, J.O. and A.C. NOBLE. Investigation of the effect of skin contact time on wine flavor. Am. J. Enol. Vitic. 34:135-8 (1983).
11. SCHRIER, P. Wine aroma composition. Identification of additional volatile constituents of red wine. J. Agric. Food Chem. 28:926-8 (1980).
12. SCHRIER, P., F. DRAWERT and K.O. ABRAHAM. Identification and determination of volatile constituents in Burgundy Pinot noir wines. Lebensm. Wiss. Technol. 13:318-21 (1980).
13. SINGLETON, V.L. and A.C. NOBLE. Wine flavor and phenolic substances. In *Phenolic, Sulfur, and Nitrogen Compounds in Food Flavors*. G. Charalambous and I. Katz (Eds.). ACS Symp. Ser 26:47-70 (1976).
14. SLINKARD, K. and V.L. SINGLETON. Total phenol analysis: Automation and comparison with manual methods. Am. J. Enol. Vitic. 28:49-55 (1977).
15. Statistical Analysis Systems. SAS User's Guide: Statistics. 584 pp. SAS Institute, Cary, NC (1982).
16. STONE, H., J. SIDEL, S. OLIVER, A. WOOLSEY and R. SINGLETON. Sensory evaluation by quantitative descriptive analysis. Food Technol. (Chicago) 28:24-34 (1974).
17. WILLIAMS, A.A., S.P. LANGRON and A.C. NOBLE. Influence of appearance on the assessment of aroma in Bordeaux wines by trained assessors. J. Inst. Brew. 90:250-3 (1984).

This research was financed by the Carneros Quality Alliance (CQA). The authors thank the members of the CQA's descriptive analysis panel, Jill Davis, Melissa Moravec, and David Rosenthal, for their active participation. The use of Dr. A. Noble's laboratory and equipment at the University of California, Davis, is also greatly appreciated.

This research conducted at the Buena Vista Winery, Sonoma, and at the University of California, Davis.

CHAPTER 4.6

DESCRIPTIVE ANALYSIS FOR WINE QUALITY EXPERTS DETERMINING APPELLATIONS BY CHARDONNAY WINE AROMA

L.P. McCLOSKEY, M. SYLVAN[1] and S.P. ARRHENIUS

At the time of original publication,
all authors were affiliated with
McCloskey, Arrhenius & Co.
29 East Napa St., Sonoma, CA 95476.

ABSTRACT

Since producer based sensory analysis of wine appellations remains in widespread use worldwide by wineries, modern sensory methods were developed for industry judge panels. U.S. appellations were studied using modern sensory analysis processes which accommodated the industry professionals' biases for perceived quality and provincial sensory language. A panel was assembled comprised of wine industry quality experts (n = 26). First, judge quality biases were determined using a multi-wine preference-testing method (n = 48) in which judges freely record their aroma terms. Subsequently QDA®type strategies were used to create the sensory language from analysis of 1100 wine × judge interactions in which free use of terms was used to describe the wines. Next, an experiment analyzed the wines using a new descriptive analysis scorecard which contained the ten most frequently used terms; and scores were computed from the number of times terms were selected by the judges (frequency of use). Groups of sixteen wines, for which the judges had no large negative bias, from the Carneros American Viticultural Areas (AVA), were compared to those from California wine growing regions (Central Coast, Napa and Sonoma). Replicate trials showed Carneros AVA and Central Coast wines clustered in the principal components (PCA) analysis of the sensory data. The detection of regional typicalness by professionals: (1) was linked to their perceived quality bias, (2) was easily detected in high quality wines linked to grape attributes more than to winemaking attributes, and (3) was not possible when judges had determined that perceived wine quality was low. ANOVA and polar spider plot analysis of the clusters indicated that the important aroma attributes of Carneros wines

---

[1]Department of Mathematics, University of California, Santa Cruz, CA 95060.

Reprinted with permission of Food & Nutrition Press, Inc., Trumbull, Connecticut.
©Copyright 1996. Originally published in Journal of Sensory Studies **11**, 49-67.

included Citrus and Green Apples/Pears. Lastly, the sensory analysis data was compared to chemical analysis of terpenes for several wines (n = 22). Chardonnay terpene (linalool) concentrations were highly correlated ($p < 0.01$) with the attribute scores for Citrus which was important in the PCA analysis of the sensory data. These sensory processes offer a descriptive analysis process that accommodates the wine industry professional. The method also offers several advantages over flavor profiling methods including the reduction of the potentially confounding problems of "quality" and "standard terms" among industry professionals as well as speed and correlation with chemical analyses.

INTRODUCTION

One of the greatest challenges for U.S. wine growers is confirmation of appellation of their bottled products by objective analysis. Wine growers could use modern consumer based sensory analysis as the litmus test for appellation since regional viticultural and enological factors must ultimately be detected among products sold in the marketplace (Stone and Sidel 1993). However, advances in modern sensory analysis used by large food manufacturers have not been systematically used by wine growers to establish appellations.

Appellations determined by producers have been important in French and German wine growing regions for a century prior to the advent of modern sensory analysis (Jackson and Lombard 1993; Kramer 1989; Larousse 1991; Moio *et al.* 1989; Morlat 1989; Peynaud 1984; Seguin 1986). In France the concepts of the *cru* and *terroir* describe tasters' detection of regionality of wine. *Terroir* is defined as a distinct region with an environment that can produce an original quality agricultural product (Larousse 1991).

In France *terroir* remains a basic tenet of regulatory controls including *appellation d'origine contrôlée* system (A.O.C.) established in 1911 (Jackson and Morlat 1993; Larousse 1991; Morlat 1989; Peynaud 1984, 1987). However, it is not strictly objective. Wine marketers have included ad hoc sensory attributes. For example, winemaking attributes such as "Brioche," or yeast-like aroma of *sur lies* Chardonnay, are often used to describe the Burgundy *terroir* (Moio *et al.* 1993). Nevertheless, *terroir* is a powerful model when used by wine industry professionals and it has remained the basis of the regulatory controls instituted to protect French regional wines and it is a focal point of enological research (Moio *et al.* 1993; Morlat 1989; Seguin 1986; Larousse 1991).

In the U.S. the first steps taken to establish tighter regulatory controls over appellations were made by the Bureau of Alcohol, Tobacco and Firearms (BATF) in 1978. The U.S. BATF established American Viticultural Areas (AVA) (Kramer 1989) based upon political factors. Once established, producers have asked in retrospect whether regional ecological factors have translated into unique and distinct wines (Guinard and Cliff 1987).

One of the great challenges for modern sensory scientists has been in establishing widespread use of descriptive analysis among traditional U.S. wine growers. Why have few wine growers used modern sensory analysis to establish appellations? A major obstacle is reconciling the traditional sensory analyses used by producers with modern wine sensory analysis. Typically, U.S. and European premium wine producers form one school which uses traditional quality assessments (TQA). Most wineries use TQA to gauge market potential (Lawless and Classen 1993). After WWII as large U.S. food and beverage companies strove to maintain the consistency of their products modern sensory analysis was introduced for wine (Amerine *et al.* 1976, 1983; Lawless and Classen 1993). "Descriptive analysis" methods were developed including quantitative descriptive analysis (QDA®) (Stone *et al.* 1974; Stone and Sidel 1993). QDA was subsequently used with consumer panels to forecast consumer acceptance of wines for large producers. Although QDA has answered questions about similarities and difference of wines for large producers, it has not been employed on any large scale basis within the premium wine industry. A third school developed at University of California at Davis called "Descriptive Analysis" (UCD DA) for the wine industry (Aiken and Noble 1984; Guinard and Cliff 1987; Noble *et al.* 1984; Noble *et al.* 1987; Noble 1988; Schmidt and Noble 1983). It is an expert system that has been used successfully by wine educators. UCD DA is similar to flavor profiling (FP) introduced in the 1950's by the Arthur D. Little Company (Caul 1957) which was based on techniques established by Sjöström and Cairncross (1954). In practice, most wine industry professionals prefer to use TQA or QDA because of several problems with FP-type methods (Williams and Langron 1984; Stone and Sidel 1993). First the judge training process, which is time-consuming and expensive, is potentially confounding for professionals with strong quality or language biases. Second the "Modified Standard Wine Aroma Terminology" (Noble *et al.* 1987) does not include many terms used by industry professionals (Moio *et al.* 1993) making it difficult to obtain complete agreement among judges. Although free-choice profiling (FCP) (Williams and Langron 1984) and QDA are available to overcome these problems, most wine producers do not yet see the merits of modern sensory analysis.

We accommodated the wine industry professional using several suggestions regarding perceived quality and language development (McCloskey *et al.* 1995; Williams and Langron 1984; Stone and Sidel 1993). We chose not to use UCD DA since it may be a difficult and potentially futile task to retrain wine quality experts to use a new language. QDA's several advantages were reviewed including: (1) reliance on 10-15 qualified subjects rather than a few specialists (2) the attributes used to characterize the products rely on a consumer based language, (3) subjects including wine industry professionals or consumers decide, by consensus or analysis, what language to use, and (4) replicate trials are required to assess subject and panel reliability (Amerine and Roessler 1976, 1983; Stone and Sidel 1993). Since our goal was to determine uniqueness of the Carneros AVA using a panel comprised completely of industry quality experts the QDA processes were used to gather terms and to develop a sensory language already in use among the judge population. A step prior to descriptive analysis was used to account for the perceived quality bias of judges (McCloskey *et al.* 1995). Preliminary chemical analysis was compared to the sensory data as part of a long range effort to create quality control techniques for regional wines (Marais 1987; Marais *et al.* 1991; Rapp 1988; Webster *et al.* 1993; Williams *et al.* 1982; Williams *et al.* 1992; Wilson *et al.* 1986).

MATERIALS AND METHODS

Wines

Forty-eight bottled wines were provided by the California producers or the Carneros Quality Alliance (CQA). All of the wines were produced in significant quantities and were available in the marketplace and were priced as premium wines.

General Practices

For all tasting sessions a leader monitored the good tasting practices including: (1) prohibition of communication among judges; (2) serving wines in coded, tulip-shaped glasses at 18C to 22C; (3) providing rinse water between samples with drinking water; (4) eliminating visual cues that may bias judges with respect to the preference testing (below) by the placement of wine glasses on yellow mats in conjunction with reducing lighting; (5) the opportunity to clarify ambiguities in oral discussions; (6) providing backup wines for objectionable pours, e.g. cork tainted wines; and (7) replication of results.

Judges

We used twenty-six wine industry professionals as judges to form a panel. Industry professional judges testified during oral discussions that they each routinely used traditional quality assessments (TQA). All the judges were familiar with UCD DA and the system of terms proposed by Noble *et al.* (1987), "A Modified Standardized System of Wine Aroma Terminology". Interjudge consistency was analyzed using Spearman's Rho and data from inconsistent judges were removed from the results.

TABLE 1.
TERMS SELECTED FOR MULTI-WINE-DESCRIPTIVE (MWD) ANALYSIS AND GROSS USE OF TERMS BY INDUSTRY PROFESSIONALS FOR CARNEROS, CENTRAL COAST, NAPA AND SONOMA CHARDONNAY WINES

Terms	Region of Origin					TOTALS
	Carneros	Carneros	Central Coast	Napa	Sonoma	
CITRUS (INCLUDE PEEL OR RIND)	27	29	23	29	25	133
OAKY, VANILLA, WOODY, TOASTY	44	52	53	44	59	252
HONEY	19	10	30	5	0	64
CARAMEL, PUMPKIN SQUASH, SWEET POTATO	8	5	10	3	6	32
FRUITY, FRUITS, ESTERY	34	41	31	38	43	187
MUSCAT, FLORAL, TERPENE	16	17	15	17	24	89
BUTTERY, MALOLACTIC	20	29	19	26	36	130
VEGETAL, HERBACEOUS, HERBAL	9	15	29	23	11	87
APPLES, RED APPLES (NOT ALDEHYDE), PEARS	11	41	28	18	17	115
NEUTRAL, VINOUS ALCOHOLS	10	5	4	6	15	40
YEASTY	0	0	0	2	1	3
H_2S	1	0	1	1	0	3
ALDEHYDE	1	1	0	0	5	7

Pre-Steps to Descriptive Analysis for Industry Professional Panels

Protocols for Selecting Wines. We analyzed wines using a preference-testing process which identifies wines for which the industry quality experts have a large negative bias. We elected to use a multi-wine preference test (MWP-test) which had previously been used by several judges (McCloskey *et al.* 1995) and was available free (MAX-Preference™ Scorecard, Enologix® 25 E. Napa St., Sonoma, CA 95776). The MWP process was efficient in selecting wines for which judges had a large negative bias, since the probability that an individual will by chance correctly identify the top eight and the bottom eight preferences of the group is 1/12870 for sixteen wines.

The preference test was simple; judges were instructed to create two equal size groups of wine by scoring wine a plus or minus. This sorting task was done by each judge individually. Scores were normalized (net score/n judges) and the MWP scores were reported from +1.0 to (−1.0). The method did not use the "no difference" or "neutral" score. The normalized scores for the forty-eight wines were used to form three equal size groups based on large (A), neutral (B) and low (C) MWP scores. Group A contained the wines for which the judges had no negative bias.

Multi-wine descriptive analysis scorecard for Chardonnay.

Describe the wines by using the terms in the column containing the "Aroma Attributes"

MY SCORE	GROUP SCORE	WINE SAMPLE	Describe each wine sample by circling two to five (no less and no more of the ten aroma terms listed below.
		301	Citrus (include Peel or Rind) Oaky, Vanilla. Woody, Toasty Honey Caramel, Pumpkin Squash, Sweet Potato Fruity, Fruits, Estery Muscat, Floral, Terpene Buttery, Malolactic Vegetal-Herbaceous, Herbal Apples,(not Aldehyde), Pears Neutral, Vinous Alcohols
		002	Citrus (include Peel or Rind) Oaky, Vanilla. Woody, Toasty Honey Caramel, Pumpkin Squash, Sweet Potato Fruity, Fruits, Estery Muscat, Floral, Terpene Buttery, Malolactic Vegetal-Herbaceous, Herbal Apples,(not Aldehyde), Pears Neutral, Vinous Alcohols
		211	Citrus (include Peel or Rind) Oaky, Vanilla. Woody, Toasty Honey Caramel, Pumpkin Squash, Sweet Potato Fruity, Fruits, Estery Muscat, Floral, Terpene Buttery, Malolactic Vegetal-Herbaceous, Herbal Apples,(not Aldehyde), Pears Neutral, Vinous Alcohols
		(064)	Citrus (include Peel or Rind) Oaky, Vanilla. Woody, Toasty Honey Caramel, Pumpkin Squash, Sweet Potato Fruity, Fruits, Estery Muscat, Floral, Terpene Buttery, Malolactic Vegetal-Herbaceous, Herbal Apples,(not Aldehyde), Pears Neutral, Vinous Alcohols
		003	Citrus (include Peel or Rind) Oaky, Vanilla. Woody, Toasty Honey Caramel, Pumpkin Squash, Sweet Potato Fruity, Fruits, Estery Muscat, Floral, Terpene Buttery, Malolactic Vegetal-Herbaceous, Herbal Apples,(not Aldehyde), Pears Neutral, Vinous Alcohols

FIG. 1. MULTI-WINE DESCRIPTIVE ANALYSIS SCORECARD WITH THE TERMS FREQUENTLY USED BY THE INDUSTRY PROFESSIONAL PANELS DURING A PRETASTING INVOLVING OVER 1100 WINE × JUDGE DESCRIPTIONS

Protocols for Selecting Terms. Applicable terms were selected from those already in use by judges, using the general strategies of QDA suggested by Stone and Sidel (1993). This process involved analyzing all terms recorded by the judges in a space on their MWP scorecard (above) during five trials. Terms were entered into a database and regression analysis was used to combine the most redundant terms. The ten most frequently used groups of clarified terms (e.g., oaky, woody, vanilla) were selected and entered into the MWDA scorecard. Table 1 lists three terms which were not included in the scorecard (Fig. 1) because they were not used but which have been used to describe French regional Chardonnay wines (Moio *et al.* 1993).

Those terms for which some ambiguity existed in the judges understanding were clarified using wines or wine extracts as reference standards to facilitate the discussion. Two terms were clarified. This involved assessing wines and applying bottled wine extracts (see below) to the inside of a glass containing a neutral wine with cotton tipped swabs followed by oral discussion. Herbal/Leafy was explained using the Grape Extract I and the standard wine aroma terms were clarified with respect to the terms proposed by Noble *et al.* (1987). Caramel/Pumpkin Squash was included *ad hoc* after discussion of Wine Extract I.

Determining Number of Terms to Describe Chardonnay Wines. Frequency analysis of the terms used by the judges on the MWP scorecard, including Poisson fit, was used to determine how many terms winemakers naturally used to describe the sixty Chardonnay wines (Tables 1 and 2).

Multi-Wine Descriptive Method

The multi-wine descriptive (MWDA) analysis uses several QDA processes except that attribute scores are determined by the rate wine attributes are selected by judges. The method uses the terms gathered by the judge panel from the MWP scorecard. Instructions to judges are to describe wines using the same number of terms judges used during the MWP process (above).

Scorecard and Scorecard Testing. The MWP scorecard includes ten terms and instructions to select two to five terms to describe Chardonnay wines (Fig. 1). In the pre-test of this scorecard (using sixteen wines) the frequency analysis showed an identical shape to that found during the tasting of sixty wines where judges freely associated terms using the MWP scorecard.

MWDA Procedure. Groups of sixteen wines were presented to judges using QDA practices and oral instructions to judges. Judges were handed a scorecard and given about one hour to assess sixteen wines. Attribute scores (β)

TABLE 2.
FREQUENCY ANALYSIS SHOWING THE AVERAGE NUMBER OF TERMS INDUSTRY PROFESSIONAL USE TO DESCRIBE CALIFORNIA CHARDONNAY WINE

REGIONS	N	Use of Terms					
		0	1	2	3	4	5
CARNEROS	13	1.4	5.7	4.5	1.3	0.1	0
CARNEROS	12	1.8	5.2	3.6	2.1	0.2	0.1
4 REGIONS	16	0	0	5.5	5.6	3.9	1

ranged from $0-n$ (n = number of judges). Scores were computed from the number of times a term was selected for a given wine by the n judges, i.e., scores were normalized by dividing by the number of judges on the panel. From this point on we followed the QDA protocols suggested by Stone and Sidel (1993).

Calculations and Statistics. Statistical analyses of the attribute scores (β) were performed using the SAS program (Statistical Analysis Systems, SAS Institute, Inc., Cary, NC 27513); SysSTAT Student (Abacus Systems, Berkeley, CA); Lotus 123 (Lotus Development, Cambridge, MA 02142) and Delta-Graph V.3.5 (Monterey, CA 93940). Interjudge consistency was analyzed using Spearman's Rho and a similarity index (Amerine and Roessler 1983). Judges were eliminated from the analysis based on the criteria suggested which improved, (1) statistical significance of the correlation between Citrus and linalool and (2) clustering in the statistical analysis of the sensory data. The statistical analysis of the sensory data included: Principal Components Analysis (PCA) of the sensory attribute scores; ANOVA; and polar plots. Multiple regression and correlation analysis was used in the analysis of the chemical and tasting data.

Extracts Used to Clarify Ambiguous Aroma Terms. Extracts for sensory processes were made from Solid Phase Extractions (SPE) using the methods proposed by Williams *et al.* (1982). SPE extracts were made of grape-stems/leaves (Grape Extract I) and a wine which judges had previously described using the term in question (Wine Extract I) which was Caramel-Pumpkin. Wine Extract-I was prepared from 500 mL of Chardonnay wine eluted with 2.5 mL ethanol (200 proof) using flash columns packed by gravity with 40μ C-18 pelicular absorbent (J.T. Baker, Cat. No. 7025-00).

Chemical Analysis. The methods used for extraction of volatile aromatic compounds were taken from the methods of Williams *et al.* (1982, 1992). Chemical analyses were performed by Enologix®, (25 E. Napa St., Sonoma, CA 95476).

RESULTS

Part I contains the results of the MWP testing using wine industry professional judges. Part II contains the results of the MWDA analysis. Preliminary chemical analysis are presented and complete analysis are found in a companion publication (Arrhenius *et al.* 1996).

Part I. Sorting Wines and Selecting Terms

The industry professional panel's quality assessments and strong technical winemaking opinions were voiced during five tastings of fifty-four different wines. For *sur lies* method wines California winemakers often had strong opinions. Six wines with acetic acid and hydrogen-sulfide aroma attributes were rejected out of hand by the industry judges. Forty-eight wines were sorted into three equal size groups of sixteen wines containing four wines per region based on the MWP scores (McCloskey *et al.* 1995). Group A contained wines with no large negative judge bias (large MWP scores). Group B contained wines with mid-range MWP scores and Group C contained wines with a large negative judge bias (low MWP scores).

Terms recorded in MWP scorecards from over 1100 wine by judge descriptions were analyzed by the sensory staff to create the judge vocabulary. The twenty-six judges consistently used one to four terms to describe Chardonnay wines and the Poisson fit indicated that 95% of the data was gathered from 1 to 4 terms. In only 10 instances did judges use more than five terms and in those cases only six to seven terms were used. Terms selected and the frequency of their use are presented in Table 1 and Fig. 1. The frequency analysis results are included in Table 2. Judges were removed from the study based on use of the similarity index as recommended (Amerine and Roessler 1983).

Part II. Multi-Wine Descriptive Analysis of Regional Wines

Three groups (A, B and C) of sixteen California Chardonnay wines each were assessed using the MWDA scorecard, Fig. 1. Aroma attribute scores, the number of terms circles on the MWDA scorecard per trial (including replications), were compared using regression analysis to analyze the relationship among the terms, Tables 3 and 4. Principal components analysis (PCA) of the

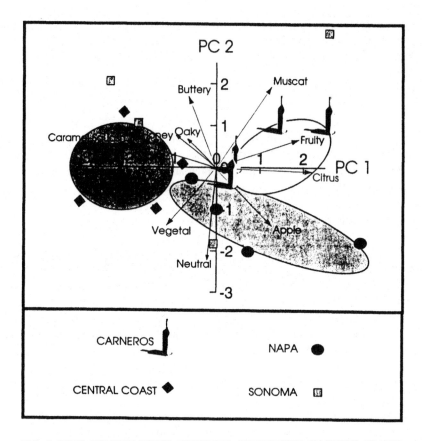

FIG. 2. HIGH QUALITY WINES: PRINCIPAL COMPONENT ANALYSIS OF THE SIXTEEN WINES FOR TEN AROMA ATTRIBUTES
The first two principal components are shown.
The attribute vectors and their origins are plotted.

sensory data was used to determine regional uniqueness, i.e., clustering (Fig. 2 and 3). Separation of clusters in the PCA plot indicated regional distinctness. The significance of the clusters was assessed by analyzing the ten aroma attributes with polar plots, Fig. 4 and ANOVA. The clustering of wines in the PCA of the sensory data were related to the industry professional's perception of quality (Group A, B and C).

Group A (Large MWP Scores). High quality Carneros wines were unique and distinct. Carneros, Central Coast and Napa wines clustered in the PCA of the sensory data, Fig. 2. Clusters were separated from each other along the first and second principal components (PC2 × PC1). Sonoma wines did not cluster.

TABLE 3.
CORRELATION MATRIX AMONG THE DESCRIPTIVE TERMS (n = 16) FOR GROUP A WINE (HIGH QUALITY)

	1	2	3	4	5	6	7	8	9	10
1. CITRUS	1									
2. OAKY, WOODY, VANILLA, TOASTY	-0.36	1								
3. HONEY	-0.31	-0.36	1							
4. CARAMEL, NOUGAT, PUMPKIN, SQUASH	-0.77*	0	0.66*	1						
5. FRUITY, FRUITS, ESTERY	0.76*	-0.18	-0.13	-0.60*	1					
6. MUSCAT, FLORAL, TERPENE	0.54*	-0.19	0.10	-0.33	0.69*	1				
7. BUTTERY, MALOLACTIC	-0.30	0.29	0.26	0.13	-0.35	0.20	1			
8. HERBAL/LEAFY, VEGETAL	-0.40	-0.03	0.14	0.42	-0.14	-0.36	-0.53*	1		
9. GREEN APPLE (not aldehyde), PEARS	-0.46	-0.44	-0.13	-0.42	0.20	0	-0.40	-0.17	1	
10. NEUTRAL, VINOUS ALCOHOLS	0.21	-0.19	-0.26	-0.07	-0.27	-0.46	-0.37	-0.05	0.29	1

* significant at p<0.05

TABLE 4.
CORRELATION MATRIX AMONG THE DESCRIPTIVE TERMS (n = 16) FOR GROUP B WINE (MIDRANGE QUALITY)

	1	2	3	4	5	6	7	8	9	10
1. CITRUS	1									
2. OAKY, WOODY, VANILLA, TOASTY	-0.17	1								
3. HONEY	-0.18	-0.30	1							
4. CARAMEL, NOUGAT, PUMPKIN, SQUASH	-0.48	-0.02	0.32	1						
5. FRUITY, FRUITS, ESTERY	0.44	-0.55*	0.23	-0.29	1					
6. MUSCAT, FLORAL, TERPENE	0.33	-0.78*	0.29	-0.10	0.74*	1				
7. BUTTERY, MALOLACTIC	-0.33	0.51*	-0.19	0	-0.39	-0.50*	1			
8. HERBAL/LEAFY, VEGETAL	-0.18	-0.07	0.06	0.40	-0.15	-0.13	-0.46	1		
9. GREEN APPLE (not aldehyde), PEARS	0.31	-0.37	-0.47	-0.54*	0.38	0.33	-0.24	-0.36	1	
10. NEUTRAL, VINOUS ALCOHOLS	-0.57*	0.14	-0.24	-0.12	-0.46	-0.33	0.47	-0.34	0.16	1

* significant at $p<0.05$

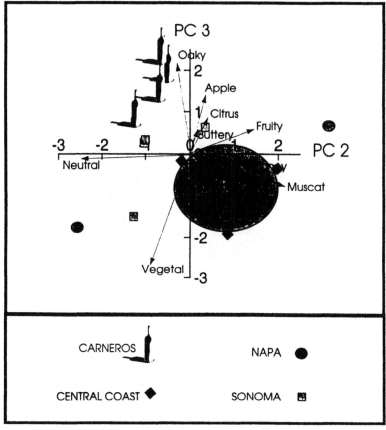

FIG. 3. MID-RANGE QUALITY WINES: PRINCIPAL COMPONENT ANALYSIS OF THE SIXTEEN WINES FOR TEN AROMA ATTRIBUTES
The second two principal components are shown.
The attribute vectors and their origins are plotted.

The first two PC's accounted for 56% of the variance and the first five PC's accounted for 90% of the variance.

PC1 was comprised of a linear combination of aroma terms which may be positively or negatively weighed (Eigenvectors in parenthesis below). The positively weighted terms of PC1, important in the description of the Carneros, wines, were: Citrus (0.51), Fruity (0.43), Muscat (0.30), and Green-Apples/Pears (0.30). Negatively weighted terms of PC1 included: Caramel-Pumpkin (-0.46) and Herbal/Leafy Vegetal (-0.23). The positively weighted terms associated with PC2 included: Muscat (0.45) and Buttery/Malolactic (0.51). Carneros wines also clustered in the PC3 × PC1 plane. The major positively weighted terms comprising PC3 included: Honey (0.58) and Caramel/Pumpkin Squash (0.31).

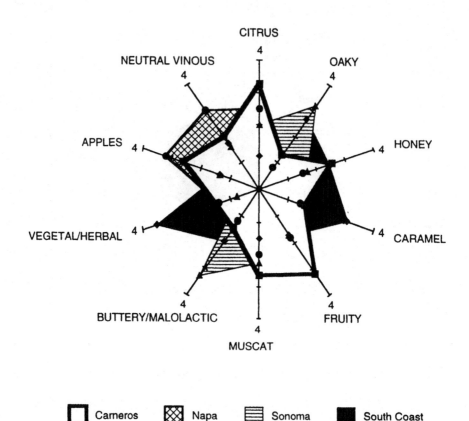

FIG. 4. POLAR PLOT ANALYSIS SHOWING REGIONAL VARIATION IN USE OF TEN AROMA ATTRIBUTES FOR GROUP A
(High quality wines, variation about the mean)

Group B (Neutral MWP Scores). Mid-range quality wines were unique and distinct. Carneros and Central Coast wines clustered in the PCA of the sensory data, Fig. 3. Carneros wines clustered tightly and separated from the other wines along the second and third principal components (i.e., in the PC3 × PC2 plane). Napa and Solomon wines did not cluster. The two PC's accounted for 57% of the variance and the first five PC's accounted for 89% of the variance. (Group B wines were assessed a second time and the results were compared with the results of the initial Group B tasting. Carneros wines again clustered in the PC3 × PC2 plane.)

The positively weighted terms of PC2 important in the description of the Carneros wines were: Muscat (0.53), Honey (0.38), and Fruity (0.32). An important negatively weighted attribute was Herbal/Leafy Vegetal (−0.20). Wines separated along PC3 based on positive weights for Oaky (0.56) and Green-Apples/Pears (0.32) and the negatively weighted Herbal/Leafy Vegetal (−0.66).

Group C (Low MWP Scores). Low quality wines did not cluster in the PCA of the sensory data. Analysis of the data indicated that aroma intensity, as reflected by frequency of use of attributes, was 20-50% of the high quality wines in Group A. Differences among the ten aroma attributes was minimal except for Caramel-Pumpkin/Squash and Honey which remained the major distinguishing attribute of the Central Coast wines.

Correlation Among Terms. The correlation matrices of the frequency of use of terms were computed and compared for the three groups of wines (Group A, Table 3). When the r-values for all attributes were compared for Groups A and B their correlation was r = 0.83 (p < 0.001). Most of the terms were not redundant and several terms had large negative correlations with each other.

DISCUSSION

The Multi-Wine Descriptive Analysis (MWDA) was used to successfully detect differences in bottled wines. The MWDA method separates itself from the QDA method since scores are derived from frequency of use of terms and not intensity. Similar to free choice profiling the judges enter terms into their MWP scorecard but differently terms are analyzed to create a final sensory language with the QDA practices for the subsequent MWDA. Although quality measures are not a part of modern sensory processes and the "clearest departure from the central dogma is in quality testing" (Lawless and Classen 1993) MWDA uses quality assessments to segment the pool of wines being studied and thereby overcomes several confounding problems.

Results showed that MWDA processes can be used with industry professionals with strong quality biases and data can be linked to uniqueness of appellations, i.e., clustering in the principal component analysis (PCA) of the sensory data. MWDA results showed that the Carneros region was unique and distinct (clearly different and not overlapping) from nearby regions including the Central Coast and Napa Valley. An important sensory result which supports the concept of *terroir* included the finding that typicalness was linked to grape based aroma attributes scores and chemical analysis for terpenes synthesized in grapes.

The MWDA method has several advantages: (1) it accounts for the quality bias of the wine industry professional which may confound the descriptive analysis, and as a result (2) clustering of wines by region in the statistical analysis of the sensory data was clearest when perceived quality was high and (3) conversely detection of regionality was reduced when the perceived quality decreased. As suggested by Stone (pers. comm.) the quality assessments (MWP) of the industry professional panel greatly affects descriptive analysis in general. The results support the view that the wine industry quality experts cannot use descriptive techniques to detect regional wines for which they also have a large negative bias with respect to quality. Apparently low quality wines confound the judges during descriptive analysis.

Group A (Large MWP Scores). The cluster formed by the wines in the PCA confirmed that the Carneros AVA was unique. Five grape aroma attributes were factors in the clustering of Group A wines. To determine which grape and wine attributes contributed to regional typicity of Carneros wines we analyzed the data using a polar plot of the ten terms and a one-way ANOVA. Citrus, Fruity, Muscat and Green-Apples/Pears were 40-75% above the mean for Carneros wines. Equally important in the description of the Carneros Chardonnay wines was the infrequent use of Caramel-Pumpkin, Herbal-Leafy Vegetal and Neutral Vinous/Alcohol (~30% below the mean). This was exemplified by the highly significant negative correlation between Citrus and Caramel/Pumpkin Squash attributes ($r = -0.77$; $p < 0.001$). This result explained why Carneros wines are distinct from Central Coast wines which are frequently described with the Pumpkin Squash attribute.

Group B (Neutral MWP Scores). The cluster formed by Carneros wines in the PCA duplicated the first results (Group A). However as the perception of quality declined, descriptions by the industry professional panel changed. Whereas Citrus and Green Apple/Pears were conserved in high and mid-quality Carneros wines, winemaking attributes replaced the grape attributes Fruity and Muscat. These attributes may have been obscured by winemaking attributes such as Oaky. Also chemical concentrations of flavorants may have changed as quality dropped from the high to the mid-range. Scores for Oaky were higher than for Citrus and Green-Apples/Pears. Together Green-Apples/Pears and Oaky were 30-67% above the mean for Carneros wines. Honey, Caramel/Pumpkin Squash and Herbal/Leafy Vegetal attributes were used infrequently with Carneros Group B wines, duplicating the Group A results.

Group C (Low MWP Scores). Low quality wines were not unique or distinct. This result strongly suggested that low quality as perceived by the industry professional panel (e.g., opinions regarding sur lies method attributes) either confounded the descriptive sensory analysis or that the chemical matrix had changed.

The clustering of Carneros Chardonnay wines in this study (and the previous study of Pinot Noirs by Guinard 1987) confirmed beyond a reasonable doubt the uniqueness of the Carneros AVA. The AVA was also distinct; the Carneros cluster did not overlap the Napa and the Central Coast clusters. Whereas the Carneros and Central Coast wines clustered at two levels of quality, Napa wines clustered only in the statistical analysis of the sensory data for the high quality wines. There was no clustering in either group for the Sonoma AVA. This is not surprising considering the size and environmental diversity of the Sonoma region.

To determine whether these sensory processes could be used to select wines for further chemical analyses wines were analyzed using Gas-Liquid-Chromatography-Mass Spectrometry. Grape aroma attribute scores, Citrus, were compared to linalool concentrations. Results were highly correlated to the attribute scores for Citrus ($P < 0.01$). As a result the MWDA scores were compared to over thirty chemical compounds and the results are reported in a companion publication (Arrhenius et al. 1996; McCloskey et al. 1996).

CONCLUSION

Accounting for the quality and language bias of industry professionals may be useful in modern descriptive analysis of regional wines particularly when research requires industry professional panels. This strategy was successfully applied to show that the Carneros AVA is unique and clearly distinct from Napa and Central Coast regions. The multi-wine descriptive analysis presented, which uses QDA language strategy and measures the quality bias, offers a practical alternative to researchers studying relationships between chemistry and producers based assessment of wines.

REFERENCES

AIKEN, J.W. and NOBLE, A.C. 1984. Comparison of the aromas of oak and glass-aged wines. Em. J. Enol. Viticult. *35*, 196-199.

AMERINE, M.A. and ROESSLER, E.B. 1976. *Wines: Their Sensory Evaluation.* W.H. & Freeman & Co., New York.

AMERINE, M.A. and ROESSLER, E.B. 1983. *Wines: Their Sensory Evaluation.* W.H. & Freeman & Co., New York.

ARRHENIUS, S.P., MCCLOSKEY, L.P. and SYLVAN, M. 1996. Chemical markers for aroma terms used to describe vitis vinifera var. Chardonnay Regional Wines. J. Ag. Food Chem. *44*, 1085-1090.

CAUL, J.F. 1957. The profile method of flavor analysis. Adv. Food Res. *7*, 1-40.

GUINARD, J.X. and CLIFF, M. 1987. Descriptive analysis of Pinot Noir from Carneros, Napa, and Sonoma. Am. J. Enol. Vitic. *38*, 211-215.

JACKSON, D.I. and LOMBARD, P.B. 1993. Environmental and management practices affecting grape composition and wine quality — A review. Am. J. Enol. Vitic. *44*, 409-430.

KRAMER, M. 1989. *Making Sense of Wine*. Marrow and Co., New York.

LAROUSSE. 1991. *Wines and Vineyards of France*. 48, pp. 576-577, 626 First U.S. Ed. Arcade Publishing, New York.

LAWLESS, H.T. and CLASSEN, M.R. 1993. Application of the central dogma in sensory evaluation. Food Technol. *47*, 139-146.

MARAIS, J. 1987. Terpene concentrations and wine quality of Vitis Vinifera L. cv. Gewurztraminer as affected by grape maturity and cellar practices. Vitis. *26*, 231-245.

MARAIS, J., VAN WYK, C.J. and RAPP, A. 1991. Carotenoid levels in maturing grapes as affected by climactic regions, sunlight and shade. S. Afr. Enol. Vitic. *2*, 64-69.

McCLOSKEY, L.P., ARRHENIUS, S.P. and SYLVAN, M. 1995. Measuring wine quality with industry professionals. Prac. Vinyard Winery *15*, (7).

McCLOSKEY, L.P., ARRHENIUS, S.P. and SYLVAN, M. 1996. Toward defining terroir with the Carneros American Viticultural Area. Carneros Quality Alliance, Vineburg, CA.

MOIO, L., SCHLICH, P., ISSANCHOU, P., ETIEVAN, X. and FEUILLAT, M. 1993. Description de la typicité aromamtiqus de vins de bourgogne issues du Cepage Chardonnay. J. Int. des Sciences de la Vigne et du Vin *27*, 179-189.

MORLAT, R. 1989. Le terroir viticole. Contribution à l'étude de sa caractérisation et de son influence sur les vins. Application aux vignobles rogues de Moyenne Vallée de la Loire. 289 pp. Thèse doctorat d'état. Université de Bordeaux II.

NOBLE, A.C. 1988. Analysis of wine sensory properties. In *Modern Methods of Plant Analysis*. Vol. 6. (H.F. Linskin and J.F. Jackson, eds.) Springer Verlag, Berlin.

NOBLE, A.C. 1994. Sensory Evaluation of Wine, Enology 100 Course Outline. pp. 13, 15.

NOBLE, A.C., ARNOLD, J., BUECHSENSTEIN, A., LEACH, E.J., SCHMIDT, J.O. and STEKRN, P.M. 1987. Modification of a standardized system of wine aroma terminology. Am. J. Enol. Vitic. *38*, 143-146.

NOBLE, A.C., WILLIAMS, A.A. and LANGRON, S.P. 1984. Descriptive analysis and quality ratings of 1976 wines from four Bordeaux communes. J. Sci. Food Agric. *35*, 88-98.

PEYNAUD, E. 1984. *Knowing and Making Wine*. John Wiley & Sons, New York.

PEYNAUD, E. 1987. *The Taste of Wine*. Macdonald & Co., London.

RAPP, A. 1988. Wine aroma from gas-chromatographic analysis. In *Modern Methods of Plant Analysis*. Vol.6. (H.F. Linsken and J.F. Jackson, eds.) Springer Verlag, Berlin.

SCHMIDT, J.O. and NOBLE, A.C. 1983. Investigation of the effect of skin contact time on wine flavor. Am. J. Enol. Viticult. *34*, 135-138.

SEGUIN, G. 1986. "Terroirs" and pedology of winegrowing. Experientia *42*, 861-873.

SJOSTROM, L.B. and CAIRNCROSS, S.E. 1954. The descriptive analysis of flavor. In *Food Acceptance Methodology*. (D.R. Peryam, F.J. Pilgrim, and M.S. Peterson, eds.) pp.25-30, Natl. Acad. Sci. Natl. Res. Couns., Washington, D.C.

STONE, H.S., SIDEL, J.L., OLIVER, S., WOOLSLEY, A. and SINGLETON, R.C. 1974. Sensory evaluation by quantitative descriptive analysis. Food Technol. *28*, 24-34.

STONE, H.S. and SIDEL, J.L. 1993. *Sensory Evaluation in Practices*. Academic Press, San Diego, CA

WEBSTER, D.R., EDWARDS, C.G., SPAYD, S.E., PETERSON, J.C. and SEYMOUR, B.J. 1993. Influence of vineyard nitrogen fertilization on the concentrations of monoterpenes, higher alcohols and esters in aged Reisling wines. Am. J. Enol. Vitic. *44*, 275-28.

WILLIAMS, A.A. and STEVENS, S.P. 1984. The use of free-choice profiling for the evaluation of commercial ports. J. Sci. Food and Agric. *35*, 558-566.

WILLIAMS, P.J., SEFTON, M.A. and FRANCIS, I.L. 1992. Glycosydic precursors of varietal grape and wine Flavor. In *Thermal and Enzymatic Conversions of Precursors to Flavor Compounds*. ACS Symposium Series No. 490, (R. Teranishi, G. Takeoka and M. Guntert, eds.) pp. 74-86, American Chemical Society, Washington, D.C.

WILLIAMS, P.J., STRAUSS, C.R., WILSON, B. and MASSY-WESTHROPP, R.A. 1982. Use of C-18 reversed-phase liquid chromatography for the isolation of monoterpene glycosides and norisoprenoid precursors from grape juice and wines. J. Chromatogr. *235*, 471-480.

WILSON, B., STRAUSS, C.R. and WILLIAMS, P.J. 1986. The distribution of free glycosidically bound monoterpenes among skin, juice and pulp fractions of some white grape varieties. Am J. Enol. Vitic. *37*, 107-111.

CHAPTER 5.0

TEXTILE MATERIALS

Although there are published documents (Rowe and Volkman 1965; Bishop 1996; Byrne *et al.* 1993) dealing with the subjective evaluation of textile materials, most of the documents were published in a non-sensory science journal. In this section, three articles are reproduced in Chapters 5.1, 5.2, and 5.3, which are considered the basic foundation for future research in the development of improved descriptive analysis techniques and their applications to textile and fabric materials.

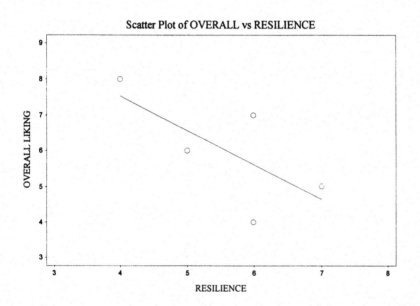

REFERENCES

BISHOP, D.P. 1996. Fabrics: sensory and mechanical properties. Textile Progress *26*(3), 1-64.

BYRNE, M.S., GARDNER, A.P.W. and FRITZ, A.M. 1993. Fibre types and end uses: A perceptual study. J. Texture Institute *84*, 275-288.

ROWE, S. and VOLKMAN, R.J. 1965. Thickness measurement of sanitary tissues in relation to softness. Tappi *48*, No.4, 54-6A.

CHAPTER 5.1

THE JUDGMENT OF HARSHNESS OF FABRICS[1]

HERMAN BOGATY, NORMAN R.S. HOLLIES
and MILTON HARRIS

*At the time of original publication
all authors were affiliated with
Harris Research Laboratories, Inc.
Washington, D.C.*

I. INTRODUCTION

The handle of cloth is utilized in many different ways as a measure of "quality" and commercial desirability. Along with cost and appearance, it is perhaps the only way consumers can assess the relative technical merit of a textile material. Throughout the course of manufacturing, from the selection of the fiber and its grading to the final chemical and mechanical finishing, a great many technical decisions are made on the basis of tactual judgments.

In view of its commercial and technical importance, it is not surprising that attempts have been made to evaluate hand instrumentally and objectively as, for example, in the classic work of Peirce [8]. More recently, with the advent of synthetics, efforts have been directed toward the understanding of fabric hand and related characteristics in terms of fiber properties [7, 9]. The basic difficulty suggested by the published work and by common experience is that handle is a complex of many properties which are integrated in the course of the subjective judgment. Thus, the A.S.T.M. has published [1] two proposed methods to assess only a few of the properties related to hand, as well as a table of eight descriptive terms of physical properties associated with it. Hoffman and Beste [7] discuss some 13 properties related to hand and Binns [3] has given a list of 28 different adjectives used by a group of observers in subjective descriptions of fabric hand.

None of the objective techniques has as yet been developed to the point where it is satisfactory to more than a very limited extent. It appears unlikely that the ability of the fingers to make sensitive and discriminating judgment and

---

[1]This work was performed under Contract QM-564 which constitutes a part of the Army quartermaster program of research on warmth, comfort, and other properties of blended fabrics.

Reprinted with permission of Textile Research Journal.

of the mind to integrate and express the result as a "single valued" judgement will be matched in the near future. In view of the widespread use of subjective evaluation of hand, the paucity of technical information on techniques and results is amazing. (In this connection the pioneering contributions of Binns [2, 3, 4] deserve more recognition and attention). This is in contrast with the considerable development in recent years of organoleptic testing in other industries (paint, food, beverage, packaging) where subjective judgments of sensory qualities of odor or flavor are now common and may even be used routinely for production control [6].

This paper describes some efforts to evaluate the subjective handle of fabric harshness in an organized and quantitative way. Some of the techniques, problems, and conclusions are thought to be applicable in other laboratories and mills. A means for classifying fabrics in a rigorous manner would appear to be a basic necessity for further development of more objective techniques. Some attempt is made to relate the handle of the fabrics studied to certain aspects of their construction-their fiber diameter and thickness.

In the work that follows, the term harshness is used to describe hand as a catchall word which seems to be most widely used and "understood" by the people involved in the test. The scale for this characteristic seems to run from harsh to soft (often "kind" in British usage), and one of its main components is the sensation of prickliness caused by contact of the surface fiber ends with the fingers. Other components appear to be stiffness and compactness.

II. METHOD OF EVALUATION

The test fabrics used in this study were a series of whipcord suitings made from wool, mohair, viscose, and nylon. Samples of these materials were supplied by the Home Economics Research Branch of the Agricultural Research Service, U.S. Department of Agriculture. The details of construction and very complete tests of the fabric are fully described by Morrison *et al.* [10]; the same sample numbers are used herein to facilitate comparisons for those interested in test results of other properties. The test materials were considered especially suitable for the following reasons: (1) They were very similar in respect to usual construction features such as yarn size and twist, fabric texture and weight. (2) Finishing processes known to alter the handle markedly—fulling and napping—were not employed in their manufacture; also the fabrics were undyed, which eliminated a psychologically distracting factor. (3) The main variable in the fabric construction was fiber composition, which permitted an assessment of the role of blend and fiber type on handle.

The chief drawback of the fabrics was that as a group, they represented fabrics somewhat stiffer and harder than those generally encountered in usual experience with garment fabrics.

Binns [2, 4] has suggested that individuals of all kinds possess a native ability to rate the feel of fiber assemblies and of fabrics. His studies involved intercomparisons among groups of materials, the entire group being offered for rating at the same time, and their arrangement in a relative order of softness. This is the common practice in evaluation of handle today. A variation of this method involves the ranking of a series of fabrics relative to one fabric in the group specifically designated as a reference "standard." An experiment along these lines is described later (Experiment III) in which fabrics were offered in pairs for relative ranking by the observers; this method is referred to hereinafter as the relative or paired method.

An alternative approach suggested itself in which fabrics would be offered one at a time for judgment. Thus, if an innate ability for such judgments does exist in the observers, they should be able to express a judgment without immediate reference to a physically present material standard. While the use of such a "mental" standard was disturbing at first to some of the judges, it is clear that previous impressions and experience are the sole basis for a great many "absolute" subjective judgments made in daily life; no "standard" is usually present when a person judges a steak to have a fine flavor or a cashmere sweater to have a soft handle. The offering of one fabric at a time for judgment of harshness against a mental standard is referred to herein as the absolute or single-fabric method.

One basis for interest in the absolute method of judgment was the possibility of evaluation of large groups of samples with fewer observations than required by the relative method. Another important reason for attempting the single rather than relative judgments was based on unfortunate experiences with "nonlinear" arrays of fabrics. Thus, it is not unusual to find in evaluating three fabrics which do not differ too widely in harshness that A rates harsher than B, B is harsher than C, but C is harsher than A. This kind of result is a consequence of the multicomponent nature of the handle characteristic.

For the present study, 8 × 8-in. specimens of the fabrics were relaxed by wetting out in warm water for several hours, followed by suspension in air till dry. The samples were offered individually (or in pairs in a later experiment) to each of 12 people in the laboratory in turn. The group was equally divided into men and women and contained some observers with considerable experience in judging the hand of wool-type cloth as well as some with very little. The order of presentation was randomized with respect to judges and fabrics in each case. No observer was asked to make a judgment more frequently than once an hour to avoid any effects due to sensory fatigue.

III. RESULTS OF HARSHNESS JUDGMENTS

A. Experiment I, Fabrics Offered One at a Time

Observers were asked "to judge with your eyes closed the handle of a sample of whipcord with respect to *harshness* or softness on a numerical scale as follows: 1-very soft, 2-soft, 3-moderately soft, moderately harsh, 5-harsh, 6-very harsh. Please do not discuss this test with anyone." The average results for the fabric (mean for 12 observers) and for the judges (mean for 12 fabrics) are given in Table 1, columns 1A. The latter average, the mean of all the fabric scores recorded for each observer, measures the *observer* in terms of his tendency to rank the samples generally harsher or less harsh than another observer. Analysis of variance showed that both observers and fabrics contribute significantly to the total variability, i.e., fabrics differ significantly in handle and independently, the observers differ in the magnitude of their ratings. (When a result is discussed herein as being significant in a statistical sense, a confidence level of 95% or greater is implied.) For the mean values shown, a difference of 0.6 units between any pair is required for significance to be asserted.

Thus, with this group of observers, differences among the fabrics can be assessed by the absolute method. There is an inbuilt scale which permits a quantitative judgment of a single fabric to be made, but this scale is somewhat different for different observers. That is, with the same external scale of values provided, some judges tended to use only the harsh end and others only the soft end of the range; some used a very restricted portion of the scale, while others spread their judgments over a wider portion of the assigned scale. Men tended to judge the harshness more severely on the average than women, but not significantly so; there was no consistent relationship between technical proficiency and level of judgment.

What is clear is that an average judgment may be made which can differentiate some of the fabrics. A single judgment by one observer, man or woman, expert or not, is of no value in discrimination by this technique. Errors in judgment, i.e., deviations from the mean, are made by all observers. For example, sample 2 is judged equal to or harsher than sample 6 by observers *C* and *H*; sample 9 is judged equal to or harsher than sample 17 by observers *L, M, B,* and *G*. Thus, averaging of the ratings of a number of observers, in addition to giving a more precise estimate of the sample through evening out the differences in scale, also evens out these momentary aberrations in judgment.

B. Experiment II, Fabrics Offered One at a Time

Since many of the observers did not use the full external range of classes of harshness provided, a second experiment was conducted with the same fabrics and observers, using fewer classes. Since, also, some of the judges were evidently using the harsh end of the scale, based on all assessment of the position of the test fabrics in the whole field of apparel fabrics in general, the question was reworded slightly to stress the desire for some limitation in the range of judgments visualized.

Observers were asked to judge ". . . the handle of a sample of whipcord with respect to harshness or softness, *for this type of fabric*, on a numerical scale as follows: 1-soft, 2-moderately soft, 3-moderately harsh, 4-harsh." The mean values are again given in Table I, columns II. To permit a comparison, the data of Experiment I were converted algebraically from a scale of 1-6 to a scale of 1-4, assuming linearity; these converted data are given in columns 1B, Table I.

Analysis of variance again showed the significant contributions of both observers and fabrics to the total variance. For the mean values shown (columns 1B and II), a difference of 0.4 units between any pair in a column is required for significance.

The change of scale and of emphasis in the question asked did not alter the main effects appreciably. The fabrics are judged in the same way and indeed the agreement in the two runs is remarkably good. There exist "harsh raters" and "soft raters" who seem to judge fabrics relatively in the same sense but differently on an absolute scale. In the same way, there are observers who use the full scale available and others who tend to use only one end.

The ability of the observers as a group to discriminate among the fabrics with respect to harshness by the absolute method is confirmed by the similarity in order of the samples and of the agreement of the numerical results in the two experiments. Every observer commits some individual errors in judgment as before so that a single rating by one observer is again deemed to be of little value in classifying the fabrics.

C. The Judgment of Paired Samples, Experiment III

In order to compare the relative efficiency in making an absolute judgment of harshness of one fabric at a time versus that in making relative judgments on pairs, six of the fabrics covering the range were selected for test. These were samples numbered 9, 2, 10, 16, 17, and 6.

The six fabrics were given in pairs to the judges who were asked to judge "the relative rank of this pair of whipcords with respect to harshness" and to answer "(a) Which of the pair is harsher? (b) Is the difference: 0-negligible, 1-small but definite, 2-moderate, 3-large." All possible pairs were tested. By using positive and negative differences in the above ratings, algebraic sums and means could be obtained, as shown in Table II.

TABLE I.
MEAN VALUES OF HANDLE JUDGMENTS
GROUPED BY FABRICS AND BY OBSERVERS

Mean of fabrics

Sample No.*	1A	1B	II	III
1	2.7	2.0	1.8	—
9	3.3	2.4	2.4	2.1
7	4.0	2.8	2.7	—
2	3.8	2.7	2.8	2.4
10	4.0	2.8	2.8	2.4
16	4.1	2.9	3.1	2.5
14	4.4	3.1	3.3	—
8	4.3	3.0	3.4	—
17	4.5	3.1	3.3	2.6
5	4.8	3.3	3.3	—
15	4.4	3.1	3.6	—
6	4.7	3.2	3.8	3.1

* Numbered as in ref. [10].

Mean of observers

Observer	1A	1B	II
A	3.9	2.8	2.8
B	4.1	2.9	3.3
C	4.1	2.9	3.3
D	4.8	3.3	2.9
E	3.8	2.7	2.6
F	2.8	2.1	2.5
G	4.4	3.1	3.8
H	4.9	3.4	3.5
J	4.3	3.0	3.1
K	3.6	2.6	2.8
L	5.0	3.4	3.3
M	3.3	2.4	2.3

The general arrangement of the fabrics with respect to harshness is the same for the paired technique (as shown by the grand average values of Table II) as those given earlier for the single judgments in Table I. Indeed the grand averages shown in Table II were converted from a scale of −3 to + 3 to a scale of 1 to 4; the results of this conversion given in column III of Table I are considered to be in good agreement with the earlier results.

TABLE II.
MEAN JUDGMENTS OF HANDLE OF PAIRS OF FABRICS

	Where X is sample:						Grand av.
	9	10	2	16	17	6	
9 is harsher than X by		-.17	-.67	-.83	-1.17	-1.58	-.88
10 is harsher than X by	.17		0	0	-.17	-1.33	-.27
2 is harsher than X by	.67	0		-.42	-.50	-.75	-.17
16 is harsher than X by	.83	0	.42		-.25	-1.17	-.03
17 is harsher than X by	1.17	.17	.50	.25		-1.42	+.13
6 is harsher than X by	1.58	1.33	.75	1.17	1.42		+1.25

A minus sign indicates fabric is less harsh. Statistical analysis indicates that a value in the body of the table of 0.8 must be reached on the average to be reliably greater than zero.

Detailed examination of the data, furthermore, indicates that better discrimination is not afforded when pairs are judged. The comparative harshness of the samples determined directly (Experiment III) and by difference in the ratings (from mean of Experiments 1B and II) are shown in Table III. This tabulation is a means for determining whether any differences among fabrics are more readily assessed by one method (relative comparison) as against the other (judgment of one fabric at a time on an absolute scale). As arranged, the total scale for both columns of data is three scale units.

The direct relative comparison is seen to result in differences of larger magnitude among the pairs of fabrics. However, in all but four cases (marked with an asterisk) the pairs are judged in the same way by the two techniques.

In only one case, pair 6-17, is a difference indicated by the results of Experiment III, where none is found in Experiments I and II; in three other comparisons (10-9, 17-2, and 17-10), no reliable difference is measured in the paired experiment, whereas a significant difference may be asserted by the results of tests of the fabrics judged one at a time. It is suggested that the paired judgments may in fact be more confusing to the observers because of the complex nature of fabric harshness. If the relative proportions of the component factors integrated by the observers into a single judgment of harshness are altered in going from pair to pair, then a difficult kind of decision is required of the judges. Thus, if a pair is presented in which one fabric is stiffer, more compact but smooth and nonprickly, and the other is compressible and lofty, but prickly on the surface, a conscious and uncomfortable decision must be reached by the observers.

This type of difficulty was clearly suggested for these fabrics by the comments of the judges. While the same factors operate in the absolute judgment of fabrics one at a time, the problem does not intrude as a conscious and distracting factor.

TABLE III.
DIFFERENCE IN HARSHNESS OF FABRICS

Sample pair	Difference in harshness	
	A. Directly (from Expt. III)	B. By difference (from Expts. I & II)
2–9	0.7	0.3
2–10	0	−.1
6–2	.8†	.8†
6–9	1.6†	1.1†
6–10	1.3†	.7†
6–16	1.2†	.5†
6–17*	1.4†	.3
10–9*	.2	.4†
10–16	0	−.1
16–2	.4	.3
16–9	.8†	.5†
17–2*	.5	.5†
17–9	1.2†	.8†
17–10*	.2	.4†
17–16	.3	.2

* Comparisons which are judged differently by the two methods and are discussed in the text.
† Statistically different from zero.

A tendency for momentary aberration in judgment was also noted in the paired series for all observers. Each judge made at least one "error" in his 15 ratings, and one observer made 4; an error was defined as the making of a judgment in the opposite sense from that given by the mean, where the latter was significantly different from zero. Where there was no significant difference in the pairs, an error was tallied where the observer noted a difference of an appreciable kind (at least two units different from the mean). The mean number of errors per judge was just under 2 per 15 observations.

It is concluded that, for these fabrics at least, the judgment of pairs or the inclusion of a standard for reference offers no clear advantage in ability to discriminate with respect to harshness. The judgment of one sample at a time by a panel of observers is at least as effective and offers economy with respect to the number of judgments that need to be made where the number of test items is large. On the other hand, with a trained panel, additional insight into the nature of the differences between a sample and a standard may possibly be obtained as a guide to research action. In either case, replication must be used, preferably by the use of a group of raters.

It will be recognized that the paired judgment technique is a special case of the common procedure in which a large group of samples may be offered for arrangement in relative rank. The main difficulties with this usual procedure are thought to be as follows:

(1) The intrusion of other psychological cues which unavoidably affect the result. The classical result of this effect is illustrated by the probably apocryphal story of the test subject who selects the blue sweater as having the best handle of group, blue being her favorite color.

(2) The onset of fatigue is well known when the array involves more than 4 or 5 items. The tendency to feel everything about the same and about in the middle of the group is a common experience for such judgments.

(3) A judgment of one array cannot be readily related to another made at a different time without the inclusion of a sufficient number of control standards covering the range. The method of judging a single sample at a time at least offers the possibility for the existence of and rating on an absolute scale.

IV. FABRIC PROPERTIES RELATED TO HARSHNESS

One of the important components of fabric harshness, especially in wool fabrics, is the resistance of surface hairs to bending. The prickle sensation given by short surface fibers acting as columns and resisting collapse by the fingers is also characteristic of some types of harshness [5]. Thus, from mechanical considerations it would be expected that harshness would be directly related to some function of fiber diameter and inversely to some function of the span length of the surface fibers. The relationship of fiber fineness to cloth softness is well known for animal fibers, particularly. Binns [4c] has shown that wool tops of various grades (i.e., fiber diameters) were readily distinguished through the sense of touch by trained or untrained people, including some who had never seen or handled top. The effects on handle caused by length of surface fibers are seen in fabrics which are made softer by a napping or brushing operation or conversely by shearing or close cropping. It was thought useful to briefly examine the subjective judgments of harshness obtained for these whipcords with reference to the fiber diameter and span length effects noted above.

In Table IV, the fabrics are grouped according to increasing fiber diameter together with their harshness ratings (mean results of Experiments I and II). The over-all effect of fiber diameter, with some interaction by effects due to length of surface fibers, is seen in the first three groups comprising fabrics made entirely of animal fibers or of wool blended with viscose. Increased harshness follows the use of coarse fiber, sometimes in spite of increased span length (group 2), sometimes concomitant with increased span length (group 3). In the last pair of fabrics of group 1, no alteration in harshness was observed with increasing mohair content, perhaps due to the counter-balancing effect of length of surface fibers. Even with the nylon blends, the tendency is noted for the fabrics containing coarse wool to be judged harsher than those containing medium wool although the differences are not statistically significant. It is suggested that these fabrics were already at the harsh end of the scale by virtue of the span length effect. Thus all of the nylon blends are substantially thinner at low pressure than the other fabrics; since the compressed thicknesses of all the fabrics are identical, it is clear that the surface fibers of these nylon blends are of very short span length, consistent with visual observation. It is interesting that some members of the panel referred to these samples as hard, stiff, boardy, and papery, as well as harsh.

The effects of span length may be more clearly seen by rearranging the samples so as to compare the viscose blends with the nylon blends in fabrics containing the identical wool and synthetic content, i.e., samples 10 vs. 17, 9 vs. 16, 8 vs. 15, and 7 vs. 14. In every case the thinner fabric, i.e., containing fibers of shorter span length at the surface, was subjectively judged harsher.

TABLE IV.
EFFECT OF FIBER DIAMETER ON FABRIC HARSHNESS

Sample no.	Nominal fiber composition*		Harshness rating†	Low pressure thickness (0.002 p.s.i.)‖ (mils)
	Wool‡ (%)	Other§ (%)		
1	100 F	—	1.9	51
2	100 M	—	2.75	53
6	75 M	25 Mohair	3.5	56
5	50 M	50 Mohair	3.3	65
9	50 M	50 Viscose	2.4	58
7	50 C	50 Viscose	2.75	69
10	75 M	25 Viscose	2.8	82
8	75 C	25 Viscose	3.2	61
16	50 M	50 Nylon	3.0	39
14	50 C	50 Nylon	3.2	42
17	75 M	25 Nylon	3.2	42
15	75 C	25 Nylon	3.35	45

* As given in [10].
† Mean of cols. 1B and II, Table I.
‡ F = fine, 64's; M = medium, 56-58's; C = coarse, 48-50's [10].
§ Mohair was 28's grade, mean diameter 31 μ; viscose and nylon were 3 den. [10].
‖ All fabrics are 28 to 30 mils at 2.0 p.s.i.

V. SUMMARY

Studies of the subjective evaluation of harshness, using a single type of fabric made from wool and blends with mohair, viscose, and nylon, are described. By means of a panel technique, fabrics may be discriminated with respect to harshness, and presumably similar techniques may be useful for other subjective properties. The evaluation of fabrics offered one at a time for judgment on an arbitrarily defined scale was found to be as efficient in discrimination as the more usual comparison with another fabric as a reference standard. A single judgment by a single observer is of little value in discerning differences in harshness; replication of the observations or of the observers is useful in smoothing out differences in judgment among observers or in minimizing a momentary error in judgment.

The harshness judgments are discussed in terms of certain structural characteristics of the fabrics tested. It is suggested that the subjective harshness is related to the diameter of the fibers used and inversely to the length of the fibers projecting from the cloth surface.

ACKNOWLEDGMENTS

The efforts of Mrs. Martha M. Kaessinger in conducting the panel tests are sincerely appreciated. A number of the ideas for judgment of subjective characteristics and methods for statistical evaluation which arose from discussions with Mr. E.B. Gasser are gratefully acknowledged.

LITERATURE CITED

1. ASTM Committee D-13, ASTM Standards on Textile Materials *41*, 610 (1954).
2. BINNS, H., Brit. J. Psychol.: (a) *16*, 237 (1925-6); (b) *27*, 404 (1936-7).
3. BINNS, H., J. Text. Inst. *25*, T89 (1934).
4. BINNS, H., J. Text. Inst. (a) *17*, T615 (1926); (b) *23*, T394 (1932); (c) *25*, T331 (1934).
5. BOGATY, H., FOURT, L. and HARRIS, M., Fourth Canadian Textile Seminar, Kingston, Ontario (September 1954).
6. FOWLER, L. and FISCHER, M.L., Monsanto Chemical Co., *The Organoleptic Panel in Production Control*, paper presented at the 127th Annual Meeting of the American Chemical Society at Cincinnati, Ohio (April 4, 1955).

7. HOFFMAN, R.M. and BESTE, L.F., *Some Relations of Fiber Properties to Fabric Hand*, Textile Research Journal *21*, T377-416 (1930).
8. PEIRCE, F.T., *The 'Handle' of Cloth as a Measurable Quantity*, J. Textile Inst. *21*, T377-416 (1930).
9. PEIRCE, J.J., *Some Wool-Like Properties of Fibers and Fabrics*, Rayon and Synthetic Textiles, 39-45 (August 1950).
10. MORRISON, B.V., WARD, R.L., DAVISON, S. and MACORMAC, A.R., Publication HNHE 104, U.S.D.A. (August 1952).

CHAPTER 5.2

MEASUREMENT OF FABRIC AESTHETICS
ANALYSIS OF AESTHETIC COMPONENTS

R.H. BRAND

*At the time of original publication,
all authors were affiliated with
E.I. du Pont de Nemours & Co., Inc.
Textile Fibers Department, Benger Laboratory
Waynesboro, Virginia.*

ABSTRACT

Effective research on the aesthetic characteristics of fabrics is difficult because explicit definitions are lacking in this field. The most reliable tool is subjective evaluation; therefore, words (*loft, clammy, hard,* etc.) become important research tools. Special meanings of these words become clear if they are logically arranged according to textile frames of reference.

Fabric aesthetic character is defined as a relationship among a minimum of six concepts: STYLE, BODY, COVER, SURFACE TEXTURE, DRAPE, and RESILIENCE. These concepts can be described by how they are subjectively perceived, by possible subconcepts (e.g., COVER can be partitioned into BOTTOM and TOP COVER), by objective tests when available, and by common word pairs used to communicate their values (e.g., *thick-thin, rough-smooth,* etc.). To illustrate application of principles, subjective scales, identified by common words, were used for analyzing the COVER concept in commercial men's suiting fabrics. These were then mathematically related to the aesthetic concept of COVER for specific fabrics.

INTRODUCTION

Most fabric properties (e.g., *weight, construction, strength*) are easily measured by physical methods. However, there is no accepted method of measuring aesthetics of fabrics. This is probably because few can agree on what fabric aesthetics is, since it is subjective and apparently means different things to different people. Experienced specialists in the textile trade, sensitive to consumer preferences, judge fabric aesthetics intuitively. On the other hand, the goal of fiber scientists in industry is to define aesthetics as a function of measurable properties P_1, P_2, P_3, etc. This requires that all elements be numerical. Previous attempts to make them numerical employed mechanical analogs such as bending length and drape projections [1] for measuring aesthetic properties. While mechanical measures and analogs are useful, they can be misleading because of a readiness to assume the analog as the aesthetic property. Fabric aesthetics is a complex response evoked from people, not machines. The only alternative is to have people measure the property. Reliable subjective evaluation of fabrics is possible, as was shown in recent studies [3, 4], and is done routinely in the food industry.

Problem

The main problem is how to question people to evoke meaningful answers. If this is done properly, then mathematical techniques and subjective test methods are available for converting these answers to the desired numerical form.

Objectives

It is proposed:
(1) To develop a definition of fabric aesthetics in terms of basic elements having the form of common words.
(2) To suggest a system for selecting rating scales. This system involves questions for subjective measurement of these basic elements.
(3) To show how basic data from judges' scores, obtained by this scheme, result in numerical definition of a specific aesthetic property, COVER, in a set of men's suitings.

TECHNIQUE

Basic Tools of Subjective Evaluation

Words, the terminology of the textile field, must be put into an orderly framework before meaningful questions can be asked. People define subjective properties of articles: so, besides the physical aspect of the object, psychological, social, and economic factors influence judgments that are expressed by words such as *lively, rough, resilient,* and *stiff.* Successful application of subjective test methods and mathematical analyses require: (i) an understanding of the words used in questions, which must be logically suited to textiles, and (ii) presence of the object to be judged—the fabric.

Jargon is often used to communicate efficiently in a specialty field. In the textile trade many common words such as *hungry, bite,* and *lively* are used in special ways. These suggest textile properties to the expert, although to most people they have no such meaning.

One way to clarify meanings is to define the words. But definitions are often either too rigid or too general to be useful. However, when the common words—the jargon—associated with fabric characteristics are ordered according to some logic, meanings can become self-evident.

Classification of Words

Useful categories for grouping words used in describing fabrics are the familiar physical sensations, such as, *touch, sight,* and *noise.* Words like RESILIENCE (which is something perceived by subjective application of force) can be grouped in a class called the kinesthetic sensation [2]. Another important set of categories is the more difficult-to-define psychological *value classes.* The most elementary of these, *desirable and undesirable,* immediately get into difficulties in terms of frames of reference.

A list of about eighty words most often used in talking about fabric aesthetics was collected[1] (first column of Appendix B). These were classified according to physical and psychological aspects of judgments. Definitions of specific classification criteria were guided by some general considerations.

DEFINITIONS

Perception of Fabric Aesthetics

When fabric aesthetics is spoken of, what is usually meant is the appearance and handle of fabrics. This is a common way of saying that fabric properties are perceived by the senses. It should be recognized that perceptions are only relative. Things are what they seem because of their relative position on scales involving contrasts, e.g., *hot-cold; high-low; good-bad; beautiful-ugly.* *Perception* is the combination and interrelationship of *sense data*, which are physical factors (*high-low; hot-cold*), and *values*, which are psychological factors (*beautiful-ugly, good-bad*).

What are concepts? Aesthetic concepts are factors and/or elements whose relationships define fabric aesthetics. Relationships in the physical sciences are symbolized by mathematical equations. These show relationships among several factors that are elements of the expression. Concepts and ideas can also be symbolized by mathematical equations [5]. A technique for doing this is called component analysis. This will be treated briefly later.

Some words associated with fabric aesthetics, for example *hand*, COVER, and BODY, do not have value polarity, that is, they are not, or need not be, good or bad, desirable or undesirable. Such words are really concepts or abstractions. They do not have a simple measurable physical reality. However, sometimes they are defined to represent sense data only. For example, *bulk*, rigidly defined as the reciprocal of fabric density, becomes technically explicit. Limiting meaning in this way often restricts communicative value. New words are then found which do not have such technical restrictions. For example, *loft* is not as technically explicit as *bulk*; so *loft* is often used when describing fabric bulk to make it clear that it is the aesthetic nature of the fabric that is being judged.

How then can these concept words be defined?

Quality Words-Polar Word Scales

Associated with concepts are quality words, their number varying depending on the complexity of the concept. These are words like *soft, cold, fuzzy*. Figure 1 shows a number of such words that are associated with the concept COVER[2]. The same quality words may be associated with different aesthetic concepts. When this happens, the meanings and values of the words appear to change. For example, in Figure 2, in association with the concept of BODY the quality *soft* implies *limp* (to indicate this symbolically the words are placed adjacent to each other); but in association with the concept of COVER, the quality word *soft* implies *fuzzy* (Figure 1). A different example will illustrate how this change in meaning comes about.

Let us consider how the concept SURFACE TEXTURE is evaluated in relation to the aesthetic and comfort characteristics of fabrics. To evaluate is to judge something on a scale that has opposite poles or a gradation. A scale for evaluating comfort concepts is the *pain-pleasure* polar scale. It is a *psychophysical* scale. Aesthetic concepts are evaluated on polar scales such as *beautiful-ugly*. These are pyscho*cultural* scales. There is nothing physical about these. So, SURFACE TEXTURE of a fabric STYLE judged by the simple polar word scale *soft-harsh* can, although it may be aesthetically beautiful, be painful from the comfort point of view. For example, a tweed may be unpleasant to the skin but pleasant to the eye.

Specific Classification Criteria

The foregoing considerations led to the following specific criteria for classifying the word collection[3]:

DESCRIPTIVE SENSORY ANALYSIS IN PRACTICE

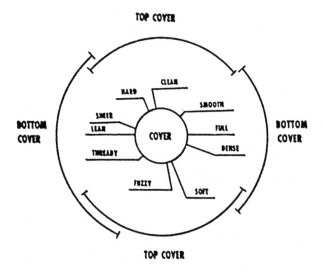

FIG. 1. COMPONENTS OF THE COVER CONCEPT,
SHOWING ASSOCIATED QUALITY WORDS
The opposite qualities that are associated with the simpler subconcepts,
TOP COVER and BOTTOM COVER, are shown in polar positions.

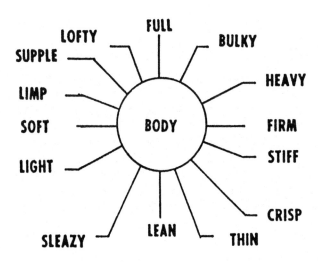

FIG. 2. COMPONENTS OF THE BODY CONCEPT,
SHOWING POLAR POSITIONS OF ASSOCIATED QUALITY WORDS

Do the Words have Value Content? Are the words as commonly used universally good or bad? Do they have polarity? Are there opposite meanings in common usage?

Are the Words Technically Explicit? For example, some technically explicit words are *bulk, covering, power, moisture absorption, weight, wrinkling,* and *pilling*. Words become technically explicit if they represent evaluations that can be produced according to specified techniques. Words can represent objective scales or subjective scales. The physical measurement of bulk is an example of the former; *wrinkling* and *pilling* evaluation methods, where standards have been defined, are examples of the latter. These scales are polar but have no value in the sense of *good-bad* or *beautiful-ugly*.

Is There a Physical Sensation Association? This criterion essentially identifies quality words. Specific words express perceived sensations—"it scratches, it is *cold, warm, heavy...*". Stimulation of sense receptors by objects brings about psychological evaluation of this experience, resulting in verbalized comment—a word symbol—which becomes recognized as negative if the experience is unpleasant or undesirable or positive if pleasant. Finally, it was asked if a specific fabric characteristic could be associated with these words.

Fabric Characteristics

Total fabric character was arbitrarily divided into four categories: AESTHETICS, COMFORT, PERFORMANCE, and CONVERTIBILITY (tailorability). In Figure 3 these are displayed in overlapping circles to show their interdependence. In everyday dealings and discussions among textile people, the so-called overlap is implied by use of the same words—*scratchy* or *wrinkly,* for example—in connection with the several characteristics even though meanings and/or values may be different for each use.

One word which ties all these characteristics together is the word symbol STYLE, taken here in its broadest sense. It defines, among other things, end-use, fabric type, and yarn and fabric structure. AESTHETICS, PERFORMANCE, COMFORT, or CONVERTIBILTY have meaning only in relation to specified fabric styles. When speaking about a fabric characteristic, one should always refer to a specific STYLE, such as PERFORMANCE *of a women's wear jersey cashmere sweater.*

Criteria for Aesthetic Concepts. AESTHETICS, one element in the framework of total fabric character, can be broken down into several aesthetic concepts defined by the following criteria (Figure 4):

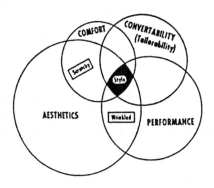

FIG. 3. FABRIC CHARACTERISTICS ARE INTERRELATED,
AS SUGGESTED IN THE ABOVE DIAGRAM, AND THEY ARE
MEANINGFUL ONLY WITH RESPECT TO SPECIFIC FABRIC STYLES
The same common quality words (*wrinkled, scratchy*)
can be used in reference to several characteristics.

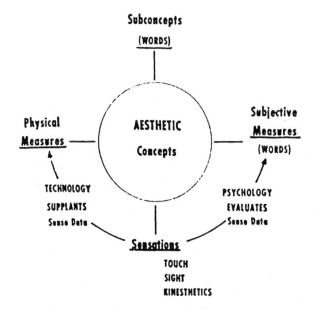

FIG. 4. CRITERIA FOR DEFINING AESTHETIC CONCEPTS

(1) The concept must be related to at least one of three main physiological sensations, the *visual, tactile,* or *kinesthetic* sensation.

(2) The concept may be a composite of subconcepts, symbolized by words which are more explicit. For example, COVER is less explicit than its component subconcepts TOP COVER and BOTTOM COVER.

(3) Some concepts may be made technically explicit by physical measurements. These measurements attempt to quantify objectively and to supplant sense data.

(4) Aesthetic concepts or subconcepts can always be evaluated subjectively. Subjective evaluation scales are represented by common words (quality words) which express the psychological value of the sense data associated with the concept.

The Aesthetic Concepts. The words BODY, COVER, SURFACE TEXTURE, RESILIENCE, DRAPE, and STYLE were selected from the classified word collection (Appendix B) as a minimum number of aesthetic concepts having to do with fabrics. The mutual dependency of these is illustrated in Figure 5. The circled words are the aesthetic concepts. The words around them illustrate either more explicit subconcepts (e.g., TOP COVER) or a relationship which is recognized by a name of its own (e.g., *liveliness* or *loft*).

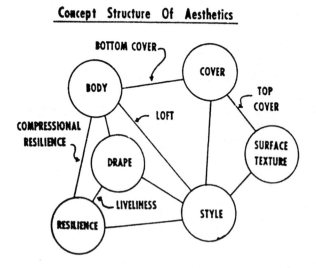

FIG. 5. THE MINIMUM NUMBER OF AESTHETIC CONCEPTS (IN CIRCLES), THEIR RELATIONSHIPS AND SUBCONCEPTS IDENTIFIED BY ASSOCIATED WORDS i.e., *loft, liveliness,* TOP COVER, etc.

Evaluation of Fabric Aesthetics

The practical consequence of the foregoing analysis was the conclusion that aesthetic character of fabrics must be defined primarily by subjective methods. Judges should be asked to evaluate qualities identified by simple word pairs whose meanings are easily recognized as opposites, words like *soft-hard*. Asking for an evaluation of DRAPE, where the judge has to decide, "Is it good or bad?" or "Are we talking about skirts or window coverings?" should be avoided. Instead, simpler qualities signified by polar words like *limp-stiff*, which are related to DRAPE, should be evaluated. More of these may be needed to define DRAPE of a fabric,[4] but they minimize uncertainty about meaning of concepts themselves. The proper word pairs represent well-understood subjective properties that can be related to concepts by mathematical analysis, as will be described. Also, simple polar quality word scales often suggest possible objective measurements, whereas concept words do not.

Defining One Aesthetic Concept-COVER. The first step in defining an aesthetic concept is to choose a set of fabrics in the specific STYLE of interest. For this illustrative example, three sets of commercial, men's suiting fabrics, *shetlands, flannels,* and *worsted,* were selected. The next step was to formulate the right questions to be asked the evaluators.

Seven polar-pair words (Figure 6), all of which had some association with the cover concept (Figure 1), were picked from the classified world list (see Table I). Some of these (e.g., *rough-smooth*) were included for evaluation

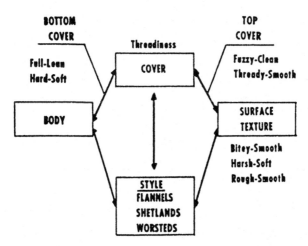

FIG. 6. POLAR-PAIR QUALITY WORDS USED FOR THE SUBJECTIVE EVALUATIONS DEFINING THE CONCEPT, COVER

TABLE I.
STRUCTURE OF AESTHETIC CONCEPTS

Concept	Principal Sensory Perceptors	Subconcepts	Possible Objective Techniques and Measures	Concept Qualities, Associated Polar Words (Subjective Value Scales)	Concept Definition (Phrase)
COVER	Sight (P) Touch (S)	Top cover Bottom cover	1. Streak meter 2. Light transmission 3. Air permeability 4. Surface contact area	Smooth-thready Fuzzy-clean Soft-hard Dense-sheer (open, sleazy) Full-lean	TOP COVER-The apparent continuity of the fabric surface. The degree of obscurity of the weave pattern due to surface fiber effects. BOTTOM COVER-The degree of obscurity of pattern in the fabric substratum.
BODY	Kinesthetic	Matter Substance Loft	1. Weight per unit area 2. Volume per unit area 3. Thickness 4. Weight density (bulk) 5. Volume density Volume fabric Volume fiber	Bulky-sleazy Full-lean Lofty-thin (crisp) Heavy-light Firm-soft Hard-limp	Technically-The substance between the selvages of woven goods - that is the "article". Semantically-The perception of the total substance of the fabric while it is being handled.
DRAPE	Sight (P) Kinesthetic	Liveliness Fit	1. Hanging heart 2. Cantilever 3. Drape meter	Lively-dead Compliant-stiff Limp-crisp Clinging-flowing Sleazy-full Boardy-supple	The manner in which a fabric *falls* when hung on a form. The perception is of the total dynamic and static behavior of the fabric. The action of "hanging" and "falling" is as important as the final "form" of folds achieved by the fabric.

DESCRIPTIVE SENSORY ANALYSIS IN PRACTICE

RESILIENCE	Kinesthetic	Compressional	1. Fabric fold compression	Lively-rubbery	Compressional resilience-The perception of resistance to *and* recovery from compression of a handful of fabric.
		Extensional	2. Tensile work recovery	Lofty-mushy	
			3. Fabric compression	Supple-compliant	Extensional resilience-The perception of the resistance to and recovery from a planar extension of the fabric.
		Liveliness	4. Vibration damping	Bounce-limp	
				Nervous-dead	Liveliness-The perception of the *rate* of recovery from small deformations.
				Snappy-stiff	
SURFACE TEXTURE	Touch	Tactility	1. Fabric friction	Dry-clammy	The visual *and* tactile perception of the fabric surface.
	Sight	Pattern	2. Friction sounds	Warm-cool	
			3. Strain gauge "feelers"	Cottony-waxy	
			4. Optical flying spot reflectance	Slick-greasy	
			5. Surface contact area	Scroopy-smooth ⎱-grainy / -rough / -pucker / -thready⎰	
				Fuzzy (nap)-clean	
				Soft-hard	
				Wiry-harsh	
				Bitey (scratchy)-picky (snaggy)	
				Shiney	
				Bloom	
				Streaky	
STYLE	Sight	Pattern	Fabric (weave) & yarn		The distinctive way in which the textile arts and technology combined to produce the perceived article.
		Fabric type	Structure analysis		
		Coloration	Fiber analysis:		
			Type		
			Linear density		
			Length		

because COVER is linked or interrelated with BODY and SURFACE TEXTURE concepts (Figure 5). Semantically, some surface characteristic is implied by the concept COVER; therefore, words such as *rough-smooth*, which are clearly associated with SURFACE TEXTURE, can conceivably also be associated with the concept of COVER. This meant that some of the same polar-pair words were listed under the different concepts in the classification (Appendix B). The time spent in evaluating these seemingly extraneous qualities was not great, and the results justified the procedure.

The subjective quality named *threadiness* and subconcepts BOTTOM COVER and TOP COVER were evaluated on a *least-most* scale. A similar approach was employed for evaluation of COVER itself. The fabrics had to be evaluated on the basis of these questions to provide validity checks on the approach to defining COVER by evaluating the associated qualities.

RESULT HIGHLIGHTS

Evaluations by all judges on any one question were analyzed and combined into a single numerical rating for each fabric for the particular question. The results, represented by the polar pair or single word used in the evaluation, could be separated into two distinct groups for each fabric set. In the flannel fabric set (Table II), e.g., it was immediately apparent that two aspects of COVER, i.e., BOTTOM COVER and TOP COVER, were meaningful, separate elements of COVER. They were in separate noncorrelated groups designated by numerals I and II. Within the grouping, the evaluations were correlated.

TABLE II.
SUBJECTIVE WORD SCALE CORRELATIONS
FOR THE FLANNEL FABRIC SET

Group I	Group II
Fuzzy-clean	Thready-smooth
Full-lean S_1	Full-lean s_2
Bitey-smooth	Hard-soft S_1
Harsh-soft	Threadiness
Rough-smooth	TOP COVER
Hard-soft s_2	Over-all COVER
BOTTOM COVER	

*See Table III for explanation of S_1 and s_2.

Over-all COVER evaluation, which was easily done by the judges in a separate evaluation session, was correlated with TOPCOVER. This meant that for the flannel fabric TOP COVER is the main consideration in the commonly used words *over-all* COVER.

The polar scale *full-lean* appeared in both groups. Mathematical analysis of judges' scores of this evaluation uncovered an ambiguity in the meaning of this word pair. One component of the evaluation was associated with fabric thickness, a second component was porosity. These evaluations reflected different interpretations of this scale by the judges.

The correlations within the two groups illustrate another point. An objective measurement is hard to define in terms of the word COVER but may be much easier in terms of a polar word pair. For example, the word pair *fuzzy-clean*, whose evaluation was correlated with COVER evaluation, suggest several experimental approaches. The scale *more* COVER-*less* COVER does not, in spite of the fact that *more* COVER-*less* COVER was an adequate basis for subjective judgement. So here is a means for identifying words suggesting objective measurements which can be expected to correlate with subjective aesthetics of interest.

Experiment Outline

Six different commercial fabrics of each of the three styles were obtained in a range of quality and cost. A multiple comparison test plan was chosen so that all fifteen possible paired comparisons among the fabrics in each STYLE were evaluated by each judge twice. Fabric sets were made up so that they could be rated quickly. Individual judges evaluated only one set according to one polar-pair word scale at each session. These sessions lasted about 4 to 8 min. No tie ratings were allowed. The raw data were processed by component-analysis computer programs developed by Dr. R.E. Seruby in our laboratory. These programs combined the judges' evaluations and calculated numerical values for each fabric on each of the subjective quality judgments. The quality ratings were identified by the polar-pair words used in the questions. They will be referred to as component polar-scale values (S_1).

Component Analysis

This technique is a means of finding a minimum number of independent factors from among a larger number.

Consider a simple case: Six text items are evaluated by two judges for a given quality. The ratings assigned to the fabrics by one judge can be plotted against the ratings given to the fabrics by the other judge. Keep in mind that paired comparisons result in distinct numerical ratings for the items. The row sum of a paired comparison preference matrix provides a rating scale in terms of the number of times each item was preferred. A best-fit line is defined for the points. The x and y axes are translated and rotated so that one axis corresponds to the best-fit line. Each point is then resolved into vectors along the new coordinates. High agreement or correlation between the two judges' ratings will be indicated by a significantly large sum of squares of all test item vectors in the major axis direction. The sum of squares of all vectors perpendicular to this principal component line will be negligible. In such a case, each test item is described completely by a single scalar value along the component axis which is the best-fit line, which if the sum of their squares is significant, indicates a second independent quality or judgment of the test items.

If six test items have a common quality on which all judges agree, this will result in only one meaningful component (dimension) when the evaluation question is right for this quality. When the six items are evaluated by many judges, the contribution of each judge's evaluation in the direction of the line of best fit is a vector in this principal axis direction. The principal scale value of the item is computed from the sum of all judges' preference ratings weighted according to the magnitude of the vector of their evaluation in the principal axis direction.

RESULTS

Component Analysis of Fabric Evaluations

Eleven judges were used in this study. Component analysis, applied to such a multidimensional system, results in components which are interpreted as majority and minority opinions or qualities. Technically, the procedure was as follows. Judges' preference scores were mean normalized. Variance-covariance matrix for judges over items was calculated and eigenvalues-eigenvectors for this matrix were determined. Components were calculated from the eigenvectors and the mean-normalized ratings. These components are then the composite scale ratings of the judges. A measure of their significance is the percentage of data variance explained by each component (Table III). For most of the evaluations, high agreement among judges was indicated by the high value of this percentage measure for the first component in each evaluation. Significant independent second components were observed in some cases. For example, the subjective quality labeled *hard-soft* for the flannel fabric set resulted in an almost even split opinion, indicated by the near equivalence of the two variance percentages. This, however, was not the case when evaluating this quality on the shetlands. The importance of the interdependence of aesthetic quality and STYLE is hereby sharply focused. Another example of this is the variance percentages for the *rough-smooth* quality in all three fabric styles.

TABLE III.
RESULTS OF COMPONENT ANALYSIS OF SUBJECTIVE EVALUATIONS*

Polar Word Scale	Flannels		Shetlands		Worsted	
	S_1	s_2	S_1	s_2	S_1	s_2
Fuzzy clean	72.3	-	83.0	-	92.3	-
Thready smooth	74.7	-	54.4	26.9	†	
Full lean	70.2	27.0	54.2	33.2	61.4	21.9
Bitey-smooth	87.6	-	90.6	-	†	
Harsh soft	73.8	-	77.7	-	†	
Rough smooth	77.8	-	87.8	-	70.6	21.3
Hard soft	41.5	39.6	76.9	-	†	
Threadiness	74	-	66.4	20.3	75.4	-
TOP COVER	78.3	-	60.2	25.1	84.0	-
BOTTOM COVER	81.7	-	88.1	-	94.3	-

* Percentage of subjective evaluations data explained by component scales (S_1) denotes majority opinion component, (s_2) denotes independent minority opinion (quality) components. Dagger indicates this question was not evaluated for the worsted fabric set. A dash represents nonsignificant percentages.

Significance of components was established by comparing component variances of the evaluations with component variances of a set of comparable random evaluations. Description of computational details is beyond the scope of this paper.

Objective Measures of a Subjective Polar Word-pair Scale, *full-lean.* Recall the ambiguous *full-lean* evaluation mentioned earlier in connection with correlated groupings in the flannel fabric set. It also appeared in the shetland fabric evaluations. The fabrics could be aligned according to two independent sets of values which resulted from component analysis of judges' evaluations (Figure 7). These are the abscissa values in the top and bottom halves of Figure 7 (the scale extends ± 2 standard deviations in standardized units). Note that fabric F, for example, was very *lean* in the meaning of component scale S_1 but relatively *full* in the meaning of component scale S_2. Here again, clear-cut reasons, indicated by the ordinates, were found for the apparent ambiguity. It was noted, during the evaluation sessions, that each judge looked at *and* felt the fabrics; but how they weighted those two independent ways of judging was not obvious. Thickness and a measure of porosity seemed good physical measures to try to correlate with the subjective data. This example illustrates the power of component analysis to separate the elements of judgment. Other significant split opinions (qualities) were not as successfully explained.

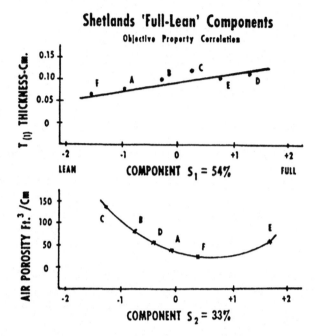

FIG. 7. OBJECTIVE MEASURES OF TWO INDEPENDENT ATTRIBUTES OF THE *FULL-LEAN* SUBJECTIVE EVALUATION ON SHETLAND FABRICS

Mathematical Definition of Concept Factors

Numerical scaling of subjective evaluations by component analysis is the first step toward mathematical definition of a concept factor. As has been indicated, evaluations, when correlated, can be grouped by inspection. But correlations are not functional relationships. The fabrics could be evaluated by the concept itself, and this was done for the COVER concept. In this case, regression analysis could be used to relate the qualities (that is, the polar-word scales) to the concept. However, for many concepts this is not easy. Let us consider the concept BODY. Semantically speaking, any object has BODY. How does one evaluate an object as being more or less of an object? Clearly this concept cannot be easily evaluated when understood in this way.

But suppose polar words associated with the concept of BODY (Figure 2) are used for evaluations. The subjective scale values S_i obtained by component analysis can then be used to get meaningful relationships for the specific STYLE being judged. The calculated scale values S_i are treated like quality values or like judges' scores in a component analysis, and this results in a general relationship of the form:

$$C_j = f_1S_1 + f_2S_2 + f_3S_3 + \ldots + f_iS_i$$

where C_j are component values, now identified as concept factors such as BODY or COVER; f_1, f_2, $f_3\ldots f_i$ are weighting coefficients for the contribution of subjective scales, S_1, S_2, $S_3\ldots S_i$ (*full-lean, harsh-soft*) to the relationship. Using this expression, numerical concept factor values can be computed for any given item.

The meaning of the concept factor C_j is given by the words used during subjective evaluations. These are the polar word pairs, the labels for the subjective scales S_i.

Meaning of the Concept Factors

The subjective component scale data for the three fabric series were analyzed as described. Generally, three significant components could be derived. In Table IV only two of these, labeled C_1 and C_2, are shown. Each concept factor is defined by weighted contributions from subjective polar-word evaluations indicated in the columns. S signifies the major subjective components scales (S_i) and s the secondary subjective component scales. For the *full-lean* subjective evaluation, these two subjective component scales were physically associated with porosity p and thickness t. The concept factors C_1 and C_2, although mathematically arrived at, have semantic content given by the words which identified the evaluation.

TABLE IV.
CONCEPT FACTOR STRUCTURE*

	C_1			C_2		
	Flannels	Shetlands	Worsteds	Flannels	Shetlands	Worsteds
Fuzzy clean	S				S	S
Thready smooth		S		S	s	
Full lean	t	t	t	P	P	P
Bitey smooth	S	S				S
Rough smooth		S	s			S
Harsh-soft	S	S				
Hard soft	s			S	S	
Threadiness				S		S

* The meaning of concept factors C_j. For each fabric set, the polar words used in the evaluations give meaning to the concept factors derived by component analysis.

The TOP COVER Concept Factor. Recall how the evaluation words were picked. They relate to the concept words, COVER, BODY, and SURFACE TEXTURE (Figure 6). COVER was further broken down into sub-concepts, TOP COVER and BOTTOM COVER (Figure 1).

In each fabric set, one of the concept factors (C_2) has meanings associated with the TOP COVER sub-concept. This is demonstrated for the worsted fabric series in Figure 8, where calculated concept factor values are plotted against actual subjective TOP COVER scale values. Scales are in standardized units.

But note in Table IV that the polar word structure of each concept factor differs for the different fabrics. For example, concept factor C_2 for the flannel fabrics does not have the same semantic content as for the worsteds. Observe the difference in the *fuzzy-clean* contribution to the concept factor in each case; it was not meaningful for flannels but was meaningful for worsteds in this concept factor. The subjective TOP COVER evaluation was correlated with C_2 concept factor values also for flannels and shetlands. The correlation coefficients for these two, 0.84 and 0.98, were as good as for the worsted fabric set.

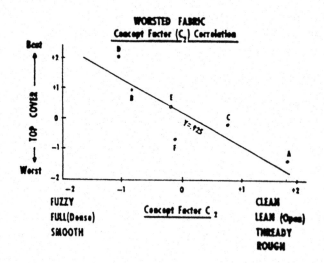

FIG. 8. THE CONCEPT FACTOR, DERIVED BY COMPONENT ANALYSIS, CORRELATED WITH INDEPENDENT SUBJECTIVE TOP COVER EVALUATION

Associated Concept Factors. The concept factor C_1 for each fabric set which could not be correlated with TOP COVER had associated polar words which were components of BODY and SURFACE TEXTURE, aesthetic concepts linked with COVER (see Figure 6) which could not be directly evaluated. The concept factors did not constitute a complete description of BODY and SURFACE TEXTURE concepts because not *all* word-pair qualities associated with them (see Table I)[5] were evaluated, but only those in common with COVER.

The BOTTOM COVER Concept Factor. This subconcept scale comes nearest to being replaceable by an objective fabric property measurement. BOTTOM COVER was for the most part evaluated visually and was identified with the *openness* of the weave. Air-flow rates through the fabric were an objective measure of this property, showing correlation coefficients between 0.88 and 0.93 for the three fabric sets.

SUMMARY

Aesthetic concepts are basically people's preferences and should be evaluated subjectively by people. Proper choice of questions, combined with mathematical analysis, leads to meaningful numerical values of these concepts.

The aesthetic concept of COVER was useful in illustrating this approach because it could be used as a basis for direct evaluation and for checking the results of the analysis. When direct concept evaluation is possible there appears to be no need to evaluate the individual concept elements on quality-word scales. This procedure however may still provide leads to physical measures. This was demonstrated in the correlation of the thickness and porosity measurements with the *full-lean* evaluation. Elemental polar-word scales are more easily related to physical properties than are concepts.

To conclude, this same approach can be used on any of the aesthetic concepts. The first step is selection of polar-word scales. The selection can be based on the word analysis as summarized in Table I. Next, a numerical scaling with reference to specific fabric styles must be carried out by suitable mathematical techniques. Third, the scales are related to the concepts by techniques such as component analysis. And finally, if possible, the word scales are replaced by or related to objective properties.

ACKNOWLEDGMENTS

Appreciation is due the many colleagues in Du Pont Textile Fibers Department in whose association, over a period of years, these ideas have borne fruit. Of special note are the debts owed to Dr. R.E. Scruby, in the mathematical treatments, and to Dr. W.A.B. Purdon and Dr. F.R. Millhiser for their sustained encouragement and constructive criticisms of this work.

FOOTNOTES

[1] Dr. O.C. Wetmore, Special Studies Manager, Marketing Research Section (assisted by technical personnel of the department) some years ago started a compilation of textile trade words and definitions. This guided the present listing. With Dr. Wetmore's permission, his collection is also appended (Appendix A).

[2] I am indebted to Dr. R.M. Hoffman for pointing out the polar nature of some words and how they can relate to concepts. This basic idea suggested structuring aesthetic concepts as illustrated in Figures 1 and 2.

[3] Sources of common word meanings were (a) *Webster's Unabridged Dictionary*, textile usages being preferred; (b) *Roget's Thesaurus*; and three textile dictionaries: (c) *Calaway*, (d) *Fairchild*, (e) *Linton*.

[4] See Table I for a suggested outline of possible polar word pairs associated with DRAPE. These were selected on the basis of a compilation such as shown in Appendix B.

[5] This table provides for the other aesthetic concepts an outline for a program, similar to the one described in detail for COVER in this paper.

LITERATURE CITED

1. CHU, C.C., PLATT, M.M. and HAMBURGER, W.J. Textile Res. J. *30*, 66 (1960).
2. DASHIELL, J.F. *Fundamentals of Psychology*, Cambridge Mass., The Riverside Press, 1937, Chapters 10 and 11.
3. HOWORTH, W.S. and OLIVER, P.H. J. Textile Inst. *49*, T540-553 (1958).
4. NAKAZOTO, H. and KAGEYAMA, I. J. Textile Mach. Soc. Japan *14*, No.2, 24-29 (1961) [Japanese Edition].
5. OSGOOD, C.E., SUCI, G.J. and TENNENBAUM, P.H. *The Measurement of Meanings*, Urbana, University of Illinois Press, 1957.

APPENDIX A

Terms Describing Hand and Resilience

Compiled by Dr. O.C. Wetmore

Bite	Fullness	RESILIENCE
Boardy	Fuzz	Sleazy
Body	Hand	Slickness
Bottom COVER	Hard	Suppleness
Bounce	Harshness	Surface
Bulk	Hungry	Tensile resilience
Cold	Lean	Thready
Compressional resilience	Liveliness	Warm
Dryness	Loft	Wrinkle
Firm	Mushy	

RESILIENCE: A term applied loosely to describe a variety of fabric recovery properties. Basically, a resilient fabric is one which yields when stressed but recovers rapidly and completely from any deformation. Another general term used synonymously is *life* (as contrasted to *liveliness*—see below).

It is more meaningful to separate fabric RESILIENCE into component parts, of which there are at least four:

(i) **Compressional resilience:** The feel of a crimped-up wad of fabric when squeezed. High RESILIENCE is shown when the fabric gives the impression that if it were squeezed harder it would compress still further. Conversely, when pressure is released, fabric expands and fills increased volume rapidly. Some people detect this property by rolling a single fold of fabric between thumb and finger.

(ii) **Liveliness:** Tendency of a fabric to revert quickly to a planar arrangement after being distorted. *Liveliness* can be evaluated by forming random, accordion-like folds with a corner or strip of fabric and observing how rapidly the arrangement jumps out like a "jack-in-the-box" when released. *Liveliness* can also be judged by waving the fabric (as a flag waves in the wind) and observing the grace and action of the ripples. Obviously, then, DRAPE and liveliness are related. The term *bounce* is used, sometimes, to mean a combination of compressional resilience (sponge rubber) and *liveliness* (jump).

(iii) **Tensile resilience:** Fabric stretch, and recovery from a planar type of deformation. Nearly all good wool fabrics have a certain amount of stretch in both warp and filling directions. Tensile resilience is judged by the amount of stretch and by the rapidity of recovery.

(iv) **Wrinkle resistance and recovery:** These terms describe behavior of fabrics in wear. The degree of deformation can vary from mild folds formed below the belt line as a result of sitting (or inside the elbow of a coat sleeve) to severe creases in the crotch or on the seat of a pair of trousers.

Surface slickness vs. dryness: Premium animal fibers such as cashmere and vicuna, have a *slick*, almost *greasy* touch which is detectable in both rawstock and in finished fabric. Fine wools (70's and 80's grade) have a *worstedy* or drier hand.

Surface bite: The drier or "worsted-like" hand of fine wool is sometimes spoken of as imparting a slight *bite* to the fabric. In coarse-textured fabrics, such as tweeds, increased fiber sizes have increased *bite*, the result of the feel of stiffer fibers on the fabric surface.

Harshness: *Bite* when bite is undesirable. Thus, a harsh tweed would be described as having bite, but a *bitey* gabardine would be described as *harsh*.

Boardy: Generally applied to an over-constructed fabric. It will be *stiff* (i.e., like a board) but not resilient because it will have a *tinny* stiffness. A *boardy* fabric is *stiff* because of fabric construction rather than because of fiber stiffness, and it lacks suppleness.

Hard: *Dense* and lacking in compressional resilience.

Firm: Firmness connotes desirability. When a house-wife buys tomatoes, she wants them *firm*, not too *soft* and not too *hard*. Likewise, fabric hand should be *firm*, not too *soft* (*mushy*) or too unyielding (*hard*). Since *firmness* is desirable, some *bulkiness* is necessary. Without *bulk*, *firmness* would probably be interpreted as *hardness*.

BODY: A generic term. A fabric always has BODY just as it always has a surface. Good BODY is a desirable combination of *loft* (see Appendix B) and *firmness*. Though low BODY is generally not desirable, this is not always so: i.e., a sport shirting might not have much BODY, but it could be desirable by virtue of being *lofty* and *fluffy*.

Bulk and loft: We use these terms as referring to the same property, but think of *bulk* as the scientific measurement of volume per unit weight, and *loft* as being the subjective impression of *bulk*.

Fullness: Fullness is more descriptive than *loft*, and includes *bulk* per se and implies that a *felting* or *fulling* has taken place. Thus, it denotes how well the structure has been filled in: a fabric could be bulky by virtue of yarn spacing, yet not be full.

Lean: A lack of *loft* and *fullness* that is generally accompanied by *threadiness*.

Sleazy: Said of a loose, badly underconstructed fabric.

Hungry: Describes in general a *thready* fabric lacking in COVER, BODY, and RESILIENCE. A polite term for a totally undesirable fabric.

A dog: A not-so-polite term for a totally undesirable fabric.

Cold hand: Said of a fabric which feels *cold* to the touch. Generally, cool fabrics are dense (lack bulk) and have a *smooth,* clear surface.

Warm hand: The opposite, of course, of a cold hand. In general, fabrics which feel warm are lofty and have considerable surface fuzz. Warmth and coolness have been shown to be essentially independent of fiber thermal properties.

APPENDIX B
Classification Scheme for Words Associated with Aesthetic Character of Fabrics

	Code		Code
Value:		**Aesthetic concept categories:**	
Good, desirable	(+)	Cover	V
Not desirable	(−)	Surface texture	X
Value changes	(O)	Eye (appearance)	E
Scaled value, most-least	(S)	Touch (tactile feel)	T
		Drape	D
		Body	B
		Resilience	R
		Style	S
Technical explicitness:		Objective measure available Y(es), N(o)	
Associated character concept category:			
Aesthetics	A		
Comfort	C		
Convertibility (tailorability)	T		
Performance (functionality)	P		

	Value	Tech. exp.	Associated character concept	Aesthetic concept
Bitey*	−		A C	T S X
Bloom				E
Boardy*	−		A C T	D R V X
Body*	+ S		A T	D R V
Bounce*	+		A	R
Bulkiness* (fabric bulk)	+	Y	A T	B S
Clammy	−		C	T X
Clean	+		A C	E S X
Clinging			C	T
Comfort		N		
Compliant	+ S		A C T	R D
Coolness*	+ S		C A	T X
Color (coloration)			A	S
Crispness	+ S		A C	B D T X
Cottony	O S		C A	B D T X
Covering power		Y	A T	V E X
COVER, TOP* (surface)	O S		A C T	V E X
COVER, BOTTOM*	O S		A C T	V E
Dead	−		A	R
DRAPE	+ S		A T P C	B R
Dryness*	+ S		A C	T X
Elasticity	O S		P A C T	R
Firmness*	+ S		A	B D
Fit	O S		A C T	D
Fullness*	+ S		A T	D B
Fuzzy*	O S		A C	V E T X
Greasy (oily)	−		A C	T
Hand		N	P A C T	B R T D X
Hardness*	O S		P A C T	B T X
Harshness*	− S		A C	T S X
Heavy		Y	P A C T	B S
Hungry*	−		A	B V
Lean*	−		A T	B V
Leathery	− ?		A C	T X
Light		Y	A C T	B
Limp	−		A T C	D R
Liveliness*	+ S		A	R
Loft*	+ S		A	B R

DESCRIPTIVE SENSORY ANALYSIS IN PRACTICE 441

Term					
Luster	+	S		A	E X
Moisture absorption			Y	P C	
Mushy*	-			A C	R T B
Mussy	-	S		P A	E X
Nap			Y	A	T E X
Nervousness	O	S	N	A C	R
Papery	-	S		A C T	B D R T X
Pattern			N	A	E T S X
Pickiness	-	S		C A	T X
Pilling	-	S	Y	P A	E X
Pucker				A T	E S X
Resilience			N	A T	R
Resilience, compressional*					
Resilience, extensional*					
Resilience, long-term					
Resilience, short-term					
Rubbery	-			C A T	R
Scratchiness	-	S		C	T X
Scroop	O	S		A	T B X
Scroopiness, cottony	O	S		A	T B X
Scroopiness, silky feel	+	S		A	T D X
Sheer	+	S		A T	E V B S
Shininess	O	S		A	E X
Sleazy*	-			A C	B D
Slickness*	O	S		A C	T X
Smoothness	+	S		A C	T E S X
Snagging	-	S		A C P	T X
Snap	+			A	R
Softness	+	S		A C T	T R S B X
Static			Y	C	
Stiffness (crispness)	O	S		A C T	R D B
Streakiness	-	S		A	E
Style			N	A	
Suppleness*	+	S		A C	D R
Texture			N	A	S X
Threadiness*	-	S		A	V E X
Threadiness, surface continuity					V E X
Threadiness, stitch deformation, knits					V E X
Thinness	-	S	Y	A C T	B
Whiteness	O	S	Y	A	S
Warmth* (tactile)	+	S		A C	T X
Waxy	-			A C	T

Weave		Y	P A C T	S X
Weight		Y	A C T	S B
Wrinkling*	– S		P	
Wiry	–		A C	T X

*Dr. O.C. Wetmore's compilation (Appendix A).

CHAPTER 5.3

DEVELOPMENT OF TERMINOLOGY TO DESCRIBE THE HANDFEEL PROPERTIES OF PAPER AND FABRICS

GAIL VANCE CIVILLE and CLARE A. DUS

At the time of original publication,
all authors were affiliated with
Sensory Spectrum, Inc.
24 Washington Avenue
Chatham, NJ 07928.

ABSTRACT

Understanding the tactile feel of paper, nonwoven, and woven products requires a valid and reliable sensory evaluation method which discriminates and describes handfeel properties. The Handfeel Spectrum Descriptive Analysis method separates the sensory tactile properties of paper and fabrics into clearly defined characteristics that are based on sound physical properties. The benefit of using a trained descriptive sensory panel is that resulting analytical sensory data allow full documentation of a sample's sensory tactile properties that can be related to consumer responses and instrumental physical tests. This benefit derives from strict protocols for manipulation and the use of precisely defined terms to discriminate and describe the qualitative properties (characteristics) and their relative intensities (strength) in each product. This paper discusses in detail the protocols for (1) sample preparation, presentation, and handling during evaluation, (2) the definition and scale range for each sensory attribute/characteristic and (3) the application of these data to address business and technical situations with consumer products.

INTRODUCTION

Hand feel properties contribute substantially to the consumer's overall acceptance or preference of paper and fabric products. For over fifty years, researchers have attempted to use the sense of touch to measure the sensory properties that contribute to the "hand," "handle" or handfeel properties of woven and nonwoven fabrics (Binns 1926). The need to assess the effects of technology on the perceived changes in handfeel and overall acceptance has stimulated several attempts to define terminology and rating scales, develop trained panels, standardize test conditions or correlate panel results with instrumental or consumer measurements (Ellis and Garnsworthy 1980).

Previously published studies involving sensory testing of handfeel properties have been limited in terminology, sample variety, scaling methods, and test protocols. Evaluation of the sensory properties of paper and fabric products generally has been limited to one or two integrated terms, that is, terms which are, in fact, combinations of singular primary terms. Integrated terms, such as, softness (Bates 1965; Gallay 1976; Hollmark 1976), harshness (Bogaty *et al.* 1956), or hand (Harada 1971; Kim and Vaughn 1979) have been measured in the past but no attempt has been made to determine what primary characteristics, which are discrete from one another, contribute to the integrated characteristic. Several studies limited evaluations to only one or two varieties of fabric or paper. Rowe and Bates studied sanitary paper products; Gallay studied paper toweling; whereas Bogaty studied wool, mohair, viscose and nylon (Rowe and Volkman 1965; Bates 1965; Gallay 1976; Bogaty *et al.* 1956).

In an attempt to rate or rank fabrics for given characteristics some studies used affective or hedonic criteria rather than a measure of intensity or strength (Kim and Vaughn 1979; Ellis and Garnsworthy 1980; Lundgren 1969). Sensory tests demand strict protocols for the handling of sample preparation, sample evaluation and subject/panelist training. In previous studies such test conditions usually have been undefined or deliberately left to the individual panelist to define.

Some of these limitations have been addressed by several research attempts to correct or expand the sensory handfeel capability. The Committee D-13 on Textile Materials of the American Society of Testing and Materials has successfully identified and defined primary characteristics. Howorth and Oliver used an array of 27 woven fabrics to study handfeel, but this type of broad product sampling is not typical for most handfeel studies (Howorth and Oliver 1958). A few attempts to quantify characteristics have used intensity rating scales, which can be more directly related to instrumental measures (Bogaty *et al.* 1956; Hollmark 1976). The use of strict test conditions has been limited to controlling the procedures for evaluation (Ray 1947) or the use of reference samples (Hollmark 1976).

In the area of sensory evaluation of foods there has been a considerable amount of success in the use of descriptive analysis techniques to qualify and quantify texture characteristics (Brandt *et al.* 1963; Szczesniak 1963; Szczesniak *et al.* 1963; Civille and Szczesniak 1973; Civille and Liska 1975; Munoz 1986; Meilgaard *et al.* 1987). These methods utilize precisely defined terminology, referenced rating scales, and strict test protocols to yield valid and reliable data. The development of the Handfeel Spectrum Descriptive Analysis (Handfeel SDA) method is based on using the Texture Profile (Brandt *et al.* 1963) and

Spectrum Descriptive Analysis (Meilgaard *et al.* 1987) methods as models. The benefits of a more analytical, comprehensive, and controlled approach to sensory analysis of foods have been demonstrated in the correlation of these descriptive data with consumer acceptance (Szczesniak 1975), instrumental methods (Szczesniak 1963; Szczesniak *et al.* 1963) and other descriptive panels.

The Handfeel SDA system was developed in the early 1980's for use in the paper industry. Terms and protocols were then refined or further developed by the Other Senses Task Group (E18.02.06.03) of ASTM Committee E18. Applications of this Handfeel SDA method to paper, nonwoven, and woven fabrics has already been successfully completed for several consumer product companies and for the Sensory Analysis Center at Kansas State University.

METHOD

Terminology

The significance of the Handfeel SDA method is that it permits valid and reliable comprehensive documentation of a product's handfeel properties. The descriptions include qualification and quantification of several mechanical, geometrical, moisture, thermal, and sound properties (Table 1) of woven, nonwoven, and paper products. The qualification of the characteristics is based on identification and definition of physically based terms, which are primary rather than integrated and neutral rather than positive or negative.

One key element in the classification and definition of characteristics is the separation of the terms by sense into two major sensory categories.

Tactile. The tactile properties include mechanical, geometrical, moisture and thermal characteristics. The mechanical characteristics which are related to the reaction of the products to stress and strain, include such terms as stiffness, resilience and force to compress. Geometrical properties, such as fuzzy and gritty, are related to the size, shape, and orientation of particles in or on the surface of the samples. The moisture property listed is moistness which is related to the perception of water and/or oil on or in the product. For some applications this term can be extended to describe the moistness more specifically as only oily or wet. To account for perception of heat transfer the thermal term, warmth, is used and it embraces products perceived as removing (cool) or contributing (warm) temperature to the hand.

Sound. The only nontactile properties included are the sound properties, which are related to the noise produced during manipulation in terms of pitch and loudness.

TABLE 1.
TERMINOLOGY USED TO DESCRIBE HANDFEEL

Mechanical Characteristics:	Related to the perception of stress and strain.
Force to Gather	Depression Resilience
Force to Compress	Tensile Stretch
Stiffness	Tensile Extension
Fullness	Hand Friction
Compression Resilience	Fabric Friction
Depression Depth	
Geometrical Characteristics:	Related to the perception of the size, shape and orientation of particles.
Roughness	Lumpiness
Grittiness	Fuzziness
Graininess	Thickness
Moisture Properties:	Related to the perception of fat or water
	Moistness
Thermal Properties:	Related to the perception of heat transfer'
	Warmth
Sound Properties:	Related to aural perception
	Noise Pitch
	Noise Intensity

In order to assure the correct use of each attribute, term definitions and corresponding rating scales with anchored ends are provided to panelists for each evaluation (Table 2).

Protocols for Evaluation

In addition to terminology and rating scales to describe the sensory characteristics of paper or fabrics, strict protocols are necessary to assure a well run analytical test. In sensory testing, it is necessary to establish strict controls for the samples, both preparation and presentation, the environment, and the panelists.

Sample Preparation. Whether treatment effects are being measured or sensory-instrumental relationships are being explored, it is essential to control the handling, preparation and presentation of the sample in order to minimize extraneous variables which may bias the data. Samples on bolts are prepared in the following manner: samples are precut from the bolt into 10 in. × 10 in. squares. The edges should be square with the machine direction for nonwoven and square with the grain for wovens. Samples are then marked with an arrow indicating machine direction (nonwoven) or weave direction (woven). For all products the following sample preparation is necessary: samples are labeled with random three digit codes to prevent panelists from being biased. Twenty-four hours before evaluation, samples are preconditioned at 73 ± 2F and 50 ± 2%RH or 73 ± 2F and 65 ± 2%RH. A given panel should use only one set of temperature and humidity conditions for all evaluations.

The conditions are chosen before the panel training takes place and are based on the conditions which best simulate those under which products are used.

Environmental Controls. The physical environment requires specific controls. The temperature of the test laboratory is set at 73 ± 2F and the humidity range of the test laboratory is set at 50 ± 2%RH or 65 ± 2%RH, again based on each product's usage. The benchtop/tabletop, used during the evaluation is similar in color to the test samples, is low in gloss, and should have no texture (Laminate or Corian is recommended). A customary feature of most sensory laboratories is the choice of colored lights to mask sample color differences. Low pressure sodium lamps also can be used for this purpose. If visual differences between samples are too large it is possible to make use of screens or boxes to conceal the samples to control visual biases.

Panelist Conditioning. In order to assure additional control over the evaluations, protocols for panelists are also defined. Prior to testing, panelists condition their hands. Twelve to fifteen hours before the panel is scheduled to meet, panelists buff fingers with the light side of an emery board (test smoothness by passing a nylon stocking over finger tips). Two hours before the panel is scheduled to meet, panelists wash their hands with a mild soap, designated by the experimenter. Immediately after washing, panelists apply 1 mL of Vaseline Intensive Care Lotion to their hands. Panelists are instructed to wear prewashed fabric lined rubber gloves when working in water between sessions.

Sample Handling. In order to reduce variability in the evaluation process it is critical that all panelists handle the samples in the same way. Therefore, for each term a clear procedure is defined for the manipulation of the samples (Table 3). These instructions define the type and direction of the forces to be used for mechanical terms, the specific fingers and the direction of movement to be used for geometrical terms and the placement of samples close to the ears for sound terms.

TABLE 2.
DEFINITIONS AND SCALES FOR TERMINOLOGY USED TO DESCRIBE HANDFEEL

Force to Gather	The amount of force required to collect/gather the sample toward the palm. [low force------>high force]
Force to Compress	The amount of force required to compress the gathered sample into the palm. [low force------>high force]
Stiffness	The degree to which the sample feels pointed, ridged, and cracked; not pliable, round, curved. [pliable/round----->stiff]
Fullness	The amount of material/paper/fabric/sample felt in the hand during manipulation. [low amount of sample/flimsy----->high amount of sample/body]
Compression Resilience	The force with which the sample presses against cupped hands. [creased,folded------>original shape]
Depression Depth	The amount that the sample depresses when downward force is applied. [no depression----->full depression]
Depression Resilience/Springiness	The rate at which the sample returns to its original position after depression is removed. [slow---------->fast/springy]
Tensile Stretch	The degree to which the sample stretches from its original shape. [no stretch----->high stretch]
Tensile Extension	The degree to which the sample returns to original shape, after tensile force is removed. [Note: This is a visual evaluation.] [no return------>fully returned]
Hand Friction	The force required to move the hand across the surface. [slip/no drag-------->drag]

TABLE 2. (*Continued*)

Fabric Friction	The force required to move the fabric over itself. [slip/no drag--------->drag]
Roughness	The overall presence of gritty, grainy, or lumpy particles in the surface; lack of smoothness. [smooth--------------->rough]
Gritty	The amount of small abrasive picky particles in the surface of the sample. [smooth/not gritty----->gritty]
Lumpy	The amount of bumps, embossing, large fiber bundles in the sample. [smooth/not lumpy------>lumpy]
Grainy	The amount of small, rounded particles in the sample. [smooth/not grainy---->grainy]
Fuzziness	The amount of pile, fiber, fuzz on the surface. [bald--------->fuzzy/nappy]
Thickness	The perceived distance between thumb and fingers [thin------------>thick]
Moistness	The amount of moistness on the surface and in the interior of the paper/fabric. Specify if the sample is oily vs. wet (water) if such a difference is detectable. [dry-------------->wet]
Warmth	The difference in thermal character between paper/fabric and hand. [cool------------->warm]
Noise Intensity	The loudness of the noise. [soft---------->loud]
Noise Pitch	Sound Frequency of the noise [low/bass--------->high/sharp]

TABLE 3.
EVALUATION TECHNIQUES

Force to Gather	Place dominant hand on top of sample; gather sample with fingers toward palm;
Force to Compress	Place dominant hand on top of sample; after gathering sample with fingers toward palm, close hand to compress.
Stiffness	Place dominant hand on top of sample; gather sample with fingers toward palm; close hand slightly and manipulate rotating sample in palm.
Fullness	Place dominant hand on top of sample; gather sample with fingers toward palm; close hand slightly and manipulate rotating sample in palm.
Compression Resilience	Place dominant hand on top of fresh sample; gather sample with fingers toward palm; contain the gathered sample between the two cupped hands. Gently compress 5 times.
Depression Depth	Place sample flat on table top. Fold sample in quarters; using fingers tips press down gently on center of folded square.
Depression Resilience/ Springiness	Place sample flat on table top. Fold sample in quarters; using finger tips press down gently on center of folded square, release the downward force.
Tensile Stretch	Grasp opposite edges (near to the edges) in hands; pull sample square across in direction 1 for 5 seconds; repeat for direction 2.
Tensile Extension	Grasp opposite edges in hands; pull sample evenly across the square in direction 1 for 5 seconds: release and wait 5 seconds; take a fresh sample and pull and release across direction 2.

TABLE 3. (*Continued*)

Hand Friction	Place palm flat on fabric; using weight of hand and forearm, move hand horizontally across the surface in all four directions parallel to the edges.
Fabric Friction	Place fabric in half, grasp open end between thumb and fingertips; move the fabric over itself with rotating motion.
Roughness Gritty Grainy Lumpy	Lay sample flat on the table; place wrist on the table top; move index and middle fingers across the surface (1 inch from the edge) lightly (left to right) using weight of hand; rotate fabric to stroke along all four directions of fabric.
Fuzziness	Lay sample flat on the table; place wrist on the table top; rotate index finger lightly on surface in small quarter size circles.
Thickness	Hold corner of sample with non-dominant hand thumb, index and middle fingers; grasp sample with thumb and index finger of dominant hand just next to non-dominant hand; run dominant hand fingers over sample using light pressure; do not go off the edge.
Moistness Warmth	Gather fabric into palm with fingers closed; flex fingers and squeeze sample gently for 3 seconds.
Noise Intensity Noise Pitch	Gather fabric into palm with fingers opened slightly; rotate sample gently while holding hand with sample next to ear.

Panelist Screening and Training

Screening. Since the Handfeel SDA method is a precise analytical sensory tool, it requires the careful selection and training of individuals for descriptive panels. Panelists must be in good health and not have any of the following physical conditions: callouses on the hands and fingers, impaired circulation in the hands and fingers, central nervous system disorders, and dry and/or chapped skin. Panelists must also not be involved in any drug therapy which affects the central nervous system. Panelists are screened for their ability to detect and describe a limited number of texture/handfeel characteristics and their intensities.

In addition, panelists are screened for their interest in the activity as well as their availability to participate on a regular basis.

Training. The philosophy of training Handfeel SDA panels is to provide a common sensory experience to the panel using reference samples to demonstrate the qualitative and quantitative differences across a broad array of products. For training purposes, a reference scale for each attribute covers the range of intensity from very low to very high (Table 4). Such scales have been developed for woven (to be published), nonwoven, and paper products. The scale used to indicate intensity can be a linear scale of 15 cm in length which is labelled with the attribute name and the end anchors or a category scale of several points with the ends also anchored. In any application of the Spectrum Descriptive Analysis method a 15 cm line or a 15 point scale (which can be expressed to tenths yielding 150 points) is used depending on the choice of data entry or previous experience of the panelists with scaling (Meilgaard et al. 1987).

The panel training requires an initial 3-4 day orientation to familiarize panelists with the terminology, procedures for evaluation and intensity scales. Panelists then practice for 4-6 weeks at 5 h per week to use the procedures, terms, and rating scales precisely. A second three day orientation introduces panelists to the use of a full evaluation of all attributes for each sample. A descriptive ballot or scoresheet is used to evaluate all the attributes for one product at a time.

Another 4-6 weeks of practice in this process enables panelists to develop expertise and confidence in the full handfeel evaluation process. During these sessions references for each attribute scale are available to increase reliability of the data.

Intensity references for all attributes of wovens are currently being developed by a Tactile SDA panel which was established at Kansas State University through the collaboration of Sensory Spectrum, Inc. and the Sensory Analysis Center at Kansas State University.

The scoresheet or ballot can be designed to evaluate all or some key attributes. All attributes do not have to be evaluated for all samples. Terms such as fabric to fabric friction may be unsuitable since some fabrics are so fuzzy there is no movement at all between the face surfaces of the fabric. The evaluation results for three woven fabrics are shown in Fig. 1.

APPLICATION TO PRODUCT EVALUATION

Handfeel SDA data provide detailed analysis of textural characteristics which document each product's handfeel. This methodology extends itself to products other than woven fabrics. With modifications to the terminology and test protocols, nonwovens, moistened nonwovens, paper and the effect of fabric treatments (detergent, fabric softeners) are also eligible for sensory evaluation.

TABLE 4.
EXAMPLES OF INTENSITY SCALES

STIFFNESS (MECHANICAL)

INTENSITY	FABRIC ID
1.3	Polyester/Cotton 50/50 Single Knit
4.7	Mercerized Cotton Print Cloth
8.5	Mercerized Combed Cotton Poplin
14.0	Cotton Organdy

FUZZINESS (GEOMETRICAL)

INTENSITY	FABRIC ID
0.7	Dacron Taffeta
3.6	Cotton Crinkle Gauze
7.0	Cotton T-Shirt
13.6	Cotton Fleece

The handfeel information can then be related to both consumer responses and instrumental analysis.

With regard to test protocols, several test conditions have been tried. The recommended protocols define the most common conditions. Other conditions can be tried and, in fact, tested against those recommended here. In addition, new terms can be developed to meet specific product needs. These terms require new protocols for evaluation and definition of terms. All the possibilities for expansion of this method and applications to diverse products are not discussed in this paper and provide opportunities for new research.

The Handfeel SDA method traces the qualitative and quantitative texture changes of a product through product development, process and material changes and/or cost reduction projects. Definition of the procedures for evaluation, the terms and the intensities, which define a "control" or "standard", allows for careful monitoring of production in a QA/QC system.

Terms identified and defined by a texture panel provide a core list of product differences which can be related to consumer acceptance and consumer attribute diagnostics. In addition, data from a descriptive texture analysis panel

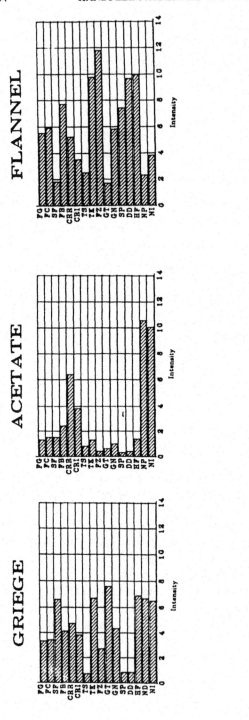

FIG. 1. HANDFEEL SPECTRUM FOR THREE FABRIC TYPES

can be used to translate or decode consumer attribute terms after a consumer study is completed. Questions regarding the possible treatment of data, in terms of statistical methods to use in analysis of the descriptive data alone or the correlation of descriptive data with instrumental and/or consumer responses for liking are not discussed in this paper.

CONCLUSION

The Handfeel SDA method is developed to provide a qualitative and quantitative description of the tactile properties of fabrics (woven, nonwoven, and paper). The advantages are (1) the flexibility of application to a wide array of products and (2) the analytical character of the data based on defined terminology, procedures for evaluation, and reference scales and (3) the capacity to use this method in relationship to both instrumental and consumer data.

REFERENCES

ASTM Committee D-13, ASTM Standards on Textile Materials.
ASTM Committee E-18, Other Senses Task Group E18.02.06.03, Unpublished Documentation.
BATES, J.D. 1965. Softness index: fact or mirage? Tappi 48, no: 4:63-4A, ABIPC 36:54.
BINNS, H. 1926. The discrimination of wool fabrics by the sense touch. Brit. J. Psychol. *16*, 237-247.
BOGATY, H., HOLLIES, N.R.S. and HARRIS, M. 1956. The judgment of harshness of fabrics. Textile Res. J. *26*, 355-360.
BRAND, R.H. 1964. Measurement of fabric aesthetics: Analysis of aesthetic components. Textile Res. J. *34*, 791-804.
BRANDT, M.A., SKINNER, E.Z. and COLEMAN, J.A. 1963. Texture profile method. J. Food Sci. *28*, 404-409.
CAUL, J.F. 1957. The profile method of flavor analysis. Adv. Food Res. *7*, 1.
CIVILLE, G.V. and LISKA, I.H. 1975. Modifications and applications to foods of the General Foods sensory texture profile technique. J. Texture Studies *6*, 19-31.
CIVILLE, G.V. and SZCZESNIAK, A.S. 1973. Guidelines to training a texture profile panel. J. Texture Studies *4*, 204-223.
ELLIS, B.C. and GARNSWORTHY, R.K. 1980. A review of techniques for the assessment of hand. Textile Res. J. *50*, 231-238.

GALLAY, W. 1976. Textural properties of paper: measurements and fundamental relationships. In *The Fundamental Properties of Paper Related to its Uses*. (Francis Bloom, ed.) pp. 684-695, British Paper and Board Ind. Fed., London.

HARADA, T., SAITO, M. and MATSUO, T. 1971. Study on the hand: part 3: The method for describing hand. J. Textile Machinery Soc. Japan. *17*, (4).

HOLLMARK, B.H. 1976. The softness of household paper products and related products. In *The Fundamental Properties of Paper Related to its Uses*. (Francis Bloom, ed.) pp. 696-703, British Paper and Board Ind. Fed., London.

HOWORTH, W.S. and OLIVER, P.H. 1958. The application of multiple factor analysis to the assessment of fabric handle. J. Textile Inst. *49*, T540-T553.

KIM, C.J. and VAUGHN, E.A. 1979. Prediction of fabric hand from mechanical properties of woven fabrics. J. Textile Machinery Soc. Japan. *32*, T47-T56.

LUNDGREN, H.P. 1969. New concepts in evaluating fabric hand. Textile Chemist and Colorist. *1*(1), 35-45.

MEILGAARD, M., CIVILLE, G.V. and CARR, B.T. 1987. *Sensory Evaluation Techniques. Volume 2*. pp. 1-23, CRC Press, Boca Raton.

MUÑOZ, A.M. 1986. Development and application of texture reference scales. J. Sensory Studies *1*(1), 55-83.

RAY, L.G., Jr. 1947. Tensile and torsional properties of textile fibers. Textile Res. J. *17*, 1.

ROWE, S. and VOLKMAN, R.J. 1965. Thickness measurement of sanitary tissues in relation to softness. Tappi 48, no: 4:54-6A, ABIPC 36:55.

SZCZESNIAK, A.S. 1963. Classification of textural characteristics. J. Food Sci. *28*, 385-389.

SZCZESNIAK, A.S., BRANDT, M.A. and FRIEDMAN, H.H. 1963. Development of standard rating scales for mechanical parameters of texture and correlation between the objective and sensory methods of texture evaluation. J. Food Sci. *28*, 397-403.

SZCZESNIAK, A.S., LOEW, B.J. and SKINNER, E.Z. 1975. Consumer texture profile technique. J. Food Sci. *40*, 1253-1256.

CHAPTER 6.0

GENERAL APPLICATIONS

In the chapters to follow, various aspects of descriptive analysis technique are reported. These include comparison of descriptive analysis methods, comparison of trained panels and their performance, methods of data analysis, and sensory evaluations of peanut flavor, noodles, and odor pungency. To advance the descriptive analysis technique, it is apparent that active research is needed. The results of the research should be published to confirm its applicability and usefulness for the benefit of the industry. Although industrial research participation is indeed necessary for its advancement, unfortunately, this type of participation is quite limited. Obviously, this is a challenge for sensory and statistics professionals to increase industrial research participation.

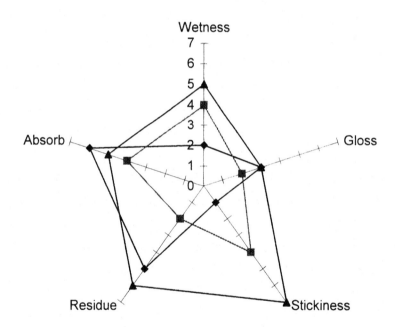

CHAPTER 6.1

THE USE OF FREE-CHOICE PROFILING FOR THE EVALUATION OF COMMERCIAL PORTS

ANTHONY A. WILLIAMS and STEVEN P. LANGRON

At the time of original publication,
author Williams was affiliated with
Food and Beverages Division
Long Ashton Research Station
University of Bristol
Long Ashton, Bristol BS189AF

and

author Langron was affiliated with
Pedigree Petfoods, Ltd
Melton Mowbray, Leicestershire.

The paper describes a new approach to profile analysis in which each assessor produces individual profiles of the products, using his or her own terms for describing them without the need to explain the meaning of such terms. The spatial configurations derived from individual profiles are rationalised by generalised Procrustes statistics. The result is a consensus configuration revealing the interrelationships between the samples for the panel as a whole. An experiment conducted on commercial ports using ten assessors, both expert and non-expert, illustrates the technique.

1. INTRODUCTION

Since the initial introduction of flavour profiling in the 1950s by the A.D. Little Corporation,[1,2] significant advances have been made with the procedure, both in the way terminology has been standardised and defined for particular products,[3-8] and in the way the data have been analysed and presented.[9-15] Flavour profiling, as it is practised in most laboratories at the moment, still suffers from two major disadvantages: it requires the development of a language for the products being assessed and a measure of agreement among the panelists about the meaning of the terms they employ. This is usually

Reprinted with permission of the J. Sci. Food Agric.
©Copyright 1984. Originally published in J. Sci. Food Agric. 35, 558-568.

achieved after several sessions wherein the products are discussed and the derived terms are defined verbally or by the production of standards. These may take the form of samples containing added ingredients[13] or chemicals, flavours or actual foods adsorbed on to paraffin wax.[6,7]

Such procedures are very time-consuming, particularly at the vocabulary development stage, if intelligible results are to be obtained. Despite much effort spent in developing vocabularies or standardising terms, judges are individuals, and invariably have their own differing sensitivities and idiosyncracies when it comes to describing foods. Therefore, complete agreement between assessors is often very difficult, if not impossible, to obtain.

The mathematical approach, based on generalised Procrustes analysis,[16-18] has been used recently to investigate reasons for panel variation[14] and to provide more reliable data for relating to external factors such as experimental variables or chemical and physical data (Langron, S.P.; Williams, A.A.; Collins, A.J., unpublished). The procedure works well by removing variation in the use of terminology in conventionally profiled data when the same terms are available for scoring by everyone. There is no logical reason, however, why the same approach could not be used to rationalise assessments made with completely different vocabularies, particularly as recent work has shown that generalised Procrustes analysis can operate on configurations of different dimensions to produce an unambiguous consensus configuration (Langron, S.P., unpublished results). Using undefined words and an assessor's own descriptors as a means of assessing foods has been applied successfully in market research.[19] In general, when analysing such data, each person's assessment has to be treated separately and can only be interpreted by relating it to external factors such as manufacturing variables or instrumental and chemical data.

This paper explores free-choice word profiling, in which each assessor uses his own privately developed vocabulary, for the assessment of ports; generalised Procrustes analysis is then used to produce information on the interrelationships between the samples, much in the same way that principal component or canonical variate analyses are used to treat conventional profile data.

2. EXPERIMENTAL

2.1. Materials

Eight commercial ports, seven tawnies and one ruby, readily available on the French market, were supplied by a French wine merchant (négociant) (Table 1).

TABLE 1.
PORTS EXAMINED IN
FREE CHOICE PROFILING EXPERIMENT

1.	Croft
2.	Cruz
3.	Sandeman
4.	Real
5.	Rozes
6.	Diez
7.	Pitters
8.	Offley

TABLE 2.
DETAILS OF TEN ASSESSORS USED FOR ASSESSING PORTS

Assessor		Qualification	Sex
1	Expert	Port producer	Male
2	Expert	Port producer	Male
3	Expert	Port producer – limited experience	Male
4	Expert	Buyer	Male
5	Expert	Buyer	Male
6	Expert	Profile panelist	Male
7	Expert	Profile panelist	Female
8	Non-Expert	Port scientist	Male
9	Non-Expert	Port scientist	Male
10	Non-Expert	Port scientist	Female

2.2. Assessors

A panel of eight men and two women was used to assess the wines (Table 2). Two were experienced tasters and producers of port; a third was in a similar category but with less experience; two were concerned with the buying of ports; two were experienced in the sensory profiling of ports; and three were scientists involved in port experiments.

2.3. Tasting environment

All assessments were conducted in a well ventilated daylight-illuminated room. After initial discussions all participants assessed samples independently in separate booths.

2.4. Presentation of samples

All samples were presented at room temperature in clear wine glasses (B.S. 5586). During initial vocabulary development all samples were available and assessors could select and assess in any order they chose. For the actual assessment samples were presented in duplicate in sets of eight; the order was randomised, and different between individuals.

2.5. Development of individual assessors' vocabularies

To introduce the concept of free-choice profiling the assessors were given a brief outline of the procedure and the ideas behind it. During this period drawings of different shapes varying in area, and solutions of different concentrations of tartaric acid and sucrose were used to introduce the panelists to the idea of scoring the magnitude of different characteristics on a 6-point scale, both in isolation and when other components were varying simultaneously.

Afterwards panelists were presented with a range of unlabelled ports, including the eight commercial samples listed in Table 1. They were asked to list sensory characteristics which described the sensory properties of the ports and to obtain an indication of the range of the characteristics they had selected in the samples. They were instructed to list only objective attributes and not things which could only be scored hedonically, e.g. the intensity of colour or redness rather than just colour.

To assist in vocabulary development participants were instructed to group terms for appearance, aroma and flavour separately. Having devised their list of terms they then had to define them to themselves so that they would always have the same meaning in subsequent assessments. They did not have to convey that meaning to anyone else, merely be certain in their own minds what they meant by a particular word and keep to it. Each individual's language was then incorporated into a 6-point optical mark readable score sheet and was ready for port assessment.

2.6. Treatment of data

To unravel the important sensory dimensions from this set of data, all results in a particular general category, appearance, aroma or taste, were subjected to generalised Procrustes analysis, the various steps in the analysis enabling each person's assessment of the ports to be rationalised and their relative agreement, as far as the final sample configuration was concerned, established.

To illustrate how the analysis works, consider one person's assessment of the appearance of the ports. If he or she chose n terms to describe the port;

it was possible to represent the scores given to each of the ports by a unique position in an n dimensional space, each of the dimensions representing one of the terms and the distance of the points along the dimension representing the score given to that term for a particular port. If this exercise is repeated for all the eight ports examined one derives eight points in this n dimensional space; the interrelationship of the points representing the differences and similarities in respect to the appearance of the ports.

A similar sample space can be obtained independently from all assessors, each space being based on an assessor's own terminology. By matching the centroids of each assessor's space (to neutralise the effect of individual scoring on different levels of the scale) rotating the individual configuration and finally expanding or shrinking each configuration as a whole, the space for each assessor is made to match as closely as possible that of other assessors yet maintaining each assessor's inter-sample relationships. Each assessor's assessment of port 1, for example, is brought as close as possible to every other assessor's assessment of port 1, and each assessor's assessment of port 2 as close as possible to every other assessor's assessment of port 2, and so on for all assessors and all ports. By taking the centroid of all the points for similar ports a consensus configuration can be obtained. This configuration may be referred to new orthogonal axes, accounting successively for decreasing amounts of variation in the data. These new axes may be interpreted in terms of each assessor's original description or related to any external factor, such as production variables, and hedonic, chemical and physical measurements of the port. The consistency or variation in agreement in respect to the relative positions of ports in the final configuration can be obtained from the scatter of the points about the centroid position. The interpoint information for each port can also be used to determine the relative agreement among the assessors. The distance between each assessor's points for each port are calculated and used as the input similarity matrix to a principal co-ordinate analysis which allows the relative position of the assessors to be determined (Langron, S.P.; Collins, A.J., unpublished).

3. RESULTS AND DISCUSSION

The descriptions developed by each of the ten assessors are listed in Table 3; these range from six to eighteen words, the least number of terms in general being used by the non-expert assessors. Plots showing the relative position of the points in respect to appearance, aroma and flavour are given in Figures 1-3, these two dimensions accounting for 96.3, 65 and 71% of the variation respectively. The interpretation of the final major axes for each assessor is presented in Tables 4-6.

TABLE 3.
ADJECTIVE SCORED IN THE ASSESSMENT OF PORTS BY THE TEN ASSESSORS

	Appearance	Aroma	Flavour
Assessor 1	Depth	Cleanness	Cleanness
	Fresh	Freshness	Fresh
	Brightness	Fruitiness	Body
		Richness	Grip
		Smoothness	Round
		Concentration	Aftertaste
Assessor 2	Tawny	Clean	Body
	Ruby	Fruity	Firmness
	Purple	Green	Coarse
	Cloudy		Tannin
			Hard
			Crisp
			Sour
Assessor 3	Red	Fruit	Tannin
	Blue	Esters	Acid
	Brown	Wood	Sweetness
	Intensity	Oloroso	Chocolate
		Spirit	Body
		Burnt	Green
Assessor 4	Mauve	Tannin	Tannin
	Red	Sweet	Sweet
	Brown	Balanced	Balanced
	Soft	Clean	Clean
	Grey	Strong	Strong
	Plum	Astringent	Astringent
Assessor 5	Ruby	Fruit	Fruity
	Tawny	Maturity	Body
	Depth	Spirit	Tannin
	Tint	Depth	Balance
		Taint	Taint
		Wood	Maturity
Assessor 6	Red	Fruity	Paper
	Purple	Pungent	Rich
	Brown	Cedar	Astringent
		Soft	Green
		Burnt	Harsh
		VA	
Assessor 7	Ruby	Spirit	Sweetness
	Tawny	Burnt	Acidity
	Intensity	Grape	Tannin
		Off-taint	Spirit
		Woody	Depth/body
		Intensity	Aftertaste
Assessor 8	Clarity	Volatile acidity	Acid
	Intensity	Fruitiness	Sweetness
	Redness	Woodiness	Smoothness
	Yellowness	Alcoholic strength	Body
		Overall intensity	
Assessor 9	Intensity	Burnt	Astringent
	Red		Acidity
	Brown		
Assessor 10	Colour	Smell	Taste
	Red		Acidity
	Brown		Tannin
			Softness

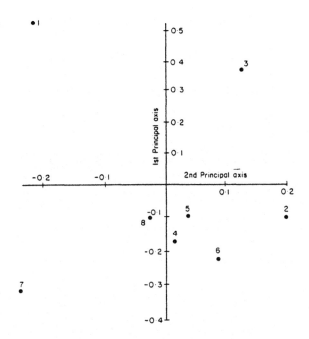

FIG. 1. SAMPLE CONSENSUS PLOT FOR APPEARANCE: FIRST TWO PRINCIPAL AXES
For identification of samples see Table 1.

TABLE 4.
DEFINITION OF FIRST TWO PRINCIPAL AXES FOR APPEARANCE IN TERMS OF ASSESSORS' ORIGINAL DESCRIPTIONS

Assessor	Principal axis 1	Principal axis 2
1	Depth of redness (0.9)[a]	−Brightness (0.93)
2	Ruby (0.59)+purple (0.47)−tawny (0.63)	Ruby (0.71)+clarity(0.40)−purple (0.51)
3	Red (0.63)+blue (0.60)	Intensity (0.75)+brown (0.65)
4	Mauve (0.6)+red (0.2)+plum (0.36)−brown (0.56)−soft (0.39)	−Soft (0.83)
5	−Tawny (0.80)	Depth (0.89)
6	Red (0.44)+purple (0.77)−brown (0.46)	Red (0.72)+brown (0.68)
7	Ruby (0.49)+intensity (0.49)−tawny (0.71)	Intensity (0.71)+tawny (0.61)
8	Red (0.81)+intensity (0.30)−yellow (0.48)	Intensity (0.71)+yellow (0.60)
9	Intensity (0.47)+purple (0.68) − brown (0.54)	Intensity (0.86)+brown (0.44)
10	Red (0.83)−brown (0.53)	Colour (0.69)+brown (0.28)

[a] Figures in parentheses refer to vector loadings for individual assessors.

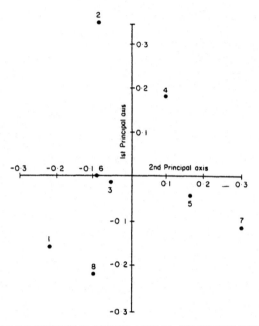

FIG. 2. SAMPLE CONSENSUS PLOT FOR AROMA: FIRST TWO PRINCIPAL AXES
For identification of samples see Table 1.

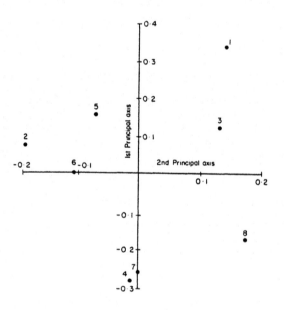

FIG. 3. SAMPLE CONSENSUS PLOT FOR FLAVOUR: FIRST TWO PRINCIPAL AXES
For identification of samples see Table 1.

TABLE 5.
DEFINITIONS OF FIRST TWO PRINCIPAL AXES FOR AROMA
IN TERMS OF ASSESSORS' ORIGINAL DESCRIPTION

Assessor	Principal axis 1	Principal axis 2
1	−Cleanness (0.56) − freshness (0.52) − richness (0.45)	−Fruitiness (0.81)
2	Fruity (0.77) − clean (0.54)	−Cleanness (0.71) − fruitiness (0.53)
3	Burnt (0.81) + estery (0.46) − waxy (0.25)	Oloroso (0.68) − fruity (0.46) − esters (0.39)
4	Astringent (0.54) + tannin (0.42) − clean (0.74)	−Strong (0.61) − astringent (0.54)
5	Taint (0.81) − fruit (0.31)	Maturity (0.57) spirit (0.38) − depth (0.76) − fruity (0.30)
6	Burnt (0.42) − fruity (−0.68) − cedar (0.52)	Cedar (0.69) − fruity (0.68)
7	Burnt (0.90) − grape (0.29)	Woody (0.74) + grape (0.50) − off-taint (0.36)
8	Vol. acidity (0.67) + alcoholic strength (0.33) − fruitiness (0.41) − woodiness (0.49)	Vol. acidity (0.34) − alcoholic strength (0.83)
9	Burnt (0.04)	Burnt (0.79)
10	Smell (−0.8)	Smell (0.57)

TABLE 6.
DEFINITION OF FIRST TWO PRINCIPAL AXES FOR FLAVOUR
IN TERMS OF ASSESSORS' ORIGINAL DESCRIPTIONS

Assessor	Principal axis 1	Principal axis 2
1	Fruit (0.62) + body (0.66) + cleanness (0.35)	Grip (0.85) + round (0.51)
2	Firmness (0.59) − coarse (0.66) − body (0.27)	Crispness (0.80) − hard (0.48)
3	Green (0.83) + tannin (0.39)	Body (0.41) + tannin (0.34) − acid (0.79)
4	Tannin (0.63) − clean (0.60) − astringent (0.30)	Astringent (0.48) − balanced (0.76)−clean (0.34)
5	Tannin (0.47) − maturity (0.83) − balance (0.317)	−Tannin (0.71) − maturity (0.52)
6	Astringent (0.73) + hash (0.48 + green (0.34) + rich (0.38)	Green (0.41) + hash (0.35) − rich (0.68) − paper (0.48)
7	Tannin (0.89) + spirit (0.39)	Acid (0.84) − spirit (0.43)
8	Acid (0.42) − sweet (0.47) − body (0.69)	Acid (0.66) + smoothness (0.73)
9	Astringent (0.93)	Acidity (0.72)
10	Tannin (0.59) − softness (0.73)	Taste (0.63) + acidity (0.48) + tannin (0.41) + softness (0.43)

Examination of the combination of descriptions used by each of the assessors in defining these axes enables one to get some indication as to their meaning in terms of each individual's own terminology. As far as appearance is concerned, the first axis seems to be a contrast between red and tawny characteristics in the port; the second, overall intensity of colour.

The meaning of the aroma dimensions are less easy to summarise, the exact interpretation varying from assessor to assessor. From examination of Table 5 it would however appear that the first axis is the contrast between burnt, bitter, astringent and possible heavy fruity notes with clean, waxy, fruity cedarwood-type characteristics. The second axis appears in general also to be the contrast between various attributes and fruitiness. As far as taste is concerned (Table 6), the first axis is largely reflected by a clean tannin content, whereas the second appears to be measuring such factors as acidity, crispness, body, balance and the ill-defined term 'grip'.

Examination of the sample residuals (derived from the sum of the squared distances from the consensus position to actual individual assessor's final sample position for each sample) (Table 7) shows greater agreement in respect to the relative position of the ports when assessing them by appearance than there was when assessing them by aroma and flavour. As far as appearance was concerned, port 1 showed the greatest disagreement and port 7 the closest agreement. In the case of aroma, port 5 showed most disagreement and port 3 the least, and in the case of flavour port 3 the greatest disagreement and port 1 the least.

TABLE 7.
SAMPLE RESIDUALS IN FINAL CONFIGURATION

Sample	Appearance	Aroma	Mouth flavour
1	0.44	0.66	0.43
2	0.25	0.51	0.50
3	0.23	0.45	0.62
4	0.28	0.69	0.42
5	0.23	0.73	0.54
6	0.25	0.51	0.48
7	0.15	0.53	0.47
8	0.20	0.60	0.48

More information on assessor conformity in respect to the relative spatial arrangement of the ports following the various manipulations can be obtained from the assessor residuals (derived from the sum of the squared distances from the consensus position to actual individual assessor's final sample position for each assessor) (Table 8) and the assessor plots (Figures 4-6). As far as aroma and flavour are concerned (Table 8, columns 2 and 3, and Figures 5 and 6), the assessors appear to be performing very similarly, all showing roughly the same degree of agreement or disagreement in the samples. No obvious groupings are emerging. As far as appearance is concerned, however, (Table 8, column 1, and Figure 4), assessors 2, 5 and 8 and in particular assessor 1, are behaving differently from the remainder of the panel.

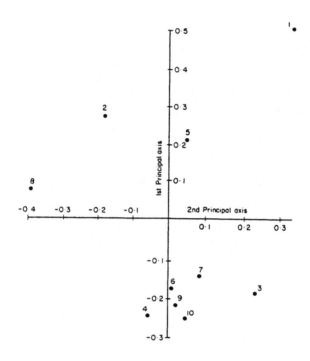

FIG. 4. ASSESSORS' PLOT FOR APPEARANCE: FIRST TWO PRINCIPAL AXES
For information on assessors see Table 2.

If one examines the individual rotated scores for each assessor for appearance, or the amount of variation accounted for by each dimension (Table 9), part of the reason for this split in the panel can be found. Assessor 1, and to some extent Assessors 2 and 8, seem to require three dimensions to describe differences in the appearance of the samples, whereas the remainder's sample variation is quite adequately accounted for by two dimensions. This third axis, in the case of Assessor 1, is associated largely with an attribute which he defines as 'beading'.

TABLE 8.
ASSESSORS' RESIDUALS IN FINAL CONFIGURATION

Assessor	Appearance	Aroma	Mouth flavour
1	0.43	0.45	0.39
2	0.25	0.49	0.43
3	0.15	0.44	0.34
4	0.17	0.44	0.38
5	0.19	0.47	0.41
6	0.17	0.47	0.45
7	0.13	0.45	0.38
8	0.27	0.46	0.37
9	0.13	0.52	0.44
10	0.15	0.50	0.36

TABLE 9.
%VARIATION ACCOUNTED FOR IN THE FIRST FOUR PRINCIPAL AXES
FOR EACH ASSESSOR IN CASE OF APPEARANCE

Assessor	Axis 1	Axis 2	Axis 3	Axis 4	Variation accounted for by next largest axis
1	32.9	24.7	41.4	0.3	0.6
2	64.9	24.5	8.4	1.8	0.1
3	80.5	14.7	2.9	0.4	1.4
4	75.1	14.0	5.4	3.8	1.1
5	80.1	17.7	1.4	0.05	0.7
6	63.6	33.9	1.9	0.1	0.3
7	68.5	26.9	4.2	0.1	0.1
8	45.4	49.1	3.2	1.9	0.16
9	84.1	14.8	0.3	0.6	0.2
10	69.6	26.6	3.7	0.1	0.02

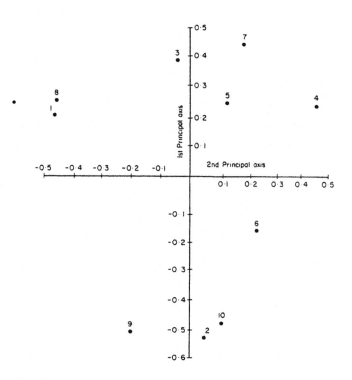

FIG. 5. ASSESSORS' PLOT FOR AROMA: FIRST TWO PRINCIPAL AXES
For information on assessors see Table 2.

4. CONCLUSION

Free-choice word profiling of the eight commercial ports, coupled with generalised Procrustes analyses, has enabled sample plots to be obtained for the ports in respect to their appearance, aroma and flavour. Although each assessor used his own descriptive term when describing the profile, the multidimensional plots showing the interrelationships of the ports demonstrated a fair degree of agreement in respect to their shape (with the exception of one assessor in respect to appearance), if not in terms of the words that they were using to define them.

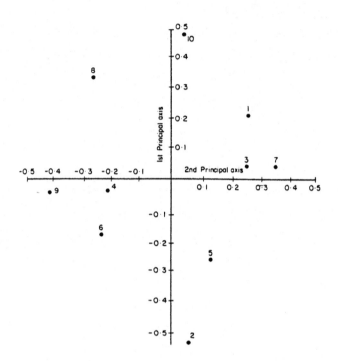

FIG. 6. ASSESSORS' PLOT FOR FLAVOUR: FIRST TWO PRINCIPAL AXES
For information on assessors see Table 2.

The consensus plot of these individual sample plots after matching also appeared sensible in terms of previous knowledge of the ports. The principal axes through the consensus space could be defined in terms of the original assessor descriptions. Using various regression approaches, they could also be related to external factors, such as laboratory colour measurements, and gas chromatographic data, as is done in similarity scaling procedures.[20] The axes too, may readily be related to hedonic measurements enabling reasons for preference to be interpreted.

The experiment proved a success and clearly showed that there is no need to develop precisely defined vocabularies for describing products in order to reveal relationships and differences between samples. Unlike the use of similarity scaling which, at present, requires many pairs of samples to be assessed before meaningful information can be obtained, this approach requires relatively few examinations and hence is much more economical of time. Allowing people to use their own terminology is a much more natural way to assess products and overcomes the frustration experienced by panel leaders in conventional profile approaches when trying to force agreement among panelists in the use of terminology.

The present exercise was conducted with people who all had some experience of sensory analysis and in the use of scales for scoring. They were all experienced in examining beverages critically and articulate in putting words to the criteria they considered important. In principle the approach should be applicable to any type of assessor, experienced or inexperienced, and could therefore find use in the field of consumer research. However, for its successful operation people must be capable of using scales for scoring, they must be objective in their assessments and that once having defined an attribute in their own minds, they must be consistent in its use. Experience indicates that consumers can use scales correctly, but we have yet to explore thoroughly their objectivity when using the words they have defined. Despite this small reservation, it is believed that free-choice word profiling, coupled with the appropriate statistical techniques for the analysis of the data, provides a valuable adjunct to present-day sensory methodology.

REFERENCES

1. CAUL, J.F. The profile method of flavour analysis. Adv. Food Res. 1957, 7, 1-4.
2. CAIRNCROSS, S.E., SJÖSTROM, L.B. Flavour profiles—a new approach to flavour problems. Food Technol. Champaign 1950, 4, 308-311.
3. VON SYDOW, E., KARLSON, G. The aroma of black currants V. The influence of heat measured by odour quality assessment techniques. Lebensm. Wiss. Technol. 1971, 152-157.
4. CLAPPERTON, J.F. Derivation of a profile method for sensory analysis of beer flavour. J. Inst. Brew. 1973, 79, 495-508.
5. CLAPPERTON, J.F. Profile analysis flavour discrimination. J. Inst. Brew. 1974, 80, 164-173.

6. WILLIAMS, A.A. The development of a vocabulary and profile assessment methods for evaluating the flavour contribution of cider and perry aroma constituents. J. Sci. Food Agric. 1975, 26, 567-582.
7. WILLIAMS, A.A., CARTER, C.S. A language and procedure for the sensory assessment of Cox's Orange Pippin apples. J. Sci. Food Agric. 1977. 28, 1090-1104.
8. SHORTREAD, G.W., RICHARDS, P., SWAN, J.S., BURTLES, S.M. The flavour terminology of Scotch whisky. Brew. Guardian 1979, 55-62.
9. CLAPPERTON, J.F., DALGLEISH, C.E., MIELGAARD, M.J. Progress towards an international system of beer flavour terminology. J. Inst. Brew. 1976, 82, 7-13.
10. CLAPPERTON, J.F., PIGGOTT, J.R. Differentiation of ale and lager flavours by principal component analysis of flavour characterization data. J. Inst. Brew. 1979, 88, 271-274.
11. WILLIAMS, A.A., BAINES, C.R., ARNOLD, G.M. Towards the objective assessment of sensory quality in inexpensive red wines. *Centennial Symposium Department of Viticulture and Enology*, University of California, Davis, Ca, USA. 1982, pp. 322-329.
12. WILLIAMS, A.A., LANGRON, S.P., ARNOLD, G.M. Sensory assessment of ciders and apple juice. In *Sensory Quality in Foods and Beverages: its Definition, Measurement and Control* (Williams, A.A.; Atkin, R.K., Eds), Ellis Horwood Ltd. Chichester, England, 1983, pp. 310-323.
13. NOBLE, A.C., WILLIAMS, A.A., LANGRON, S.P. Descriptive analysis and quality rating of 1976 wines from the Bordeaux communes. J. Sci. Food Agric. 1984, 1, 88-98.
14. WILLIAMS, A.A., BAINES, C.R., LANGRON, S.P., COLLINS, A.J. Evaluating tasters' performance in the profiling of foods and beverages. In *Flavour '81* (Schreiber, P., Ed.) Walter de Gruyter & Co., Berlin. New York, 1981, pp. 83-92.
15. STONE, H., SIDEL, J., OLIVER, S., WOOLSEY, A., SINGLETON, R.C. Sensory evaluation by quantitative descriptive analysis. Food Technol. 1974, 28, 24-34.
16. GOWER, J.C. Generalised Procrustes analysis. Psychometrica 1975, 40, 35-51.
17. GOWER, J.C. Procrustes rotational fitting problems. Math. Sic. 1976, 1, 12-15 (Suppl.)
18. TEN BERG, J.M.F. Orthogonal Procrustes rotation for two or more matrices. Psychometrica 1977, 42, 267.

19. MOSKOWITZ, H.R., CHANDLER, J.W. New uses of magnitude estimation. In *Sensory Properties of Foods* (Birch, G.G., Brennan, J.G. and Parker, K.J., Eds). Applied Science Publishers Ltd., London, 1977, pp. 189-211.
20. SCHIFFMAN, S.S., REYNOLDS, M.L., YOUNG, F.W. *An Introduction to Multi-dimensional Scaling: Theory Methods and Application.* Academic Press, London, New York, 1981, pp. 413.

CHAPTER 6.2

A COMPARISON OF THE AROMAS OF SIX COFFEES CHARACTERISED BY CONVENTIONAL PROFILING, FREE-CHOICE PROFILING AND SIMILARITY SCALING METHODS

ANTHONY A. WILLIAMS and GILLIAN M. ARNOLD

*At the time of original publication,
all authors were affiliated with
Food and Beverages Division and Biometrics Section
Long Ashton Research Station
University of Bristol
Long Ashton, Bristol BS18 9AF.*

A selected panel, trained in sensory procedures, used a previously developed vocabulary to profile the aromas of six ground coffees. The same coffees were also examined by a different panel of assessors not familiar with coffees, using free-choice profiling and, on a separate occasion, similarity scaling procedures. The data from the three experiments were independently subjected to the appropriate statistical techniques to produce two-dimensional diagrams showing the inter-relationships of the samples. All three procedures produced similar information both in respect to the way the coffees grouped and differed from one another. The paper concludes by discussing the assessor variation associated with each of the procedures and shows that free choice profiling compares very favourably with the other two.

1. INTRODUCTION

Multivariate statistical techniques have been used to show the inter-relationships between samples assessed by conventional profile procedures by extricating the most significant dimensions from the multidimensional data provided by such techniques and plotting the information they contain as a two- or three-dimensional diagram.[1-6] Similarity scaling[7-10] and more recently free-choice profiling[11-13] have also been used to provide data from which similar sample spaces can be devised.

Free-choice profiling has advantages over both conventional profiling and similarity scaling.[11] It is faster than conventional profiling because it lacks the need for the elaborate training of assessors, and it does not require every sample to be compared with every other sample, a current requisite of similarity scaling approaches. Because words are used to assess the samples, the dimensions extracted can be provided with meaning. Whether these three approaches for examining the inter-relationships of samples provide the same information needs to be assessed, particularly with respect to the newer free-choice profiling approach.

The following examination was undertaken to determine just how similar were the sample spaces derived by all three procedures.

2. EXPERIMENTAL

2.1. Materials

Coffee beans of five types, Ethiopian Yirgachefe Arabica, Ivory Coast Robusta, Costa Rican Arabica, Angolan Robusta and Brazilian Arabica, were medium-roasted as single batches (3 kg) using a Probat (LG3) roaster, then stored in airtight containers and medium ground in a Birzerba coffee grinder (grind 8) 24 h prior to examination. The sixth sample was an instant coffee (Nestlé Gold Blend) purchased through a retail outlet and ground to similar particle size as the beans using the same grinder.

2.2. Conventional profile assessment

2.2.1. Assessors

The 15 assessors (four men, 11 women) were selected from staff of the International Coffee Organisation for their coffee profiling suitability. Following a period of training in general sensory assessment, panelists were presented with a wide range of both fresh and instant coffees over six sessions during which time they developed a defined vocabulary and assessment procedure for describing the various sensory characteristics of coffee both as ground bean and as brewed coffee.

2.2.2. The assessment procedure

All assessments took place in individual assessment booths illuminated by artificial daylight. At each session assessors were presented with three randomly-coded samples of ground coffee beans (10 g) in a wine glass (BS

5586) covered with a watch glass, and they assessed each for the intensity of the attributes listed in Table 1 using a six-point continuous line scale. All assessors attended six sessions; each coffee was assessed in triplicate and the order of presentation followed a balanced experimental design.

TABLE 1.
TERMS AND SCALE USED IN CONVENTIONAL PROFILING OF COFFEES

Terms	
Intensity	Beef broth-like
Complexity	Toasted bread-like
Depth/Body	Burnt/Charcoal-like
Minty/Acid/Citrus	Rubber-like
Blackcurrant-like	Motor oil/Tar-like
Fragrant/Floral	Phenolic
Winy	Cereal-like/Nutty
Milky/Creamy	Grassy/Stalky
Sugary/Caramel	Earthy
Fruit cake-like/Liquorice	Papery/Woody/Ashy
Spicy	Animal-like
Chocolate-like	Rancid/Sour/Cheesy
Malty	

Scale

0	1	2	3	4	5
None of the attributes	Very weak	Weak	Moderate	Strong	Very strong

2.3. Similarity scaling

2.3.1. The assessors

The 10 assessors (two men, eight women) were members of the staff of Long Ashton Research Station; eight had previous experience in profile analysis, two were unfamiliar with descriptive sensory analysis and none had any experience in the sensory assessment of coffees.

2.3.2. The assessment procedure

All coffees were compared with each other in both orders by all assessors. Five pairs of samples were presented at each session and each pair was scored on the similarity of their aromas using a six-point scale (0 = no similarity to 5 = identical). All samples were presented in randomly coded wine glasses (BS 5586) covered with watch glasses and were assessed in individual booths. Identical samples could not be identified visually and no identical samples were directly compared.

2.4. Free-choice profiling

2.4.1. The assessors

The panel was the same as that used for similarity scaling (Section 2.3.1) except that one of the panelists not familiar with profiling did not take part.

2.4.2. The assessment procedure

Samples were assessed using the general procedure as previously described for ports.[12] All six coffees were first presented to the assessors in a similar way to that used for conventional profiling. Samples were used to develop individual vocabularies for assessing differences and similarities in the aromas of the ground beans. The terms devised by each assessor are given in Table 2. The assessors then examined the same coffees (randomly coded) in triplicate, scoring each adjective using a six-point scale. Samples were presented in groups of three in the form of a balanced experimental design and, as with conventional profiling, the complete experiment took six sessions.

2.5. Statistical analysis

Mean profile data were subjected to principal component analysis,[14] similarity data to INDSCAL analysis[7] and free-choice profiling data to generalised Procrustes analysis.[15] Sample plots were obtained showing the inter-relationships of the coffees over the first two principal axes. To assist in comparing the overall sample spaces by these three methods, coordinates from each of the procedures were used as input to a further generalised Procrustes analysis.

To explore how well each assessor conformed to the panel means, individual sample spaces were superimposed on the consensus spaces. In the case of the principal component analysis of the conventional profile data individual spaces were compared following translation of the data so that scores for each assessor were equally distributed about a common zero.

Individual assessor plots[12,16] were also obtained for each of the experiments. Such plots were obtained as previously described for free-choice profiling[12] and similarity scaling:[16] for conventional profile data, assessor plots were obtained by taking every possible pair of assessors, calculating the mean distance between their assessments of the same coffees and using this as input

TABLE 2.
TERMS USED IN FREE-CHOICE PROFILING OF COFFEES

	Assessor 1	Assessor 2	Assessor 3	Assessor 4	Assessor 5	Assessor 6	Assessor 7	Assessor 8	Assessor 9
1.	Bitter	Black treacle	Bitter	Cardboard	Bitter	Aromatic	Almond	Burnt	Acid/sharp
2.	Chocolate	Cardboard	Chocolate	Musty/dusty	Cardboard	Burnt/roasted	Black treacle	Cardboard	Burnt/scorched
3.	Coffee	Chemical	Chocolate/bitter	Nutty	Mocha	Chocolate	Bitter	Dry	Caramel
4.	Drains	Chocolate	Cooked	Oily	Nutty	Closed up	Butter	Green	Chocolate
5.	Germanic	Earthy	Fusty	Roasted/burnt	Powdery	Fishy	Cardboard	Malty	Creamy
6.	Plastic	Mousy	Fragrant	Sawdust	Roasted	High roast Carma	Chocolate (plain)	Musty	Lemon/fragment/winy
7.	Roasted	Nutty	Mothballs	Sharp	Scented	Musty/dirty	Fishy	Nutty	Malty
8.	Smoky/tobacco	Sawdust	Musty/dusty	Smoky/tobacco	—	Scented	Greasy	Powdery	Musty
9.	Stale	Sweet	Nutty	Stale	—	Tobacco	Malty	Roasted	Wood/papery
10.	Sweet	Tobacco	Powdery	Vanilla	—	—	Roasted	Scented/fragrant	Tobacco
11.	Burnt	Urinals	Smoky/tobacco	Scented	—	—	Scented	Sweet	—
12.	Choking	Vanilla	Drains	Cooked	—	—	Woody	Tea-leaves	—
13.	—	Volatile	—	Mouldy	—	—	Caramel	Toast	—
14.	—	Cigarette ash	—	Chocolate/bitter	—	—	Tobacco	Woody	—
15.	—	—	—	—	—	—	Vanilla	Tobacco	—

to a principal coordinate analysis, much in the same way as assessor plots are obtained following Procrustes analysis of free-choice profiling data.

A generalised Procrustes analysis was performed on the conventional profile data also and sample and assessor plots obtained to determine how much rationalisation of the terms and scales improved the results obtained from this approach.

3. RESULTS

Principal component analysis of the conventional profile data showed the first two principal axes to account for 80.6% variation and provided the sample plots shown in Figure 1a. The two robusta coffees were closely grouped and the instant coffee was completely different from the other five. The arabicas were also grouped but less closely than the robustas. Of these the Brazilian arabica was closest to the robustas.

Examination of vector loadings showed the two principal axes to be largely the contrast of (i) intensity, complexity, depth of aroma, minty/acid/citrus and winy with malty, rubbery, cereal-like, earthy, papery and rancid/sour/chemical, and (ii) milky, sugary and chocolate with beefy/fruit/charcoal and phenolic notes.

Correlation of the INDSCAL solutions with individual assessor's mean similarities (Table 3) showed that two dimensions were required to explain variations in the data. Correlation of mean similarities with the two dimensional solution shows that these two dimensions account for approximately 61% of the variation. Sample plots are given in Figure 1b. Inter-relationships were virtually identical to those obtained by conventional profiling except that the space was rotated through 90° about an axis 90° to the two shown, and through a further 180° around the second axis.

TABLE 3.
CORRELATION BETWEEN INDIVIDUAL ASSESSORS' SAMPLE SIMILARITIES AND RESULTS OF ONE-, TWO- AND THREE-DIMENSIONAL INDSCAL SOLUTIONS

Assessor	One-dimensional solution	Two-dimensional solution	Three-dimensional solution
1	0.734**	0.855***	0.887***
2	0.392	0.671**	0.937***
3	0.686**	0.757**	0.849***
4	0.382	0.838***	0.972***
5	0.833***	0.850***	0.951***
6	0.690**	0.802***	0.802***
7	0.377	0.521*	0.668**
8	0.606*	0.634*	0.922***
9	0.864***	0.896***	0.945***
10	0.378	0.893***	0.963***

*Significant at 5% level.
**Significant at 1% level.
***Significant at 0.1% level.

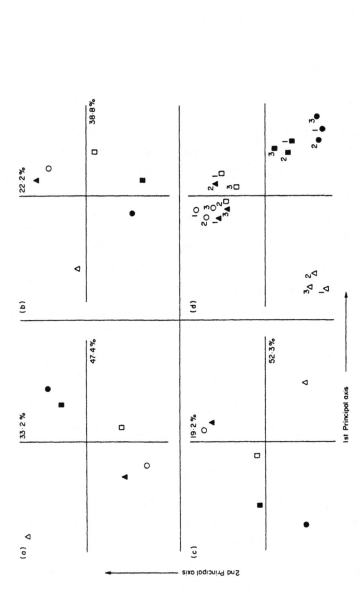

FIG. 1. *a* Mean sample plot on first two principal axes following conventional profiling; *b* mean sample plot following similarity scaling and INDSCAL analysis (two dimensional solution); *c* mean sample plot following free-choice profiling and generalised Procrustes analysis (first two principal axes); *d* sample plots for conventional profiling, similarity scaling and free-choice profiling following generalised Procrustes analysis. Conventional profiling (1); similarity scaling (2); free-choice profiling (3). ○. Angolan Robusta; •. Yirgachefe Arabica (Ethiopian); □. Brazilian Arabica; ■. Costa Rican Arabica; △. Instant Coffee (Nestle Gold Blend); ▲. Ivory Coast Robusta.

Examination of the space obtained by free-choice profiling indicated that the first two principal axes accounted for 71.5% of the variation and produced the sample plot shown in Figure 1c which is similar to those produced by conventional profiling except that the space is rotated through 180° about an axis 90° to the two shown. Interpretation of these axes for each of the nine assessors is given in Table 4; the first dimension appears to be the contrast between cardboardy, smoky-tobacco and powdery-type aromas with chocolate scented notes: the second to be the contrast between bitter; nutty; stale and roasted notes with sweet scented-type aromas. When the rotation is taken into account they show similarity with the descriptions used in conventional profiling.

The comparison of these three spaces by generalised Procrustes analysis (Figure 1d) emphasises the similarity obtained by the three procedures, all similar samples being grouped together.

Examination of individual sample-points of the respective spaces from each of the three assessments (Figures 2, 3 and 4) showed the extent of agreement amongst assessors, free-choice profiling and similarity scaling giving the tightest clusters. Selecting the best 10 of the 15 assessors from the profiling exercise removed outlying points but only improved the tightness of the groupings marginally, clearly demonstrating the need to take account of individual assessor variation in use of terms and scoring procedure, even when analysing conventionally devised profile data. When allowance was made for individual variations by applying generalised Procrustes techniques to the conventional profile data, mean sample inter-relationships were only marginally affected but individual assessor agreement (Figure 5) was brought more into line with that obtained by free-choice profiling.

Assessor plots[12] derived from each of the three procedures showed some clustering with different assessors grouping together depending on the procedure. Examination of the general spread of the assessor plots derived from conventional profiling and free-choice profiling again showed the overall grouping and basic agreement of the assessors in free-choice profiling to be better than that for the whole panel when using conventional profiling, confirming the information deduced from Figures 2 and 3.

Application of generalised Procrustes analysis to the conventional profile data improved the overall scatter of assessors bringing the results more in line with those obtained with free-choice profiling.

Because the results of the similarity scaling exercise are derived by an entirely different approach, it is impossible to make similar direct comparisons, although comparison of Figures 2 and 4 would seem to indicate a marked improvement in the agreement between assessors when compared to conventional profiling.

TABLE 4.
INTERPRETATION OF PRINCIPAL AXES FOLLOWING
FREE CHOICE PROFILING

Assessor	Principal Axis 1	Principal Axis 2
1	Plastic (0.50) +smoky/tobacco (0.37) − chocolate (0.61)	Bitter (0.43) +coffee (0.41) − sweet (0.43)
2	Cardboard (0.53) − volatile (0.57) − sweet (0.38)	Nutty (0.32) +earthy (0.22) +chocolate (0.20) − cardboard (0.58) − volatile (0.53)
3	Smoky/tobacco (0.40) − chocolate (0.59) − fragrant (0.45)	Fragrant (0.30) +fusty (0.27) +mothballs (0.24) − powdery (0.59) − chocolate (0.52)
4	Musty/dusty (0.54) − vanilla (0.54) − oily (0.43)	Stale (0.51) +sawdust (0.46) − musty/dusty (0.43)
5	Powdery (0.65) − roasted (0.48) − scented (0.48)	Cardboard (0.55) +Mocha (0.48) +bitter (0.45) − powdery (0.34) − scented (0.31)
6	Musty/dirty (0.6) − scented (0.49) − high roast Carma (0.46)	Burnt/roasted (0.45) − scented (0.6)
7	Tobacco (0.49) +cardboard (0.43) − almond (0.42) − black treacle (0.34)	Chocolate, plain (0.45) − black treacle (0.55)
8	Toast (0.36) +dry (0.34) − green (0.41) − malty (0.40)	Cardboard (0.45) +burnt (0.36) − tea leaves (0.39) − sweet (0.38) −scented/fragrant (0.35)
9	Creamy (0.35) +musty (0.33) +tobacco (0.31) −lemony/fragrant/winy	Burnt/scorched (0.47) +woody/papery (0.36) − tobacco (0.43) − malty (0.43) − acid/sharp (0.42)

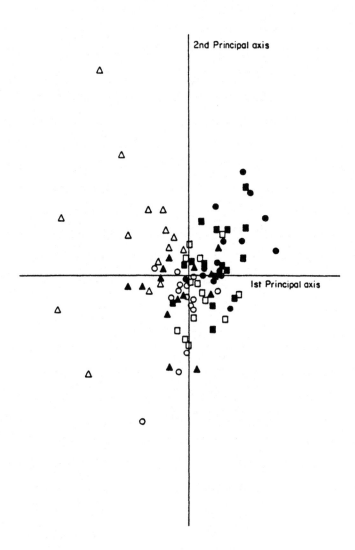

FIG. 2. INDIVIDUAL SAMPLE PLOTS FOR CONVENTIONAL PROFILING FOLLOWING PRINCIPAL COMPONENT ANALYSIS (15 ASSESSORS)
For identities of coffees see Figure 1.

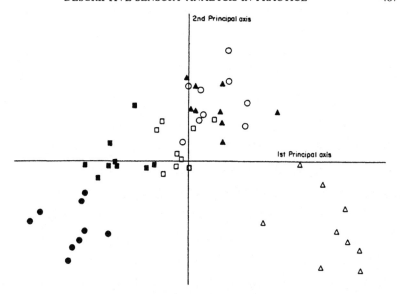

FIG. 3. INDIVIDUAL SAMPLE PLOTS FOR FREE-CHOICE PROFILING
FOLLOWING GENERALISED PROCRUSTES ANALYSIS
For identities of coffees see Figure 1.

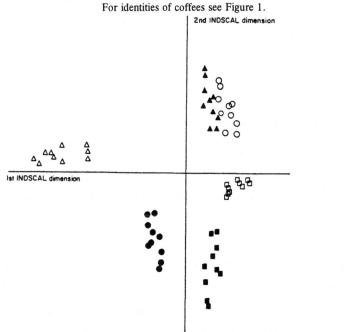

FIG. 4. INDIVIDUAL SAMPLE PLOTS FOLLOWING SIMILARITY SCALING
AND INDSCAL ANALYSIS
For identities of coffees see Figure 1.

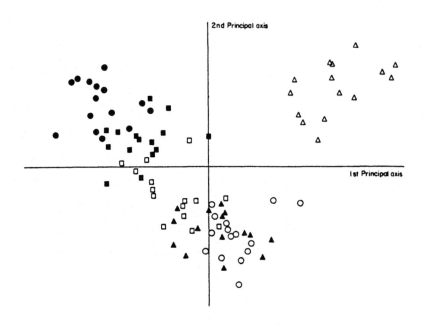

FIG. 5. INDIVIDUAL SAMPLE PLOTS FOLLOWING TREATMENT OF CONVENTIONAL PROFILING DATA WITH GENERALISED PROCRUSTES ANALYSIS
For identities of coffees see Figure 1.

4. CONCLUSIONS

The results of the three assessments described in this paper clearly show that conventional profiling, similarity scaling and free-choice profiling produced similar relationships between samples. They validate conclusions drawn from free-choice profiling and demonstrate the advantages of taking individual assessor variations into account when evaluating conventional profile data. Despite the different conceptual approach to profiling and similarity scaling, in this example they have produced very similar information on the inter-relationships of samples. Chauhan *et al.*, when examining soft drinks, came to similar conclusions.[9]

All three sensory procedures described in this paper require the assessors to be familiar with the use of scaling and to be consistent in the use of criteria once established. Both free-choice profiling and similarity scaling, however, are simpler to operate than conventional profiling, requiring much less training. Similarity scaling obviously requires no vocabulary development and that required in free-choice profiling is much less rigorous than in conventional profiling. In operation, free-choice profiling has two advantages over similarity scaling. In multiple sample assessments it requires far fewer assessments, as samples are assessed individually and not as all possible paired comparisons. As descriptors are used in devising the sample space, the dimension of the space, unlike similarity scaling, can be interpreted directly from the data.

Unless account is taken of individual variation when using conventional profiling approaches, free-choice profiling would appear to be a more precise tool providing greater agreement between assessors in the positioning of samples within the sample space. The improvement can be checked statistically[17] but in the example described in this paper it is clearly demonstrated from inspection of Figures 2 and 3.

Because free-choice profiling attempts to rationalise the use of words when people are perceiving the same intersample differences, it highlights individuals or sub-groups who are responding differently to the main panel and are possibly more sensitive to particular attributes.[12] Such information is lost when conventional profile data are treated in the normal way, thus making the technique inherently more sensitive. The analysis of conventional profile data by generalised Procrustes techniques would similarly highlight outliers.

An individual's reproducibility in free-choice profiling is similar to that in other procedures, all of which are influenced by order effects in multiple sample experiments, but because assessors are using their own words they are more at ease with their use.

With these advantages, and as free-choice profiling would appear to provide compatible results to more conventional procedures particularly with assessors who are familiar with profiling techniques,[12] there are many instances, therefore, where it could provide a more appropriate, faster and cheaper alternative to conventional profile assessments.

ACKNOWLEDGEMENTS

The authors thank the International Coffee Organisation for permission to publish the conventional profiling part of this paper. This was carried out at the conclusion of a panel training exercise commissioned by that organisation.

REFERENCES

1. CLAPPERTON, J.F., PIGGOTT, J.R. Differentiation of ale and lager flavours by principal component analysis of flavour characterisation data. J. Inst. Brew. 1979, 88, 271-274.
2. WILLIAMS, A.A., BAINES, C.R., ARNOLD, G.M. *Towards the objective assessment of sensory quality in inexpensive red wines.* Centennial Symposium, Dept. of Viticulture and Enology, University of California, Davis, USA, 1982, pp. 322-329.
3. WILLIAMS, A.A., LANGRON, S.P., ARNOLD, G.M. Objective and hedonic sensory assessment of ciders and apple juices. In *Sensory Quality in Foods and Beverages; its Definition, Measurement and Control* (Williams, A.A.; Atkin, R.K., Eds.), Ellis Horwood Ltd, Chichester, England, 1983, pp. 310-323.
4. NOBLE, A.C., WILLIAMS, A.A., LANGRON, S.P. Descriptive analysis and quality rating of 1976 vines from the Bordeaux communes. J. Sci. Food Agric. 1984, 35, 88-98.
5. WILLIAMS, A.A., BAINES, C.R., LANGRON, S.P., COLLINS, A.J. Evaluating tasters' performance in the profiling of food and beverages. In *Flavour '81* (Schreier, P., Ed.), Walter de Gruyter, Berlin, New York, 1981, pp. 83-92.
6. STONE, H., SIDEL, J., OLIVER, S., WOOLSEY, A., SINGLETON, R.C. Sensory evaluation by quantitative descriptive analysis. Food Technol. 1974, 28, 24-34.
7. SCHIFFMAN, S.S., REYNOLDS, M.L., YOUNG, F.W. *An Introduction to Multidimensional Scaling: Theory, Methods and Application.* Academic Press, London, New York, 1981, pp. 413.
8. MacFIE, H.G.H., THOMSON, D.M.H. Multidimensional scaling methods. In *Sensory Analysis of Foods* (Piggott, J.R., Ed.). Appl. Sci. 1984, pp. 351-375.
9. CHAUHAN, J., HARPER, R., KRZANOWSKI, W. Comparisons between direct similarity assessment and descriptive profiles of citrus soft drinks. In *Sensory Quality in Foods and Beverages; its Definition, Measurement and Control* (Williams, A.A.; Atkin, R.K., Eds.), Ellis Horwood, Chichester, 1983, pp. 297-309.
10. THOMSON, D.M.H., MacFIE, H.J.H. Is there an alternative to descriptive sensory assessment? In *Sensory Quality in Foods and Beverages; its Definition, Measurement and Control* (Williams, A.A.; Atkin, R.K., Eds), Ellis Horwood, Chichester, 1983, pp. 96-107.
11. WILLIAMS, A.A., LANGRON, S.P. A new approach to sensory profile analysis. In *Flavour of Distilled Beverages* (Piggott, J.R., Ed.), Ellis Horwood, Chichester, 1983, pp. 219-224.

12. WILLIAMS, A.A., LANGRON, S.P. The use of free choice profiling for the examination of Commercial Ports. J. Sci. Food Agric. 1984, *35*, 558-568.
13. WILLIAMS, A.A., LANGRON, S.P. The influence of colour on the assessment of Ports. J. Food Technol. In press.
14. CHATFIELD, C., COLLINS, A.J. *Introduction to Multivariate Analysis.* Chapman and Hall, London, 1980, pp. 57-81.
15. GOWER, J.C. Generalised Procrustes analysis. Psychometrica, 1975, *40*, 35-51.
16. WILLIAMS, A.A., LANGRON, S.P., NOBLE, A.C. The influence of visual cues in the assessment of aroma in Bordeaux wines by trained judges. J. Inst. Brew. 1984, *90*, 250-253.
17. LANGRON, S.P. The Statistical Treatment of Sensory Analysis Data. Ph.D. Thesis, University of Bath, 1981.

CHAPTER 6.3

EVALUATION AND APPLICATIONS OF ODOR PROFILING

M.A. JELTEMA and E.W. SOUTHWICK

*At the time of original publication,
all authors were affiliated with
Philip Morris, U.S.A.
P.O. Box 26583
Richmond, Virginia 23261.*

ABSTRACT

An odor profiling procedure was developed based on the ASTM odor profiling method. This modified procedure involved using approximately twenty panelists. Panel sessions and data collection were controlled by computer. The results obtained by this panel compared favorably to results obtained by the ASTM panel for which 150 panelists evaluated each compound, indicating that a small panel can be used to produce replicable results. Statistical methods of finding similarities and dissimilarities among compounds using profile data are discussed and compared to results from a multidimensional scaling (MDS) study in which degrees of differences among compounds were judged directly. These results indicate that profile data can be used to define and map the degree of similarity/dissimilarity among compounds, as well as to define the sensory dimensions on which these compounds differ. The use of factor analysis to study the underlying sensory dimensions of the odor space is also discussed. It is hoped that this type of research will lead to a better understanding of the underlying dimensions used to describe odorants.

INTRODUCTION

The major objectives of an odor profiling program are to provide a database of odor attributes for pure chemicals and well-defined mixtures and to be able to determine the sensory differences between compounds.

An odor profile is a mathematical construct derived from a subjective assessment of the presence (or absence) of defined odor qualities, e.g., peanutty, rose like, and an estimate of the apparent intensity of that quality in the stimulus presented. Most often the profile is obtained by combining assessments of several panel members such that the result is an "average" profile. This technique is used widely in the flavor and fragrance industry as a means of systematically recording subjective data in a retrievable form.

There are numerous methods of obtaining profiles of odorants. The more usual methods include using trained perfumers either to obtain anecdotal descriptors on each compound or to identify odor aspects in relation to defined odorant references. The first approach makes statistical comparisons between compounds difficult and relies on superb panelist recall. With the latter approach, the profile is restricted to the choice of reference odors, and nuances may not be apparent. Also, the amount of sniffing required for comparison may be prohibitive.

The recently completed ASTM profile study (Dravnieks 1982) used a ballot with 146 descriptors. Each odorant was presented in a foil packet containing a balsa wood chip impregnated with the odorant. In a few instances the odorant was contained on scratch and sniff blotters. This method of presentation was used principally because the ballots were distributed by mail to 15 different laboratories. The profiles were generated by 10 panelists at each location and the "average" profiles were very stable and reproducible. Panelists received little, if any, training although most were probably involved in odor evaluation to some extent in their daily work.

Profiles obtained in this manner provide data on different compounds which can be statistically compared. The method, however, in the format used by the ASTM, would be unusable by most sensory laboratories due to the amount of time which would have to be spent obtaining and recording the data. This paper describes the approach taken by our laboratory to overcome these difficulties and compares the type of information which can be obtained by the profiling technique to that obtained by other sensory techniques.

METHODS

Mechanics of the Program

The ASTM procedure of profiling (Dravnieks 1982) was used with several modifications. The number of panelists was decreased from 150 to approximately 20. The number of descriptors was increased from 146 to 147 by adding an intensity rating. To avoid the need for laborious panel sessions and data input, a computer controlled sensory laboratory was developed (Jeltema *et al.* 1984).

The odorants to be evaluated are presented to the panelist as a dilute solution (usually 1%) in either dipropylene glycol or diethyl phthalate contained in an amber, wide-mouth bottle with a screw cap. Each sample bottle is identified with a code number. The panelists sniff each sample ad lib.

The data are collected by direct entry into the computer using touch-sensitive terminals. Panelists are free to do the evaluations at any time during a 48 h period; three samples per session are done. There are two sessions per week. Provisions have been made for panelists to make up sessions they have missed. No formal panel sessions are held.

A panelist logs in and the computer responds by displaying the code number of the sample to be evaluated. The order of presentation is randomized for each panelist. When a panelist indicates that he is ready, a ballot of 147 descriptors is presented (8 at a time) and the panelist records his responses by touching the screen. When the profile is complete, the terminal displays the name of the odorant and a brief description pertaining to the compound. There is a two minute pause before the terminal displays the next sample to be evaluated.

The raw data obtained from each panelist are in the form of scores (0-5) for the 147 descriptors. An "average" analysis for each compound is determined by calculating the percent frequency of use of every descriptor (percentage of scores > 0) and the applicability of each descriptor as described by Dravnieks (1982).

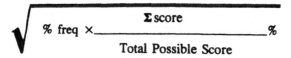

$$\sqrt{\% \text{ freq} \times \frac{\Sigma \text{ score}}{\text{Total Possible Score}}} \%$$

For each evaluation, the data from each panelist are compared to the "average" for the panel to provide a correlation of each individual with the average. This provides a method of monitoring individual panelists. Intensity scores are not included in this analysis but are simply recorded to indicate whether the compound needs to be redone at a different concentration. All of these analyses are performed by computer programs. The finished data are then stored in a searchable database.

RESULTS AND DISCUSSION

Estimation of Panel Size

To estimate an appropriate panel size, the results of the ASTM evaluation of acetophone were analyzed. Both the "average" evaluation and the evaluations from each of the 15 labs which participated in the study were kindly provided by A. Dravnieks. The results from pairs of labs (10 panelists per lab) were randomly combined and correlated with the "average" evaluation. The mean correlation using one lab (10 people) to compare with the "average" was 0.80; with randomly combined data from two labs (20 people) 0.90; and with three labs (30 people), 0.93. Since the replicability of samples was found in the ASTM study to be approximately 0.95, it was determined that 20-30 panelists could be used to obtain data which would agree fairly well with data obtained from 150 panelists. An initial panel of 25 people, who have an interest in this kind of research, was recruited.

Tests of Reproducibility

Thirty-five odorants, including acetophenone, which spanned a broad range of subjective response were selected from the ASTM list to use as a test data set (Table 1). Eight of these were profiled as dilute solutions in both diethyl phthalate and dipropylene glycol to determine whether solvent had any effect. Fourteen of these 35 were repeated to evaluate the reproducibility of our panel. These are presented in Table 2. There was no indication that the solvent had any effect on the profiles.

The mean correlation of all compounds with the ASTM data was 0.85, which was in the expected range base on the original ASTM data. Data from acetophenone showed a correlation of 0.93 with the ASTM data, which is in agreement with the data used to estimate panel size. These results confirmed that a panel size of 20 would be satisfactory.

Reproducibility of the panel was slightly better (0.90) than was the correlation with the ASTM data; it was similar to the reproducibility cited by the ASTM study. The reproducibility of the profile of a compound was related to the overall consistency of the panel for that compound. When the mean of the correlations of each panelist with the "average" profile was high, i.e., the panelists were consistent with each other in their evaluations, the repeat evaluation of the same odorant gave a profile very similar ($r = 0.81$) to the initial profile. This information provides a means of determining whether an odorant should be repeated to obtain more stable results.

The 95% confidence limit around the correlation of one compound with a different compound was fairly stable over the range of correlations. For most, a correlation could be expected to fall in the range ± 0.07. At very high correlations (> 0.8) the range of a value was smaller (± 0.03).

While the data obtained from our small panel correlated well with the data obtained from ASTM, it was found that our panel rated applicable descriptors more often that did the ASTM panel, i.e., if a descriptor was found to be applicable, a higher percentage of our panel indicated that the descriptor applied.

TABLE 1.
ODORANTS FOR INITIAL PROFILING

Sample No.	Name	Sample No.	Name
A101	Acetophenone	A158	Dipropylene Glycol
A102	Anethole	A159	Skatole
A103	1-Butanol	A162	2-Ethylpyrazine
A104	l-Carvone	A164	Indole
A107	1-Heptanol	A165	
A108	1-Hexanol	A167	2,3-Dimethylpyrazine
A110	Pyridine	A169	Diethylphthalate
A114	Methyl Salicylate	A186	d-Limonene
A116	Thymol	A189	Isovaleric Acid
A127		A192	Diethyl Sulfide
A131	2,5-Dimethylpyrazine	A196	2-Acetylpyridine
A134	Vanillin	A197	Furfuryl Mercaptan
A135	Menthol	A200	1-Octen-3-ol
A145	DL-camphor	A202	Isoamyl Acetate
A147	Dibutylamine	A215	3-Hexanol
A150	Beta-Ionone	A239	Cyclotene
A152	Alpha-Pinene	A265	Phenylacetic Acid
A153	1-Octanol	A266	Dihydrocoumarin

Evaluation of Panelists

Data obtained for 25 compounds were used to determine whether there were any panelists who were performing differently from the rest. The correlation of each panelist's profile with the "average" profile was used for the analysis.

Analysis of variance showed that there were significant effects due to panelists and odorants. While panelists were performing differently, the differences depended on the odorant. Since not all panelists had evaluated each odorant, a regression analysis was done to estimate the effect of each panelist and each odorant on the correlations. For this model, each panelist's correlation with the "average" profile was used as the dependent variable. There was one independent variable for each compound ($=1$ if the dependent variable was based on that compound, $=0$ if not), and there was one independent variable for each panelist ($=1$ if the dependent variable was obtained from that panelist, $=0$ if not). By this analysis, coefficients were estimated for all compounds and panelists. Panelists with a high coefficient had done a better job of agreeing with the "average" profile than panelists with low coefficients. In this way, panelists could be ranked according to how well their data conformed to the total evaluation.

TABLE 2.
RESULTS OF THE PANEL FOR 14 SELECTED COMPOUNDS[a]

Compounds	Reproducibility (r)		Correlation with ASTM
	Our Panel	ASTM	
1-Heptanol[b]	.90	.92	.87
1-Hexanol	.92	.91	.85
1-Butanol[b]	.83	.93	.79
A127	.95		.94
Vanillin	.99		.93
A165	.92[c]		.92
Phenylacetic acid	.89[c]		.83
Pyridine	.93	.96	.93
2-Acetylpyridine	.91		.79
Furfuryl Mercaptan	.96 .95[d] .86[e]		.60
Thymol	.88		.81
Indole	.87[c]		.86
Dibutylamine	.77		.55
Acetophenone	.92	.96	.93
Mean ± 2σ	.90 ± .03		.83 ± .06
Mean ± 2σ over 35 compounds			.85 ± .04

a. 2 reps
b. 3 reps
c. 2 x's conc of 1st
d. 1/100 conc of 1st
e. 1/1000 conc of 1st

The odor profiles of some individual panelists were found to be in poor agreement with the "average" panel evaluation. However, no panelist provided an evaluation which correlated poorly with the "average" for every sample examined. In general, any statistical effect arising from individual profiles which differed from the "average" profile was ameliorated by an increase in the number of panelists. It was gratifying to learn that an increase in the number of evaluations for an odorant did improve the correlation of the data with that of the ASTM study. The performance of individual panelists continues to be monitored.

Effect of Concentration of Odorant

A valid odor profile for a particular compound can only be obtained if the quality of the perceived odor is not strongly concentration dependent. An earlier MDS study performed at this laboratory of four compounds over a 100-fold concentration range failed to show any dramatic concentration dependency. As an additional safeguard, several compounds have been profiled at different concentrations, usually varying by a factor of two. Furfuryl mercaptan was investigated over a 1000-fold dilution range. Only at the lowest concentration (2 ppm) did the correlation diminish slightly (0.86), probably because some panelists were having difficulty detecting the odor. This does not imply that we will never find compounds which do not differ in odor quality as the concentration is altered. For this reason the concentration of each compound is listed along with the descriptive analysis.

We have added an intensity score to the ballot to monitor perceived intensity. For cases in which the panel results indicate either too strong or too weak a stimulus, the odorant is repeated at an adjusted concentration.

Comparisons Among Odorants

Correlations. The correlation of profiles of two compounds over the 146 descriptors (not including intensity) has been used in the ASTM procedure as an indication of the similarity of the two odorants. This criterion provides a good first approximation of the similarity. However, since many of the 146 descriptors are intentionally redundant, the usefulness of this type of correlation is mitigated by the degree of redundancy and the fact that some descriptors will be more important to the overall differences among compounds than others.

Factor analysis. Factor analysis is a statistical technique which can be used to reduce a large number of variables (descriptors) to a smaller subset of variables (factors). Factor loadings can be used to identify those descriptors which are highly correlated (i.e., are used by panelists interchangeably). In this way, the compounds can be compared on a much smaller groups of independent variables. Although the strength of individual terms will still differ with the compound analyzed, this makes it much easier to determine on which descriptors a set of compounds vary. For factor analysis to provide meaningful information, the number of samples must be very large and cover the entire range of interest (in this case all possible odor types).

Factor analysis of the 146 descriptors was performed on 415 profiles from the ASTM data and our panel. While the results of factor analysis differed slightly depending on the compounds analyzed, the underlying terms have remained fairly constant. Table 3 lists the descriptors which contribute most of each factor. Grouping terms in one factor does not imply that the terms are identical, but that panelists are using those terms to describe some underlying sensory dimension. After a fairly large number of compound have been analyzed, a more precise grouping of terms should be achieved along with information concerning terms which are not useful to sensorially describing odorants. This should provide us with more information concerning the number of underlying sensory variables used in odor description (assuming that all possible sensory dimensions are covered by this list of terms). From this initial analysis approximately 17 nonredundant factors emerged each of which explained greater than 2% of the variation in the data. The total variation accounted for was 89%.

TABLE 3.
MAJOR DESCRIPTORS* ASSOCIATED WITH EACH FACTOR

Factor 1 Animal, Foul	Factor 2 Solvent	Factor 3 Non-citrus Fruit	Factor 4 Green	Factor 5 Nutty
rancid	cleaning fluid	non-citrus fruit	fresh green vegetable	popcorn
sweaty	paint	apple	crushed grass	nutty
dirty linen	varnish	pear	green pepper	peanut butter
cat urine	gas, solvent	strawberry	raw potato	grainy
putrid, foul	kerosene	peach	crushed weeds	meaty
sickening	chemical	melon	raw cucumber	fresh smoke
fecal	nail-polish remover	grape juice	beany	yeasty
sour milk	etherish	pineapple	cooked vegetables	---
animal	disinfectant	cherry	herbal, green	burnt, smoky
cadaverous	turpentine	raisins	musty, earthy, moldy	mouse
urine	alcoholic	fermented fruit	celery	
sour, vinegar	metal	banana		
sewer	tar	sweet		
wet wool	---	---		
blood, raw meat	medicinal			
sharp, pungent	raw rubber			
cheesy	mothballs			
oily, fatty	creosote			
heavy				
---				
fishy				
smoked fish				
stale				
sauerkraut				
mouse				

Factor 6 Brown	Factor 7 Floral	Factor 8 Cool, Minty	Factor 9 Burnt	Factor 10 Citrus
caramel	Lavender	Eucalyptus	burnt paper	lemon
maple	Violet	camphor	burnt candle	grapefruit
molasses	Rose	cool	sooty	citrus
buttery	perfumery	minty	burnt rubber	orange
honey	floral	medicinal	burnt, smoky	
chocolate	cologne	mothballs	burnt milk	
malty	fragrant		creosote	
coffee				
vanilla				
bakery				

Factor 11 Sulfidic	Factor 12 Woody	Factor 13 Spicy	Factor 14 Coconut, Almond	Factor 15 Leather, Rubber
garlic, onion	cedarwood	clove	coconut	leather
sulfidic	woody	cinnamon	almond	---
household gas	bark	spicy	---	rubber
---	cork		vanilla	
sewer	oakwood, cognac			

Factor 16 Fishy	Factor 17 Caraway, Anise
Fishy	Caraway
Smoked Fish	Anise

*Descriptors below the dashed lines indicate descriptors have factor loadings $\geq$ .4 but are in common with other factors.

Two sets of structurally related compounds have been profiled to test the discriminative ability of the odor profiling method. One set consisted of the simple aliphatic alcohols shown in Table 4. For alcohols differing only by one methylene unit (1-hexanol and 1-heptanol), analysis of the ASTM data showed they were only slightly different (0.89); the correlation for repeating the same odorant was 0.95/0.96. Data from our panel results gave similar results, (0.90 for the correlation between the homologues and 0.90/0.92 for the same compound). All of the other comparisons showed a greater degree of difference. Therefore, simple correlation data alone were not sufficient to differentiate reliably between 1-hexanol and 1-heptanol.

Results of factor analysis showed, however, that while 1-heptanol tended to be more citrus, 1-hexanol was considered more floral. For 1-butanol, scores for citrus and floral were low and high scores for solvent were given. For compounds which are very similar, repeat analyses need to be performed to provide a measure of the stability of the differences.

TABLE 4.
CORRELATIONS FOR ALCOHOLS

	1-Butanol	1-Hexanol	1-Heptanol	1-Octanol	3-Hexanol
1-Butanol	0.83				
1-Hexanol	0.76	0.92			
1-Heptanol	0.79	0.90	0.90		
1-Octanol	0.51	0.60	0.59	--	
3-Hexanol	0.83	0.72	0.76	.77	--

Differences Derived from Odor Profiling Versus Differences Derived from MDS

A more stringent test of the ability of odor profile data to differentiate between compounds is to compare the data gathered by odor profiling to a method which directly measures and maps the dissimilarity between compounds. Multidimensional scaling (MDS) is one method which has been used extensively for this purpose (Schiffman *et al.* 1981). In this method, panelists evaluate all pairs of compounds for similarity/dissimilarity. While this provides a direct measurement of the overall differences among samples, the underlying reasons for the differences are not elucidated. Also, while the MDS map may indicate that two compounds are very different along one dimension, this dimension may cover multiple sensory dimensions. For example, while two compounds are very different, it may be because one is fruity and the other is nutty rather than simply a difference on one of these terms.

Dissimilarity measures such as chi-square and euclidean distance using all descriptors as dimensions, have previously been successfully correlated to direct dissimilarity measurements of selected odorants having varying degrees of dissimilarity (Dravnieks et al. 1978). In this study, the profile data were used to estimate the differences among compounds in a set of very similar compounds. Factor analysis was first used to define orthogonal dimensions and determine on which dimensions the compounds differed.

To compare the data obtained by MDS to odor profiling, 14 compounds, which had previously been evaluated by MDS (Ennis and Seeman 1985), were also evaluated by the odor profiling panel. These compounds are listed in Table 5.

TABLE 5.
COMPOUNDS COMPARED BY MDS AND ODOR PROFILING

1	1,3-di-t-butylbenzene
2	1,3-diethylbenzene
3	2,6-di-t-butylpyridine
4	2,6-dimethylpyridine
5	2-ethyl-6-methylpyridine
6	2-isopropyl-6-methylpyridine
7	2-methylpyridine
8	2-t butyl-6-isopropylpyridine
9	1-ethyl-3-methylbenzene
10	meta-xylene
11	2,6-di-isopropylpyridine
12	2-t butyl-6-methylpyridine
13	2-t butyl-6-ethylpyridine
14	2,6-diethylpyridine

While the MDS data had indicated that the differences among these compounds could be explained by two dimensions (Fig. 1), factor analysis of the odor profiling data indicated that these compounds differed on five underlying variables. A factor was chosen if at least one compound received a high score on that factor (>2). These included: animal/foul, solvent, green, burnt, and rubber. Maps of the compounds on these dimensions are shown in Fig. 2. Similarity between the methods was estimated by calculating the correlation between data obtained by odor profiling and MDS. The correlations among compounds from odor profiling and direct similarity scores from MDS showed a correlation of 0.56 ($p < .0005$) while the euclidean distances from the MDS dimensions and the five factor scores on which the compounds differed showed a correlation of 0.74 ($p < .0005$). These correlations are as good as those found between the raw and massaged data from either method (similarity versus MDS distances = 0.76; correlation between compound and distances from factor scores = 0.79).

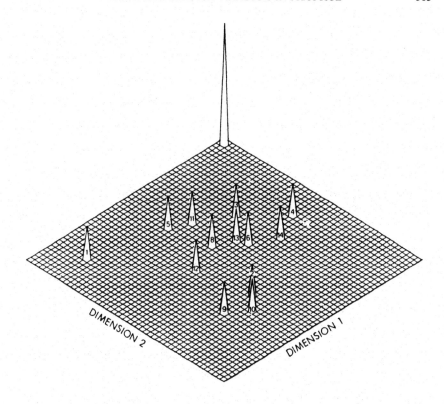

FIG. 1. MDS PLOT (ENNIS AND SEEMAN 1985) OF COMPOUNDS LISTED IN TABLE 5
This map represents a subset of 22 compounds which were originally compared in an MDS study. Compound 7 was deleted from this analysis due to a large standard deviation in the distance estimate.

The similarities between the two techniques are evident by comparing Fig. 1 and 2. For example, compound one is dissimilar from all other samples on the MDS map (Fig. 1) and on the rubber dimension (Fig. 2A). Compounds 2, 9 and 10 are similar to each other and set apart from other compounds (particularly 3, 4 and 5) on all dimensions of both figures. However, while the MDS map shows agreement with the odor profiling maps in representing degree of compound dissimilarity, it is evident, as previously suggested, that the two MDS dimensions cover multiple sensory dimensions which were pulled apart by factor analysis of the odor profiling data. Therefore, when sensory data is being used to relate to other measurements (e.g., analytical data), the MDS dimensions may be found to provide less useful information than data from odor profiling.

While this comparison of the two methods does not provide a validation of the data obtained by odor profiling, it does show that the two methods, with different ingoing assumptions, generated similar results.

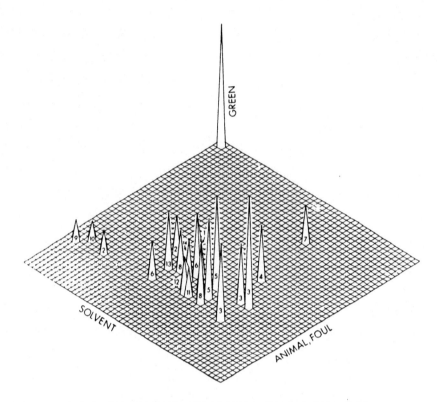

FIG. 2. FACTOR SCORE PLOTS OF THE FOURTEEN COMPOUNDS
WHICH HAD BEEN PREVIOUSLY COMPARED BY MDS
Names of each are listed in Table 5.

SUMMARY

The odor profiling procedure using a panel of approximately 20 was found to provide a good reliable method of defining the odor qualities of odorants. The information obtained by the odor profiling method can be statistically compared to determine compound similarities and differences. Statistical techniques such as factor analysis can be used for data simplification as well as a means of studying the underlying sensory dimensions of the odor space. The validity of this approach was tested by a comparison of results obtained by MDS. Therefore, the data obtained from odor profiling can be used to provide stable sensory descriptions of a compound and to estimate the degree of difference between compounds as well as the underlying sensory reasons for those differences.

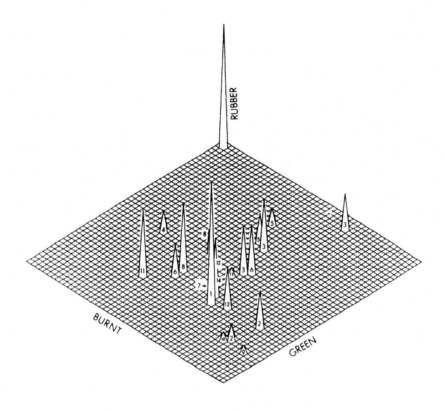

FIG. 2A.

ACKNOWLEDGMENTS

The authors would like to thank everyone who contributed to the collection of these data. This includes the panelists who dedicated their time, the chemists who supplied the compounds, and the computer specialists who aided in setting up this program. We would like to express special thanks to Lisa Eby and Archie Williams for collecting the MDS data which were used in the evaluation of the odor profiling procedure, to Drs. Daniel Ennis and Jeffrey Seeman for making available their unpublished results, and to John Tindall for his valuable advice and support.

REFERENCES

DRAVNIEKS, A., BOCK, F.C., POWERS, J.J., TIBBETTS, M. and FORD, M. 1978. Comparison of odors directly and through profiling. Chemical Senses and Flavour 3, 191.

DRAVNIEKS, A. 1982. Odor quality: Semantically generated multidimensional profiles are stable. Science 218, 799.

ENNIS, D. and SEEMAN, J. 1985. Unpublished results—personal communication. Philip Morris, Richmond, VA.

JELTEMA, M.A., JELTEMA, B.D. and SOUTHWICK, E.W. 1984. Automation of the sensory laboratory. Presented at 38th TCRC. Atlanta, GA.

SCHIFFMAN, S.S., REYNOLDS, M.L. and YOUNG, F.W. 1981. Introduction to Multidimensional Scaling. Academic Press, New York.

CHAPTER 6.4

COMPONENT AND FACTOR ANALYSIS APPLIED TO DESCRIPTORS FOR TEA SWEETENED WITH SUCROSE AND WITH SACCHARIN

NANCY M. ROGERS, ROLF E. BARGMANN and JOHN J. POWERS

*At the time of original publication,
author Rogers was affiliated with
Quik-to-Fix Products, Inc.,
Garland, Texas*

and

*authors Bargmann and Powers
were affiliated with
University of Georgia
Athens, GA 30602.*

ABSTRACT

Component and factor analysis were compared as means of reducing scaled values for 10 descriptors of tea to two dimensions so that relations among the attributes could be visualized. The association diagrams yielded by the two methods were generally similar, and there were significant differences between the diagrams for English Breakfast, an 80:20 blend of English Breakfast and Sencha teas, Ceylon and Darjeeling teas. Bartlett's test for equality of variance-covariance matrices was significant. When tea was equivalently sweetened with sucrose and with saccharin, the association diagrams were different. Preference and sweetness were highly correlated with each other and they had high communality with the other attributes when saccharin was the sweetening agent. When sucrose was the agent, sweetness still had high communality with the other attributes, but preference did not, indicating preference was the result of interplay of all the attributes rather than a dominant one. The correlation matrices were not significantly at $p = 0.05$, but they were at 0.10.

INTRODUCTION

Sensory profile analysis has grown apace during the last decade. While not an essential part of the sensory or statistical analysis, graphical representation of relations among the attributes of the products is useful for various purposes. When the sensory technologist merely wants to make differences readily apparent upon quick inspection, a spider-web diagram suffices. When the investigator wishes to depict more realistically relations which probably exist among the attributes, diagramming based upon some form of multivariate analysis is more likely to approach that goal. The purpose of this investigation was twofold; one objective was to compare, for the purpose of plotting, component and factor analysis (FA) as a means of reducing to two dimensions attributes originally examined in 10 (or 12) dimensions, and the second was to compare the effects of sucrose and saccharin on perception of attributes of tea, other than sweetness, the perceived sweetness level having been set to be equivalent.

Although spider-web diagramming is by far the most common graphical procedure, increasingly investigators are utilizing principal component analysis (PCA) of FA as a basis for depicting relations among variables. Jounela-Eriksson (1981), for example, constructed a spider-web diagram to illustrate differences in the profiles of Scotch, Bourbon and Canadian whiskies, but she applied PCA to study the relations of sensory attributes to certain types of compounds. Yoshikama *et al.* (1970), Harries *et al.* (1972), Baines (1976), Fjeldsenden *et al.* (1981), McLellan *et al.* (1983, 1984), and Brehn and Schultz (1986), among others, employed FA. Powers (1984) has cited several of the studies where either PCA or FA was utilized.

Component analysis and PA operate upon two different premises. In component analysis, the intent is to represent measurements made on many variables as a lesser number of "components" with as little change in the original data as possible. In FA, each variable is represented by two parts, the "common factor" and "unique factor" loading. The only requirements of the unique components is that they be uncorrelated. Discrepancies are not minimized (Thurstone 1947). Component analysis minimizes the differences in the residuals between the observed and reproduced scores; thus especially when reduction in rank from 10 (or 12) dimensions to 2 is effected, the reproduced correlations between attributes are likely to be changed. Additional steps should then be taken to make the reproduced correlations approximate the observed correlations. The customary technique of component analysis employs least squares, i.e., PCA, but in this study a mimimax approach (Baker 1976; Bargmann and Baker 1977) was employed. The mimimax process minimizes the largest discrepancy and results in a considerably smaller number of component errors than the least squares procedure.

MATERIALS AND METHODS

Materials and Terms

Initially 10 kinds of tea were used in the training of panelists to educate their palates to different sensory notes and to give them an idea of the intensity different attributes might have. Terms to describe attributes were taken from various glossaries and from Palmer (1974). The prospective panelists likewise suggested terms which might be appropriate. The terms ultimately chosen are listed in Table 1.

For the first phase of the trial, four teas were used: English Breakfast, an 80:20 blend of English Breakfast and Sencha teas, Ceylon and Darjeeling. For the second phase, sucrose or saccharin was added to the tea. The triangular test was used to determine the amounts of sucrose and of saccharin which would be perceived as being equivalent in sweetness. For sucrose, 4.5 g was added to 200 mL of freshly brewed tea; for saccharin, 12.3 mg of sodium saccharin was added. For this phase of the study, sweetness and preference were added as terms.

Preparation of Tea

The tea was made by steeping 2 g of tea in 100 mL of freshly boiled water for 3 min. The infusion were then immediately separated from the leaves by filtering through four layers of cheesecloth. The tea was held at 80C in stoppered flasks until used, but no longer than 1 h. The panelists received 20 mL of each sample.

Sensory Procedure

Only after a panelist had arrived were his (her) samples prepared. The panelists were instructed to commence evaluation immediately. Samples of the hot tea thus varied by no more than 1C. Neither cream nor lemon was added to the tea. The samples were coded with three-digit numbers. At each of the sessions one of the tea samples served as a reference as well as being included among the coded samples. For the phase involving a comparison of the four types of tea, the panelists thus evaluated four coded teas against the reference sample. For the trial involving unsweetened tea and tea sweetened with either saccharin or with sucrose, three samples were submitted to the panelists plus one of them as the reference tea. A balanced design was used so that for each of the two trials each tea under investigation eventually served as the reference an equal number of times. An objection to using a single product as the reference is that it is unlikely to be midway among the products for all the attributes being examined. Using different teas for the reference necessitated that adjustments be made in the scores. Bargmann (1981) devised a procedure (to be published separately) which compensated for different references having been used and for error induced by panelists assigning a different score to the coded reference than should have been assigned. Scaling was from -4 to +4 (Wu et al. 1977). The panelists were instructed to check off for each descriptor whether the intensity of that attribute was the same as the reference tea, stronger or weaker. The panelists did not know the identity of the reference sample nor that an identical sample was being inserted as a coded sample. For statistical analysis the scale was transformed to 1 to 9 with 0 on the original scale becoming 5.

For the comparison of the four kinds of tea, 20 panelists were used. For the sweetening-agent trial, 10 of the most effective of the 20 panelists were selected. Evaluation of the four kinds of tea was replicated eight times; the sweetener trial was replicated nine times.

Distilled water at 23C was provided for oral rinsing and Matso° crackers were provided to remove flavor from the mouth if the panelist so wished. Panelists were permitted to use either of both of the mouth-clearing substances, but they were instructed they had to consistently use the same procedure throughout the trial.

Statistical Analysis

Before proceeding to multivariate analysis all results were subjected to univariate analysis. In actuality, the MUDAID program (Applebaum and Bargmann 1967) which encompasses uni- and multivariate analysis in one pass was employed, but analysis of variance (ANOVA) and correlation results were examined first to search for outliers. The results of one panelist were deleted before the multivariate analysis phases because that panelist was an outlier for two of the terms.

Once the data had been examined univariately, the original data were reduced to a rank-two representation using the minimax procedure. The lengths of the vectors were then reduced without changing direction so that the scalar products between vectors approached the observed correlations. A computer program (Bargmann 1981) was used to decompose the rank-two matrix into its common and unique parts. In that sense, the differences between the observed and reproduced correlations were minimized by a weighted least squares procedure. For FA, the FCAN program of Bargmann (1975) was utilized. Hotelling's T^2 test was employed to test for differences over all the attributes. Bartlett's test was used to test for equality among the variance-covariance matrices.

RESULTS

When the Roy-Scheffe test for mean comparisons was applied to the means of the various attributes, the Darjeeling tea was clearly different from the other teas. English Breakfast tea was likewise different from the other teas. The 80:20 blend of English Breakfast and Sencha teas and the Ceylon tea differed in only five of the 10 attributes. Bartlett's test established that the variance-covariance matrices of the four teas differed significantly at $p \leq 0.05$. Table 1 lists the vector representations for the four teas when the correlation matrices were subjected to the minimax-component-communalities procedure. To save space, diagrams are not given. Were they, they would indicate that the attributes are grouped together in different ways. Comments relative to interpretation of association diagram will be reserved until the diagrams for the sweetened tea have been presented.

Figure 1 shows a comparison of relations among the attributes when the diagrams were derived via the component-communalities technique and by FA and when the tea was sweetened with saccharin, sucrose and was unsweetened. Application of Hotelling's T^2 test to the means of the attributes for the tea sweetened with saccharin and with sucrose indicated that there was not a significant difference between the teas. ANOVA had indicated however that the aroma and the fishy note of the teas did differ significantly and that some of the other attributes approached significance. When Bartlett's test was applied to the variance-covariance matrices, the level of significance was between 0.10 to 0.05; thus while not significant, there was some evidence that the equality of the correlation matrices was not the same. The component and FA analysis bore out that conclusion. Figure 1 shows that the association diagrams for the saccharin- and the sucrose-sweetened teas were different.

Unlike the spider-web diagram where the length of the vector is a measure of the magnitude of the mean, long vectors in the case of component analysis indicates that the attribute possesses certain properties in common with other attributes. The length of a vector is governed by the communality or coefficient of determination of the attribute. Short vectors show that the attribute has little communality with other attributes. The direction of short vectors should thus be ignored. Vectors which cluster together are highly correlated. The same applied to those at approximately 180°. Vectors at right angles are uncorrelated (orthogonal). The uppermost diagram to the left (an association diagram based on component analysis) indicates that preference and sweetness possess communality with other attributes and they are highly correlated with each other. A similar conclusion can be drawn as to flavor and pungency.

The second diagram from the top and to the left illustrates a different state of affairs. Preference no longer has a high communality with any one of the other attributes. Sweetness is still orthogonal to aroma; unlike preference,

TABLE 1.
VECTOR REPRESENTATION (X AND Y COORDINATES) FOR ENGLISH BREAKFAST, ENGLISH BREAKFAST-SENCHA, CEYLON AND DARJEELING TEAS USING THE MINIMAX COMPONENT ANALYSIS AND COMMUNALITIES PROGRAMS

Attribute	English-Breakfast		English Breakfast-Sencha		Ceylon		Darjeeling	
	x	y	x	y	x	y	x	y
1. Aroma	0.16	0.52	0.23	0.34	0.13	0.32	0.55	0.53
2. Flavor	0.84	-0.08	0.64	0.32	0.68	0.73	0.38	0.22
3. Pungency	0.90	0.02	0.74	0.19	0.62	-0.02	-0.21	0.62
4. Briskness	0.67	0.26	0.46	0.11	0.45	0.43	-0.16	0.26
5. Bitterness	0.88	-0.17	0.86	-0.21	0.89	-0.45	-0.65	0.47
6. Astringency	0.74	-0.07	0.80	-0.43	0.15	0.47	0.70	0.38
7. Color	0.15	-0.72	-0.21	0.24	0.06	-0.22	-0.23	0.05
8. Brightness	0.50	0.09	0.24	0.51	-0.01	0.02	0.30	0.14
9. Fishiness	0.71	-0.19	0.19	-0.20	0.08	0.24	0.11	0.45
10. Smokiness	0.67	0.32	0.30	-0.33	0.31	-0.11	-0.35	0.29

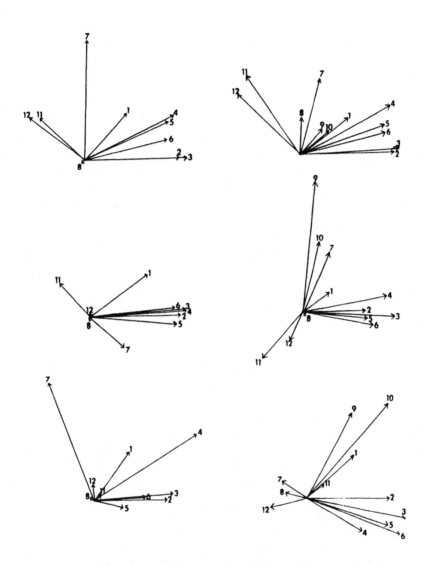

FIG. 1. ASSOCIATION DIAGRAMS FOR ATTRIBUTES OF TEA

The three diagrams at the left were derived via the minimax-components-communalities procedure. The topmost diagram is for saccharin-sweetened tea, the middle one for sucrose-sweetened tea and the bottom one for unsweetened tea. The three diagrams to the right were derived via factor analysis. From top to bottom, the identities of the tea are the same as for those at the left.

The attributes are coded as: 1 = aroma, 2 = flavor, 3 = pungency, 4 = briskness, 5 = bitterness, 6 = astringency, 7 = color, 8 = brightness, 9 = fishiness, 10 = smokiness, 11 = sweetness and 12 = preference.

it has high communality with the other attributes. The role of the sucrose seems to be more a melding agent for other attributes than a dominant factor in determining preference. The lowest diagram to the left shows that preference for unsweetened tea has low communality indicating that in the absence of a sweetener, preference is not highly correlated with any one of the other attributes. In other words, the interplay among all of them is the force that determines preferences. It may be noted that flavor, pungency and astringency have high communality and are highly correlated with each other.

The diagram to the right at the top is based on FA. When the component analysis was conducted, two of the attributes (fishy and smoky) were dropped because the minimax program can handle only 10 factors. They are included in the FA diagrams since the FCAN program has no limitation as to the number of variables allowed. The FA patterns bear out the component patterns. Sweetness and preference are highly correlated and orthogonal to the attributes, fishy, smoky and aroma. By the component procedure, preference and sweetness were likewise orthogonal to aroma. The FA graph in the middle to the right tells almost the same story for sucrose as its component counterpart to the left. The FA association diagram for unsweetened tea at the bottom to the right differs somewhat from its component counterpart to the left, but there are several similarities.

DISCUSSION

While there are some discrepancies between the components and the FA association diagrams, both have the merit of probably representing relations among the attributes more truly than any spider-web diagram can do. There is without doubt some distortion between the "true" relations and the rank-two representations. A reduction in rank for 10 to 2 is drastic and likely to distort relations among the variables some. Provided the same procedure and the same variables are used, departures from the "true" relations between variables probably will be comparable; thus the resulting patterns, if distorted at all, are likely to be equally affected.

In a lucid and probing presentation of elements affecting associations among products and attributes, Vuataz (1977) pointed out how correlations over the tasters and the disturbances may affect the information about the product. He considered that the planar projections of the actual configurations and structures do not provide a distorted but only incomplete information. He explained that the information can be completed by obtaining projections on other planes. Later, Vuataz (1981) discussed the subject further.

One advantage the minimax-component-communalities procedure has is that the error is less (Table 2). When compared with the original correlation

TABLE 2.
MAXIMUM DISCREPANCIES FOR PANELS I AND II (OBSERVED RATINGS - RANK TWO APPROXIMATIONS)

	Product	Maximum Error (Minimax)	Maximum Error (Principal Component)
Panel I	English Breakfast	.788	1.220
	English Breakfast-Sencha blend (80:20)	.822	1.316
	Ceylon	.994	1.835
	Darjeeling	.999	1.794
Panel II	Tea sweetened with saccharin	.928	1.535
	Tea sweetened with sugar	1.235	1.650
	Unsweetened tea	.966	1.849

matrices, the components-communalities process resulted in relationships between attributes being closer than by the FA procedure. A disadvantage the minimax program has is that it permits only 50 experimental units and 10 variables to be handled. FA programs are more flexible; they can encompass a greater number of variables and any FA computer program can be used to obtain factor loading to plot the association diagrams. For interpretation they must be used in conjunction with the correlation matrices.

If general statistical packages are used instead of the minimax program employed here, association diagrams may still be derived but the error involved in reducing the rank will be greater.

Although the chief objective of the study was to compare the component and the FA procedures, the study did bear out a situation often encountered by those seeking to reformulate foods. Even though the sweetness level was set the same (the sweetness scores for saccharin- and sucrose-sweetened tea were almost identical), the panelist's perception of the tea was different. The bitterness score was higher (Helgren et al. 1955; Hyvonen et al. 1978) for the saccharin-treated tea but not significantly so. Neither from Hotelling's T^2 test nor the fact only two attributes differed significantly would one claim that the two sweetened teas differed in any major way; nonetheless the panelist's perception of them was considerably different as the association diagrams indicate. In reformulating foods, instances are encountered where panelists can detect a difference between the products but none of the attributes being quantitatively scaled are significantly different from each other. One explanation is that the panelists are homing in on an attribute other than the ones being quantified. Another is that the replacement ingredient causes changes for reasons other than its own sensory characteristics. Here, for example because of the difference in the amounts of sweeteners added, there could have been differences in the vapor pressure. In fact, ANOVA revealed that the means for total aroma and fishiness were significantly different. Examination of the diagrams derived by FA (top and middle diagrams to the right) shows that the communalities and the correlations between total aroma, fishiness and smokiness did differ. Although none of the attributes arising from nonvolatiles were significantly different, the association diagrams for them varied too. Note that among the attributes, flavor, pungency, briskness, bitterness and astringency of the sucrose-sweetened tea, the attributes are more highly correlated with each other and their communalities vary somewhat from the corresponding attribute of the tea sweetened with saccharin. Though the means themselves say otherwise, minor differences in the ratios of the intensity values seem to have caused the panelists to perceive the teas differently. The correlation matrices, the basic source of the association diagrams, while not significantly different at the 0.05 level, were at the 0.10 level. Upon that basis, the differences observed in the association diagrams probably are not spurious relations but represent real differences in perception.

ACKNOWLEDGMENTS

Appreciation is expressed to the Coca-Cola Company, Atlanta, GA, for providing a graduate fellowship under which this study was conducted.

REFERENCES

APPLEBAUM, M. and BARGMANN, R.E. 1967. A Fortran 11 program for MUDAID, multivariate, univariate and discriminant analysis of irregular data. Tech. Rept. NONR 1834 (39), Univ. of Illinois, Urbana.

BAINES, E. 1976. Evaluation of flavours in dental creams. J. Soc. Chem. 27, 271-287.

BAKER, F.D. 1976. A minimax approach to data analysis. Ph.D. dissertation, University of Georgia, Athens, GA 30602.

BARGMANN, R.E. 1975. FCAN user guide. Dept. of Statistics, Univ. of Georgia.

BARGMANN, R.E. 1981. Communalities program. Data set COMUN on UGREBI on Cyber. Dept. of Statistics, Univ. of Georgia, Athena, GA.

BARGMANN, R.E. and BAKER, F.D. 1977. A minimax approach to component analysis. In *Applications of Statistics*, (P.R. Krishnaiah, ed.) pp. 55-69.

BRUHN, C.M. and SCHUTZ, H.C. 1986. Consumer perceptions of dairy and related-use foods. Food Technol. 40, 79-85.

FJELDSENDEN, B., MARTENS, M. and RUSSWURM, H. JR. 1981. Sensory quality criteria of carrots, swedes and cauliflower. Lebensm. Wiss. u. Technol. 14, 237-241.

HARRIES, J.M., RHODES, D.N. and CHRYSTALL, B.B. 1972. Meat texture. I. Subjective assessment of the texture of cooked beef. J. Text. Studies 3, 101-114.

HELGREN, F.J., LYNCH, M.J. and KIRCHMEYER, F.J. 1955. A taste panel study of the saccharin "off-taste". J. Am. Pharm. Assoc. 44, 353-355, 422-446.

HYVÖNEN, L., KRUKELA, R., KOIVISTOINEN, P. and RATILAINEN, A. 1978. Sweetening of coffee and tea with fructose-saccharin mixtures. J. Food Sci. 43, 1677-1685.

JOUNELA-ERIKSSON, P. 1981. Predictive values of sensory and analytical data for distilled beverages. In *Flavour '81'*, (P. Schreier, ed.) pp. 145-164, Walter de Gruyter & Co., Berlin.

LYON, B.G. 1980. Sensory profiling of canned boned chicken; sensory examination procedures and data analysis. J. Food Sci. 45, 1341-1346.

MCLELLAN, M.R., CASH, J.N. and GRAY, J.I. 1983. Characterization of the aroma of carrots (Daucus carote L.) with the use of factor analysis. J Food Sci. *48*, 72-74.

MCLELLAN, M.R., WIND, L.R. and KIME, R.W. 1984. Determination of sensory components accounting for intervarietal variation in apple sauce and slices using factor analysis. J. Food Sci. *49*, 751-755.

PALMER, D.H. 1974. Multivariate analysis of flavour terms used by experts and non-experts for describing tea. J. Sci. Fd. Agric. *25*, 153-164.

POWERS, JOHN J. 1984. Current practices and applications of descriptive methods. In *Sensory Analysis of Foods*, (J.R. Piggott, ed.) pp. 179-242, Elsevier Applied Sci. Publishers, London.

THURSTONE, L.L. 1947. Multiple Factor Analysis. 535 pp. University of Chicago Press.

VUATAZ, L. 1977. Some points on methodology in multidimensional data analysis as applied to sensory evaluation. Nestle Research News 1976/1977, pp. 57-71. Tech. Documentation Center, Lausanne, Switzerland.

VUATAZ, L. 1981. Information about products and individuals in multicriteria description of food products. In *Criteria of Food Acceptance*, (J. Solms and R.L. Hall, eds.) pp. 429-446, Forster Verlag AG, Zurich, Switzerland.

WU, L.S., BARGMANN, R.E. and POWERS, J.J. 1977. Factor analysis applied to wine descriptors. J. Food Sci. *42*, 944-952.

YOSHIKAWA, S., NISHIMARUA, S., TASHIRO, T. and YOSHIDA, M. 1970. Collection and classification of words for description of food texture. J. Text. Studies *1*, 452-463.

CHAPTER 6.5

INTENSITY VARIATION DESCRIPTIVE METHODOLOGY: DEVELOPMENT AND APPLICATION OF A NEW SENSORY EVALUATION TECHNIQUE

HARVEY H. GORDIN

*At the time of original publication,
the author was affiliated with
R.J. Reynolds Tobacco Company
Bowman Gray Technical Center
Winston-Salem, NC 27102.*

ABSTRACT

The Intensity Variation Descriptive Methodology (IVDM) is a modification of the descriptive analysis procedure commonly used in the sensory evaluation of foods. The IVDM technique provides multiple attribute evaluations within specified sections of a product sample. In this study, the IVDM technique was used to quantify changes in attribute intensity as a cigarette sample was consumed.

A touch sensitive Hewlett Packard 150 Personal Computer was used as the data collection instrument. Proprietary software was developed which created a graphic representation of an unstructured line scale on the computer screen. Subjects indicated the intensity of sensory attributes, within various sections of tobacco rod, by touching the computer screen anywhere along the line scale. Data obtained from IVDM evaluations indicated that this method was sensitive to the quantitative changes in cigarette smoke components during consumption of the product.

The program developed for IVDM testing may be modified for conventional descriptive testing. The application of computers as sensory evaluation data collection instruments was explored.

INTRODUCTION

The descriptive analysis technique used in the sensory evaluation of foods and other consumer products provides a detailed profile of product attribute intensities. The information provided by descriptive testing is useful in comparing different product formulations, and competitive products. The conventional descriptive technique does not however, provide the sensory analyst with any information concerning the changes in attribute intensity which occur as a product is consumed. The Time-Intensity method described by Larson-Powers and Pangborn (1978) and later by Guinard *et al.* (1985) provides multiple attribute intensity measurements at various time intervals during product evaluation. These methods generally focus on a single product characteristic and the decay or increase in its' intensity over time.

The complete evaluation of a cigarette product requires a unique sensory testing methodology. It is known that the quantity of cigarette smoke components change as the product is consumed (Davis *et al.* 1973). Conventional descriptive tests which offer singular evaluation of the product do not reveal the perceptual changes which occur during the smoking process. Human smoker variability causes subjects to smoke products very differently (Thorton 1978), for this reason a time-based method such as Time-Intensity would not be appropriate for sensory evaluation of cigarette products, as different portions of the sample would be evaluated by the various tests subjects within the same time interval. A method was developed that concentrated the subjects evaluations on specific locations of the product. The method used a touch-sensitive computer screen as both the ballot and data collection device. Software was developed for this application. Modifications of the software would allow it to be used for conventional descriptive testing, or time-intensity testing. The major advantage of the computerized test method is the elimination of the paper ballot used in conventional tests. Replacing the paper ballot with a touch screen computer increased the efficiency of the testing operation since the digitizing of sensory data was not required. In time or section intensity studies, a computer ballot removes the bias present when subjects are allowed to view their previous attribute evaluations.

MATERIALS AND METHODS

Cigarette Sample

The product evaluated in the study was an 85 mm Full Flavor cigarette, consisting of a typical American blend, and a conventional filtration system. The product was commercially available.

Prior to sensory evaluation, the product was divided into six sections by drawing lines on the tobacco rod with a marking pen. The first section of the product began at the fire-end of the cigarette and included the first 10 mm of the tobacco rod. Each subsequent section comprised 7 mm of the tobacco rod.

Analytical Method

The cigarette sample was conditioned and smoked by the standards FTC method (35 cc puff, 2 second duration, 1 puff per minute) described by Pillsbury (Pillsbury et al. 1969). The smoke was collected on Cambridge pads. The Cambridge pads were extracted and analyzed for moisture and nicotine by gas chromatography. Total dry particulate matter (TPM) was calculated as wet TPM minus nicotine and water. Carbon monoxide (CO) was determined by collecting the gas phase of the mainstream smoke and analyzing for CO by nondispersive infrared spectrophotometry (Williams 1980). Moisture was determined by accurately weighing samples prior to and after heating the samples in a forced-draft oven for 3 h at 214F. Moisture values were determined as the loss in weight exhibited by the samples, based on the methods of Moseley et al. (1951). Tobacco nicotine was determined by a method very similar to that of Harvey and Handy (1981). Puff-by-puff determinations of 'tar' and nicotine were determined by a method similar to Davis (Davis et al. 1973).

Sensory Method

Panel Development. The IVDM technique is a modification of the Quantitative Descriptive Analysis Method (QDA) described by Stone et al. (1980). The IVDM panel consisted of twenty experienced descriptive subjects. The subjects had previously participated in language development sessions and had received training in the use of the unstructured line scale. The software used for the IVDM test was designed to display line scales for four sensory attributes at a time. Additional attributes could appear on subsequent screens. During panel development, the subjects decided that because of the limited amount of product evaluated within a section of the product, the number of attributes to be evaluated should be limited to four. Training sessions were held with the subjects to determine the attributes most appropriate for quantifying the changes in sensory perceptions which occur during the smoking experience. The attributes selected by the panel were: Concentration of Smoke, Tobacco Taste, Impact, and Drying Sensation. In addition to providing the needed language, training sessions allowed the IVDM subjects to become familiar with the testing procedure and the computer ballot.

Software Requirements. A proprietary software system was developed specifically for IVDM testing. The software was written for the Hewlett Packard 150 Personal Computer in the C language. The IVDM software provided the following capabilities:

Ballot Creation. The program generated a graphic representation, on the computer screen, of the line scale used in conventional descriptive tests (Stone and Sidel 1985). Additionally the software allowed the user, i.e. the individual(s) who are designing the study, to title each line scale with an attribute name, and place descriptors at each anchor, thereby creating a complete ballot (Fig. 1). The number of line scales which may appear on the computer screen at one time was limited to four. The user may request any number of additional ballot screens to allow for additional attribute evaluations, or as was done in this study, to obtain replicate evaluations of the attributes in the various sections of the product.

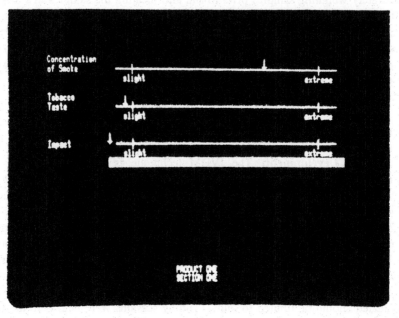

FIG. 1. COMPUTER BALLOT

In addition to creating the ballot, this portion of the program allowed the user to create a screen of text. This was used to provide the sensory subject with test instructions.

Test Design. The program facilitated the design and administration of the sensory test. The user entered the subjects name and the appropriate product serving order. The program generated unique three digit code numbers which correspond to the particular subject and product. At the beginning of a test session, a touch activated key pad was displayed on the computer screen. The subject entered the code number of the product into the computer using the graphic key pad (Fig. 2). Once the three digit code was entered, the program verified that the code number was valid for the particular subject. The key pad display may be repeated, allowing numerous products to be evaluated during a testing session.

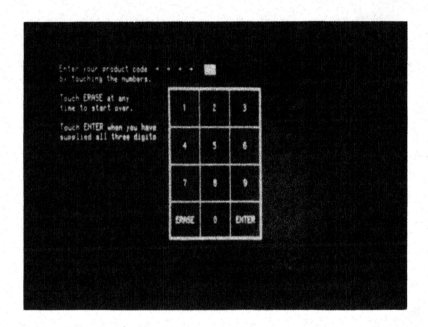

FIG. 2. GRAPHIC KEY PAD

Data Files. When a subject evaluated the intensity of the product, for one of the four sensory characteristics, by touching the line scale, the program converted the location touched to a value from 1 (slight) to 60 (extreme). The numerical values of attribute intensity were recorded in a data file. Also contained in the data file were the subjects' name, the date and time of the test session, and the identity of the product(s) evaluated.

Timing Device. The ability to time certain activities during product evaluation was also provided by the program. This was useful for timing the rest period between the evaluation of aftertaste dimensions. The IVDM software displayed a "clock" in the form of a circle (Fig. 3). Sections of the clock were removed (Fig. 4) until at the end of a specified time interval, the clock was completely removed from the screen. The length of the time delay, the size and location of the "clock" were determined by the user. The insertion of the clock is allowed anytime during a test session.

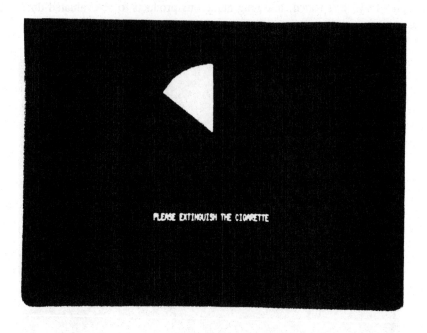

FIG. 3. TIMING DEVICE

Test Procedure. The subjects were situated in temporary booths constructed of a foam core material, which exposed only the terminal touchscreen (Fig. 5). The booths prevented the subjects from observing the evaluations of others. For this study, four test stations were available in the sensory facility. The subjects were given a sample, which was divided into six sections. The sample was contained in a capped, glass, screw-topped test tube. A 3-digit product code number, printed on an adhesive label was affixed to the test tube.

At the beginning of a session the subject entered the product code number into the computer by using the graphic key pad. Once the validity of the sample code had been determined, the first line scale for the evaluation of Concentration of Smoke was displayed. After entering the intensity of Concentration of Smoke, the line scale for Tobacco Taste was displayed. This sequence was followed through the evaluation of Drying Sensation.

FIG. 4. TIMING DEVICE

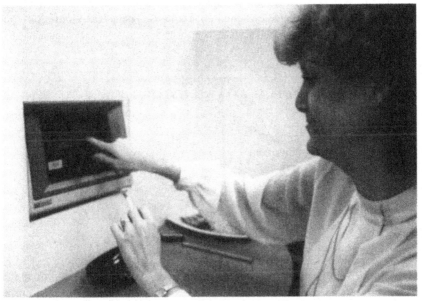

FIG. 5. SENSORY SUBJECT IN TEST BOOTH

After all four attributes were evaluated, the screen was cleared, when the fire-cone reached the second section of the product, the product was evaluated again with reference to the same four attributes. This sequence was repeated until all six sections of the sample were evaluated. The evaluation of a second sample followed a three minute rest period. The IVDM software provided a three minute clock to enforce this rest period. During the break, the subjects were provided with breadsticks and water to cleanse their mouths prior to evaluation of the second sample.

Data Analysis. The IVDM software recorded all test responses on a 3.5 in. computer disc. This information was transferred to the IBM computer system for statistical analysis using the Statistical Analysis System (SAS). The SAS procedure treated the individual sections of the products as main effects in the analysis of variance model, Duncan's multiple range test was employed to determine specific section differences. The SAS analysis also provided mean scores and standard deviations for each subject within each product section.

RESULTS

Cigarette Analysis

The data obtained from the analysis of the sample, as determined by the standard FTC methods, is contained in Table 1. The puff-by-puff 'tar' and nicotine data is contained in Table 2.

TABLE 1.
ANALYTICAL DATA FOR TEST CIGARETTES

Puff Count	9.00
TPM (mg/Cigarette)	20.50
Nicotine (mg/Cigarette)	2.28
"Tar" (mg/Cigarette)	16.60
CO (mg/Cigarette)	13.85
Moisture (%)	12.08

Sensory Data Analysis

Within each attribute, an analysis of variance test was performed on the individual product sections, specific differences were determined by the Duncan test. The results of the statistical analysis, for each of the attributes follows:

TABLE 2.
PUFF BY PUFF ANALYSIS OF TEST CIGARETTES

Puff Number	Nicotine (mg/Cigarette)	'Tar' (mg/Cigarette)
1	0.0480	0.630
2	0.0850	1.230
3	0.1040	1.410
4	0.1090	1.520
5	0.1190	1.670
6	0.1300	1.810
7	0.1390	1.960
8	0.1440	1.990
9	0.1520	2.210

Concentration of Smoke. The sections of the product were significantly different at $p=0.12$. Section one of the product had significantly less Concentration of Smoke than section six.

Tobacco Taste. No significant differences ($p=0.63$) were noted among the sections of the product.

Impact. Significant differences in the intensity of Impact were noted among the sections of the product ($p=0.0001$). Section one was perceived as having significantly less Impact than sections three, four, five, and six. Section two was perceived as having significantly less Impact than sections four, five, and six.

Drying Sensation. Significant differences in the Intensity of Drying Sensation were perceived among the sections of the product ($p=0.003$). Sections one and two were perceived as having significantly less Drying Sensation than section six.

DISCUSSION

The sensitivity of the IVDM technique to changes in attribute intensity which occur during the smoking experience was demonstrated. For the attri-

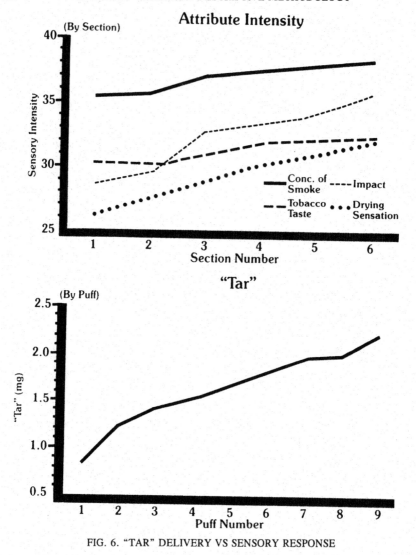

FIG. 6. "TAR" DELIVERY VS SENSORY RESPONSE

butes, Concentration of Smoke, Impact, and Drying Sensation, the increase in attribute intensity from first to final puffs, corresponded to increases in puff-by-puff 'tar' and nicotine delivery. The intensity of Tobacco Taste did not change significantly from the initial to final puffs. It is assumed that the perception of this attribute is not related to the quantity of 'tar' and nicotine available in the mainstream cigarette smoke.

In Fig. 6 and 7 the mean sensory response to the four attributes is compared to the smoke chemistry data.

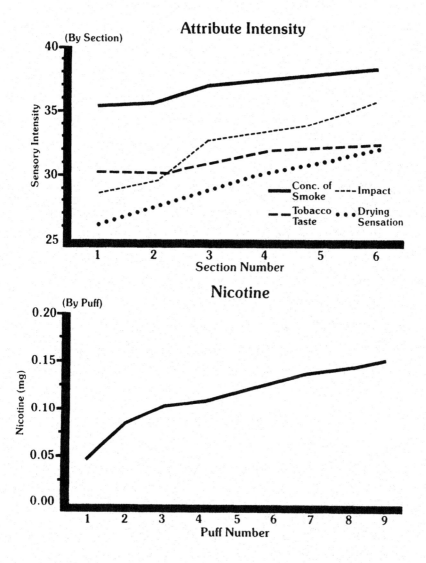

FIG. 7. NICOTINE DELIVERY VS SENSORY RESPONSE

The IVDM technique provided attribute evaluations within specific sections of the cigarette tobacco rod, unlike the Time-Intensity method which provides evaluations during specific time intervals. Since the amount of a cigarette consumed by an individual smoker is highly variable, a time based method of evaluation, such as Time-Intensity, would not have linked sensory judgments to specific sections of the product.

The computer program developed for IVDM may be modified for use in conventional descriptive tests, such as QDA. The use of the touch screen computer as the data collection device eliminates the time and cost involved with preparing paper ballots. Data entry is automatic, therefore the digitizing of sensory data prior to statistical analysis is not required. As indicated in this study, a computerized method would also facilitate timing between samples, and provided subject/sample verification.

The use of computers in conducting time-intensity studies would prevent the subjects' from viewing previous evaluations, therefore, the anticipation of the directional movement in attribute intensity would be minimized.

Numerous applications within the sensory laboratory been improved by the use of computers (Savoca 1984). The IVDM approach is another example of how the use of computers can increase the accuracy and efficiency of a modern sensory facility.

ACKNOWLEDGMENTS

The author wishes to thank Ms. Anne Syvarth, Ms. Linda Hill, and Mr. Cliff Keith (R.J. Reynolds Tobacco Co.), Dr. Herb Stone, and Mr. Joel Sidel (Tragon Corp.), Mr. Jack Armstrong (Armstrong Assoc.) for their assistance in the development of the IVDM technique.

REFERENCES

DAVIS, B.R., HOUSEMAN, T.R. and RODERICK, H.R. 1973. Studies of cigarette smoke transfer using radioisotopically labeled tobacco constituents. Part III: The use of Dotriacontane-16, 17-14C as a marker for the deposition of cigarette smoke in the respiratory system of experimental animals. Beitrage zur tabakforschung Int. 7(3), 148.

GUINARD, J.X. and PANGBORN, R.M. 1985. Computerized procedure for time-intensity sensory measurements. J. Food Sci. 50, 543-546.

HARVEY, W.R. and HANDY, B.M. 1981. On-Line generation of cyanogen chloride as a replacement reagent for cyanogen bromide in the total alkaloids determination. Tobacco Sci. 45, 131.

LARSON-POWERS, N. and PANGBORN, R.M. 1978. Paired comparison and time-intensity measurements of the sensory properties of beverages and gelatins containing sucrose of synthetic sweeteners. J. Food Sci. 43(1), 41.

MOSELEY, J.M., HARLAN, W.R. and HAMMER, H.R. 1951. Burley tobacco–relation of the nitrogenous fractions to smoking quality. Ind. Eng. Chem. 43, 2342.

PILLSBURY, H.C., BRIGHT, C.C., O'CONNOR and IRISH, F.W. 1969. "Tar" and nicotine in cigarette smoke. J. A.O.A.C. *52*(3), 458.

SAVOCA, M.R. 1984. Computer applications in descriptive testing. Food Technol. *38*(9), 74.

STONE, H., SIDEL, J.L. and BLOOMQUIST, J. 1980. Quantitative descriptive analysis. Cereal Foods World *25*(10).

STONE, H. and SIDEL, J.L. 1985. *Sensory Evaluation Practices*. Academic Press, San Diego, CA.

THORNTON, R.E. (Ed.) 1978. *Smoking Behavior – Physiological and Psychological Influences*. Churchill-Livingstone, London.

WILLIAMS, T.B. 1980. The determination of nitric oxide and gas phase cigarette smoke by nondispersive infrared analysis. Beitrage zur Tabakforschung Int. *10*(2), 91.

CHAPTER 6.6

DEVELOPMENT OF A LEXICON FOR THE DESCRIPTION OF PEANUT FLAVOR

PETER B. JOHNSEN[1], GAIL VANCE CIVILLE[2], JOHN R. VERCELLOTTI[3],
TIMOTHY H. SANDERS[4] and CLARE A. DUS[2]

*At the time of original publication,
author Johnsen was affiliated with
Monell Chemical Senses Center
3500 Market Street
Philadelphia, PA 19104.*

*Authors Civille and Dus were affiliated with
Sensory Spectrum, Inc.
44 Brentwood Drive
East Hanover, NJ 07936.*

*Also, author Vercellotti was affiliated with
USDA-ARS-Southern Regional Research Center
P.O. Box 19687
New Orleans, LA 70179*

and

*author Sanders was affiliated with
USDA-ARS-National Peanut Research Laboratory
1011 Forrester Drive
Dawson, GA 31742.*

ABSTRACT

A lexicon of terms to describe desirable as well as undesirable flavors in peanuts has been developed. The lexicon and an intensity rating scale was developed by a 13 member panel of flavor and peanut specialists representing industry and the USDA-Agricultural Research Service. This system is intended to provide definitive, common terminology for use in communicating differences in peanut flavor variables among all phases of peanut research and industry.

---

[1]Send correspondence to: Peter B. Johnsen, USDA-ARS-Southern Regional Research Center, P.O. Box 19687, New Orleans, LA 70179.

*Reprinted with permission of Food & Nutrition Press, Inc., Trumbull, Connecticut.
©Copyright 1988. Originally published in* Journal of Sensory Studies *3, 9–17.*

INTRODUCTION

In response to the designation of "off-flavor" as a significant researchable problem critical to the peanut industry, USDA-ARS initiated flavor quality research programs at the Southern Regional Research Center, New Orleans, LA (SRRC) and the National Peanut Research Laboratory, Dawson, GA (NPRL). A critical element for these programs was the establishment of a concise, industry acceptable, expandable peanut flavor lexicon. The development of that lexicon is reported here.

This lexicon, used to document the sensory characteristics of peanuts, is based on the sensory descriptive analysis method (Cairncross and Sjostrom 1950; Caul 1957; Brandt et al. 1963; Civille and Szczesniak 1973; Stone et al. 1974; Meilgaard et al. 1987).

Earlier methods used to define peanut flavor quality lack sufficient descriptive capabilities. In 1971, the CLER (Critical Laboratory Evaluation Roast) Method was introduced as a quality measurement method (Holaday 1971). This procedure combines good flavor plus off-flavor on one continuum. It also combines qualitative and quantitative scales on one continuum (that is, the use of "good", "bad" and "low" on one scale). Similarly the roast scale does not reflect directly the roasted flavor of a sample because it mixes hedonics (excellent and good) with roast level (under-roasted and over-roasted).

Since its 1971 introduction, the CLER method has been revised (Fletcher 1987) and these discrepancies have been eliminated. Although the revised CLER method does provide a measure of quality, as an integrated system it only provides an indication of specific flavor changes in a supplemental comments section.

Previous descriptive lexicons for peanut flavor are incomplete in some additional aspects. Oupadissakoon and Young (1984) developed a lexicon for describing roasted peanut flavor which is lacking in terms relating to degree of roast. Because all peanuts in a lot do not roast in the same manner, it is necessary to separate the degree of roast component (light roast, medium roast and dark roast) in order to respond to the presence of each of the roast levels mixed in a sample of nuts.

Syarief et al. (1985) used principal component analysis to study the underlying dimensions of the sensory flavor characteristics of roasted peanuts and peanut butter. The off-flavor terminology that they developed is limited to oxidized, mold, earthy, and petroleum. Unfortunately, the single term "oxidized" does not differentiate between the distinct flavor perceptions that are generated at earlier and later stages of the oxidation process.

Syarief et al. (1985) also suggest reducing the descriptive terminology developed by Oupadissakoon and Young (1984) for peanut butters to seven words. Missing from this limited lexicon, however, are descriptors for the sweet/caramel character as well as terms describing the various off flavors.

Civille and Lawless (1986) outline the characteristics of a complete lexicon. Terminology must be orthogonal, based on underlying structure, based on a broad reference set, precisely defined, and primary. It was believed, therefore, that existing lexicons required additional terms and clarification of others.

The peanut flavor descriptive lexicon is intended to provide a means of communication within the peanut industry. For researchers the terminology may be used to associate flavor attributes with both biological treatment variables and the subsequent chemical analyses. For growers and various handlers/processors the lexicon presents a means to communicate quality issues related to flavor beyond the hedonic "good/bad" or "like/dislike" responses. For manufacturers the lexicon can provide a common language for communication with suppliers, within the company (R&D to Marketing) and ultimately with the consumer.

MATERIALS AND METHODS

As Johnsen and Civille (1986) outlined in their work with warmed-over flavor of beef, an effective method for establishing a lexicon is to gather a panel of flavor experts in the industry to contribute terminology for an array of both desirable and "off-flavor" samples.

Panel Formation

Contracted by the USDA-ARS-SRRC, the Monell Chemical Senses Center convened a panel to develop the peanut lexicon. The panel was comprised predominantly of industry personnel who are involved in the daily assessment of peanut flavor for consumer products made with peanuts. Other panelists included USDA-ARS scientists involved in peanut quality research. Table 1 lists the panel members and their affiliations.

Sample Identification

Eighteen peanut samples (*Arachis hypogea*, L. cultivar Florunner) were obtained through efforts made by the National Peanut Council, NPRL, SRRC and private industry. Samples were prepared by the NPRL and SRRC following the methods described in "Sample Preparation".

TABLE 1.
PANEL PARTICIPANTS

Representative	Affiliation
Karen Carter	Beatrice
Karen Heuther	Nabisco Brands
Judy Heylmun	Nabisco Brands
Martha Holland	Azar Nut
Jack Pearson	USDA-ARS-NPRL
Tim Sanders	USDA-ARS-NPRL
Art Schmidt	Procter & Gamble
Fred Smith	Snackmaster, M&M/Mars
Lisa Thompson	Best Foods, CPC
Margaret Twomey	Hershey Chocolate
John Vercellotti	USDA-ARS-SRRC
Carolyn Vinnett	USDA-ARS-SRRC
Suzanne Whitlock	Hershey Chocolate

All samples were screened by P. Johnson (Monell), G. Civille (Sensory Spectrum), T. Sanders and J. Pearson (NPRL), and J. Vercellotti (SRRC). The screening team identified a roast level which optimized flavor expression, as well as, four roast levels to represent "very light", "light", "dark" and "very dark" samples needed to generate baseline peanut words. Also identified by the team were eleven samples representing an array of off flavors.

Sample Preparation

Peanut samples for this project were prepared either at the NPRL or SRRC. Table 2 gives the sample number, general description, roast conditions, and the color values for the peanut samples used in the development of the peanut flavor lexicon.

Samples 1-5, 15 and 17 were prepared at the SRRC. Roasting was done with a pilot scale (.1 to 10 kg capacity) gas heated surface combustion dryer (Midland-Ross, Toledo, OH). Venturis were adjusted to deliver hot air at 325F. Roasting variations were based on time while degree of roasting was established by color evaluations using a Hunter Lab D25-PC2 colorimeter. The samples were immediately cooled by forced air, blanched and then stored at 40F until processed into butters. Peanut butters (unstabilized peanut slurry, without added salt, sweeteners, emulsifiers, and stabilizers) were made with a commercial food processor and all samples were stored at −4F until used.

TABLE 2.
DESCRIPTION/ORIGIN, ROAST CONDITIONS AND COLOR VALUES[1] FOR PEANUT SAMPLES USED IN DEVELOPMENT OF PEANUT FLAVOR LEXICON

Sample	Description/Origin	Commercial Size	Roast Temp. (°F)	Roast Time (min.)	Peanut Butter Color Values		
					L	a	b
1	Very light roast	Medium	325	7.5	59.0	4.0	20.8
2	Light Roast	Medium	325	10	53.7	7.3	22.4
3	Medium Roast (Ref.)	Medium	325	12.5	49.2	9.7	22.7
4	Dark Roast	Medium	325	15	45.8	11.0	22.3
5	Very Dark Roast	Medium	325	17	40.8	12.6	21.2
6	Immature	Other Edible	320	26	48.4	10.8	22.9
7	Freeze Damaged	No. 1	320	29	47.9	10.5	23.0
8	Excess Methyl Bromide	Medium	320	37	47.0	11.4	23.0
9	Warehouse Damaged	Medium	320	37.5	47.4	11.1	22.7
10	Rancid Oil Added	Medium	320	38	47.7	11.1	23.1
11	Immature (1984 crop)	No. 1	320	26	47.6	11.1	23.1
12	1980 Crop (drought)	Medium	320	34	42.3	12.5	21.6
13	1980 Crop (drought)	Medium	320	31	48.9	10.6	22.9
14	Improper Curing (1983 Crop)	Medium	320	37	40.9	13.0	21.2
15	Blanched	Medium	325	12.5			
16	Commercial Roasted Peanuts (rancid)	------					
17	Peanuts & 20% Hearts	Medium	325	12.5	49.2	9.7	22.7

[1]Color values were obtained with a Hunter Lab D25-PC2 Colorimeter

Samples 6-14 were prepared at the NPRL. The roaster was a Blue-M mechanical convection oven with horizontal air flow. The heating chamber was fitted with a 15.5 in. × 6 in. cylindrical, four compartment sample holder which rotated at ca. 7 rpm. The cylinder was constructed of 16 gauge stainless steel with 1/8 in. × 3/4 in. staggered slots. Approximately 1 kg of peanuts were roasted at one time. After roasting, peanuts were cooled with forced ambient air and blanched with a small laboratory blancher. Peanut butter was made with a Bauer mill and all samples were kept frozen.

Sample 16 was a commercial sample for which preparation information is not available.

Sample Presentation

Peanut samples were presented to the panel members as both blanched splits and peanut butter at room temperature. Samples were coded with random three digit numbers. Specific reference samples were identified as such. Nonchlorinated commercial spring water was provided to the panelists as rinse water.

RESULTS

Lexicon Development

The first step of this process involved the complication of a list of terms to characterize the aromatics, basic tastes, and feeling factors typically found in peanuts. The panel did this by evaluating the peanut samples which represented the optimum roast, over and under roast, very over and very under roast. This session generated terminology for desirable peanut flavor as well as terminology for roast variations. The panel then evaluated off-flavor samples and suggested terms to characterize the off-flavors which were present. Discussion of the proposed terminology led to consensus agreement on the final terminology presented in Table 3.

To rate the intensities of peanut flavors, a ten point scale was utilized. This scale was established in reference to flavor intensities that are assigned to specific characteristics apparent in several commercially available food products (Table 4). The panel agreed that the intensity of the "roasted peanutty" flavor note of a reference sample would be a 6.5 on the scale. Evaluation of a reference sample as a blind coded sample yielded a mean score of 6.1 confirming this expectation (Table 5).

To become more familiar with the flavor terms and intensity scales, the panel evaluated 50:50 mixes of the different roasts demonstrating that the intensities of specific flavor notes persisted in mixes and were not averaged by blending.

Lexicon Validation

Once the lexicon was established, the final step was to validate the terminology. This was accomplished by allowing the panel to evaluate samples using both the lexicon and the established intensity scale. The number coded samples evaluated included a reference, a rancid sample (rancid oil had been added to the sample), and a sample that had been stored in a warehouse which was involved in a fire. Data collected from these evaluations are presented in Table 5.

TABLE 3.
LEXICON OF PEANUT FLAVOR DESCRIPTORS

AROMATICS

Roasted peanutty
The aromatic associated with medium-roast peanuts (about 3-4 on USDA color chips) and having fragrant character such as methyl pyrazine.

Raw bean/peanutty
The aromatic associated with light-roast peanuts (about 1-2 on USDA color chips) and having legume-like character (specify beans or pea if possible.)

Dark roasted peanut
The aromatic associated with dark-roasted peanuts (4+ on USDA color chips) and having very browned or toasted character.

Sweet aromatic
The aromatics associated with sweet material such as caramel, vanilla, molasses, fruit (specify type).

Woody/hulls/skins
The aromatics associated with base peanut character (absence of fragrant top notes) and related to dry wood, peanut hulls, and skins.

Cardboard
The aromatic associated with somewhat oxidized fats and oils and reminiscent of cardboard.

Painty
The aromatic associated with linseed oil, oil based paint.

Burnt
The aromatic associated with very dark roast, burnt starches, and carbohydrates, (burnt toast or espresso coffee.

Green
The aromatic associated with uncooked vegetables/grasstwigs, *cis*-3-hexanal.

Earthy
The aromatic associated with wet dirt and mulch.

Grainy
The aromatic associated with raw grain (bran, starch, corn, sorghum).

Fishy
The aromatic associated with trimethylamine, cod liver oil, or old fish.

Chemical/plastic
The aromatic associated with plastic and burnt plastics.

Skunky/mercaptan
The aromatic associated with sulfur compounds, such as mercaptan, which exhibit skunk-like character.

TASTES

Sweet
The taste on the tongue associated with sugars.

Sour
The taste on the tongue associated with acids.

Salty
The taste on the tongue associated with sodium ions.

Bitter
The taste on the tongue associated with bitter agents such as caffeine or quinine.

CHEMICAL FEELING FACTORS

Astringent
The chemical feeling factor on the tongue, described as puckering/dry and associated with tannins or alum.

Metallic
The chemical feeling factor on the tongue described as flat, metallic and associated with iron and copper.

TABLE 4.
INTENSITY REFERENCES

0	
1	
2	Na_2CO_3 in a saltine
3	
4	Apple in Motts apple sauce
5	
6	Orange in Minute Maid orange juice
6.5	Roasted Peanutty in reference sample
7	
8	Grapes in Welches Grape Juice
9	
10	Cinnamon in Big Red Gum

TABLE 5.
INTENSITIES OF FLAVOR CHARACTERISTICS
AS RATED FOR PEANUT SAMPLES[1]

	REFERENCE (651)	RANCID (109)	SMOKE DAMAGED (426)
Roasted Peanutty	6.1±0.6	2.7±1.6	1.1±1.3
Raw Bean/Peanut	1.7±1.0	1.4±1.3	0.2±0.6
Dark Roasted Peanut	2.1±1.3	1.3±1.7	2.2±2.6
Sweet Aromatic	3.5±1.3	1.3±1.5	0.7±1.0
Woody/Hulls/Skin	1.2±1.2	2.1±1.8	
Cardboard		3.4±2.7	
Painty		4.7±1.8	
Burnt			2.7±3.2
Sweet	2.9±0.9	1.4±1.2	0.3±0.8
Bitter	1.2±1.3	2.0±1.8	2.2±2.0
Astringency	1.6±1.2	2.0±1.4	2.3±1.6

[1]Values are means and standard deviations of 13 individual judgments

The data indicate that the panel was able to use the lexicon to describe basic flavor differences among the samples. Sample number 109 (rancid) was described as being both cardboard-like and painty. The panel also indicated a reduction in the roasted peanutty character, sweet aromatic, and the sweet taste when compared to the reference sample.

Sample 426 (warehouse smoke damaged) was described as having burnt characteristics. The panel also suggested including the term "smoke" to the lexicon to account for smoke damaged samples as opposed to burnt or phenolic (burnt plastic) character. As with sample 109, the results showed a reduction in roasted peanutty character, sweet aromatic, and sweet taste.

It should be noted that these data can not be construed as the definitive flavor descriptions of the samples. They are intended only to confirm the appropriateness of the lexicon and intensity scale for describing desirable and off-flavor peanut flavors.

CONCLUSIONS

This lexicon of peanut flavor descriptors with definitions provides a comprehensive, nonredundant list of terms to evaluate and communicate flavor characteristics of peanuts and peanut products. The terminology can be used in correlations with instrumental data, product development, shelf-life, quality control as well as basic research. While this lexicon is comprehensive, at this time, the lexicon can be expanded as other "off-character" peanuts are identified from new peanut sources and treatments.

REFERENCES

BRANDT, M.A., SKINNER, E.Z. and COLEMAN, J.A. 1963. Texture profile method. J. Food Sci. 28(4), 404.

CAIRNCROSS, S.E. and SJOSTROM, L.B. 1950. Flavor profiles—a new approach to flavor problems. Food Technol. 4, 308.

CAUL, J.F. 1957. The profile method of flavor analysis. Adv. Food Res. 7, 140.

CIVILLE, G.V. and SZCZESNIAK, A.S. 1973. Guidelines to training a texture profile panel. J. Texture Stud. 4, 204-223.

FLETCHER, M.M. 1987. Evaluation of peanut quality. In *Peanut Quality—Its Assurance and Maintenance from the Farm to End-Product* (Tech. Bull. 874) (E.M. Ahmed and H.E. Pattee, eds.) pp. 60-72, Agric. Exp. Sta., Inst. Food and Agric. Sci., Univ. of Florida, Gainesville.

HOLADAY, C.E. 1971. Appendix III. Report of the peanut quality committee. J. Amer. Peanut Res. Educ. Soc. 3(II), 239-241.

JOHNSEN, P.B. and CIVILLE, G.V. 1986. A standardized lexicon of meat WOF descriptors. J. Sensory Studies 1, 99-104.

MEILGAARD, M., CIVILLE, G.V. and CARR, B.T. 1987. *Sensory Evaluation Techniques,* Vol. 2, pp. 1-23. CRC Press, Boca Raton, Florida.

OUPADISSAKOON, C. and YOUNG, C.T. 1984. Modeling of roasted peanut flavor for some Virginia type peanuts from amino acid and sugar contents. J. Food Sci. *49*, 52-58.

STONE, H., SIDEL, J., OLIVER, S., WOOLEY, A. and SINGLETON, R.C. 1974. Sensory evaluation by quantitative descriptive analysis. Food Technol. *28*(11), 24-34.

SYARIEF, H., HAMANN, D.D., GIESBRECHT, F.G., YOUNG, C.T. and MONROE, R.J. 1985. Interdependency and underlying dimensions of sensory flavor of selected foods. J. Food Sci. *50*, 631-638.

CHAPTER 6.7

SENSORY MEASUREMENT OF FOOD TEXTURE BY FREE-CHOICE PROFILING

RICHARD J. MARSHALL[1] and SIMON P.J. KIRBY

*At the time of original publication,
all authors were affiliated with
AFRC Institute of Food Research, Reading Laboratory
(University of Reading)
Shinfield, Reading, RG2 9AT, UK.*

ABSTRACT

Panelists were trained to measure texture of processed cheese analogues of known composition by free-choice profiling. Panel scores were analyzed by generalized Procrustes (a multivariate method). Plots of the Procrustes scores for the first 2 principal component axes from analysis of separate replicates reflected the experimental design. To compare replicates, scores from Procrustes analysis of the original data were subjected to Procrustes analysis. The results showed that panelists' assessment of replicates was consistent. Regression of the scores from this second Procrustes analysis showed that they were well explained by the linear effects of either moisture in nonfat solids or fat content. Panelists who performed differently from the rest could be identified. Panelists were more sensitive to changes in moisture in nonfat solids content than fat. It was concluded that the sensory texture of the cheese analogues could be measured satisfactorily by free-choice profiling.

INTRODUCTION

Sensory analysis of food texture has often been carried out using texture profiling (Szczesniak 1963; Civille and Szczesniak 1973; Bourne *et al.* 1975). The training of assessors for conventional profiling is long because of the need to develop a descriptive vocabulary and scales by consensus (Bourne 1982; Williams and Langron 1984). Panelists may also have difficulty in understanding definitions as has been observed in the texture assessment of Cheshire cheese (R.J. Marshall unpublished observations). The use of common numerical scales can also give rise to problems such as the reluctance of panelists to use the full range of the scale (Stevens and Galanter 1957).

Recently a modification of the standard profiling method has been used to assess the appearance, flavor and aroma of commercial ports (Williams and Langron 1984) and it has been compared with conventional profiling and similarity scaling for the assessment of coffee aromas (Williams and Arnold 1985). During training the assessors are presented with a range of samples similar to those to be measured finally and each individual develops his own unique list of descriptors and definitions. While identical terms may be used by different panelists they may have totally different meanings. During an experiment, each panelist scores the degree or intensity of each descriptor on his list and scores from all panelists are collated and then analyzed by generalized Procrustes analysis (Gower 1975; Langron 1981). In this procedure the results from each assessor are considered as coordinates in a multi-dimensional space. These are then transformed by translation, rotation/reflection and scaling to obtain a consensus configuration that minimizes the residual variation between the assessors (Arnold and Williams 1986; Banfield and Harries 1975; Vuataz 1977).

Williams and Arnold (1985) showed that this Free-choice Profiling with the scores analyzed by generalized Procrustes analysis gave results similar to conventional profiling and similarity scaling. The panelists required less training, fewer assessments and the method avoided the tendency of panelists to use restricted and differing scoring ranges and overcame minor individual variation while highlighting those panelists who responded differently from the rest.

The objectives of the present work were to validate Free-choice profiling and the generalized Procrustes analysis of the data for the sensory analysis of the texture of unflavored processed cheese analogues by using a designed experiment. It was found that the data fell into a pattern similar to that of the compositional variation between analogues confirming that the differences detected were genuine.

EXPERIMENTAL

Experimental Design

Five cheese analogues in which percentage (w/w) fat (butter oil) and moisture in nonfat solids (MNFS) (Lawrence and Gilles 1980) were varied (Fig. 1) were made on the same Friday by a standard method to be described elsewhere.

The following day each cheese was divided into 10 equal-sized blocks which were vacuum-packed individually in food-grade polythene bags. Packs for daily use were made from one portion of each of the 5 cheeses vacuum packed into second larger bags. During the experiment one pack of 5 samples was opened on each of 10 consecutive days (excepting the middle Saturday and

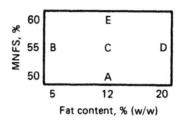

```
           A                              B
          BLOCK                          BLOCK
        1  2  3  4  5                 1  2  3  4  5
       | A  B  C  D  E               | A  B  C  D  E
  ROW  |                             |
       | B  E  D  A  C               | C  D  B  E  A
```

FIG. 1. EXPERIMENTAL DESIGN AND RANDOMIZATION
OF SAMPLES PRESENTED TO PANEL

a) Butter fat content (%, w/w) and Moisture in nonfat solids (%, w/w)(MNFS) of cheeses calculated from quantities of ingredients. A,B,C,D,E: the 5 compositions.

b) Randomization derived from this 2(2×5) incomplete Latin square; each cheese occurred with every other cheese only once. The squares were randomized for each taster to preserve this property.

Sunday) starting the Monday following preparation of the cheeses. One randomized pair of cheeses was presented to each panelist in sterile plastic Petri dishes on each occasion. Within each dish were 4 pieces of the cheese of approximately 30 × 15 × 5mm size. The randomization was derived from an extended 2 × 5 incomplete Latin square (Cochran and Cox 1957) (Fig. 1). This extended incomplete square has the property that each treatment occurs with each of the other treatments once in the same column (daily session in this case) and also that each treatment occurs equally often in the two rows (corresponding to order of tasting). The squares were randomized for each panelist in such a way as to preserve these properties. Each Petri dish was labelled with a unique, random three digit code to avoid bias. The training and the experimental trial were carried out in a panel room with separate booths for each panelist. Surplus samples from the first and last sessions were analyzed for composition.

Preliminary Panel Sessions

The panelists were staff of the IFR, Reading Laboratory, selected on the basis of answers to a questionnaire about likes, dislikes, health, availability and a short test of extent of textural vocabulary. Fifteen were selected for training and vocabulary development (although one subsequently missed all experimental sessions due to illness) which consisted of 10 × 0.5h sessions at which they were given samples of the cheese analogues of varying compositions and textures including the range of fat and moisture contents that were used in the final experiment. Samples were presented and labelled as in the experiment described above. For one session, panelists were given samples of different varieties of 'real' cheeses to help the development of their vocabulary. They were encouraged to use whatever terms they wished to describe the texture.

```
MANUAL
                    ┌ Feel with fingers without squashing or moulding
        1 piece    ┤  Snap
                    └ Mould one half of snapped piece

                    ┌ Compress very gently
        1 piece    ┤  Compress until first fracture
                    └ Squash completely

ORAL
                    ┌ 1st bite with incisors
                    │
                    │ 1st chew with molars
        1 piece    ┤
                    │ Chew until ready to swallow
                    │
                    │ Feel with tongue (against hard palate, cheek, teeth)
                    └ after chewing

        1 piece      Feel unchewed sample in mouth
```

FIG. 2. ASSESSMENT SCHEME USED BY PANELISTS
DURING SENSORY EVALUATION OF THE TEXTURE OF MODEL CHEESES

A standard scheme in which texture was assessed first manually then orally was developed jointly by all panelists (Fig. 2). Panelists were also introduced to scoring on a continuous scale consisting of a 12cm horizontal line with anchors 1cm from each end labelled "Low" at the left-hand end and "High" at the other. After training session 6, panelists were asked individually to define their descriptors and, where possible, to eliminate synonyms. The descriptors and definitions were compiled into personal lists with descriptors in order of appearance during assessment following the standard scheme. During the remaining training sessions these lists were refined further as the panelists became more familiar with the procedure. Personal score sheets were prepared with the same order of descriptors, with a score line beside each and subheadings "Manual" and "Oral" to indicate to which stage of the scheme terms belonged.

Measurement of Scores and Statistical Analysis

The distances of the scoring marks along the score lines from the "Low" anchor were measured in mm using a MOP3 digitiser (Kontron Electronics Ltd, Watford, UK) connected directly to the host computer. To detect any statistically significant session and order effects an analysis of variance of the panel scores was calculated for all descriptors for each assessor. Generalized Procrustes analyses were carried out separately for the manual and oral descriptors for each replicate using the Genstat macro described by Arnold (1986). The calculated consensus and individual assessor scores for all dimensions obtained from these analyses were then themselves used as input to generalized Procrustes analysis to assess the agreement between replicates and to obtain overall consensus and individual assessor results. To interpret the major axes from the Procrustes analysis of the manual and oral data for each replicate, correlations between the calculated Procrustes scores for each assessor and his descriptor scores were calculated. Also calculated were the correlations between descriptor scores for each assessor.

To relate compositional variables to sensory data, the calculated consensus scores for each cheese for the first two principal component axes of the consensus configuration, derived from the generalized Procrustes analysis of the calculated consensus scores for each replicate, were regressed separately against percentage fat and percentage MNFS. Terms up to the quadratic were included in the model if they were significant at the 10% level.

RESULTS

Analyses of Variance of the Sensory Data

Analysis of variance of the panel scores for all 276 descriptors developed by all 14 panelists showed 4 with a statistically significant session effect ($0.001 < p < 0.01$). Two descriptors showed a statistically significant order effect at this same level and one at $p < 0.001$. The number of descriptors with significant session or order effects was so low that no allowance was made for such effects in the generalized Procrustes analyses.

Each panelist had a number of descriptors that were highly correlated with each other.

Procrustes Analyses of the Sensory Data

Absences from the panel meant that only for replicates 1 and 2 were there full results. Assessor 13 had some replicate number 3 results missing and assessors 5, 9 and 12 had some replicate number 4 results missing. These assessors were excluded from the generalized Procrustes analysis of the replicates for which they had missing data.

The experimental design based on actual sample composition is shown in Fig. 3. The plots of the calculated consensus scores for the oral descriptors (Fig. 4) clearly displayed the experimental design in terms of the relationship between consensus points. Similarly the design was also evident in the plots for the analyses of the manual descriptors (Fig. 5) although less clear. For the manual data, the percentage variation of the consensus configuration for each replicate accounted for by the first 2 principal axes was approximately 80% (Table 1). For the oral data, the percentage variation of the consensus configuration for each replicate accounted for by the first 2 principal axes was greater than 80%. For both manual and oral data, the generalized Procrustes analysis of the calculated consensus scores for each replicate showed good agreement between replicates (Fig. 6). The percentage variation of the consensus of the consensuses accounted for by the first 2 principal axes was 79.2 and 88.6 for the manual and oral data, respectively.

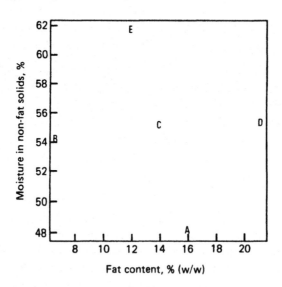

FIG. 3. FAT AND MOISTURE IN NONFAT SOLIDS CONTENT OF PROCESSED CHEESE ANALOGUES DERIVED FROM CHEMICAL ANALYSES
A,B,C,D,E: the individual cheeses.

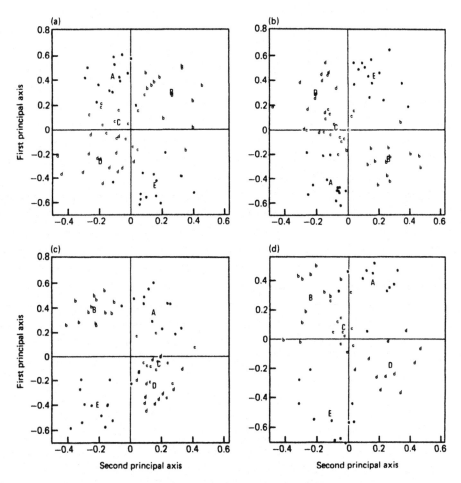

FIG. 4. CONSENSUS AND INDIVIDUAL SCORES FOR THE
FIRST 2 PRINCIPAL COMPONENT AXES FROM GENERALIZED PROCRUSTES
ANALYSIS OF SEPARATE REPLICATES, ORAL DATA

a) Replicate 1; b) Replicate 2; c) Replicate 3; d) Replicate 4.

A, B, C, D, E: Consensus points for cheeses A, B, C, D, E corresponding to composition in Fig. 3.
a, b, c, d, e: Individual assessor points for cheeses A, B, C, D, E, respectively.

The generalized Procrustes analyses of each individual's calculated scores derived from generalized Procrustes analyses of all replicates gave information about each assessor's sensitivity to compositional change. For example, Fig. 7

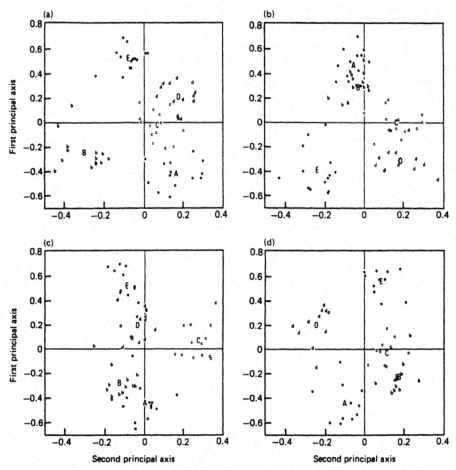

FIG. 5. CONSENSUS AND INDIVIDUAL SCORES FOR THE FIRST 2 PRINCIPAL COMPONENT AXES FROM GENERALIZED PROCRUSTES ANALYSIS OF SEPARATE REPLICATES, MANUAL DATA

a) Replicate 1; b) Replicate 2; c) Replicate 3; d) Replicate 4.

A, B, C, D, E: Consensus points for cheeses A, B, C, D, E corresponding to compositions in Fig. 3.

a, b, c, d, e: Individual assessor points for cheeses A, B, C, D, E, respectively.

shows the plot of the calculated consensus scores for the first 2 principal axes of the consensus configuration derived from a generalized Procrustes analysis of panel scores for each replicate for the manual descriptors of assessor Number 2. This assessor placed cheeses D and E closer together than cheeses A and B.

TABLE 1.
PERCENT VARIATION ACCOUNTED FOR BY THE FIRST TWO PRINCIPAL AXES
OF THE PROCRUSTES CONSENSUS CONFIGURATIONS

	Replicate 1	Replicate 2	Replicate 3	Replicate 4	Consensus of consensuses
Manual Data					
	82.0	79.2	79.9	82.4	79.2
Oral Data					
	86.8	82.3	83.4	85.6	88.6

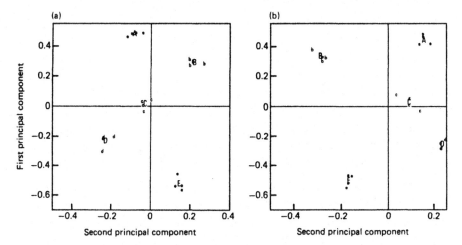

FIG. 6. CONSENSUS AND INDIVIDUAL REPLICATE CONSENSUS SCORES
FOR THE FIRST 2 PRINCIPAL COMPONENT AXES FROM GENERALIZED
PROCRUSTES ANALYSIS OF CONSENSUS SCORES FROM
ANALYSIS OF SEPARATE REPLICATES

a) Manual data; b) Oral data.

A, B, C, D, E: Overall consensus points for cheeses A, B, C, D, E corresponding to compositions in Fig. 3.

a, b, c, d, e: Individual replicate consensus points for cheeses A, B, C, D, E, respectively.

Relationships Between Sensory Data and Sample Composition

The actual composition of the cheese was very close to the composition estimated from the quantities of ingredients (Fig. 1a and 3). The latter were used in the regressions rather than the theoretical values.

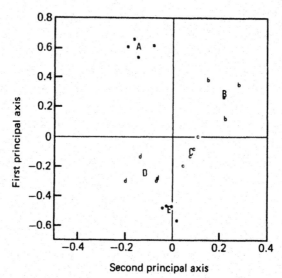

FIG. 7. CONSENSUS AND INDIVIDUAL REPLICATE SCORES FOR THE FIRST 2 PRINCIPAL COMPONENT AXES FROM GENERALIZED PROCRUSTES ANALYSIS OF MANUAL DATA FROM ONE PANELIST

A, B, C, D, E: Consensus points for the 4 replicates, corresponding to cheeses of compositions A, B, C, D, E given in Fig. 3.

a, b, c, d, e: Individual replicate points for cheeses A, B, C, D, E, respectively.

Regression of the consensus scores derived from the generalized Procrustes analysis of the calculated consensus scores for each replicate showed that for both manual and oral descriptors, the calculated scores on the first principal component axis were well explained by the linear effect of MNFS (Table 2). The calculated scores for the second principal component axis were similarly well explained for both manual and oral descriptors by the linear effect of percentage fat.

For the manual and oral descriptors, those terms that correlated most highly with the calculated scores of individual assessors for the first 2 principal component axes of the consensus configuration, derived from the generalized Procrustes analysis of each replicate, tended to be similar for all panelists (Tables 3 and 4). For manual data, the first principal axes were generally described by terms relating to deformability of the cheeses. The second axes contained more terms relating to fracture. For the oral data, terms were more

TABLE 2.
BEST SIMPLE LINEAR REGRESSIONS FOR THE CALCULATED SCORES FOR THE FIRST TWO PRINCIPAL AXES FOR THE CONSENSUS, DERIVED FROM THE PROCRUSTES MATCHING OF THE CONSENSUS CONFIGURATIONS FOR ALL REPLICATES, ON FAT OR MNFS (STANDARD ERRORS IN BRACKETS)

Manual data

First principal axis

Score = 4.22 − 0.0770 MNFS
(1.04)   (0.0190)

Percentage variance accounted for : 79.5

Second principal axis

Score = 0.4522 − 0.03283 fat
(0.0707)   (0.00485)

Percentage variance accounted for : 91.8

Oral data

First principal axis

Score = 4.00 − 0.0730 MNFS
(1.22)   (0.0222)

Percentage variance accounted for : 71.1

Second principal axis

Score = −0.539 + 0.03908 Fat
(0.123)   (0.00843)

Percentage variance accounted for : 83.7

---

variable but the first principal axes seemed to be described by terms relating to firmness and viscosity and the second principal axes by terms relating to behavior during chewing.

Panelists who performed differently from the rest could be identified by examining their residual sums of squares in the analysis of variance of the transformed data for each replicate (Table 5). In 4 out of the 8 analyses, panelist number 14 had a relatively high residual but he was not excluded from the results because the other 4 residuals were similar to those of most other assessors.

DISCUSSION

Previous free-choice profiling of flavors and aromas (Williams and Langron 1984; Williams and Arnold 1985) used random samples which were not all replicated, unlike our approach of designed differences with replication of all

TABLE 3.
LARGEST CORRELATIONS OF SCORES ON THE FIRST TWO PRINCIPAL AXES OF THE CONSENSUS CONFIGURATION FOR EACH INDIVIDUAL DERIVED FROM MANUAL DATA WITH CORRESPONDING DESCRIPTORS

Assessor	First Axis Description	Correlation	Second Axis Description	Correlation
First replicate				
1	Shapeability	0.99	Springy	0.77
2	Moist	0.98	Clean fracture	-0.84
3	Moist	0.98	Breaks cleanly	-0.96
4	Elasticity	-0.97	Crumbliness	0.88
5	Sticky	0.98	Elastic	-0.93
6	Deformability	0.96	Rubberiness	0.93
7	India rubber like	-0.98	Ease of breaking	-0.90
8	Deforms before fracture	0.99	Smoothness-graininess	0.84
9	Elasticity	-0.97	Hardness-softness	-0.78
10	Bounciness	-0.95	Strength on moulding	0.97
11	Dryness	-0.95	Shear	-0.88
12	Softness	0.90	Resilience	-0.80
13	Rubbery	-1.00	Springy	0.87
14	Deformability	0.97	Springiness	-0.98
Second replicate				
1	Moist	-0.98	Shapeability	-0.63
2	Mouldability	-0.98	Fractures easily	-0.67
3	Hard	0.95	Breaks	-0.78
4	Compressability	0.99	Flakey surface	-0.92
5	Paste	-1.00	Resistance to tear	0.98
6	Deformability	-0.96	Granular	-0.92
7	Softness	-0.99	Ease of breaking	-0.97
8	Pliability/mouldability	-1.00	Smoothness-graininess	0.88
9	Flakiness	0.93	Tearability	-0.90
10	Squashiness	-1.00	Breakability	-0.96
11	Softness-firmness	0.94	Shear	0.82
12	Crumbliness	0.96	Dryness	0.98
13	Sticky	-0.98	Smooth	-0.91
14	Springiness	0.99	Deformability	0.96

TABLE 3. (*continued*)

Assessor	First Axis Description	Correlation	Second Axis Description	Correlation
Third replicate				
1	Breakability	0.99	Snapability	0.93
2	Greasy	0.99	Granular	-0.57
3	Rough	-0.94	Grainy	-0.56
4	Resistance to fracture	-0.99	Mouldability	-0.95
5	Malleability	1.00	Stretch	0.85
6	Fluidness	0.98	Elasticity	0.70
7	Firmness, tackiness	-0.97, 0.97	Jelly-like	-0.85
8	Smoothness-graininess	-1.00	Dryness	-0.54
9	Greasiness	0.99	Malleable-brittle	0.94
10	Deformability	0.99	Strength	0.71
11	Dryness	-0.93	Springiness	0.79
12	Brittleness	-0.96	Crumbliness	-0.93
13				
14	Stickiness	0.99	Springiness	0.56
Fourth replicate				
1	Breakability	0.78	Powderiness	-0.96
2	Fractures easily	0.96	Rubbery	-0.91
3	Elastic-springy	0.97	Dry	-0.99
4	Compressability	-0.97	Ease of snapping	0.79
5				
6	Hardness	-1.00	Mouldability	-0.76
7	Shininess	0.99	Malleability	-0.83
8	Hardness	0.75	Lumpiness	-0.94
9				
10	Squashiness	0.98	Springiness	0.87
11	Softness-firmness	-0.96	Springiness	0.86
12				
13	Rigid	-0.98	Springy	0.99
14	Springiness	-0.98	Peelability	-0.63

TABLE 4.
LARGEST CORRELATIONS OF SCORES ON THE FIRST TWO PRINCIPAL AXES OF THE CONSENSUS CONFIGURATION FOR EACH INDIVIDUAL DERIVED FROM ORAL DATA WITH CORRESPONDING DESCRIPTORS

Assessor	First Axis Description	Correlation	Second Axis Description	Correlation
First replicate				
1	Firm-rubbery	1.00	Smoothness	0.87
2	Firm-tough	0.98	Requires saliva to dissolve	-0.99
3	Moist	-0.98	Elastic-springy	0.99
4	Ease of chewing	-0.98	Stickiness	-0.63
5	Rough	1.00	Creamy	-0.83
6	Clean bite	-0.98	Squashiness	0.91
7	Smoothness	-1.00	Gluiness	0.86
8	Hardness	0.99	Paste formation	-0.98
9	Viscosity	-0.99	Grittiness	-0.70
10	Smoothness	-0.99	Brittleness	-0.80
11	Solubility	-0.97	Lumpiness	0.93
12	Stickiness	-1.00	Springiness	-0.95
13	Glossy mouthfeel	-1.00	Granular	0.95
14	Pastiness	-1.00	Jelly-like	0.99
Second replicate				
1	Firm	-0.84	Smoothness	0.88
2	Firm-tough	-0.99	Requires saliva to dissolve	-0.50
3	Moist	0.96	Elastic-springy	0.91
4	Rough	-0.80	Hardness	-0.69
5	Wetness	-0.99	Lard-like	0.76
6	Goes smooth	0.99	Solubility	-0.65
7	Hardness	0.99	Cheese-like	-0.60
8	Viscosity	-0.99	Paste formation	-0.96
9	Squashiness	0.96	Dryness	-0.38
10	Firmness	0.99	Solubility	-0.56
11	Brittleness	-0.91	Lumpiness	0.68
12	Tough	-0.78	Rubberiness	-0.89
13	Graininess	-0.92	Granular	0.84
14		-0.94	Jelly-like	0.93

TABLE 4. (continued)

Assessor	First Axis Description	Correlation	Second Axis Description	Correlation
Third replicate				
1	Granular	0.99	Rubbery	-0.89
2	Moist	-0.97	Clings to teeth	0.73
3	Smooth	-0.98	Elastic-springy	0.60
4	Solubility	-0.98	Hardness	0.90
5	Wet	-1.00	Chewy	-0.96
6	Chewability	0.99	Solubility	0.83
7	Goes smooth	-0.99	Cheese-like	0.85
8	Dryness	0.98	Stickiness	0.99
9	Hardness-softness	-0.98	Granularity	-0.92
10	Crumbliness	0.98	Firmness	0.95
11	Slipperiness	-0.94	Stickiness-clinginess	0.88
12	Stickiness	-0.88	Springiness	-0.91
13				
14	Jelly-like	-0.98	Stickiness	0.98
Fourth replicate				
1	Watery	-1.00	Smoothness	0.98
2	Requires saliva to dissolve	1.00	Firm-tough	0.50
3	Hard	0.95	Rubbery	0.93
4	Ease of biting	-0.99	Stickiness	0.80
5				
6	Softness	-1.00	Solubility	0.88
7	Creaminess	-0.96	Granularity	-0.87
8	Hardness	0.97	Paste formation	0.96
9				
10	Smoothness	-0.98	Crumbliness	-0.90
11	Firmness	0.87	Stickiness-clinginess	0.85
12				
13	Tough	0.97	Pastey	1.00
14	Jelly-like	-0.95	Pastiness	0.91

TABLE 5.
ASSESSOR RESIDUAL SUM OF SQUARES
FOR GENERALIZED PROCRUSTES ANALYSES

Manual

Assessor	Replicate 1 Residual	Replicate 2 Residual	Replicate 3 Residual	Replicate 4 Residual
1	0.166	0.092	0.218	0.150
2	0.117	0.053	0.122	0.073
3	0.077	0.091	0.099	0.348
4	0.100	0.105	0.123	0.053
5	0.100	0.125	0.090	-
6	0.176	0.219	0.085	0.075
7	0.069	0.058	0.068	0.218
8	0.084	0.115	0.091	0.131
9	0.110	0.076	0.386	-
10	0.045	0.084	0.069	0.062
11	0.124	0.104	0.163	0.154
12	0.139	0.373	0.250	-
13	0.107	0.150	-	0.043
14	0.374	0.348	0.106	0.063

Oral

Assessor	Replicate 1 Residual	Replicate 2 Residual	Replicate 3 Residual	Replicate 4 Residual
1	0.186	0.137	0.212	0.191
2	0.159	0.154	0.149	0.175
3	0.106	0.106	0.080	0.316
4	0.250	0.211	0.170	0.158
5	0.154	0.133	0.201	-
6	0.233	0.147	0.128	0.092
7	0.150	0.088	0.112	0.229
8	0.170	0.174	0.262	0.162
9	0.165	0.141	0.503	-
10	0.117	0.096	0.151	0.128
11	0.246	0.103	0.111	0.152
12	0.189	0.349	0.224	-
13	0.253	0.233	-	0.278
14	0.426	0.405	0.157	0.160

samples. This enabled us to measure the consistency of the panel results by doing a generalized Procrustes analysis for each replicate and then a generalized Procrustes analysis of the resulting consensus scores. We were also able to relate the calculated consensus scores from the analysis of the consensus results for each replicate to the known MNFS and fat contents of the cheeses.

The almost total absence of statistically significant session or order effects in the analysis of variance of the panel scores indicated that panelists' level of scoring was similar between sessions and was unaffected by the order of presentation of the cheeses. The plots of the calculated consensus scores of the first 2 principal component axes from generalized Procrustes analysis of each replicate further confirmed that the panelists were generally in quite good agreement (Fig. 4 and 5). Although panelists who performed differently from the rest could be identified by examining their residual sums of squares (Table 5) deciding which had an unusually large residual was subjective. No panelist had residuals that were consistently greater than the rest.

The high correlations between some descriptors for each panelist indicated that terms were either synonyms or antonyms. Thus the number of descriptors used by each panelist could be reduced to simplify their profile, for example by using the method suggested by Krzanowski (1987).

The reflection of the experimental design in the plots of the calculated consensus scores for the first 2 principal component axes of the consensus configuration derived from the generalized Procrustes analyses of the manual and oral data for each replicate strongly suggested that the method was measuring a real effect. The high variance explained by simple linear regressions of the generalized Procrustes consensus scores for the first 2 principal component axes of the consensus configuration derived from the calculated consensus scores for each replicate on MNFS and fat confirmed the closeness of these scores to the experimental design.

Regression of the consensus scores from generalized Procrustes analysis of the derived consensus scores from each replicate showed that for both the manual and oral data the first principal axis was describing the MNFS content of the cheeses and the second the fat content. This suggested that the panelists were most sensitive to the moisture content of the cheeses and slightly less sensitive to the fat content. The correlations between the final consensus scores and the descriptors further suggests that panelists determined changes in texture manually by assessing deformability and fracture and orally by assessing firmness, viscosity and behavior during chewing.

These results show that Free-choice profiling can be used to measure differences in the texture of food in an objective and systematic way similar to conventional profiling (Bourne 1982). The method enabled us to assess the performance of the panelists and showed they were more sensitive to changes in the MNFS than the fat content of the cheeses.

ACKNOWLEDGMENTS

We thank Emma C. Boult and Francis Cassera for skilled technical assistance, our panelists and the Analytical Services section of IFR, Reading Laboratory. We also thank Drs. Margaret L. Green, John C. Gower and Alan Huitson for helpful discussions.

REFERENCES

ARNOLD, G.M. 1986. A Generalised Procrustes macro for sensory analysis. In *Genstat Newsletter No. 18.* pp. 61–80. Numerical Algorithms Group Ltd: NAG Central Office, 256 Banbury Road, Oxford, U.K.

ARNOLD, G.M. and WILLIAMS, A.A. 1986. The use of generalised Procrustes techniques in sensory analysis. In *Statistical Procedures in Food Research*. (J.R. Piggott, ed.) pp. 233-253, Elsevier Applied Science, London.

BANFIELD, C.F. and HARRIES, J.M. 1975. A technique for comparing judges' performance in sensory tests. J. Food Technol. *10*, 1-10.

BOURNE, M.C. 1982. Sensory methods of texture and viscosity measurement. In *Food Texture and Viscosity: Concept and Measurement*. pp. 247-279, Academic Press, New York.

BOURNE, M.C., SANDOVAL, A.M.R., VILLALOBOS, C.M. and BUCKLE, T.S. 1975. Training a sensory texture profile panel and development of standard rating scale, in Colombia. J. Texture Studies *6*, 43-52.

CIVILLE, G.V. and SZSZESNIAK, A.S. 1973. Guidelines to training a texture profile panel. J. Texture Studies *4*, 204-223.

COCHRAN, W.G. and COX, G.M. 1957. *Experimental Designs*. 2nd edition. p. 538, John Wiley & Sons, New York.

GOWER, J.C. 1975. Generalized Procrustes analysis. Psychometrika. *40*, 33-51.

KRZANOWSKI, W.J. 1987. Selection of variables to preserve multivariate data structure, using principal components. J. Royal Statistical Soc. (Series C) Appl. Stat. *36*, 22-33.

LANGRON, S.P. 1981. The Statistical Treatment of Sensory Analysis data. PhD thesis University of Bath.

LAWRENCE, R. and GILLES, J. 1980. The assessment of the potential quality of young Cheddar cheese. New Zealand J. Dairy Sci. and Technol. *15*, 1-12.

STEVENS, S.S. and GALANTER, E.H. 1957. Ratio scales and category scales for a dozen conceptual continua. J. Experimental Psychology *53*, 377-411.

SZCZESNIAK, A.S. 1963. Classification of textural characteristics. J. Food Sci. *28*, 385-409.

VUATAZ, L. 1977. Some points of methodology in multidimensional data analysis as applied to sensory evaluation. Nestle Res. News 1976/77, 57-71.

WILLIAMS, A.A. and ARNOLD, G. 1985. A comparison of the aromas of six coffees characterized by conventional profiling, free-choice profiling and similarity scaling methods. J. Sci. Food and Agric. *36*, 204-214.

WILLIAMS, A.A. and LANGRON, S.P. 1984. The use of free-choice profiling for the evaluation of commercial ports. J. Sci. Food and Agric. *35*, 558-568.

CHAPTER 6.8

TASTE DESCRIPTIVE ANALYSIS: CONCEPT FORMATION, ALIGNMENT AND APPROPRIATENESS

M. O'MAHONY, L. ROTHMAN, T. ELLISON, D. SHAW and L. BUTEAU

*At the time of original publication,
all authors were affiliated with
Dept. Food Science and Technology
University of California
Davis, California 95616.*

ABSTRACT

The commonly used approach to psychophysical taste descriptive analysis called 'taste profiling' was assessed to determine whether it produced concept alignment with an agreed set of labels; it did not. After judges had been trained using a concept alignment procedure involving fifty-nine standards, the procedure was assessed to see whether broadly defined concepts of 'sweet', 'sour', 'salty' and 'bitter' generalized to various stimuli; they did not. It was concluded that the technique is inferior to currently used methods of descriptive analysis.

INTRODUCTION

The descriptive analysis of flavor is a widely used technique in the sensory analysis of foods. There are several standard techniques available (Amerine *et al.* 1965; Cairncross and Sjöström 1950; Caul 1957; Meilgaard *et al.* 1987; Stone *et al.* 1974). These use a system whereby particular sensory characteristics of a food are defined for trained panelists, using physical standard stimuli. In this way, an ad hoc language is created for the communication of the sensory characteristics of the food product; various scaling techniques are used to measure the strengths of these characteristics. To understand fully the mechanisms involved in the formation of the ad hoc descriptive language it becomes necessary to consider the information processing taking place in the brain during this process. This, in turn, leads to a consideration of mechanisms involved in the formation of sensory concepts (Civille and Lawless 1986).

*Reprinted with permission of Food & Nutrition Press, Inc., Trumbull, Connecticut.
© Copyright 1990. Originally published in Journal of Sensory Studies 5, 71-103.*

According to current theory of sensory concept formation (Hull 1920; Miller and Johnson-Laird 1976; Millward 1980; Nelson 1974; Pikas 1966), a sensory concept is formed by a two-part process: abstraction and generalization. For simplicity, consider the color concept of 'redness'. For the first part, the concept is abstracted from red and nonred stimuli. For the second, this concept is generalized or broadened beyond those sensations used in the abstraction process. Then, stimuli colored shades of red that have not been seen before, can be categorized as falling within the concept; they have 'redness'. The concept is given a label, 'red', for purposes of communication between those who share the concept. The same reasoning can be applied to the formation of taste or flavor concepts like 'sour', 'sweet', 'crunchy', 'fruity', 'bitter', 'sour', etc.

If a group of untrained judges were to list the sensory characteristics of a food, there would be little agreement. Firstly, untrained judges do not have a common language with which to communicate sensations; a given sensation will elicit different descriptive labels from different judges. This can be simply rectified by agreeing on a set of labels. The second reason is more difficult to resolve; untrained judges do not have an agreed system of categorization. For example, a sensation that may be categorized as 'salty' by one judge, may be categorized as something other than 'salty' by another. This is similar to the phenomenon of a 'greenish-blue' color being categorized as 'green' by one judge and 'blue' by another. In the terms of those who study the information processing that takes place for such categorizations (Miller and Johnson-Laird 1976), the sensory concept for saltiness for one judge is not exactly aligned with that of the other judge. A sensation may fall within one judge's concept of saltiness, while for a different judge with a slightly different saltiness concept, the sensation may fall outside his concept. Without some type of concept alignment procedure, sensory concepts will be idiosyncratic (Miller and Johnson-Laird 1976).

For a panel of judges to be used as an analytical tool for descriptive analysis, the judges will need to have their concepts aligned. That is, judges must agree on which range of sensations are to be included within a given concept and which are to be excluded. If not, panelists might disagree on the reporting of a given sensory characteristic, not because of differences in sensitivity to that characteristic but because the sensation elicited fell within the particular concept of some judges but not others. This would be analogous to using a set of instruments that were not calibrated in the same way.

The techniques of descriptive analysis commonly used in sensory evaluation (Amerine *et al.* 1965; Cairncross and Sjöström 1950; Caul 1957; Meilgaard *et al.* 1987; Stone *et al.* 1974) use standard physical stimuli to align the concepts of their judges during training for panel participation. Refinement of concept alignment for sensory characteristics may be gained by the use of multiple standards (Ishii and O'Mahony 1987a, 1990).

For descriptive analysis of taste characteristics, variations on an approach that has been called 'taste profiling' are sometimes utilized generally by psychophysicists. Taste profiling has been applied to various food constituents like sweeteners, monosodium glutamate (MSG), KCl and sodium benzoate (Acton and Stone 1976; Du Bois et al. 1977, 1981; Kier 1980; Peryam 1960, 1963). The approach requires judges to describe tastes in terms of numerical estimates of their 'sweetness', 'sourness', 'saltiness' and 'bitterness'; either some or all of these categories are available to the judge. The numerical estimates have been given in terms of category scales (Peryam 1960, 1963; Smith and Halpern 1983), line scales (Lawless 1979), percentages (Acton and Stone 1976; Acton et al. 1970; Du Bois et al. 1977, 1981; Kier 1980) or magnitude estimation (Smith and McBurney 1969). The latter version of this technique is used widely by taste psychophysicists (Bartoshuk 1978, 1979; Bartoshuk et al. 1972, 1974a, 1974b, 1978, 1988; DeSimone et al. 1980; Kuznicki and Ashbaugh 1982; Lawless 1979; McBurney 1972; McBurney and Bartoshuk 1973; McBurney and Gent 1978; McBurney and Shick 1971; McBurney et al. 1972; Meiselman and Halpern 1970a, 1970b; Murphy et al. 1981). Sometimes additional terms like 'other' have also been added for tastes that did not fit into the traditional four (Bartoshuk 1974, 1975; Bartoshuk and Cleveland 1977; Bujas et al. 1974; Kuznicki 1978; Kuznicki and Ashbaugh 1979; Settle et al. 1986) and for sensations other than taste (Gent 1979; Gent and McBurney 1978). Sometimes nongustatory sensation categories are provided (Cowart 1987).

Those who use taste profiling sometimes report giving standards beforehand to judges to define the terms: 'sweet', 'sour', 'salty', and 'bitter' (Smith and McBurney 1969; Riskey et al. 1982); this would be expected to bring about some degree of the necessary concept alignment. Another approach has been to screen judges by giving stimuli like NaCl (salty), citric acid or HCl (sour), quinine or KCl (bitter) and sucrose (sweet) and selecting only those judges who give the appropriate descriptions (Cardello 1978, 1979; Cardello and Murphy 1977; Dzendolet and Meiselman 1967a, 1967b; McCutcheon and Saunders 1972; Meiselman and Halpern 1970a, 1970b; Meiselman and Dzendolet 1967; Sandick and Cardello 1983). Kuznicki (Kuznicki 1978; Kuznicki and McCutcheon 1979) required judges to be consistent in their screening responses over several experimental sessions and in a further study (Kuznicki and Ashbaugh 1979) made the screening more appropriate to the experimental task by including binary mixtures among the screening stimuli. Yet, many studies have been reported where no screening or concept alignment procedures were used, although practice in magnitude estimation of line lengths or colors has sometimes been given (Settle et al. 1986).

To the sensory analyst, used to developing a pragmatic descriptive language using judges trained with standards, two problems would seem evident. Firstly, taste profiling often employs judges with no procedures for prior presentation of standards to define the descriptive terms. This would seem unlikely even to achieve agreement on descriptive labels, let alone concept alignment. The rationale for adopting this approach is that it is assumed that judges will understand what is meant by the four taste adjectives so well that training or presentation of standards would not be necessary. The second problem is that even if concept alignment were achieved, with agreement on the descriptive terms to be used for each concept, the limited number of descriptive terms might render the resulting conceptual structure inappropriate for stimuli like sodium benzoate, M.S.G. or KCl which evoke sensations that might not fall into combinations of the four categories.

Sensory evaluation has adopted several techniques that are widely used in psychophysics. Sometimes this has been appropriate, while on other occasions the differences in goals of the two disciplines has rendered such adoption inappropriate. Furthermore, it is as well to test the validity of any technique borrowed from another discipline. Because taste profiling has been used by some sensory analysts and because there has been no direct validation study of this method, the current study examined the technique from a concept formation, alignment and appropriateness point of view. The first experiment examined whether judges using the taste profiling technique had their concepts aligned with agreed descriptive terms for these concepts. The second study trained judges with standards to form aligned concepts with agreed labels for the four tastes and then tested whether these concepts were appropriate for descriptive analysis of various taste stimuli.

EXPERIMENT I

This first experiment tested the degree of concept alignment and common labelling of the concepts achieved by untrained judges, in standard taste profiling procedures, when untrained judges were used with no prior presentation of standards. If concept alignment had been achieved with a common set of descriptive labels and the concepts stabilized by sufficient training, certain requirements would be fulfilled in the description of a set of taste stimuli. These are as follows:

(1) Different stimuli would all be given different descriptive labels by a given judge.
(2) A judge would give the same descriptive label to a stimulus presented twice, in the absence of cross-adaptation effects.
(3) Judges would agree on the label chosen for a given stimulus.

Further, if tastes did not blend into new unitary tastes then a mixture of tastes would feel like a mixture. This is the assumption that taste is analytic rather than synthetic (Erikson 1982; Erikson and Covey 1980; O'Mahony *et al.* 1983) and is an assumption sometimes made in the interpretation of descriptive analytic data. Should it be true, two further requirements would be fulfilled:

(4) Descriptive labels would reflect the number of stimuli added to a mixture.
(5) Mixtures would be given labels that were composites of the labels used for the component stimuli.

As a comparison with the standard taste profiling procedure where responses were restricted to combinations of the available labels, judges were first allowed to use their own descriptive labels with no restriction and no suggestion of labels (O'Mahony and Thompson 1977). This comparison allowed the effect of restriction of labels per se to be studied, besides the effect of lack of prior agreement on labelling.

METHOD

Judges

28 judges (14M, 14F, 18-30 years), students at U.C.D. and their friends were used. Twenty had participated in taste experiments before but none had any experience with or specific training in taste description.

Stimuli

Single solute solution stimuli and two solute mixtures were prepared as follows: 300mM NaCl, 529mM fructose, 4.4mM citric acid, 62μM quinine hydrochloride (QHCl), 363mM monosodium glutamate (M.S.G.), 192mM KCl. Binary mixtures of these stimuli: NaCl/fructose, NaCl/citric acid, citric acid/fructose were prepared so that the molar concentration of each solute was the same as for the single solute stimuli. Preliminary experiments had established that the stimuli were easily distinguishable.

NaCl, citric acid and KCl were analytical reagent grade (Mallinkrodt Inc., Paris, KY), fructose was standard grade (Sigma Chemical Co., St. Louis, MO), M.S.G. was standard grade (U.S. Biochemicals Corp., Cleveland, OH), QHCl was reagent grade (BDH Chemicals, Poole, U.K.). The solvent was Millipore purified water (Milli-RO15 in series with Milli-Q system; reverse osmosis followed by ion exchange and activated charcoal; Millipore Co., Bedford, MA).

Ten mL samples of stimulus were dispensed (Oxford Adjustable Volume Dispensor, Model L; Oxford Laboratories Corp., Athy, Co. Kildare, Republic of Ireland) in 1 oz. unwaxed paper cups (Lily Portion Cups, Type 100; Lily Tulip Inc., Toledo, OH). All stimuli were presented at constant room temperature: 19-28C.

Procedure

Judges tasted all 9 solutions, the M.S.G., KCl and NaCl/fructose mixture were presented twice. This gave a total of 12 stimuli presented in random order. Judges were required to describe the tastes of these stimuli using their own descriptive labels without training or prior presentation of standards. Two conditions were used.

Condition 1. Firstly, judges described tastes using an 'unrestricted response' descriptive procedure. They were required to describe the taste of each stimulus in their own words; no words were supplied nor were they suggested. To allay anxieties, judges were informed that the stimuli were solutions of food ingredients, either single stimuli or mixtures of two. They were told that they might perceive one taste or as many as two, three, four or more. They were told that they may or may not be given a sample twice. Care was taken to avoid subjects having prior expectancies.

Condition 2. Secondly, for Condition 2, the experiment was repeated in a 'restricted response' condition. It was the same as Condition 1 with the modification that judges were only allowed to use combinations of the descriptive labels: 'sweet', 'sour', 'salty', 'bitter', 'other'. This gave 31 possible combinations. 'Tasteless' was also an allowed response but was never used.

After completing these two conditions, judges returned for a further session for control experiments.

Control Condition 1. This experiment was performed to ensure that judges could differentiate the tastes of the stimuli. Judges who described the taste of two different stimuli with the same label in Condition 1, returned to perform six triangle tests to establish that they could distinguish the stimuli significantly (criterion 5/6; binomial $p < 0.018$). This was found necessary for 12 judges (one stimulus pair for 10 judges, 3 pairs for 2 judges); they all performed the triangle tests successfully.

Control Condition 2. This control condition was performed to demonstrate that all the five requirements could be fulfilled if judges used a descriptive language, where standard stimuli were presented beforehand to define taste concepts which were given agreed descriptive labels.

The training procedure was as follows. All judges tasted the single solute solutions and were given a taste descriptive label for each one. The descriptive label chosen was its chemical name. For example, the label for NaCl was 'sodium chloride', for fructose it was 'fructose'. Judges were then allowed to repeat this as often as desired. They then tasted the six stimuli in random order, not knowing their identity, and were required to respond with the appropriate descriptive label. If they responded with the wrong label, they were corrected immediately. This was continued until all six single solute stimuli were given correct labels.

Judges then tasted and were required to name correctly, all 12 stimuli given in random order. They were told that the stimuli would be either single solutions or binary mixtures. No feedback was given. All subjects made at least one mistake. For training, they were then given the stimuli which they had misnamed and given the correct names. They went on to taste and attempt to name correctly with feedback, all twelve stimuli until this could be done without error. They were then tested as before with no feedback and if they did not perform perfectly, training was given in the same manner. All judges achieved perfect performance. This demonstrated that the five required conditions could be fulfilled perfectly if a descriptive system were employed which made use of prior presentation of physical standards to define the descriptive labels and sufficient training to learn the descriptive terms.

For all experiments, judges rinsed the mouth a minimum of 3 times (ranging 3-9) with tapwater, before tasting each stimulus. This minimized the effects of adaptation to residual amounts of stimulus left over from prior tastings (O'Mahony 1979a). Five rinses were taken at the beginning of each session. Conditions 1 and 2 were run during an initial experimental session (session lengths ranged 17-44 min). Control experiments were performed in a separate session (one subject needed 2 sessions); total time for control experimentation ranged 11-93 min.

RESULTS

All judges were completely successful in fulfilling all five requirements in Control Condition 2. Thus, the required concept alignment and common labelling of concepts was achieved when judges had been given standards to define the concepts and a common set of labels for those concepts. This indicated that the five requirements were possible to fulfill given suitable protocols. The measure of success of the taste profiling method and the modification using unrestricted selection of labels will be considered requirement by requirement.

To determine whether judges had fulfilled the first requirement, the data in Conditions 1 and 2 were examined to determine whether for each judge any of the 9 separate stimuli (including mixtures) were given the same label (facsimile labellings). The difference tests given in Control Condition 1 indicated that facsimile labellings were not caused by lack of perception of differences in taste between the stimuli. Because there were 9 solutions, there were a total of 8 possible facsimile labellings per judge (total 224 over 28 judges). A solution presented twice and given the same label was not scored as a facsimile labelling, yet such a stimulus does have twice the opportunity to score facsimile labellings with other stimuli.

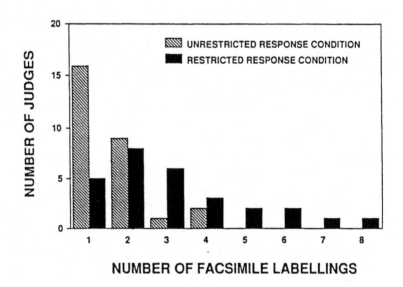

FIG. 1. HISTOGRAM INDICATING THE NUMBER OF SUBJECTS WHO GAVE VARIOUS NUMBERS OF FACSIMILE LABELLINGS FOR UNRESTRICTED AND RESTRICTED DESCRIPTIVE RESPONSE CONDITIONS

Facsimile labelling occurred both in the unrestricted (Condition 1) and restricted (Condition 2) response conditions; the frequencies of occurrence are given in Fig. 1. In the unrestricted response condition facsimile labelling tended to occur few times per subject. High frequencies (>3), indicating a more serious lack of descriptive differentiation between the separate stimuli, occurred only in the restricted response condition. Facsimile labelling was significantly less common in the 'unrestricted response' than in the 'restricted response' condition (for 12/28 vs. 22/28 subjects; for 17/224 vs. 59/224 possible facsimile labellings; binomial comparison of proportions $p \leq 0.008$). It could be argued that the fact that judges knew the stimuli to be food ingredients encouraged judges to use more descriptive labels in the unrestricted response condition and so reduce the number of facsimile labellings. For about half the judges, the descriptions had generally the same numbers of labels as were used in prior studies when no mention was made of the stimuli being food ingredients (O'Mahony and Thompson 1977; O'Mahony et al. 1976a, 1976b). It could further be argued that freedom to use texture descriptions in the unrestricted response condition could account for the fewer facsimile labellings. Texture labels, which do not necessarily indicate textural sensations, represented only 4.5% of the responses; this was too few to account for the differences. On the other hand, textural responses may have been described using taste labels.

To examine fulfillment of the second requirement, the labels given to the three stimuli that were presented twice (M.S.G., KCl, NaCl/fructose mixture) were examined to see whether they were labeled consistently by each judge. Table 1 indicates the number of judges who were consistent in their labelling of the repeated stimuli. The majority of subjects were inconsistent for at least one stimulus. The total possible number of inconsistent responses for the 28 subjects and 3 stimuli is 84 (3 × 28); there were significantly more in the unrestricted response condition (67 vs. 38: binomial comparison of proportions, $p < 0.00006$). However, there were not significantly more subjects giving inconsistent responses in the unrestricted response condition (27 vs. 24, $p = 0.9$).

TABLE 1.
NUMBER OF SUBJECTS WHO WERE CONSISTENT IN THEIR LABELLING OF THE THREE STIMULI WHICH WERE PRESENTED TWICE

Number of stimuli for which subjects were consistent	Number of Subjects	
	Unrestricted response condition	Restricted response condition
3	1	4
2	1	13
1	12	8
0	14	3

TABLE 2.
NUMBER OF SUBJECTS GIVING UNIQUE DESCRIPTIONS FOR EACH STIMULUS

Stimulus	Response Condition	
	Unrestricted	Restricted
NaCl	10	2
QHCl	18	2
fructose	17	3
citric acid	25	5
M.S.G. 1st presentation	28	4
M.S.G. 2nd presentation	24	2
KCl.1st presentation	21	2
KCl.2nd presentation	26	1
citric acid/fructose	16	1
citric acid/NaCl	22	5
fructose/NaCl.1st presentation	16	3
fructose/NaCl.2nd presentation	14	2

For the third requirement, the data were examined to see whether there was inter-subject agreement on the use of labels. For each stimulus, the number of subjects (out of 28) who gave a unique label (not agreeing with any other subject) were counted. The picture is complicated for M.S.G., KCl and the fructose/ NaCl binary mixture, because each stimulus was presented twice; the opportunity for uniqueness was halved compared with other stimuli. To allow simple comparison, the first and second presentations of each of these stimuli were regarded as presentations of separate stimuli. The number of judges giving unique descriptions are given in Table 2 for each stimulus. Over the 12 stimuli and 28 subjects, the total number of possible unique descriptions was 336. In the unrestricted response condition unique descriptions formed a significant majority (237/336, binomial $p < 0.00006$) while in the restricted response condition they were a significant minority (32/336, binomial $p < 0.00006$). The proportion of unique descriptions was significantly higher in the unrestricted response condition (binomial comparison of proportions, $p < 0.00006$). Thus, for inter-subject agreement, the unrestricted response condition was not satisfactory,

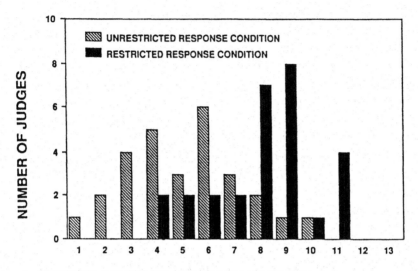

FIG. 2. HISTOGRAM INDICATING THE NUMBER OF JUDGES WHO GAVE LABELS FOR WHICH THE NUMBER OF WORDS MATCHED THE NUMBER OF SOLUTES IN THE STIMULUS

while the restricted response condition gave essentially good intersubject agreement for several stimuli: citric acid/fructose mixture (23 subjects called it 'sour-sweet'), QHCl (22, 'bitter') fructose/NaCl (25, 'salty-sweet', on the first and more consistent presentation), fructose (20, 'sweet'), NaCl (19, 'salty') citric acid/NaCl (18 'sour-salty'), KCl (12, salty-bitter; 9 bitter) citric acid (11, sour; 9 sour-bitter). Note that good inter-subject agreement was obtained for mixtures.

To test fulfillment of the fourth requirement, labels were examined to determine how they reflected the number of solutes present in the stimulus. As a strict test, the data were examined to see how frequently a single solute stimulus was given a one-word label and a binary mixture was given a two-word label. No judges fulfilled this requirement for all 12 stimuli in either Condition 1 or 2. The numbers of subjects giving labels for which the numbers of words matched the number of solutes in the stimulus are given in Fig. 2. The most stimuli for which the number of words matched the number of labels was eleven; four subjects achieved this in the restricted response condition. Inspection of the histogram indicates that such responses tended to be more common in the restricted response condition. The 12 stimuli and 28 subjects give a possible total of 336 responses in which the number of words in a label can match the number of solutes present; significantly more occurred in the restricted response condition (199 vs. 116; binomial comparison of proportions, p<0.00006). Matching the number of words in a label to the number of solutes occurred more for the binary mixtures. Over both response conditions the proportion of such responses was significantly higher for binary mixtures than for single solute stimuli (149/224 vs. 166/448; binomial comparison of proportions, p<0.00006).

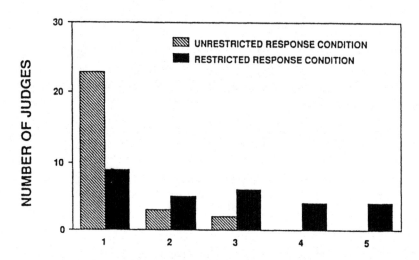

NUMBER OF RESPONSES TO BINARY MIXTURES MATCHING THE SUM OF THE COMPONENT MIXTURES

FIG. 3. HISTOGRAM INDICATING THE NUMBER OF SUBJECTS WHO GAVE LABELS TO BINARY MIXTURES WHICH WERE COMPOSITES OF THE LABELS GIVEN TO THE COMPONENT SINGLE SOLUTE STIMULI

To test fulfillment of the fifth and final condition, the data were examined for consistency to determine whether binary mixtures were given the sum of the labels of the components of the mixture, regardless of the number of words involved in the labels. Figure 3 gives the number of subjects who described various numbers of mixtures in terms of their component solute descriptions. It can be seen that although the trend was not strong, it was stronger for the restricted response condition. With the four binary mixtures and 28 subjects, there are a total of 112 possible responses to binary mixtures in each response condition. Although numbers were low, the total proportion of such matched responses was significantly higher in the restricted response condition (47/112 vs. 7/112; binomial comparison of proportions $p < 0.00006$).

The taste profiling procedure fulfilled the five conditions poorly, while a procedure akin to current food science descriptive analysis using physical standards to define descriptive labels fulfilled all the five conditions perfectly. Using the taste profiling procedure, only a significant minority of judges (binomial $p = 0.001$) gave all different stimuli separate descriptive labels. Such duplication of labels for different tastes leads to confusion in communication, confirming earlier work (Ishii and O'Mahony 1987b). Relaxing the restriction on available labels (unrestricted response condition) improved the situation significantly. Using the taste profiling procedure only a significant minority (binomial $p < 0.001$) of judges managed to give the same descriptive label to a given stimulus presented twice, while there was far from perfect agreement between judges on the descriptive labels to be applied to the stimuli. In both cases, the restriction in the number of descriptive labels available was an advantage, limiting the range of disagreement possible. Thus, the method of taste profiling failed to align judges' concepts with agreed descriptive labels, a necessary prerequisite for descriptive analysis.

For some uses of taste profiling it would be necessary to assume that the descriptive labels represented in some consistent way, the number and identity of the solutes present in the stimuli. Failure to fulfill the last two conditions indicated that this assumption cannot be made. In the unrestricted response condition, failure was even greater. Thus, a multi-adjectival descriptive term cannot necessarily be interpreted as indicating the presence of more taste sensations than a single adjective description. This is not surprising considering the partially synthetic nature of taste (Erickson 1982; Erickson and Covey 1980; O'Mahony *et al.* 1983).

EXPERIMENT II

Although the standard method of taste profiling would seem to have little success in aligning judges' concepts, it could be argued that training with standards would achieve this end. The question then becomes one of whether taste concepts that would be appropriate for the labels 'sweet', 'sour', 'salty', and 'bitter' have any general applicability in the descriptive analysis of tastes. Might not a restriction to only a four concept descriptive system be inadequate, there being tastes that might not fit such a categorization system? If this were so, responses to such tastes in terms of the four descriptive labels would give spurious data, being a response more to task demand characteristics than to perceived sensation.

This proposition was tested in the following manner. Judges were given standards and extensive training to define, learn, stabilize and align four taste concepts: 'sweet', 'sour', 'salty' and 'bitter'. The concepts were defined using 59 standard stimuli, including many mixtures. In this way, very broad concepts were defined for each of the four labels. For example, a sweet taste stimulus, only slightly modified by the addition of a small amount of NaCl, would be included not only in the 'sweet' concept but also in the 'salty' concept; it would be labeled 'sweet-salty.' Preliminary experimentation indicated that this gave much broader than usual definition of the four concepts, to give the experiment the greatest chance of succeeding. Use of mixture stimuli as standards also allowed for possible blending of tastes caused by any synthetic properties (Erickson 1982; Erickson and Covey 1980; O'Mahony et al. 1983).

Having aligned the concepts of the trained judges, new stimuli were presented. Should the concepts be appropriate to the tastes elicited by these new stimuli, the sensations caused by them would fall within the judges' concepts. As the judges' concepts were aligned, a given sensation would fall into identical concepts for each judge; the judges would agree on their descriptions. If the concepts were not appropriate, the judges would need to modify their concepts. Such modification without the control afforded by new standards would allow the judges' concepts to vary in an uncontrolled way, resulting in loss of alignment of the concepts. The descriptions would then disagree. In this manner, the appropriateness of the four concept system was tested.

METHOD

Judges

Five students from U.C.D. (1M, 4F, 20-29 years) were used; they were naive to taste psychophysical testing. All were tasters of phenylthiocarbamide (PTC), having distinguished without difficulty $67\mu M$ PTC from purified water on five successive (binomial $p = 0.004$) 3-AFC tests (Green and Swets 1966).

The sensitivity for each judge to the tastes of NaCl, fructose, QHCl and citric acid was known and available for reference. Sensitivities were determined after the main experiment using the R-index rating procedure (Brown 1974; O'Mahony 1979b, 1986). The procedure for these determinations was as follows. After practice, familiarization with the stimuli, and thorough rinsing (12 rinses), judges tasted twenty 10 mL samples each of 5 mM, 3 mM and 1 mM NaCl and purified water (signal stimuli), as well as twenty further purified water samples (noise stimuli), in random order. The low stimulus concentrations did not necessitate interstimulus rinsing (O'Mahony 1979a). Stimuli were rated as 'salt', 'salt-not sure', 'water', 'water-not sure'. From such responses, R-indices, which give estimates of the probability of distinguishing between water and 5 mM, 3 mM or 1 mM NaCl were calculated; such indices were measures of sensitivity. The calculated probability of distinguishing water (signals) from water (noise), provided an estimate of the variance of chance levels (O'Mahony 1988).

For other stimuli, the procedure was the same except that stimulus concentrations were: fructose 30 mM, 18 mM and 6 mM; QHCl 30μM, 18μM and 6μM; citric acid 0.1 mM, 0.06 mM and 0.02 mM.

The procedure was then repeated except that sensitivity to NaCl, fructose, citric acid and QHCl was not measured in water. Instead, it was measured in a medium comprised of the other three stimuli at their weaker concentrations, namely: 300 mM NaCl, 529 mM fructose, 4.4 mM citric acid, 90μM QHCl. The concentrations of stimuli to be detected were higher than in the pure aqueous solutions; they were: NaCl, 15, 10, 5 mM; fructose 78, 60, 42 mM; citric acid 0.3. 0.2, 0.1 mM; QHCl 40, 30, 20μM.

Stimuli

Fifty-nine stimuli were used as standards to define the taste concepts. These were solutions of NaCl, fructose, citric acid, QHCl and their mixtures, detailed in Table 3.

Five final test single solute stimuli were prepared to test the appropriateness of the concepts formed: 363 mM M.S.G., 192 mM KCl, 171 mM sodium benzoate, 431 mM NaHCO$_3$, 208 mM sodium citrate.

The 529 mM fructose, 4.4 mM citric acid, 62μM QHCl, and the five test single stimuli were previously matched in intensity to 300 mM NaCl; the 958 mM fructose, 8.8 mM citric acid and 120μM QHCl were matched to 500 mM NaCl. This was done using 11 judges with a double-blind modification of the double staircase procedure (Cornsweet 1962), with random alternations and permitted equality judgements. The judges had been screened for matching skill by matching NaCl solutions to 300 mM NaCl; these judges were not used in the main experiment, in case their experience with the five test stimuli biased their descriptions.

The 12 mixtures at the bottom of Table 3 consisted of a first solute with a trace of the second solute, sufficient only to alter the taste of the first solute slightly. To include the label for the trace solute in the description of this mixture ensured that the associated concept was broad. Preliminary experiments had indicated that such a broad concept included more stimuli than an everyday language concept, a concept used prior to any training. To ensure that such solutions were distinguishable from the solution comprising of the stronger solute alone, the judges performed difference tests; judges were required to

TABLE 3.
STIMULI USED AS STANDARDS TO DEFINE THE FOUR TASTE CONCEPTS IN EXPERIMENT II

<u>Single stimuli</u>: 300mM NaCl, 259mM fructose, 4.4mM citric acid, 62μM QHCl
500mM NaCl, 958mM fructose, 8.8mM citric acid, 120μM QHCl

<u>Total = 8</u>

Mixtures indicated below, made up to concentrations: 300mM NaCl, 529mM fructose, 4.4mM citric acid, 90μM QHCl.

| NaCl/fructose | NaCl/citric acid | NaCl/QHCl |
| fructose/citric acid | fructose/QHCl | citric acid/QHCl |

NaCl/fructose/citric acid
NaCl/QHCl/citric acid

NaCl/fructose/QHCl
QHCl/fructose/citric acid

NaCl/fructose/QHCl/citric acid

<u>Total = 11</u>

Mixtures in the same combinations as above made up to the same concentrations, except that in every combination the mixture was made up with one of the stimuli having a higher concentration of 500mM NaCl, 958mM fructose, 8.8mM citric acid or 120μM QHCl.

<u>Total = 28</u>

<u>Binary mixtures</u>:

300mM NaCl with 105.8mM fructose, 0.88mM citric acid or 36μM QHCl
529mM fructose with 30mM NaCl, 0.44mM citric acid or 18μM QHCl
4.4mM citric acid with 60mM NaCl, 105.8mM fructose or 36μM QHCl
62μM QHCl with 30mM NaCl, 52.9mM fructose or 0.44mM citric acid

<u>Total = 12</u>

TOTAL STIMULI = 59

distinguish stimuli on five successive 3-AFC tests (binomial $p = 0.004$). Judges performed these tests after the stimuli had been introduced into the concept learning session. All judges distinguished the stimuli; had they not, they would have been eliminated from the study.

All stimuli were prepared from pure non-odorous chemicals. As a further check, to ensure that there were no odorous impurities present, the stimuli were tested for odor. Highest concentration single solute solutions were given to the subjects in ten paired comparisons, to be distinguished by odor from purified water. No judges could distinguish them from water better than chance (binomial $p > 0.17$). There was one exception. One judge detected sodium benzoate by odor on all ten trials ($p < 0.001$). Because of this, noseclips were used for sodium benzoate tasting for all subjects. It is hypothesized that the odor may have been due to a trace volatile impurity.

NaCl, citric acid, sodium citrate, $NaHCO_3$ and KCl were analytical reagent grade (Mallinckrodt Inc., St. Louis, Missouri); sodium benzoate, β-D(-)fructose were standard grade from Sigma Chemical Company (St. Louis, Missouri); M.S.G. was standard grade from United States Biochemical (Cleveland, Ohio); QHCl was laboratory reagent grade (B.D.H. Chemicals Ltd., Pool, Dorset, U.K.). Solutions were made up in water, purified by passing deionized water through a Milli-R015 filtration and reverse osmosis system (conductivity $< 10^{-6}$ mho/cm; surface tension ≥ 71 dyne/cm).

Fructose solutions were allowed to stand at least four hours at room temperature to mutarotate. All solutions were freshly prepared at least each week (NaCl every fortnight) and stored in steamed pyrex bottles. Fructose, citric acid, M.S.G., and sodium citrate were refrigerated to inhibit microbial growth and were warmed for presentation at room temperature, 20-28C (± 0.5C), immediately prior to experimentation. Approximately 20 mL of stimulus was presented for tasting in a 1 oz. unwaxed paper cup (Dixie No. 4100: American Can Company, Greenwich, Connecticut). The appropriate mixtures were all prepared beforehand except for standards with two or three solutes. These were made up during experimentation, by mixing appropriate volumes of more concentrated stock solutions from dispensors (Oxford Adjustable Volume Dispensor, Model L; Oxford Laboratories International Corp., Athy, Co. Kildare, Republic of Ireland).

Procedure

Judges learned the four taste concepts by tasting the 54 standard stimuli and learning to attribute to them the appropriate labels. Any solution containing NaCl was labeled 'salty', citric acid: 'sour', fructose: 'sweet' and QHCl: 'bitter'. (Thus, a mixture of NaCl, citric acid and fructose was labelled 'salty-sour sweet'). The four labels allowed a possible of 15 word combinations for response. Having learned to give the appropriate response to each of the stimuli without error, the judges were then required to taste the five new stimuli, believing them to be further mixtures of four stimuli, and to describe them in terms of which of the four taste concepts their sensations fell into. Consistency of labelling between judges was noted to determine whether the taste concepts generalized to these new stimuli in a consistent manner. If the four concepts were appropriate for communicating the sensations elicited by the new stimuli, the concepts should generalize to them in a consistent manner.

To learn the standards, judges sipped and expectorated approximately half the sample of solution, responding verbally. If they responded with the appropriate label, as defined, they were told that they were correct and presented with the next solution. If they did not respond with the appropriate label, they were told the appropriate response and required to taste the remaining portion of the sample. They were then allowed to taste more of that sample if they so desired.

All eight single solute stimuli were available as labelled standards, for tasting at anytime during the session. The judge could also call for any of the stimuli for comparison by naming the appropriate labels. Judges were informed that any number of the solutes used for the single solute stimuli might be present in a mixture and that the order of presentation of the stimuli was random with the possibility of repetition.

Initially all the single, solute stimuli and solutions with two solutes (but not the 12 binary mixture stimuli at the bottom of Table 3) were presented to judges. Once these had been learned, solutions with three and four solutes were introduced. Finally, once these had been learned, the 12 binary mixtures at the foot of Table 3 were introduced. Subjects proceeded at their own pace, taking 3-5 approx. 40 mL purified water mouth rinses between tastings (always five after a stimulus with a QHCl solute), to minimize the effects of adaptation to residuals from priorly tasted stimuli (O'Mahony 1979a). Session lengths were determined by the judge, each session starting with the judge tasting all the single solute standards. Training was continued informally in this manner until the experimenter estimated that the judges had learned the concepts sufficiently well to be able to respond to all 59 training solutions with the appropriate labels.

Judges were then tested under stricter test conditions to ensure that they could respond to all 59 training solutions appropriately, before being presented with the nonprimary stimuli. A criterion of 100 correct successive responses to stimuli, including all 59 training solutions, was required. Under test conditions, only the eight solutions with single solutes were available as labeled standards; other solutions could not be requested for comparison. The experimenter was isolated behind a screen and indicated a correct response using a 'click' apparatus. Judges were allowed to taste the test solution and standards as often as they wished before responding. If the judge responded with the wrong label, he was corrected and the session was continued as an informal concept learning session. Further training was given as necessary, until the judge demonstrated sufficient skill to be able to attempt a further session under test conditions.

Once the criterion of 100 successive correct responses had been achieved under test conditions, four of the five final test stimuli (sodium benzoate omitted) were immediately presented three times each, in random order, under the same test conditions. Judges were told that the nonprimary stimuli would be either the same single solutes or mixtures of two, three or four of them, in new proportions. If the responses of a judge to each of the three tastings of a given nonprimary stimulus were inconsistent, the judge was required to taste the stimulus again, having been told his responses to the stimulus beforehand. He was then asked to make a final decision. The judge then donned a noseclip, to eliminate olfactory cues, and tasted sodium benzoate as often as desired, arriving at a final judgement once again. The judge was told that the reason for using the noseclip was that some researchers, in error, had suggested that this particular mixture of stimuli might react to give an odor. The experimenters explained that they felt that although this was impossible, they were issuing the judges with noseclips to allay any criticism by other researchers along these lines. All judges appeared completely convinced and later confirmed this.

During the learning of the concepts; the final 12 binary mixtures were also introduced to the subjects under test conditions, to allow assessment of whether they fell within the concepts already formed up to that point.

During the learning of the concepts, subjects came to the laboratory regularly over periods of up to six weeks. The total experimental times needed to learn the concept, to a criterion of 100 successive correct responses, were for the five subjects: 67.75, 35.0, 27.25, 14.0 and 13.5 h. The final part of the experiment, in which nonprimary stimuli were presented, ranged 24-39 min.

RESULTS

Table 4 indicates the labels given by the five judges to the five final test stimuli. It is evident that there was little agreement between judges on the labels chosen for the stimuli; no more than two judges could agree on the label for a given stimulus. Furthermore, a judge sometimes gave different stimuli the same label while reporting that they elicited different sensations. Subjective reports indicated that the new stimuli were harder to label because they did not appear to elicit sensations falling within the four concepts. Judges reported that the four labels were inadequate for describing them. There was little consistency in the labelling of these stimuli. For only eight of the total 25 descriptions was there intrajudge consistency. Interestingly, all but two of the 25 descriptions included the label 'salty'. All subjects reported a stimulus in which an extra taste was present which, contrary to experimental instructions, did not fall within the four concepts. Subjects #3 and #5 reported that although they used two taste labels, only one taste was present (#3: 'sour-salty'; #5: all 'salty-bitter' descriptions).

TABLE 4.
LABELS GIVEN TO THE TASTES OF THE NONPRIMARY STIMULI AFTER COMPLETION
OF CONCEPT LEARNING

Subject	Stimuli				
	M.S.G.	sodium benzoate	KCl	NaHCO$_3$	sodium citrate
1.	sour-salty	sweet-salty-bitter	sour-salty	sour-salty-bitter	sweet-salty-bitter
2.	sweet-salty-(+)	sweet-salty-bitter	salty-bitter-(+)	salty-bitter-(+)	sweet-salty-bitter
3.	sour-salty	sweet-(+)	salty-bitter*	sweet-salty	sweet-salty-(+)*
4.	salty*	sweet-salty	sour-salty	salty-bitter*	salty-bitter
5.	salty-bitter*	sweet-bitter	salty-bitter*	salty-bitter*	salty-bitter*

(+) denotes the presence of an extra taste which could not be described
* denotes consistency of response over three replications: not applicable to sodium benzoate

A similar effect was found when subjects first tasted the final 12 mixtures; only 52% of the responses had the appropriate labels. This indicated that the concepts did not even generalize consistently to mixtures in which the stimuli were not mixed in the same proportions as the standards.

There are alternative explanations for the lack of generalization of the four taste concepts to the five test stimuli. One is that the concepts were appropriate to the tastes of the new stimuli, but that the range of sensations elicited by the new stimuli were not all equally perceived by the subjects, because of differences in sensitivity. Thus, inconsistency in labelling could have been caused by the differences in sensitivity of subjects to the defined 'salty', 'sweet', 'sour' and 'bitter' tastes, elicited by the new stimuli. In this way, Subject #1 was more sensitive to the defined sour taste concept than Subject #4, because the former labeled M.S.G. as 'sour-salty', whereas the latter labelled it as 'salty'. However, the subjects' taste sensitivities to the stimuli on which the concepts were based (NaCl, fructose, QHCl, citric acid) did not account for the lack of inter-subject consistency, encountered for the new stimuli. Furthermore, the inter-subject differences were so inconsistent between stimuli, that it would have been impossible to infer any consistent pattern of sensitivity difference among the subjects.

It could still be argued that the four taste concepts were sensations elicited by more than one taste receptor type. Thus, the sourness perceived by Subject #3 and not by #4 for M.S.G. was the result of one sour system. The sourness perceived by Subject #4 but not by #3 for KCl would have been the result of a separate sour system. Following this argument, it would need to be concluded from the results of only five subjects and five stimuli that there were two separate taste systems serving the defined sour taste, two separate systems for the sweet taste and two for the bitter. This argument is logically possible yet lacks parsimony and evolutionary sense, when so many separate systems result in so few sensations.

DISCUSSION

The present study indicated that the method of taste profiling failed to align judges' concepts and establish a common set of descriptive labels. Clearly, the assumption that untrained judges already have a set of aligned concepts with agreed labels is unjustified. Further, when judges had their concepts aligned by training with physical standards, the four concept system had limited applicability. It did not generalize to stimuli that were not mixtures of the standards. There were even limitations in generalization to mixtures whose components had intensities very different from those of the standards. This means that application of the method, using trained judges for assessment of food products like

M.S.G., or KCl might result in data more dependent on demand characteristics than taste sensation. It might have limited applicability to mixtures of the standards or perhaps to sweeteners with side tastes like bitter aftertastes (Acton and Stone 1976; DuBois *et al.* 1977, 1981; Kier 1980). Yet, even here it would seem sensible to interpret data with caution. Earlier work has indicated that for a set of taste concepts, limitation in the labels available for their description to 'sweet', 'sour', 'salty', 'bitter' and 'other' causes distortions caused by facsimile labelling (Ishii and O'Mahony 1987b). It would seem wiser to use current methods for descriptive analysis and not adopt taste profiling.

Psychophysicists have used taste profiling for the descriptive analysis of stimuli that might be regarded as having 'nonprimary' tastes: M.S.G. (Bartoshuk *et al.* 1974a; Peryam 1963), sodium benzoate (Bartoshuk *et al.* 1988; Peryam 1960, 1963), KCl (Bartoshuk *et al.* 1988; Kuznicki and Ashbaugh 1979; McBurney and Shick 1971; McBurney *et al.* 1972; Murphy *et al.* 1981; Peryam 1963; Smith and McBurney 1969). The current study demonstrates the questionability of trying to describe the taste of M.S.G. in terms of the four primary tastes (Bartoshuk *et al.* 1974a; Cairncross 1948; Dorn 1949; Galvin 1948; Hanson *et al.* 1960; Lockhart and Gainer 1950; Mosel and Kantrowitz 1952, 1954; Peryam 1963; Sjöström and Crocker 1948), rather than treat its taste as conceptually different, as do the Japanese with the label 'umami' (Cagan *et al.* 1979; Kuninaka 1965, 1981; O'Mahony and Ishii 1986; Torii and Cagan 1980; Yamaguchi 1979; Yamaguchi and Kimizuka 1979; Yamaguchi *et al.* 1971, 1982) or 'Ajinomoto taste' (O'Mahony and Ishii 1986). The only feasible way to describe a taste to a person, is to give it to them and name it. In the same way, 'red' cannot be described to a blind man. Redness can only be communicated by showing red objects to a person and labelling them 'red'.

Similar conclusions about labelling of taste sensations would be made from multivariate studies which indicated the spatial locus of M.S.G. to be distinct from those of the four primaries (Schiffman *et al.* 1980; Yoshida 1963). The same was also found for KCl (Schiffman and Erickson 1971) and other stimuli (Schiffman and Dackis 1975; Schiffman *et al.* 1975, 1980). Similar conclusions were also made from taste sorting experiments (Ishii and O'Mahony 1987b).

The restriction of taste descriptive labels to only 'sweet', 'sour', 'salty' and 'bitter' has been defended by assuming that all other oral sensations are nongustatory. If a stimulus like M.S.G. does not appear to fit the descriptive scheme, the assumption is that this occurs because it also elicits tactile stimulation. This may well be true but it is irrelevant to the results and conclusions of this experiment.

There are reports of tactile stimulation caused by NaCl (Green and Gelhard 1989). It would certainly seem likely for stimuli like acids, KCl and perhaps M.S.G. Introspection would not be able to determine where taste finished and tactile or common chemical sense stimulation began. In Erickson's (1983) words, we cannot be exactly sure what we mean by a 'taste'. Yet, the goal of descriptive analysis is to communicate sensations; a descriptive term does not indicate what senses are involved in producing the sensation. It is naive to assume a direct relationship between language and physiological mechanisms. The important feature for descriptive analysis is that concepts are aligned; which senses are involved in the maintenance of that sensory concept is not an essential province of descriptive analysis.

The only sure way to be sure of the senses involved in a sensation is to block sensory input to eliminate cross-sensory interference (O'Mahony and Goldstein 1987a, 1987b; O'Mahony et al. 1983, 1985a, 1985b, 1985c). Yet, this is not always possible. The only way to differentiate between taste and tactile stimulation would be to use judges with blocked input channels like severed nerves. Merely to ask judges to ignore tactile stimulation and respond only to taste makes the assumption that they have the skill to do this. Such an introspective technique could only be considered if the judges were well trained and validation studies had shown that they had the required skills.

The long training times in Experiment II, and the information to be processed with each judgement (Attneave 1959; Beebe-Center et al. 1955; Garner 1952) force a consideration of whether taste has a 'synthetic' nature (blending like color) or whether it is purely 'analytic' (not blending like sound). These concepts have been explored by Erickson (1977). Should taste be perfectly analytic, the task for subjects during training in Experiment II would be merely to search for the four component tastes in each mixture and report which of the four were present; four stimuli only amount to 2 bits of information. Should there be complete synthesis, the judge would need to learn the unique tastes of 59 stimuli, a 5.88 bit task. This would stretch the channel capacity for taste (Beebe-Center et al. 1955). The possibility that taste may be only partially synthetic (O'Mahony et al. 1983), as well as the logical possibility of tastes being suppressed in mixtures, makes it impossible to estimate channel capacities exactly. The long training times and subjective reports suggest at least a degree of synthesis for taste. However, McBurney and Gent (1979) state that taste is analytic. Yet, reports of changes in taste quality of stimuli when mixed with other stimuli (Cairncross 1948; Erickson 1981, 1982; Kuznicki and Ashbaugh 1979, 1982; Moskowitz 1972; Schiffman and Erickson 1980) suggest some degree of synthesis. Appropriate assignment of suitable descriptive terms, as was encountered after the judges had been trained, has been taken to imply that taste is analytic (Kuznicki and Ashbaugh 1982; McBurney and Gent 1979). This is a logical error. Subjects could merely have learned to assign the correct labels to new synthetic experiences; long training times would be compatible with such a learning task. Bartoshuk (1975) argued that taste could be classified either as synthetic or analytic, because the idea of fusion of taste qualities is not clearly defined. Direct judgements on the singularity of taste mixtures (Erickson 1981,

1982; Erickson and Covey 1980) have suggested that taste is partly synthetic partly analytic, but the categorization method used was prone to response bias; more recently a rating scale method has been used (Erickson 1982). O'Mahony et al. (1983) using signal detection measures to avoid response bias, reported that taste is sufficiently analytic to distinguish single solute stimuli from many solute stimuli, but sufficiently synthetic to confuse the number of solutes present in multiple solute stimuli. Further, all single solute stimuli did not taste equally singular. The synthetic/analytic nature of the sensory characteristics of the components of foods is under-researched but is of vital importance for understanding the cognitive mechanisms involved in descriptive analysis. Synthetic senses would require presentation of mixtures as standards for the training of judges.

The long training times required for subjects to learn the appropriate descriptive terms in Experiment II confirm the result of a previous study (O'Mahony and Buteau 1982). Kuznicki and Ashbaugh (1979), however, found that training had little effect on taste profiling; the profiles from untrained and trained subjects were essentially very close. Their study differs from this one in that their subjects were only required to label single and binary solute stimuli. Their untrained subjects had already had considerable experience, having been selected on the basis of having shown consistency in naming 'primary' taste stimuli. The training given in their study consisted only of five fairly brief sessions, involving the differentiation and naming of single 'primary' stimuli and mixtures of two. Thus, the difference between trained and untrained subjects was not great, the profiling task was relatively simple, and the training was not intense. The lack of agreement with the present study is not surprising.

Before conclusion, it is worth considering how a method like taste profiling evolved. The choice of the four descriptive taste labels is the result of tradition based on the contentious assumption that there are four basic or primary tastes. This widely discussed issue has been thoroughly reviewed by Erickson (Erickson 1977, 1981, 1982, 1983, 1984a, 1984b, 1985a, 1985b; Erickson and Covey 1980; Erickson and Schiffman 1975; Schiffman and Erickson 1980) as well as many other authors (Andersen 1970; Dethier 1978; Faurion 1983; Frings 1954; Kuznicki and Ashbaugh 1979; McBurney 1974; McBurney and Gent 1979; O'Mahony and Ishii 1987; Price and DeSimone 1977; Scott and Chang 1984). The issue is confused by a lack of differentiation between physiological and psychological levels of argument, as well as by a lack of definition of 'primacy'. The number of primary tastes could refer to the number of taste receptors, the number of transduction mechanisms, the number of neural codes, channels or fibre types or the number of gustatory cortical processes, areas or cell types. In the absence of sufficient evidence, it would seem unwise to assume a priorly specified number of any of these physiological entities. Considerations of some type of 'psychological primacy' also suffer from lack of definition. A more useful approach would be to examine specific physiological or psychological processes per se. To award any of these processes with the attribute 'primary' is irrelevant to the understanding of taste mechanisms.

It may be hypothesized that the choice of these particular four tastes is more dependent on language and culture than physiology. There is a need to communicate about food and an appropriate language is developed for this purpose; the language will depend on the cuisine of that culture. There is the possibility that these are the wrong categories; Erickson (1985a) has discussed the implications of this. Williams (1977) has discussed how a spurious hypothesis like the concept of four basic tastes can gain such wide acceptance.

The main effect of the primacy issue has been the creation of traditions. It has often resulted in the number of stimuli used in an experiment being restricted to four. It has restricted the number of labels used for taste concepts to four (or five with an 'other' label), despite reports of difficulties with such a restriction (Murphy 1971; O'Mahony and Thompson 1977; Schiffman and Erickson 1980) and despite the tendency for subjects to use more labels when the restriction is lifted (O'Mahony and Thompson 1977; O'Mahony et al. 1976a, 1976b).

In their discussion of the use of restricted labels for taste descriptive analysis, Dzendolet and Meiselman (1967) cite a technique of color naming. Here, only the labels 'red', 'yellow', 'green' and 'blue' are allowed, with an indication of relative strength (Boynton and Gordon 1965; Boynton et al. 1964; Jacobs and Gaylord 1967; Jameson and Hurvich 1959; Kintz et al. 1969). The reported success with this technique is not surprising. The few judges used in these studies were only required to label relatively simple spectral colors; colors like 'brown' were not considered. For more general naming of colors, color chips or color standards are required (Munsell 1976). Even though everyday language concepts only constitute an approximate tool, the constant communication regarding colors in western culture would ensure greater commonality for color labels than for taste labels; this is certainly true for pre-school children (O'Mahony et al. 1978). There are certainly many examples of taste 'confusions' like the 'sour-bitter' confusion (Amerine et al. 1965; Gregson and Baker 1973; McAuliffe and Meiselman 1974; Meiselman and Dzendolet 1967; Myers 1904; Pangborn 1961; Robinson 1970) which can be eliminated merely by definition of the labels and concepts (O'Mahony et al. 1979).

Argument has been made for the restriction of taste labels to 'sweet', 'sour', 'salty' and 'bitter' from the fact that when given these labels, judges rarely use any others. However, O'Mahony and Thompson (1977) demonstrated how the little use of other labels can be caused by task demand characteristics and suggestion. Argument has also been made for the restriction of labels to four, from the fact that subjects can reliably respond to tastes using only these labels. Yet, this does not imply validity. Subjects can reliably describe days of the week and states of the U.S.A. using only these labels (O'Mahony 1983) and there can be no implication that days or states have tastes. The fact that Friday and Saturday are sweet and Monday is bitter has obvious nongustatory interpretations.

Finally, it is worth noting that scaling and descriptive analysis are essentially introspective techniques and as such are less objective than difference testing, which requires the judge to demonstrate behaviorally that he can distinguish differences. Introspective techniques should be validated before use. If, in descriptive analysis, a judge reports the presence of a given sensation, this report can be trusted if that judge has demonstrated that he can reliably report the sensation when it is introduced during training and not report it when it is not introduced. For this, sufficient concept alignment is essential.

ACKNOWLEDGMENTS

The authors would like to thank the following for their assistance: Melanie Almen, Sanah Sheldon, Claudia Carhart, Jeanine Corea-Prado, Steve Dorfman, Deborah Dreyer Michael Feno, Corie Taira, Lisa Irskens, Jane Killebrew, Natalie Lum, Caroi Morikawa, Nancie Odbert, Mary Rice, Lee Ann Sucht, Elaine White, Joanne Wong, and Sau-Yin Wong.

REFERENCES

ACTON, E.M., LEAFFER, M.A., OLIVER, S.M., and STONE, H. 1970. Structure-taste relationships in oximes related to perillartine. J. Agr. Food Chem. *18*, 1061-1068.

ACTON, E.M. and STONE, H. 1976. Potential new artificial sweetener from study of structure-taste relationships. Science *193*, 584-586.

AMERINE, M.A., PANGBORN, R.M., and ROESSLER, E.B. 1965. *Principles of Sensory Evaluation of Food*, pp. 267-270, Academic Press, New York.

ANDERSEN, H.T. 1970. Problems of taste specificity. In *Taste and Smell in Vertebrates* (G.E.W. Wolstenholme and J. Knight, eds.) pp. 71-82, J. & A. Churchill, London.

ATTNEAVE, F. 1959. *Application of Information Theory to Psychology*, Holt, Rinehart and Winston, New York.

BARTOSHUK, L.M. 1974. NaCl thresholds in man: Thresholds for water taste or NaCl taste. J. Comp. Physiol. Psychol. *87*, 310-325.

BARTOSHUK, L.M. 1975. Taste mixtures: Is mixture suppression related to compression? Physiol. Behav. *14*, 643-649.
BARTOSHUK, L.M. 1978. The psychophysics of taste. Amer. J. Clin. Nutr. *31*, 1068-1077.
BARTOSHUK, L.M. 1979. Bitter taste of saccharin related to the genetic ability to taste the bitter substance 6-n-propylthiouracil. Science *205*, 934-935.
BARTOSHUK, L.M., CAIN, W.S., CLEVELAND, C.T., GROSSMAN, L.S., MARKS, L.E., STEVENS, J.C., and STOLWIJK, J.A.J. 1974a. Saltiness of monosodium glutamate and sodium intake. J. Amer. Med. Assoc. *230*, 670.
BARTOSHUK, L.M., GENTILE, R.L., MOSKOWITZ, H.R., and MEISELMAN, H.L. 1974b. Sweet taste induced by miracle fruit (synsepalum dulcifium). Physiol. Behavior *12*, 449-456.
BARTOSHUK, L.M., LEE, C-H. and SCARPELLINO, R. 1972. Sweet taste of water induced by artichoke *(cynara scolymus)*. Science *178*, 988-990.
BARTOSHUK, L.M. and CLEVELAND, C.T. 1977. Mixtures of substances with similar tastes. A test of a psychophysical model of taste mixture interactions. Sensory Processes *1*, 177-186.
BARTOSHUK, L.M., MURPHY, C. and CLEVELAND, C.T. 1978. Sweet taste of dilute NaCl: Psychophysical evidence for a sweet stimulus. Physiol. Behav. *21*, 609-613.
BARTOSHUK, L.M., RIFKIN, B., MARKS, L.E. and HOOPER, J.E. 1988. Bitterness of KCl and benzoate: related to genetic status for sensitivity to PTC/ PROP, Chem. Senses *13*, 517-528.
BEEBE-CENTER, J.G., ROGERS, M.S. and O'CONNELL, D.M. 1955. Transmission of information about sucrose and saline solutions through the sense of taste. J. Psychol. *39*, 157-160.
BOYNTON, R.M. and GORDON, J. 1965. Bezold-Brucke hueshift measured by color naming technique. J. Optical Soc. Amer. *55*, 78-86.
BOYNTON, R.M., SCHAFER, W. and NEUN, M.E. 1964. Hue-wavelength relation measured by color-naming method for three retinal locations. Science, *146*, 666-668.
BROWN, J. 1974. Recognition assessed by rating and ranking. Brit. J. Psychol. *65*, 13-22.
BUJAS, Z., SZABO, S., KOVACIC, M. and ROHACEK, A. 1974. Adaptation effects on evoked electrical taste. Percept. Psychophys. *15*, 210-214.
CAGAN, R.H., TORII, K. and KARE, M.R. 1979. Biochemical studies of glutamate taste receptors: The synergistic taste effect of L-glutamate and 5'ribonucleotides. In: *Glutamic Acid: Advances in Biochemistry and Physiology* (L.J. Filer, S. Garattini, M.R. Kare, W.A. Reynolds, and R.J. Wartman, eds.) pp 1-9. Raven Press, New York.

CAIRNCROSS, S.E. 1948. The effect of monosodium glutamate on food flavor. In *Flavor and Acceptability of Monosodium Glutamate* pp. 32-38. Food and Container Inst., Chicago, IL.

CAIRNCROSS, S.E. and SJÖSTRÖM, L. B. 1950. Flavor profiles—A new approach to flavor. Food Technol. *4*, 308-311.

CARDELLO, A.V. 1978. Chemical stimulation of single human fungiform taste papillae: Sensitivity profiles and locus of stimulation. Sensory Processes, *2*, 173-190.

CARDELLO, A.V. 1979. Taste quality changes as a function of salt concentration in single human taste papillae. Chem. Senses Flav. *4*, 1-13.

CARDELLO, A.V. and MURPHY, C. 1977. Magnitude estimates of gustatory quality changes as a function of solution concentration of simple salts. Chem. Senses Flav. *2*, 327-339.

CAUL, J.F. 1957. The profile method of flavor analysis. Adv. Food Res. *7*, 1-40.

CIVILLE, G.V. and LAWLESS, H.T. 1986. The importance of language in describing perceptions. J. Sensory Stud. *1*, 203-215.

CORNSWEET, N. 1962. The staircase-method in psychophysics. Amer. J. Psychol. *75*, 485-491.

COWART, B.J. 1987. Oral chemical irritation: Does it reduce perceived taste intensity? Chem. Senses. *12*, 467-479.

DE SIMONE, J.A., HECK, G.L. and BARTOSHUK, L.M. 1980. Surface active taste modifiers: A comparison of the physical and psychophysical properties of gymnemic acid and sodium lauryl sulfate. Chem. Senses *5*, 317-330.

DETHIER, V.G. 1978. Other tastes, other worlds. Science *201*, 224-228.

DORN, H.W. 1949. Role of monosodium glutamate in seasoning. Food Technol. *3*, 74.

DUBOIS, G.E., CROSBY, G.A., LEE, J.F., STEPHENSON, R.A. and WANG, P.C. 1981. Dihydrochalcone sweeteners. Homoserine-dihydrochalcone conjugate with low aftertaste, sucrose-like organoleptic properties. J. Agric. Food Chem. *29*, 1269-1276.

DUBOIS, G.E., CROSBY, G.A., STEPHENSON, R.A. and WINGARD, R.E. 1977. Dihydrochalone sweeteners: Synthesis and sensory evaluation of sulfonate derivatives. J. Agric. Food Chem. *25*, 763-772.

DZENDOLET, E. and MEISELMAN, H.L. 1967a. Gustatory quality changes as a function of solution concentration. Percept. Psychophys. *2*, 29-63.

DZENDOLET, E. and MEISELMAN, H.L. 1967b. Cation and anion contributions to gustatory quality of simple salts. Percept. Psychophys. *2*, 601-604.

ERICKSON, R.P. 1977. The role of 'primaries' in taste research. In *Olfaction and Taste VI*. (J. LeMagnen and P. MacLeod, eds.) pp. 369-376, I.R.L., London.

ERICKSON, R.P. 1981. A new direction in taste psychophysics: Aristotle and Henning. Abstract: 3rd Annual Meeting: Association for Chemoreception Sciences. Sarasota, FL.

ERICKSON, R.P. 1982. Studies on the perception of taste: Do primaries exist? Physiol. Behav. 28, 57-62.

ERICKSON, R.P. 1983. Taste: A time for re-evaluation. E.C.R.O. Newsletter, No. 27, 281-282.

ERICKSON, R.P. 1984a. Öhrwall, Henning and von Skramlik; The foundations of the four primary position in taste. Neuroscience and Biobehavioral Reviews 8, 105-127.

ERICKSON, R.P. 1984b. On the neural bases of behavior. Amer. Scientist 22, 233-241.

ERICKSON, R.P. 1985a. Grouping in chemical senses. Chem. Senses. 10, 333-340.

ERICKSON, R.P. 1985b. Definitions: A matter of taste. In *Taste, Olfaction and the Central Nervous System*, (D.W. Pfaff, ed.) pp. 129-150, Rockefeller Univ. Press, New York.

ERICKSON, R.P. and COVEY, E. 1980. On the singularity of taste sensations: What is a primary? Physiol. Behav. 25, 527-533.

ERICKSON, R.P. and SCHIFFMAN, S.S. 1975. The chemical senses: a systematic approach. In *Handbook of Psychobiology*, (M.S. Gazzaniga and C. Blakemore, eds.) pp. 303-426, Academic Press, London.

FAURION, A. 1983. Taste—A time for re-evaluation—Reply. E.C.R.O. Newsletter No. 28, 291-293.

FRINGS, H.W. 1954. Gustatory stimulation by ions and the taste spectrum. Abstract Papers A.C.S., 126th Meeting. Div. Agric. Food Chem. p. 14A.

GALVIN, S.L. 1948. The taste of monosodium glutamate and other amino acids in dilute solutions. In *Flavor and Acceptability of Monosodium Glutamate*, pp. 25-31, Food and Container Inst., Chicago, IL.

GARNER, W.R. 1962. *Uncertainty and Structure as Psychological Concepts*. John Wiley & Sons, New York.

GENT, J.F. 1979. An experimental model for adaptation in taste. Sensory Processes 3, 303-316.

GENT, J.F. and McBURNEY, D.H. 1978. Time course of gustatory adaptation. Percept. Psychophys. 23, 171-173.

GREEN, B. and GELHARD, B. 1989. Salt as an oral irritant. Chem. Senses 14, 259-271.

GREGSON, R.A.M. and BAKER, A.F.H. 1973. Sourness and bitterness: confusions over sequences of taste judgements. Brit. J. Psychol. 64, 71-76.

HANSON, H.L., BRUSHWAY, M.J. and LINEWEAVER, H. 1960. Monosodium glutamate studies. 1. Factors affecting detection of and preference for added glutamate in foods. Food Technol. 14, 320-327.

ISHII, R. and O'MAHONY, M. 1987a. Defining a taste by a single standard: aspects of salty and umami tastes. J. Food Sci. *52*, 1405-1409.

ISHII, R. and O'MAHONY, M. 1987b. Taste sorting and naming: Can taste concepts be misrepresented by traditional psychophysical labelling systems? Chem. Senses *12*, 37-51.

ISHII, R. and O'MAHONY, M. 1990. Use of multiple standards to define sensory characteristics for descriptive analysis: aspects of concept formation. J. Food Sci., In Press.

JACOBS, G.H. and GAYLORD, H.A. 1967. Effects of chromatic adaptation on color naming. Vision Research *7*, 645-653.

JAMESON, D. and HURVICH, L.M. 1959. Perceived color and its dependence on focal surrounding and preceding stimulus variables. J. Optical Soc. Amer. *4*, 890-898.

KIER, L.B. 1980. Molecular structure influencing either a sweet or bitter taste among aldoximes. J. Pharm. Sci. *69*, 416-419.

KINTZ, R.T., PARKER, J.A. and BOYNTON, R.M. 1969. Information transmission in spectral color naming. Percept. Psychophys. *5*, 241-245.

KUNINAKA, A. 1965. Flavor potentiators. Oregon State Univ., 4th Symp. Foods pp. 515-535.

KUNINAKA, A. 1981. Taste and flavor enhancers. In *Flavor Research, Recent Advances* (R. Teranishi, R.A. Flath, and H. Sugisawa, eds.) pp. 305-353, Marcel Dekker, New York.

KUZNICKI, J.T. 1978. Taste profiles from single human taste papillae. Percept Motor Skills *47*, 279-286.

KUZNICKI, J.T. and ASHBAUGH, N. 1979. Taste quality differences within the sweet and salty taste categories. Sensory Processes *3*, 157-182.

KUZNICKI, J.T. and ASHBAUGH, N. 1982. Space and time separation of taste mixture components. Chem. Senses *7*, 39-62.

KUZNICKI, J.T. and McCUTCHEON, N.B. 1979. Cross-enhancement of the sour taste on single human taste papillae. J. Expt. Psychol. *108*, 68-89.

LAWLESS, H.T. 1979. Evidence for neural inhibition in bittersweet taste mixtures. J. Comp. Physiol. Psychol. *93*, 538-547.

LOCKHART, E.E. and GAINER, J.M. 1950. Effect of monosodium glutamate on taste of pure sucrose and sodium chloride. Food Res. *15*, 459-464.

McAULIFFE, W.K. and MEISELMAN, H.L. 1974. The roles of practice and correction in the categorization of sour and bitter taste qualities. Percept. Psychophys *16*, 242-244.

McBURNEY, D.H. 1972. Gustatory cross adaptation between sweet-tasting compounds. Percept. Psychophys. *11*, 225-227.

McBURNEY, D.H. 1974. Are there primary tastes for man? Chem. Senses Flav. *1*, 17-28.

McBURNEY, D.H. and BARTOSHUK, L.M. 1973. Interactions between stimuli with different taste qualities. Physiol. Behav. *10*, 1101-1106.

McBURNEY, D.H. and GENT, J.F. 1978. Taste of methyl-α-D-mannopyranoside: Effect of cross adaptation and gymnema sylvestre. Chem. Senses Flav. *3*, 45-50.

McBURNEY, D.H. and GENT, J.F. 1979. On the nature of taste qualities. Psychol. Bull. *86*, 151-167.

McBURNEY, D.H. and SHICK, T.R. 1971. Taste and water taste of twenty-six compounds for man. Percept Psychophys. *10*, 249-252.

McBURNEY, D.H., SMITH, D.V. and SHICK, T.R. 1972. Gustatory cross adaptation: sourness and bitterness. Percept. Psychophys. *11*, 228-232.

McCUTCHEON, N.B. and SAUNDERS, J. 1972. Human taste papilla stimulation: stability of quality judgements over time. Science, *175*, 214-216.

MEILGAARD, M., CIVILLE, G.V. and CARR, B.T. 1987. *Sensory Evaluation Techniques*, Vol. 2, C.R.C. Press, Boca Raton, FL.

MEISELMAN, H. and DZENDOLET, E. 1967. Variability in gustatory quality identification. Percept. Psychophys. *2*, 496-498.

MEISELMAN, H.L. and HALPERN, B.P. 1970a. Human judgements of gymnema sylvestre and sucrose mixtures. Physiol. Behav. *5*, 945-948.

MEISELMAN, H.L. and HALPERN, B.P. 1970b. Effects of gymnema sylvestre on complex tastes elicited by amino acids and sucrose. Physiol. Behav. *5*, 1379-1384.

MILLER, G.A. and JOHNSON-LAIRD, P.N. 1976. Language and Perception, Cambridge Univ. Press, London.

MOSEL, J.N. and KANTROWITZ, G. 1952. The effect of monosodium glutamate on acuity to the primary tastes. Amer. J. Psychol. *65*, 573-579.

MOSEL, J.N. and KANTROWITZ, G. 1954. Absolute sensitivity to the glutamic taste. J. Gen. Physiol. *51*, 11-18.

MOSKOWITZ, H.R. 1972. Perceptual changes in taste mixtures. Percept. Psychophys. *11*, 257-262.

MUNSELL BOOK OF COLOR. 1976. MacBeth Division of Kollmorgen Corp., Baltimore, MD.

MURPHY, C., CARDELLO, A.V. and BRAND, J.G. 1981. Tastes of fifteen halide salts following water and NaCl: Anion and cation effects. Physiol. Behav. *26*, 1083-1095.

MURPHY, W.M. 1971. The effect of complete dentures upon taste perception. Brit. Dental J. *130*, 201-205.

MYERS, C.S. 1904. The taste names of primitive peoples. Brit. J. Psychol. *1*, 117-126.

O'MAHONY, M. 1979a. Salt taste adaptation: The psychophysical effects of subadapting solutions and residual stimuli from prior tastings on the taste of sodium chloride. Perception *8*, 441-476.

O'MAHONY, M. 1979b. Short-cut signal detection measurements for sensory analysis, J. Food Sci. *44*, 302-303.

O'MAHONY, M. 1983. Gustatory responses to nongustatory stimuli. Perception, *12*, 627-633.

O'MAHONY, M. 1988. Sensory difference and preference testing: the use of signal detection measures. In *Applied Sensory Analysis of Foods*. Vol. 1, (H. Moskowitz, ed.) pp. 145-175, C.R.C. Press, Boca Raton, FL.

O'MAHONY, M., ATASSI-SHELDON, S., ROTHMAN, L. and MURPHY-ELLISON, T. 1983. Relative singularity/mixedness judgements for selected taste stimuli. Physiol. Behav. *31*, 749-755.

O'MAHONY, M., AUTIO, J., HEINTZ, C. and GOLDENBERG, M. 1978. Taste naming by preschool children compared to colour naming: preliminary examination. I.R.C.S. Medic. Sci *6*, 208.

O'MAHONY, M. and BUTEAU, L. 1982. Taste mixtures: Can the components be readily identified? I.R.C.S. Medic. Science *10*, 109-110.

O'MAHONY, and GOLDSTEIN, L.R. 1987a. Sensory techniques for measuring differences in California navel oranges treated with doses of gamma-radiation below 0.6 Kgray. J. Food Sci *52*, 348-352.

O'MAHONY, M. and GOLDSTEIN, L.R. 1987b. Methods for sensory evaluation of navel oranges treated with electron beam irradiation. Lebensm.- Wiss u.-Technol. *20*, 78-82.

O'MAHONY, and ISHII, R. 1986. A comparison of English and Japanese taste languages: Taste descriptive methodology, codability and the umami taste. Brit. J. Psychol. *77*, 161-174.

O'MAHONY, M. and ISHII, R. 1987. The umami taste concept: Implications for dogma of four basic tastes. In *Umami: A Basic Taste,* (Y. Kawamura and M.R. Kare, eds.) pp. 75-93, Marcell Dekker, New York.

O'MAHONY, M. and THOMPSON, B. 1977. Taste quality descriptions: Can the subjects' response be affected by mentioning taste words in the instructions? Chem. Senses Flav. *2*, 283-298.

O'MAHONY, M., BUTEAU, L., KLAPMAN-BAKER, K., STAVROS, I., ALFORD, J., LEONARD, S.J., HEIL, J.R. and WOLCOTT, T.K. 1983. Sensory evaluation of high vacuum flame sterilized clingstone peaches, using ranking and signal detection measures with minimal cross-sensory interference. J. Food Sci. *48*, 1626-1631.

O'MAHONY, M., GOLDENBERG, M., STEDMON, J. and ALFORD, J. 1979. Confusion in the use of the taste adjectives 'sour' and 'bitter'. Chem. Senses. Flav. *4*, 301-318.

O'MAHONY, M., HOBSON, A., GARVEY, J., DAVIES, M. and BIRT, C. 1976a. How many tastes are there for low concentration 'sweet' and 'sour' stimuli?—Threshold implications. Perception *5*, 147-154.

O'MAHONY, M., KINGSLEY, L., HARJI, A. and DAVIES, M. 1976b. What sensation signals the salt taste threshold? Chem. Senses. Flav. *2*, 177-188.

O'MAHONY, M., WONG, S-Y. and ODBERT, N. 1985a. Sensory evaluation of navel oranges treated with low doses of gamma-radiation. J. Food Sci. *50*, 639-646.

O'MAHONY, M., WONG, S-Y. and ODBERT, N. 1985b. Initial sensory evaluation of Bing cherries treated with low doses of gamma-radiation. J. Food Sci. *50*, 1048-1050.

O'MAHONY, M., WONG, S-Y. and ODBERT, N. 1985c. Sensory evaluation of Regina freestone peaches treated with low doses of gamma radiation. J. Food Sci. *50*, 1051-1054.

PANGBORN, R.M. 1961. In discussion for Pilgrim, F.J. Interactions of suprathreshold taste stimuli. In *The Physiological and Behavioral Aspects of Taste,* (M.R. Kare and B.P. Halpern, eds.) p. 72, Univ. Chicago Press, Chicago, IL.

PERYAM, D.R. 1960. The variable taste perception of sodium benzoate. Food Technol. *14*, 383-386.

PERYAM, D.R. 1963. Variability of taste perception. J. Food Sci. *28*, 734-740.

PRICE, S. and DESIMONE, J.A. 1977. Models of taste receptor cell stimulation. Chem. Senses Flav. *2*, 427-456.

RISKEY, D.R., DESOR, J.A. and VELLUCCI, D. 1982. Effects of gymnemic acid concentration and time since exposure on intensity of simple tastes: A test of the biphasic model for the action of gymnemic acid. Chem. Senses *7*, 143-152.

ROBINSON, J.O. 1970. The misuse of taste names by untrained observers. Brit. J. Psychol. *61*, 375-378.

SANDICK, B. and CARDELLO, A.V. 1983. Tastes of salts and acids on circumvallate papillae and anterior tongue. Chem. Senses, *8*, 59-69.

SCHIFFMAN, S.S. and DACKIS, C. 1975. Taste of nutrients: amino acids, vitamins, and fatty acids. Percept. Psychophys. *17*, 140-146.

SCHIFFMAN, S.S. and ERICKSON, R.P. 1971. A psychophysical model for gustatory quality. Physiol. Behav. *7*, 617-633.

SCHIFFMAN, S.S. and ERICKSON, R.P. 1980. The issue of primary tastes versus a taste continuum. Neurosci. Biobehav. Rev. *4*, 109-117.

SCHIFFMAN, S.S., McELROY, A.E. and ERICKSON, R.P. 1980. The range of taste quality of sodium salts. Physiol. Behav. *24*, 217-224.

SCHIFFMAN, S.S., MOROCH, K. and DUNBAR, J. 1975. Taste of acetylated amino acids. Chem. Senses Flav. *1*, 387-401.

SCOTT, T.R. and CHANG, F-T.T. 1984. The state of gustatory neural coding. Chem. Senses *8*, 297-314.

SETTLE, R.G., MEEHAN, K., WILLIAMS, G.A., DOTY, R.L. and SISLEY, A.C. 1986. Chemosensory properties of sour tastants. Physiol. Behav. *36*, 619-623.

SJÖSTRÖM, L.B. and CROCKER, E.C. 1948. The role of monosodium glutamate in the seasoning of certain vegetables. Food Sci. *2*, 317-321.

SMITH, D.V. and McBURNEY, D.H. 1969. Gustatory cross-adaptation: Does a single mechanism code the salty taste? J. Exptl. Psychol. *80*, 101-105.

SMITH, V.V. and HALPERN, B.P. Selective suppression of judged sweetness by ziziphins. Physiol. Behav. *30*, 867-874.

STONE, H., SIDEL, J., OLIVER, S., WOOLSEY, A. and SINGLETON, R.C. 1974. Sensory evaluation by quantitative descriptive analysis. Food Technol. *28*, 24-34.

TORII, K. and CAGAN, R.H. 1980. Biochemical studies of taste sensation. IX. Enhancement of L-[^{3}H] glutamate binding to bovine taste papillae by 5'-ribonucleotides. Biochem. Biophys. Acta *627*, 313-323.

WILLIAMS, D.I. 1977. The social evolution of a fact. Bull. Brit. Psychol. Soc. *30*, 241-243.

YAMAGUCHI, S. 1979. The umami taste. In *Food Taste Chemistry*, (J. Boundreau, ed.) pp. 33-51, Amer. Chem. Soc. Symp. Ser. No. 115, Washington, D.C.

YAMAGUCHI, S., FURUKAWA, H. and TAKAHASHI, C. 1982. Flavor interaction between monosodium glutamate and sodium chloride. Abstr. Assoc. Chemorecept. Sci., Sarasota, FL.

YAMAGUCHI, S. and KIMIZUKA, A. 1979. Psychometric studies on the taste of monosodium glutamate. In *Glutamic Acid: Advances in Biochemistry and Physiology*, (L.J. Filer, S. Garattini, M.R. Kare, W.A. Reynolds and R.J. Wurtman, eds.) pp. 1-9, Raven Press, New York.

YAMAGUCHI, S., YOSHIKAWA, T., IKEDA, S. and NINOMIYA, T. 1971. Measurement of the relative taste intensity of some 1-2-amino acids and 5'-nucleotides. J. Food Sci. *36*, 846-849.

YOSHIDA, M. 1963. Similarity among different kinds of taste near the threshold concentration. Jap. J. Psychol. *34*, 25-35.

CHAPTER 6.9

CONTROL CHART TECHNIQUE: A FEASIBLE APPROACH TO MEASUREMENT OF PANELIST PERFORMANCE IN PRODUCT PROFILE DEVELOPMENT

MIFLORA M. GATCHALIAN, SONIA Y. DE LEON and TOSHIMASA YANO

At the time of original publication,
authors Gatchalian and DeLeon were affiliated with
Department of Food Science and Nutrition
University of Philippines, Diliman
Quezon City, Philippines

and

author Yano was affiliated with
Department of Agricultural Chemistry
Tokyo University, Bunkyu-ku Tokyo, Japan.

ABSTRACT

A quality control chart technique was tested for feasibility in measuring panelist performance in product profile development. The descriptive sensory score sheet evolved by Filipino trained panelists for fresh-frozen young coconut meat and water at "malakanin" (7-8 months old coconut fruit) stage was utilized for this study. A comparison of panelist performance in terms of their congruence, within themselves and between each other, in identifying the intensity or presence of an attribute was made. Results indicated that a similar measurement of significant differences could be obtained using both the analysis of variance (ANOVA) approach and the control chart (c.c.) technique. Considering the volume of data to be handled, the c.c. technique appeared to be more manageable than ANOVA. Besides, the c.c. was found by the panelists and the researchers to be more interesting to analyze, easier to apply, to interpret and to use for improvement follow through compared to the ANOVA.

Reprinted with permission of Food & Nutrition Press, Inc., Trumbull, Connecticut.
© *Copyright 1991. Originally published in* Journal of Sensory Studies 3, 239–254.

INTRODUCTION

Development of a descriptive sensory profile for young coconuts is a major concern of countries growing coconuts. In the last decade, the young coconut or "buko," as it is called in the Philippines, gained international demands either as fresh or as processed product (Alvarez 1989; Anon. 1983, 1989; Celso 1989). Unfortunately, there is very little known about the characteristics of "buko" for use in product development or for quality control purposes. This prompted the start of a research project in practically all aspects of this young coconut fruit, including the development of its sensory profile.

Several publications were reviewed relative to approaches in product profiling. Much work had been done in texture profiling (Bourne 1968; Baramesco and Lester 1990; Cardello *et al.* 1982) and on quantitative descriptive analysis or QDA (Johnsen and Kelly 1990; Sokolow 1988; Stone *et al.* 1974; Stone *et al.* 1980; Stone and Sidel 1985; Zook and Pearce 1988; and Zook and Wessman 1977). Their detailed presentations of approaches and applications were very encouraging. Some, however, were too sophisticated for proper application in developing countries. But with some modifications to suit the country situation, the combined good points of these approaches could be tried out.

Efforts towards development of the "buko" profile in the Philippines started with panelist recruitment or selection, and training. The descriptive sensory language development (DSLD) became the second major activity, followed eventually by the preparation of the descriptive score sheet (DSS). This became the tool for testing the panelist's precision in the use of evolved descriptive terms to characterize the sensory attributes of "buko" at the "malakanin" stage. The young coconut is generally known in the Philippines to be 6-9 months of age in the tree. At this stage, it has three maturity levels: (a) "maluhog" or mucus-like at 6-7 months; "malakanin" or cooked rice-like at 7-8 months; and "malakatad" or leather-like at 8-9 months (Adriano and Manahan 1931; Del Rosario and Malijan 1985). Of the three levels, "malakanin" stage is generally prepared as a refreshing beverage containing coconut meat and water (Gonzales *et al.* 1984).

A major consideration during the process of product profile development is the need for a fast, yet efficient, accurate and easy to interpret approach to data analysis. Even in this computer age, the right choice of statistical methods can mean a great deal in facilitating data analysis. It is important, in a study, to know if the trained panelists has a common perception of either the intensity or the presence of the attribute they have agreed to use as a description of "buko." This means analyzing each descriptive term in the DSS per trial, per panelist per sample evaluated. The amount of work involved in data analysis can be enormous, especially if there are many descriptive terms. Furthermore, just to measure the individual panelist's performance per trial per descriptive term per sample will be a very tedious task. Thus, the urgent need to search for alternative means for data analysis.

Analytical researches and reviews utilizing statistical methods and other approaches to data analysis are very helpful in the search for alternatives (Gacula 1987; McDaniel *et al.* 1987; O'Mahony 1986). Furthermore, the straight forward presentation of Netter and Wasserman (1974) on the analysis of variance applications make this multivariate approach appear appropriate for the data on hand. However, despite the strengths of the approaches presented, the complexity of computations and the difficulty in results analysis and interpretation, compound the need to identify simple methods.

The control chart (c.c.) technique is a statistical quality control tool for product/process monitoring (Besterfield 1986). In particular, the c.c. for variables allows for following-through the efficiency of a process when the attributes analyzed are measurable in terms of weights, heights, temperature, scores, etc. Briefly, a control chart consists of the number of sample observations on the abscissa and the value of the variable being monitored on the ordinate. A central line, representing the means of all observations of the variable can be drawn horizontally and then the control limits (upper and lower) can be computed and likewise drawn as broken lines above and below the central line. When individual means are plotted on the chart, any point outside of the limits is known as an "outlier."

In product profile development, a parallel can be done for panel performance measurement. A panelist, in sensory evaluation, is actually an instrument who measures a product attribute based on his own perceptions. But with training, the variability between panelists can be reduced to a minimum, so that it becomes possible for them to act as a precise measuring tool. The intensity score for a descriptive term in a DSS is a measure of product attribute. The three trials done by the panelist form the sub-group of a series of observations about the product. Each observation is done by a panelist and the mean of three trials is an intensity score for the product attribute being measured or the variable to be used in the c.c. The range of scores for the three trials per panelist will be the individual variations per observation.

On the basis of the above considerations, the c.c. for variables can be constructed. Panelist performance can, therefore, be analyzed relative to each other and to the overall observations. This can be done when control limits are set-up and average panelist performance are plotted. The extent of variation in mean scores of panelists will be determined by the distance between the upper and lower control limits. The amount of panelist variability will be shown by the number of "outliers" or points outside the limits per descriptive term analyzed.

This report will focus largely on the use of the c.c. technique as an approach to measurement and monitoring of panelist performance in "buko" profile development. Specifically, the objectives of the study are: (a) To measure panelist performance in product profile development through QDA and ANOVA approaches; (b) to determine the feasibility of the c.c. technique as an alternative approach to DSS data analysis; and (c) to identify the comparative advantages/disadvantages in the use of c.c. over ANOVA as a tool for measurement of panelist performance.

MATERIALS AND METHODS

Materials

Samples evaluated by multiple comparison test were "buko" water and meat which were either fresh, fresh-frozen, blanch-frozen and canned. The fresh sample was used as reference. All tests were done at the "Sensorium" or the sensory evaluation laboratory of the University of the Philippines College of Home Economics. There were 13 trained panelists who participated in the evaluation sessions. They were themselves participants in the descriptive sensory language development (DSLD) program which evolved the descriptive terms for young coconut. These terms were then utilized in the preparation of the descriptive score sheet (DSS) used for evaluation of samples under study.

Methods

The completed DSS obtained from the three trials of the multiple comparison tests were collated in preparation for data analysis. This report will present only the data obtained from three trials on the fresh frozen samples. Tables presenting individual panelists' score for each descriptive term per trial were prepared. This served as the raw data for which further statistical analyses were to be done.

From the panelists' mean scores per descriptive term per trial, a quantitative descriptive analysis (QDA) was done. To construct the QDA spider-web chart, first compute the means of all trials by all panelists for a given descriptive term. When each of the means is plotted on the spokes of the radiating lines criss-crossing on a common central point, then connected sequentially together, a spider-web configuration appears. The QDA figure would be expected to show the extent of variations between means per trial. Where differences between trials for a descriptive term appeared probable, a t-test was done.

For an overall analysis, the ANOVA was used (Gatchalian 1981; O'Mahony 1986). Differences between trials and between panelists were tested using the F-test at both the 1% and 5% levels of significance. Where significant differences were shown, the Duncan's multiple range test (DMRT) was done to identify the factor causing the difference (Gatchalian 1981; Larmond 1985). A summary of the ANOVA results for all descriptive terms from the DSS was then prepared adding a column which identified the panelist causing the difference.

Another approach to data analysis was tested utilizing the same DSS results. The control chart (c.c.) technique required only the data on mean score and range of the scores in the three trials per panelist for each of the descriptive term. The average of the means for all trials for all panelists per descriptive term was called the "mean of means" in c.c. parlance. This value served as the central line or $\bar{\bar{X}}$ in the c.c. To compute for the upper (UCL) and lower (LCL) control limits, the average of the ranges of all panelists was first obtained. The average of ranges is R and when multiplied by a factor A2 based on the subgroup size (Gatchalian 1989; Besterfield 1986) will be equivalent to a 3 sigma or 3 standard deviations spread above or below $\bar{\bar{X}}$. Thus, the LCL will be $\bar{\bar{X}} - A_2R$ and the UCL will be $\bar{\bar{X}} + A_2R$. A 99% confidence interval (C.I.) will be the distance between LCL and UCL. Any point (i.e., panelist score) outside the limits will be considered significantly different and the panelist may be identified as an "outlier."

To show sample applications, two control charts were constructed for the descriptive term which either had the greatest number of outliers or none at all. A table showing the data needed for the construction of the c.c. was also prepared for illustrative purposes. For an overall assessment of panelist performance relative to all the descriptive terms, a summary table showing the $\bar{\bar{X}}$, R, control limits and the panelist "outliers" was made. When desired, the panelists themselves could construct their own c.c. for any descriptive term. This way, they would be able to analyze their own position relative to the other panelists.

Using results from the c.c. analysis and tables, discussions with panelists focused on both the descriptive terms with many "outliers" and on those with none but had a wide C.I. For those with wide C.I., further definitions and comparison of terms describing samples were done until consensus on intensity of the attribute was reached. A discussion again on the concepts of attribute intensity was done, for the terms with narrow C.I. but had many outliers. Actual samples of the fresh and fresh frozen "buko" meat and water were utilized for reference during discussions.

Succeeding sensory evaluation sessions had the major objective of reducing the variations in attribute intensity between panelists and within the individual panelist himself. Hence, a narrower C.I. with fewer "outliers" would be the goal for the next sensory evaluation sessions using the DSS and analyzed by c.c. technique.

RESULTS AND DISCUSSIONS

The descriptive score sheet (DSS) utilized for evaluating the fresh frozen "buko" water and meat is shown in Fig. 1. Terms in the DSS were those evolved during the descriptive sensory language development (DSLD) sessions and then pilot tested on fresh samples. These descriptive terms (Fig. 1) were still subject to further assessment relative to their intensities or presence in the product.

Table 1 presents a sample data obtained from the completed DSS. The raw data presents only the scores for "greyish-white," one of the descriptive terms three trials per panelist for a given descriptive term. Means for each trial for every descriptive term were also obtained.

The means of panelist scores for each of the three trials per descriptive term were superimposed in a QDA spider-web configuration shown in Fig. 2. On face value, involving the means alone, closeness of means for each descriptive term per trial could imply repeatability. It seems that the panelists in three instances were able to give a consistent product profile. Perhaps there would not even be any need to make a t-test to measure differences between trials. However, the study also needed to measure panelist performance to determine extent of variability between each other and within individual panelists. It was, therefore, important to know whether there was agreement among the panelists relative to the intensities of each descriptive term as they were used to describe the product. To achieve this, further data analysis had to be done.

Shown in Table 2 is the summary of ANOVA results from the 16 descriptive terms in the DSS for "buko". The last column presents the panelists identified through DMRT. Results show that of the 16 attributes, 7 had means which differed significantly between panelists at 1% level and 2 at the 5% level of significance. Using the DMRT to identify the panelist causing the significant difference, each was actually identified per descriptive term as shown in Table 2. Furthermore, in support of the QDA presentation in Fig. 2, only one attribute (sweet taste) actually appeared to have differed significantly between trials. These findings show that the ANOVA can provide the needed information to measure panelist performance. For measurement of improvement in their performance in succeeding sensory evaluation sessions, however, another series of ANOVA computations would again be required. Much efforts need to be exerted to achieve this end.

Using the ANOVA approach to determine differences between panelists and between replications is a very cumbersome process. For those who had manually computed the ANOVA, one can easily appreciate the efforts expended to produce the summary data shown in Fig. 2. to be able to do a t-test for each of

Descriptive Score Sheet for Product Profiling
of
Young Coconut Water and Meat

Name: _____   Date: _____

INSTRUCTIONS: Presented below are the descriptive terms the trained panelists used to describe young coconut water and meat. Evaluate the sample and indicate by a slash mark (/) the intensity of the attribute which best describes the product. Take sufficient time to evaluate each characteristic of the coded sample. After you have completed the assessment return both the sample and the scoresheet, then proceed to assess the next sample. Be sure to rinse your mouth before evaluating the next sample.

Product Attribute	Intensity of the Attribute

Coconut Water Appearance

```
                        0                                           12cm.
Greyish white           ---------------------------------------------
                           slightly      moderately      extremely

Cloudy                  ---------------------------------------------
```

Coconut Meat Appearance

```
Opaque white strips     ---------------------------------------------
                           Few            Many            All

Smooth surface          ---------------------------------------------
                           slightly      moderately      extremely

Rough surface           ---------------------------------------------

Sour odor               ---------------------------------------------
                           Absent         Faint          Distinct

Sweet odor              ---------------------------------------------

Oily odor               ---------------------------------------------

Sour taste              ---------------------------------------------

Sweet taste             ---------------------------------------------

Oily taste              ---------------------------------------------
```

Coconut Water Consistency

```
Thin                    ---------------------------------------------
                           Like           Like very       Like
                           Water          Thin Syrup      Thick syrup
```

Coconut Meat Texture

```
Smooth (Mouthfeel)      ---------------------------------------------
                           slightly      moderately      extremely

Tender (first bite)     ---------------------------------------------

Tough (first bite)      ---------------------------------------------

Mealy (before           ---------------------------------------------
  swallowing)
```

Remarks:

Thank You !

FIG. 1. DESCRIPTIVE SCORE SHEET FOR PRODUCT PROFILING OF YOUNG COCONUT MEAT AND WATER

TABLE 1.
SAMPLE RAW DATA COLLECTED FROM THE DESCRIPTIVE SCORE SHEET FOR EVALUATING "BUKO" IN THREE TRIALS BY 13 PANELISTS FOR THE INTENSITY SCORE OF "GREYISH-WHITE" APPEARANCE

Panelist number	Trial 1	Trial 2	Trial 3
1	36	21	31
2	31	41	18
3	20	61	40
4	22	21	20
5	7	8	9
6	5	5	5
7	46	35	41
8	27	26	16
9	18	17	17
10	21	21	24
11	51	50	58
12	68	55	62
13	12	23	23
Total	364	364	362
$\bar{X}$ (mean)	28.0	29.5	27.8

the 16 descriptive terms where the scores were obtained in three trials by 13 panelists, 16 ANOVA computations were done to make 16 ANOVA tables. Add to this the DMRT analysis for every term with significant difference, the efforts could be tremendous, even with the use of personal computers, the same volume of data still needs to be handled and analyzed.

A search for a shorter approach to data analysis which would be easy to interpret and interesting to the panelists was necessary. Furthermore, there was a need to monitor progress in panelists' agreement in their observed intensity of each descriptive term in succeeding evaluation sessions. It is probable that levels of variability between panelists could be better observed through evaluation of confidence intervals (C.I.). The wider the interval wherein scores would vary, the greater could be the panelists' differences in perceptions relative to the intensity of a given attribute. Conversely, the narrower the C.I. the smaller would be the variability.

The approach utilizing the principles of the control chart (c.c.) technique was deemed feasible for the study. A sample data shown in Table 3 was used to illustrate the approach to computations of confidence limits. Using A_2R to represent 3 standard deviations, the LCL and UCL were computed for each descriptive term.

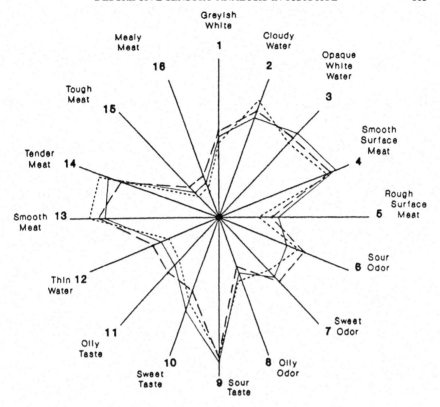

FIG. 2. PROFILE BY QDA OF FRESH FROZEN YOUNG COCONUT MEAT AND WATER IN THREE TRIALS (AVERAGE OF PANELIST SCORES PER TRIAL FOR EACH DESCRIPTIVE TERM)

Data presented in Table 3 are samples of two extreme situations: (a) a descriptive term "sweet taste" where no "outlier" was observed; and (b) "mealy" where several "outliers" were shown. Fig. 3 and 4 show the sample control chart using data in Table 3.

Figure 3 presents a c.c. where no point (or panelist) was shown to be an "outlier." Yet, at the same time, it can be observed that the C.l. is wide. This could imply high variability within panelists which collectively resulted in a wide spread of the overall mean score. It may also be implied that there was a poor understanding about the intensity of the attribute as perceived by individual panelists. Thus, even when there was no outlier, it would still be necessary for the panelists to discuss in-depth what they meant by close to absence of a "sweet taste" (score of 18) to a faint one (score of 57). This difference would need a resolution if a reduction of individual variability is to be achieved in the next evaluation sessions.

TABLE 2.
SUMMARY OF ANALYSIS OF VARIANCE TESTS FOR DIFFERENCES BETWEEN
REPLICATIONS AND BETWEEN PANELISTS RELATIVE TO DESCRIPTIVE TERMS
FOR FRESH FROZEN YOUNG COCONUT MEAT AND WATER

Descriptive terms for meat (M) and water (W)	Error mean square	F Value Between replications	F Value Between panelists	Identified panelists (P) causing differences
Appearance				
1. Greyish White (M)	66.67	0.171	12.38 **	P_3 P_5 P_6 P_7 P_{11} P_{12}
2. Cloudy (W)	214.74	0.086	2.01	None
3. Opaque (M)	112.32	0.518	6.70 **	P_2 P_4 P_5
4. Smooth surface (M)	218.39	0.499	1.62	None (P_6 P_7)a
5. Rough surface (M)	135.14	0.419	2.26 *	P_{12}
Odor				
6. Sour odor (M&W)	526.31	0.146	1.216	None
7. Sweet odor (M&W)	101.93	0.234	6.71 **	P_5 P_7 P_9 P_{12}
8. Oily odor (M&W)	172.36	0.292	3.91 **	P_4 P_6 P_{10} P_{12}
Taste				
9. Sour taste (M&W)	98.68	0.097	4.66 **	P_2 P_5 P_7
10. Sweet taste (M&W)	175.30	3.655*	1.73	None
11. Oily taste (M&W)	150.04	0.826	4.46 **	P_3 P_6 P_7 P_{11}
Kinesthetic				
12. Thin (W)	154.97	2.712	1.78	None (P_5)a
13. Smooth (M)	324.36	0.193	1.19	None (P_1)a
14. Tender (M)	451.20	0.579	1.22	None (P_2)a
15. Tough (M)	85.80	0.920	2.61 **	P_1 P_3
16. Mealy (M)	19.21	0.621	11.93 **	P_1 P_3 P_4 P_6 P_7 P_9 P_{10} P_{12}

* Significant at 5% level.
** Significant at 1% level.
a Panelists who were outliers in the control chart.

A peculiar situation is shown in Fig. 4. This is an example of another extreme case where variations within panelists were generally so small that the panelists performance showed a narrow variability level. However, the overall observation indicated significant differences between panelists resulting in several becoming "outliers." There were 5 (P_4, P_7, P_9, P_{10}, P_{12}) panelists who rated "mealy" texture to be less than slight (score of less than 6.3), 2 (P_3, P_6) considered it to be more than slightly "mealy" and 1 (P_1) thought it was half way between slight to moderate (score of 30.7). Although variations within

TABLE 3.
SAMPLE DATA FOR CONTROL CHART ANALYSIS UTILIZING SCORES FOR
"SWEET TASTE" (NO OUTLIERS) AND SCORES FOR "MEALY" TEXTURE
WITH MANY OUTLIERS

Panelists number	Scores for sweet taste					Scores for mealy texture				
	T1	T2	T3	$\bar{X}$	R	T1	T2	T3	$\bar{X}$	R
P_1	36	37	35	36.0	2	36	23	33	30.7	13
P_2	43	13	13	23.0	30	10	16	9	11.7	7
P_3	38	22	42	34.0	20	25	35	14	24.7	21
P_4	42	21	42	35.0	21	5	6	5	5.3	1
P_5	27	20	73	40.0	53	7	8	8	7.7	1
P_6	40	39	39	39.3	1	21	21	23	21.7	2
P_7	76	35	37	49.3	41	5	6	5	5.3	1
P_8	72	34	31	45.7	41	9	10	0	6.3	1
P_9	20	28	18	22.0	10	10	4	3	5.7	7
P_{10}	36	19	27	27.3	17	4	5	6	5.0	2
P_{11}	50	34	24	36.0	26	10	9	8	8.0	1
P_{12}	55	57	50	54.0	7	3	4	3	3.3	1
P_{13}	28	23	21	24.0	7	9	6	15	10.0	9
Total				456.3	276				153.1	68

Upper control limit = UCL = $\bar{\bar{X}} + A_2\bar{R}$

Lower control limit = LCL = $\bar{\bar{X}} - A_2\bar{R}$

Where: $A_2 = 1.02$, and $D_4 = 2.57$ at subgroup size of 3

$\bar{\bar{X}} = 456.3 / 13 = 35.8$ $\bar{\bar{X}} = 153.1 / 13 = 11.8$

$\bar{R} = 276 / 13 = 21.2$ $\bar{R} = 68 / 13 = 5.2$

panelists were relatively small, still, the presence of outliers made it imperative to reassess their perceptions of the attributes. Panelists who were outliers and who varied greatly within themselves needed to analyze their degree of understanding of the word "mealy."

A summary of the control limits for all descriptive terms including the identified "outlier" panelists is shown in Table 4. A panelist who saw himself as an "outlier" generally wanted to know what happened. This created deeper interest in the product profile development. The desire to improve one's standing in the c.c. made the panelists even more serious in their tasks of evaluation. In the next evaluation sessions, better results would be expected such that the C.l. should be smaller while the number of "outliers," fewer.

It is interesting to observe that the outliers in the c.c. were also the panelists identified through DMRT to cause the significant difference in the ANOVA results (Table 2). However, there were some instances in the ANOVA

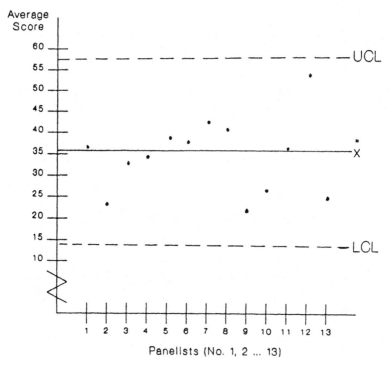

FIG. 3. SAMPLE CONTROL CHART WHERE THERE ARE NO "OUTLIERS" FOR A GIVEN DESCRIPTIVE TERM (SWEET TASTE)

where no significant difference was observed, yet 1 or more "outliers" were shown by the c.c. In general, results from data analysis utilizing the c.c. technique show that it can be a feasible measure of panelist performance in product profile development.

Based on these presentations, it may be observed that the use of c.c. technique may have certain advantages. First of all, when handling voluminous data, it can lend itself to fewer errors because of fewer mathematical computations, particularly when compared to ANOVA. Secondly, because fewer computations are needed, results are obtainable faster with much less effort expended. Thirdly, the use of confidence intervals as a measure of variability makes analysis more vivid and easier to comprehend compared to other methods. Fourthly, the fact that panelists can observe their relative variations could help sustain interest and/or enthusiasm for the development study. Finally, the use of the c.c. for monitoring a process is a unique approach to determination of progress or improvement in attaining panelists agreement relative to intensity of scores applied to a given descriptive term.

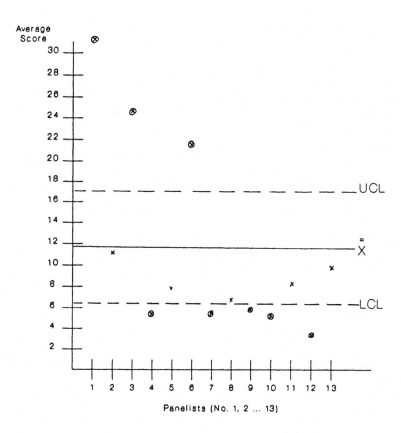

FIG. 4. SAMPLE CONTROL CHART SHOWING THE MOST NUMBER OF OUTLIERS FOR A DESCRIPTIVE TERM (MEALY)

Maybe the major disadvantage of the c.c. technique is the probable impression it might create on the panelists that because they are not "outliers" their performance is good. It may be recalled that "outliers" may not occur when variability within panelists (between trials) is high. As such the confidence

TABLE 4.
MEAN OF MEANS ($\bar{\bar{X}}$ AND MEAN OF RANGE $\bar{R}$) UTILIZED IN CONTROL CHART TECHNIQUE OF DATA ANALYSIS IN PROFILE DEVELOPMENT OF FROZEN YOUNG MEAT AND WATER

	$\bar{\bar{X}}$	$\bar{R}$	Control Limit Lower	Control Limit Upper	Outlier panelist
Appearance					
Coconut Water					
Greyish white	28.5	10.9	17.4	39.5	P_3 P_5 P_6 P_7 P_{11} P_{12}
Cloudy	45.1	33.8	10.6	79.6	None
Coconut Meat					
Opaque white	43.5	16.5	26.7	60.3	P_2 P_4 P_5
Smooth surface	61.8	19.7	41.7	81.9	P_5 P_7
Rough surface	19.0	14.2	4.5	33.5	P_{12}
Odor					
Sour	47.9	23.8	23.6	72.2	None
Sweet	37.9	14.1	23.5	52.3	P_5 P_7 P_9 P_{12}
Oily	24.5	13.7	10.5	38.5	P_4 P_6 P_{10} P_{12}
Taste					
Sour	64.3	13.6	50.4	78.2	P_4 P_5 P_7
Sweet	35.8	21.2	14.2	57.4	None
Oily	23.8	17.2	6.3	41.3	P_3 P_6 P_7 P_{11}
Texture					
Water					
Thin	25.2	15.2	9.7	40.7	P_5
Meat					
Smooth	69.0	18.0	50.6	87.4	P_1
Tender	67.5	22.4	44.7	90.3	P_2
Tough	12.7	12.7	-0.3	25.7	P_1 P_3
Mealy	11.8	5.2	6.5	17.1	P_1 P_3 P_4 P_6 P_7 P_9 P_{10} P_{11}

interval may become so wide that no one becomes an outlier, yet this may, in fact, show that there is a wide disagreement on the descriptive term intensity score. This problem may, however, be resolved by standardizing the scores.

CONCLUSIONS AND RECOMMENDATIONS

Results from sensory evaluation of young coconut or "Buko" in the Philippines, utilizing the locally evolved descriptive score sheet were analyzed. Using the quantitative descriptive analysis (QDA) approach comparative means per trial were plotted to form the spiderweb configuration. Results indicated no significant difference between trials except for the "sweet taste" descriptive term. Without analyzing panelist variability, this presentation could indicate repeatability of panelists' assessment of the product. But, since it was also desired to know the panelist's performance, another test was necessary.

Based on the analysis of variance (ANOVA) results after a series of computations for each of the 16 descriptive terms, it was observed that there were attributes where panelist means were significantly different at the 99% confidence level and 2 at the 95% level. When analyzed further using the Duncan's multiple range test (DMRT) the panelists causing the significant differences were identified for each of the descriptive terms. Panelists' relative performance was successfully measured using the ANOVA/DMRT. But the efforts expended in computations, plus the abstract nature of results, made the whole approach both cumbersome and uninteresting.

The control chart (c.c.) technique, utilized in statistical quality control for process monitoring, was tested for feasibility of use for measurement of panelist performance in product profile development. Results indicated that the c.c. can be a relatively fast measure for panelist performance, particularly with regards to computational requirements and ease in data interpretation. The use of the c.c. for panelist performance analysis made the results more interesting to the panelists while at the same time easy to understand. Besides, to follow-up improvement in performance in succeeding evaluation sessions, the c.c. could better present improvements, particularly when compared with the ANOVA. Five advantages and one disadvantage were presented in the use of the c.c. as a technique for panelist performance measurement in product profile development.

REFERENCES

ADRIANO, F.T. and MANAHAN, M. 1931. The nutritive value of green, ripe and sport coconuts, "buko", "niyog" and "makapuno". Phil. Agric. *20*, 195-198.

ALVAREZ, R.C. 1989. "Buko" market still largely untapped. Manila Bulletin. Feb. 21.

ANON. 1983. The tree of life potential of the coconut. Coconuts Today. *1*(2), 17.

ANON. 1989. Taiwan firm wins "buko" bid. Manila Bulletin. Feb. 21.

BARAMESCO, S.P. and LESTER, C.S. 1990. Application of texture profiling to baked products: Some considerations for evaluation, definition of parameters and reference points. J. Texture Studies *21*(3),235.

BESTERFIELD, D.H. 1986. *Quality Control*, 2nd ed., 358 pp., Prentice Hall, Englewood Cliffs, NJ 07682.

BOURNE, M.C. 1968. Texture profiles of ripening pears. J. Food Sci. *3*(2), 223.

CARDELLO, A.V. *et al.* 1982. Perception of texture by trained consumer panelists. J. Food Sci. *47*(4), 1186-1197.

CELSO, A. 1989. Government urged to offer buko as counter trade to Taiwan. Malaya, Feb. 15.

GACULA, M. JR. 1987. Some issues in the design and analysis of sensory data: revisited. J. Sensory Studies *2*(3), 169-185.

GATCHALIAN, M.M. 1981. *Sensory Evaluation Methods with Statistical Analysis*, pp. 376-404, U.P. College of Home Economics, Diliman, Q.C., Philippines.

GATCHALIAN, M.M. 1989. *Sensory Evaluation Methods for Quality Assessment and Development*, 470 pp. U.P. College of Home Economics, Diliman, Q.C. Philippines.

GONZALEZ, O.N., JULIETA, N., ALEJO, V., BRILLANTES, J. and VALDECANAS, M. 1984. A process for preparing noncarbonated and carbonated coconut water beverages. Nat. Inst. Sci. Technol. J. Oct-Dec. 4.

JOHNSEN, P.B. and KELLY, C.A. 1989. A technique for quantitative sensory evaluation of farm-raised catfish. J. Sensory Studies *4*, 189-199.

LARMOND, E. 1985. Methods of Sensory Evaluation of Food. Publication 1284, Dept. of Agric, Canada.

McDANIEL, M., HENDERSON, L.A., WATSON, B.T. JR. and HEATHERBELL, D. 1987. Sensory panel training and screening for descriptive analysis of the aroma of Pinot Noir Wine fermented by several strains of Malolactic bacteria. J. Sensory Studies *2*(3), 149-167.

NETTER, J. and WASSERMAN, W. 1974. *Applied Linear Statistical Models*, 842 pp., Richard D. Irwin, Homewood, IL 60430.

O'MAHONEY, M. 1986. *Sensory Evaluation of Food Statistical Methods and Procedures*, 487 pp. Marcel Dekker, 270 Madison Ave., NY 10016.

SOKOLOW, H. 1988. Qualitative Methods for Language Development, In Applied Sensory Analysis of Foods, Vol. I, (H. Moskowitz, ed.) p. 3, CRC Press, Boca Raton, FL.

STONE, H. and SIDEL, J.L. 1985. Descriptive analysis. In Sensory Evaluation Practices. Academic Press, Orlando, Florida. Chapter 6.

STONE, H., SIDEL, J.L. and BLOOMQUIST, J. 1980. Quantitative descriptive analysis. Cereal Foods World *25*, 642.

STONE, H., SIDEL, J., OLIVER, S., WOOLEY, A. and SINGLETON, R.C. 1974. Sensory evaluation by quantitative descriptive analysis. Food Technol. *28*(24).

ZOOK, K.L, and PEARCE, J.H. 1988. Quantitative descriptive analysis. In *Applied Sensory Analysis of Food*, Vol. 1, (H. Moskowitz, ed.) p. 43, CRC Press, Boca Raton, FL.

ZOOK, K. and WESSMAN, C. 1977. The selection and use of judges for descriptive panels. Food Technol. *31*(11), 56-61.

CHAPTER 6.10

COMPARISON OF THREE DESCRIPTIVE ANALYSIS SCALING METHODS FOR THE SENSORY EVALUATION OF NOODLES

FLOR CRISANTA F. GALVEZ and ANNA V.A. RESURRECCION

*At the time of original publication,
all authors were affiliated with
Department of Food Science and Technology
University of Georgia, Agricultural Experiment Station
Griffin, GA 30223.*

ABSTRACT

Unstructured line scales (ULS), semi-structured line scales (SLS), and category scales (CS) were compared in evaluating two samples of mungbean starch noodles. Results indicated that panelists discriminated better between noodle samples when ULS were used. When identical sets of descriptors were used, more terms were found to be discriminatory with ULS but a smaller group of descriptors was sufficient to discriminate between samples compared to those needed for SLS or CS. It was also found that the use of ULS was most reliable and most sensitive to product differences. Increased anchoring of the line scales with the intensity standards did not result in increased sensitivity and reliability of the scale. Results also showed that different scaling methods may be used in succession at intervals of at least 4 h by a trained panel in evaluating the sensory characteristics of noodles using descriptive analysis.

INTRODUCTION

Designing a sensory test includes the choice of a scaling method most appropriate to use for the particular product to be analyzed. Scaling methods involve the use of either numbers or words to express the intensity of a perceived attribute and the acceptability of products (Meilgaard *et al.* 1987; Lawless and Malone 1986b). There are, at present, three types of scales that are in common use for the descriptive analysis of products. These include category scales (CS), linear scales (LS), and magnitude estimation (ME) or ratio scales (Meilgaard *et al.* 1987). Lawless and Malone (1986b) stated that considering scaling as a measuring instrument, it is but appropriate to ask how sensitive a technique is or how well small physical differences can be distinguished by observers using that instrument. The literature is inconclusive relative to which method is more sensitive to product differences. Comparisons of the methods yielded different results for different products tested.

In scoring the tenderness of beef cuts from different grades, similar results were obtained using a nine-point CS and a LS that was modified to have nine marked divisions (Raffensperger *et al.* 1956). On the other hand, in assessing apple quality, the use of LS resulted in larger t-values than those using CS (Baten 1946). Similarly, larger treatment F-values for panelists' evaluations of intensities of fat in milk and sucrose in lemonade were obtained using LS as opposed to ME (Giovanni and Pangborn 1983).

A series of experiments to compare ME and nine-point CS were conducted on food acceptance (Moskowitz and Sidel 1971). Both evaluation procedures were found to be equally sensitive to differences in food acceptability with ME quantifying the ratios among different items and CS providing numerical and verbal categories of acceptance. The effectiveness of these same methods for determining "created differences" in whisky sour formulations were evaluated using a trained panel (McDaniel and Sawyer 1981a). Results likewise indicated no differences between the two methods regarding scaling mix differences. However, when consumer preferences for the same products were determined, more statistically significant differences in preference were found using ME (McDaniel and Sawyer 1981b).

The three scaling techniques (CS, LS, ME) were used to evaluate the eating quality of cooked beef steaks (Shand *et al.* 1985). Results showed that CS was most sensitive and LS was least sensitive in detecting differences in the quality attributes while ME was as sensitive as CS to most treatment differences. Lawless and Malone (1986a) likewise found a modest advantage for CS when CS, LS, ME, and a hybrid of CS and LS were compared for their ability to discriminate differences among products using consumer tests. However, approximate parity was observed among these methods in their ability to differentiate small physical differences using central location tests with untrained judges (Lawless and Malone 1986b). Similarly, studies conducted to compare the relative effectiveness of ME and CS in the measurement of hedonic response to a controlled stimulus showed no clear superiority of any one method in reliability, precision, or discrimination (Pearce *et al.* 1986).

Some of the studies conducted to compare the ability of the different rating scales to discriminate product differences also polled the subjects as to their opinions of the ease of use of the methods. In all of these studies, ME was the least preferred by judges in evaluating the sensory properties of products (Shand *et al.* 1985; Lawless and Malone 1986a; McTigue *et al.* 1989). In this regard, Giovanni and Pangborn (1983) stated that between LS and ME, LS was simpler to perform.

There is a need to qualitatively and quantitatively describe the sensory characteristics of mungbean starch noodles using the most suitable scaling method. However, comparative studies examining the use of different rating scales in noodle research are lacking. The present investigation was conducted with the following objectives:

(1) To determine differences in unstructured line scales (ULS), semi-structured line scales (SLS), and category scales (CS) in evaluating the sensory characteristics of noodles, and
(2) To determine the most suitable scaling method among three listed above, for evaluating the sensory characteristics of noodles.

MATERIALS AND METHODS

Experimental Design

Three different scaling methods were used to evaluate the noodle samples. Nine panelists evaluated two noodle samples using one method at a time during three different periods. The three periods were separated by at least 4 h to eliminate residual effects. Four replicates were conducted. In each replicate, a cross-over design, presented in Table 1, was used as described by Cochran and Cox (1957). This design is commonly used in experiments wherein the different treatments are applied in successive periods to the same subject.

Preparation of Samples

Commercial samples of mungbean starch noodle were purchased from an oriental store. The two available brands were very different from each other in appearance and were considered suitable for the purpose of the study. The samples were stored at ambient temperature until needed for the test.

Fifty grams of noodle samples were pre-soaked in 750 mL water for 5 min, drained through a wire kitchen strainer, and cooked in 1.5 L boiling water for 5 min, as determined by preliminary tests for doneness. The cooked noodles were immediately cooled under cold running water for 1 min. They were cut into 2-3 cm lengths and presented in 1 oz plastic cups in 20-g amounts. Preparation of samples was done 2 h before evaluation time.

TABLE 1.
EXPERIMENTAL DESIGN
(A) UNSTRUCTURED LINE SCALE
(B) SEMI-STRUCTURED LINE SCALE
(C) CATEGORY SCALE

REP	PERIOD	JUDGES								
		1	2	3	4	5	6	7	8	9
1	1	A	B	C	A	B	C	A	B	C
	2	B	C	A	B	C	A	B	C	A
	3	C	A	B	C	A	B	C	A	B
2	1	B	A	C	B	A	C	B	A	C
	2	A	C	B	A	C	B	A	C	B
	3	C	B	A	C	B	A	C	B	A
3	1	C	B	A	C	B	A	C	B	A
	2	B	A	C	B	A	C	B	A	C
	3	A	C	B	A	C	B	A	C	B
4	1	A	C	B	A	C	B	A	C	B
	2	C	B	A	C	B	A	C	B	A
	3	B	A	C	B	A	C	B	A	C

Sensory Evaluation of Samples

Panel. The same panel of judges evaluated the noodle samples using each of the three methods. Nine judges, previously trained on descriptive analysis, were selected to participate in this panel. The panel was composed of faculty, staff, and graduate students of the department. Judges were given additional training in perceiving and rating the sensory characteristics of mungbean starch noodles using the three scaling methods.

Terms. Descriptive terms were offered by the panel after evaluating samples in a one-hour session with a panel leader. A total of 28 descriptors were suggested for appearance, aroma, taste, and mouthfeel. Redundant terms were subsequently grouped and named with appropriate terms which reduced the number of descriptors to 14. During training, the panel decided on the definitions of the terms, anchors for the scales, and the standards to be used for the tests most of which were selected from those suggested by Meilgaard and coworkers (1987). Identical terms were used with each scaling method. The final list of descriptors used in the study is presented in Table 2.

Scoresheets. The scoresheets used by the panelists for the line scales utilized an interactive program designed to ask for responses to one attribute per sample at a time viewed by each judge on a computer monitor. The line scales

TABLE 2.
DEFINITIONS OF TERMS USED BY TRAINED SENSORY PANEL
TO EVALUATE MUNGBEAN STARCH NOODLES

TERMS	DEFINITIONS
Appearance	
Color	The actual color name or hue.
Glossiness	The amount of shine on the surface of the noodle strands.
Transparency	The extent of visibility through the noodle strands of objects lying behind them.
Speckledness	The presence of specks or visible particles within the noodle strand.
Stickiness	The degree of adherence of the noodle strands with each other.
Aroma	
Beany/green	The aroma associated with freshly cut green beans.
Musty	The aroma associated with paper that was stored in a warm and moist place for a long time.
Cooked starch	The aroma associated with starch that was heated or boiled with water and then cooled.
Taste	
Bitter	The taste on the tongue associated with caffeine.
Mouthfeel	
Slipperiness	The mouthfeel associated with boiled okra.
Moistness	The mouthfeel associated with the presence of water on the surface of the noodle strands.
Hardness	The force required to bite through the noodle strands.
Chewiness	The length of time required to masticate a sample at a constant rate of force application to reduce it to a consistency suitable for swallowing.
Elasticity	The capability of the noodle strands to recover its size and shape after application of force or pressure as in biting a rubber band.

consisted of 150-mm lines with a marker that can be moved from end to end with the cursor keys. The unstructured line scales were anchored at 12.5 mm from both ends while the semi-structured line scales were anchored with the intensity standards. Judges used the marker to indicate the intensities of a particular attribute. Responses using the 15-point category scales were written on paper ballots.

Environmental Conditions. Panelists evaluated the two noodle samples individually under white incandescent lights in partitioned booths equipped with computer terminals in a sensory evaluation laboratory. Samples in 1 oz plastic cups, prepared as described earlier, and coded with three-digit random numbers obtained using the table of random numbers (Meilgaard *et al.* 1987) were presented on white plates. Several sample cups were provided for each sample. The order of presentation of samples to panelists was randomized at every session. Panelists were asked to place the entire contents of the sample cup in their mouths when evaluating for intensities of attributes for taste and mouthfeel. The judges were provided with the reference standards, water, and water biscuits.

Statistical Analysis

The scaling methods were evaluated based on reliability, sensitivity to product differences, performance of judges, and effectiveness of terms. The data from the study were analyzed using the Statistical Analysis System (SAS Institute Inc. 1985). Significant differences between replications and between samples and significant period effects were determined using analysis of variance (ANOVA). Performance of judges was evaluated using one-way ANOVA. Results of multivariate analysis of variance (MANOVA), and stepwise discriminant analysis were used to determine whether descriptors were discriminatory of product differences.

RESULTS AND DISCUSSION

Period Effect

Table 3 shows the effect of the sequence by which the panelists used the three scaling methods on the evaluations made on the products which was found, generally, to be not significant at $p \leq 0.05$. The panel was able to evaluate noodle samples using the three scaling methods in succession without their performance in the first method significantly affecting the next. Only one of the 14 attributes evaluated, moistness, was significantly affected by the sequence of the methods during the first replicate. This result showed that the cross-over design was effective in eliminating residual or carry-over effects when treatments were separated by a time period of at least 4 h.

TABLE 3.
PROBABILITY LEVELS FOR THE DIFFERENT PERIODS ACROSS ALL PRODUCTS
BY INDIVIDUAL ATTRIBUTE IN FOUR REPLICATES

ATTRIBUTES	REPLICATION			
	1	2	3	4
Appearance				
Color	0.946	0.854	0.504	0.807
Glossiness	0.209	0.612	0.846	0.335
Transparency	0.446	0.213	0.335	0.504
Speckledness	0.636	0.458	0.869	0.292
Stickiness	0.977	0.691	0.297	0.881
Aroma				
Cooked starch	0.429	0.543	0.697	0.598
Musty	0.330	0.418	0.656	0.682
Beany	0.502	0.393	0.170	0.807
Taste				
Bitter	0.890	0.916	0.537	0.676
Mouthfeel				
Hardness	0.188	0.538	0.688	0.537
Moistness	0.029	0.697	0.781	0.235
Slipperiness	0.060	0.568	0.163	0.341
Chewiness	0.264	0.611	0.946	0.702
Elasticity	0.886	0.598	0.341	0.999

Reliability of Methods

Reliability was defined by Giovanni and Pangborn (1983) as the reproducibility of response using the F-values for replications as the criterion. Presented in Table 4 is the effect of replication on the evaluations made by the panelists using the three scaling methods.

In measuring the intensities of sensory characteristics of noodles, the use of ULS was found to be the most reliable because no significant differences were obtained for replication for any of the attributes evaluated using this scaling technique. The use of CS was the least reliable. However, for evaluating the intensities of mouthfeel and taste characteristics of the noodles, the other two methods were just as reliable to use as the unstructured line scaling technique. Also, the use of SLS were as reliable as the use of ULS for the measurement of intensities of appearance attributes of the noodles. It was apparent, that in

TABLE 4.
PROBABILITY LEVELS FOR THE DIFFERENT REPLICATIONS ACROSS ALL PRODUCTS BY INDIVIDUAL ATTRIBUTE AND SCALING METHOD

ATTRIBUTES	METHOD		
	ULS	SLS	CS
Appearance			
Color	0.096	0.144	0.504
Glossiness	0.065	0.096	0.002
Transparency	0.807	0.846	0.140
Speckledness	0.335	0.786	0.292
Stickiness	0.869	0.297	0.011
Aroma			
Cooked starch	0.697	0.009	0.273
Musty	0.656	0.006	0.032
Beany	0.170	0.682	0.881
Taste			
Bitter	0.537	0.115	0.150
Mouthfeel			
Hardness	0.568	0.688	0.598
Moistness	0.203	0.337	0.514
Slipperiness	0.114	0.946	0.341
Chewiness	0.163	0.082	0.702
Elasticity	0.781	0.611	0.999

general, increased anchoring of the line scales with the intensity standards did not increase their reliability specifically for evaluating intensities of flavor attributes.

Performance of Panelists

One-way analysis of variance was used to assess performance of panelists as suggested by Stone and coworkers (1974) who stated that any assessor whose F-value for performance has a probability ≤ 0.50 is contributing to discrimination. Figure 1 shows the number of significant attributes for each panelist at $p \leq 0.50$ using the three different scaling methods.

It was observed that more panelists had more significant attributes when using ULS. The point, however, is not whether the panelists had more significant attributes using one method compared to the others, but whether the number of attributes for which the panelist is significant exceeds chance

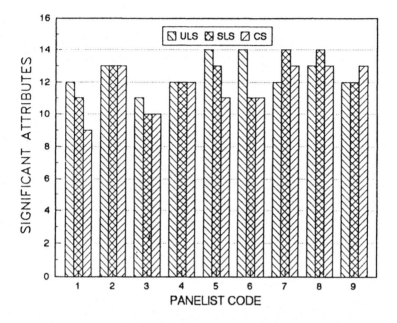

FIG. 1. NUMBER OF SIGNIFICANT ATTRIBUTES FOR EACH PANELIST AT $P < 0.50$ USING THE THREE DIFFERENT SCALING METHODS

indicating, therefore, a good performance by the assessor. In this regard, Powers (1988) used the cumulative binomial table which he found to have worked well. Based on the relationship he presented between the total number of attributes required to be significant at $p \leq 0.50$ if chance performance is to have a probability of less than or equal to 5%, the panelists in the present study should have significant performances in 11 of the 14 attributes evaluated. Results showed that all the nine panelists performed very well using ULS because all of them had 11 or more significant attributes. Only 8 panelists had good performances using SLS, and only 7 when CS were used.

Effectiveness of the Terms

The ability of a panel to use the descriptors presented in a scoresheet can be assessed in terms of the ability to discriminate among products, the reliable use across replicates, nonredundancy with other terms, prediction of consumer acceptance, and correlation with instrumental measures (Lawless 1989). To determine whether the descriptors utilized in this study were discriminatory of product differences, multivariate analysis of variance was conducted on the data. Palmer (1974) pointed out that not all descriptors are discriminatory. If the F-value for the term is significant, then the term is discriminatory and therefore,

TABLE 5.
PROBABILITY LEVELS FOR THE DIFFERENT PRODUCTS ACROSS ALL
REPLICATIONS BY INDIVIDUAL ATTRIBUTE AND SCALING METHOD

ATTRIBUTES	METHOD		
	ULS	SLS	CS
Appearance			
Color	0.0001	0.0001	0.0001
Glossiness	0.0143	0.0143	0.0772
Transparency	0.0001	0.0005	0.0128
Speckledness	0.0001	0.0001	0.0001
Stickiness	0.0001	0.0001	0.0001
Aroma			
Cooked starch	0.0023	0.2096	0.0113
Musty	0.0001	0.0001	0.0001
Beany	0.0018	0.0717	0.0001
Taste			
Bitter	0.0428	0.0468	0.2793
Mouthfeel			
Hardness	0.0001	0.0001	0.0001
Moistness	0.9627	0.9385	0.9816
Slipperiness	0.6629	0.5730	0.1063
Chewiness	0.0018	0.0001	0.0003
Elasticity	0.0001	0.0001	0.0001

effective for discriminating between samples; otherwise, the term can only be characterized as a descriptor.

Results presented in Table 5 showed that at 5% level of significance, 12 of 14 terms were discriminatory when ULS were used whereas the SLS and CS each found only 10 out of 14. All three methods indicated that moistness and slipperiness could be used only as descriptors for the noodles. In addition to these two terms, panelist using SLS classified cooked starch, and beany as descriptors and using CS found glossiness and bitterness also to be descriptors.

The reliable use of the descriptors by the panelists across replicates is indicated in Table 4. At 5% level of significance, the panelists used all the terms most consistently when using ULS.

Most Efficient Combination of Terms

Stepwise discriminant analysis was performed to seek out subsets of descriptors most useful to discriminate between the noodle samples using the different scaling methods (Powers and Ware 1986). Results showed that four descriptors were sufficient for discriminating between samples using ULS while each of the other two methods required five at $p \leq 0.05$ (Table 6). The terms 'stickiness' and 'cooked starch' were required in order to discriminate between samples when any of the scaling methods was used. In addition to these two terms, it was also found necessary to include 'elasticity' and 'bitterness' when evaluating noodle samples using ULS. 'Speckledness' together with 'slipperiness' and 'transparency' should be included when using SLS, and 'speckledness' together with 'color' and 'chewiness' should be included when using CS.

TABLE 6.
RESULTS OF STEPWISE DISCRIMINANT ANALYSIS ON THE SENSORY CHARACTERISTICS OF THE NOODLES EVALUATED USING UNSTRUCTURED LINE SCALES (ULS), SEMI-STRUCTURED LINE SCALES (SLS), AND CATEGORY SCALES (CS)

METHOD	STEP	VARIABLE ENTERED	F STATISTIC	LEVEL OF SIGNIFICANCE
ULS	1	Stickiness	180.85	0.0001
	2	Elasticity	24.94	0.0001
	3	Cooked starch	14.67	0.0003
	4	Bitterness	6.98	0.0103
SLS	1	Stickiness	234.57	0.0001
	2	Cooked starch	21.58	0.0001
	3	Speckledness	18.78	0.0001
	4	Slipperiness	4.58	0.0360
	5	Transparency	4.24	0.0435
CS	1	Stickiness	183.72	0.0001
	2	Color	19.58	0.0001
	3	Speckledness	9.59	0.0028
	4	Cooked starch	16.34	0.0001
	5	Chewiness	4.16	0.0454

Sensitivity of Methods to Product Differences

Results presented in Table 5 also show that ULS were most sensitive to product differences as indicated by the 12 significant F-values obtained at $p \leq 0.05$ from the 14 attributes evaluated when this scaling method was used. Using SLS and CS, products were found to be significantly different in only 10 attributes. These results likewise show that increased anchoring of the line scales with the intensity standards did not increase their sensitivity to product differences.

SUMMARY

Unstructured line scales (ULS), semi-structured line scales (SLS), and category scales (CS) were compared in evaluating mungbean starch noodles. Nine panelists evaluated two noodle samples using one method at a time during three different periods in four replicates.

Results indicated that the different scaling methods may be used in succession at intervals of at least four hours by a trained panel in evaluating the sensory characteristics of noodles without the performance of the panel in the first method significantly affecting the next. Results likewise showed that for the measurement of the sensory characteristics of noodles, the use of ULS was most reliable and required a smaller group of descriptors. It was also found that the panelists were better able to discriminate between noodle samples and that more terms were discriminatory of product differences when ULS were used. Increased anchoring of the line scales with the intensity standards did not result in increased sensitivity and reliability specifically for the flavor attributes of noodles.

ACKNOWLEDGMENT

The authors wish to acknowledge Aminah Abdullah, Lynn Cheney, Jih-Shiang Chern, Janey Dunn, Marilyn Erickson, Diana Hao, Chan Lee, and Kathleen Muego for participating in the sensory panel.

REFERENCES

BATEN, W.D. 1946. Organoleptic tests pertaining to apples and pears. Food Res. *11*, 84-90.

COCHRAN, W.G. and COX, G.M. 1957. *Experimental Designs*. 2nd ed. John Wiley & Sons, New York. Chapman & Hall, Limited, London.

GIOVANNI, M.E. and PANGBORN, R.M. 1983. Measurement of taste intensity and degree of liking of beverages by graphic scales and magnitude estimation. J. Food Sci. *48*, 1175-1182.

LAND, D.G. and SHEPHERD, R. 1988. Scaling and Ranking Methods. In *Sensory Analysis of Foods*. (J.R. Piggott, ed.) pp. 155-186, Elsevier Applied Science, London and New York.

LAWLESS, H.T. 1989. Editor's Corner: More thoughts on terminology development. Sensory Forum. Fall 1989, Number 46, pp. 5-6.

LAWLESS, H.T. and MALONE, G.J. 1986a. The discriminative efficiency of common scaling methods. J. Sensory Studies *1*, 85-98.

LAWLESS, H.T. and MALONE, G.J. 1986b. A comparison of rating scales: sensitivity, replicates and relative measurement. J. Sensory Studies *1*, 155-174.

McDANIEL, M.R. and SAWYER, F.M. 1981a. Descriptive analysis of whiskey sour formulations: magnitude estimation versus a 9-point category scale, J. Food Sci. *46*, 178-181, 189.

McDANIEL, M.R. and SAWYER, F.M. 1981b. Preference testing of whiskey sour formulations: magnitude estimation versus the 9-point hedonic. J. Food Sci. *46*, 182-185.

McTIGUE, M.C., KOEHLER, H.H. and SILBERNAGEL, M.J. 1989. Comparison of four sensory evaluation methods for assessing cooked dry bean flavor. J. Food Sci. *54*(5), 1278-1283.

MEILGAARD, M., CIVILLE, G.V. and CARR, B.T. 1987. Sensory Evaluation Techniques. Vol. 1 and 2. CRC Press, Boca Raton, Florida.

MOSKOWITZ, H.R. and SIDEL, J.L. 1971. Magnitude and hedonic scales of food acceptability. J. Food Sci. *36*, 677-680.

PALMER, D.H. 1974. Multivariate analysis of flavor terms used by experts and nonexperts for describing tea. J. Sci. Food Agric. *25*, 153-164.

PEARCE, J.H., KORTH, B. and WARREN, C.B. 1986. Evaluation of three scaling methods for hedonics. J. Sensory Studies *1*, 27-46.

POWERS, J.J. 1988. Current practices and applications of descriptive methods. In *Sensory Analysis of Foods*. (J.R. Piggot, ed.) pp. 187-266, Elsevier Applied Science, London and New York.

POWERS, J.J. and WARE, G.O. 1986. Discriminant Analysis. In *Statistical Procedures in Food Research*. (J.R. Piggott, ed.) pp. 125-180, Elsevier Applied Science. London and New York.

RAFFENSPERGER, E.L., PERYAM, D.R. and WOOD, K.R. 1956. Development of a scale for grading toughness-tenderness in beef. Food Technol. *10*(12), 627-630.

SHAND, P.J., HAWRYSH, Z.J., HARDIN, R.T. and JEREMIAH, L.E. 1985. Descriptive sensory assessment of beef steaks by category scaling, line scaling, and magnitude estimation. J. Food Sci. *50*, 495-500.

STONE, H. and SIDEL, J.L. 1985. *Sensory Evaluation Practices*. Academic Press, Orlando, Florida.

STONE, H., SIDEL, J., OLIVER, S., WOOSLEY, A. and SINGLETON, R.C. 1974. Sensory evaluation by quantitative descriptive analysis. Food Technol. *28*, 24-34.

CHAPTER 6.11

A COMPARISON OF FREE-CHOICE PROFILING AND THE REPERTORY GRID METHOD IN THE FLAVOR PROFILING OF CIDER

J.R. PIGGOTT and M.P. WATSON

*At the time of original publication,
all authors were affiliated with
University of Strathclyde
Department of Bioscience and Biotechnology
131 Albion Street
Glasgow G1 1SD
Scotland.*

ABSTRACT

Free-choice profiling has been used to collect descriptive sensory data from both trained and untrained assessors, though untrained assessors may have difficulty in generating sufficient and adequate descriptors. The more structured repertory grid method has been used as an alternative procedure, and is thought to help assessors in vocabulary development. To compare these two procedures, panels of assessors used them to describe the sensory properties of 25 ciders. Generalized Procrustes analyses of the two data sets provided broadly similar results, but the repertory grid method yielded more descriptors and interpretation of the resulting product space was slightly easier. However, the repertory grid method took slightly longer to carry out, and it was concluded that neither method was clearly superior.

INTRODUCTION

Descriptive analysis in its various forms has been a popular sensory technique for many years. The original Flavour Profile Method, described by Cairncross and Sjöström (1950), used a group of highly trained assessors working with an agreed vocabulary to examine and describe characteristics of products. Since then the technique has been modified by a number of authors, and current practice is most often based on the Quantitative Descriptive Analysis procedure of Stone et al. (1974). These procedures emphasize the importance of sufficient assessor training, correct experimental design and statistical evaluation of results, and it is assumed that the assessors are using the descriptors in the same way. However there are a number of sources of variation which might not be eliminated completely by training (Arnold and Williams 1986):

(1) Assessors vary in their overall level of scoring;
(2) Assessors use descriptors in different ways;
(3) Assessors vary in their range of scoring;
(4) Assessors vary in their use of terms and scales between sessions;
(5) Assessors might perceive different stimuli in the same products.

To overcome some of these problems the Free Choice Profile (FCP) procedure was developed (Williams and Langron 1984; Williams and Arnold 1985). This allows the assessors to choose their own vocabularies, thus eliminating the need for extensive training in descriptor use, and in combination with Generalized Procrustes Analysis (GPA) it eliminates the first three of the sources of variation listed above. GPA consists of three logically distinct steps: firstly, the centroids of each assessor's data are matched so as to eliminate the effect of use of different parts of the scales; secondly, isotropic scale changes remove the differences in the scoring range used by different assessors; thirdly, the configurations are matched as closely as possible by rotation and reflection of the axes (Arnold and Williams 1986). This process produces a perceptual space for each assessor, which is matched as closely as possible with the other assessors. A consensus configuration is then calculated as the average configuration for all the assessors. This is usually simplified as a reduced dimensional plot by principal components analysis (PCA). The residual errors (distances between the assessors' individual configurations and the consensus) can then be used to calculate coordinates for plotting the assessors, in order to identify outliers or groups. Oreskovich et al. (1991) have described the principles and applications of GPA in more detail.

FCP permits the use of untrained assessors, because there is no need for training in the use of descriptors. However, this may lead to misuse of descriptors and difficulties in interpretation of results (Guy et al. 1989). Some assessors also appear to have difficulty in generating sufficient descriptors (e.g., Piggott et al. 1991), and McEwan et al. (1989) suggested that assessors, in the isolation of sensory booths, find it difficult to describe their perceptions. Thus a more structured approach was developed incorporating the repertory grid method (RGM) (Kelly 1955; McEwan and Thomson 1988). This involves presenting each assessor with a sample triad, and an interviewer then asks the assessor to describe the similarities and differences between them. The interviewer must be present in order to obtain the optimum amount of information about the product (McEwan et al. 1989). A variation of the RGM involves presenting the samples in the form of a dyad grid, the purpose being to eliminate any confusion in the mind of the untrained assessor caused by triadic presentation (Ryle and Lunghi 1970; McEwan et al. 1989). It is not clear whether the original FCP procedure or the RGM followed by profiling provides better data, in the form of easier interpretation, a more complete description of the sample space, and better discrimination between samples. Both procedures have been successfully used in many studies, and McEwan et al. (1989) concluded on the basis of a limited study that neither procedure was clearly superior. Piggott et al. (1991) commented that the free-choice technique did not work successfully for all respondents in a study of perceptions of alcoholic beverages.

The aim of the work presented here was to investigate the differences, if any, between free-choice profiling and the repertory grid method for descriptive analysis of a set of ciders.

MATERIALS AND METHODS

Samples

Twenty-five ciders were used, mainly purchased through UK retail outlets (Table 1). These particular ciders were selected, as it was considered that they satisfactorily covered the products available on the U.K. market. The samples were stored and served to the assessors at ambient temperature (approximately 20C). Samples (30 mL) were presented in 150-mL clear glass tumblers covered by watch-glasses and identified by 3-digit random numbers.

Assessors

Two groups of 15 assessors were recruited from among the staff and students of the University. The two groups were roughly balanced in terms of age and sex and previous extent of sensory training and experience. There were 13 male and 17 female assessors, 18 had previously received sensory training, and ages ranged from approximately 20 to 50 years. One group used the FCP procedure, while the other used the RGM.

TABLE 1.
DETAILS OF THE SAMPLES USED

Abbreviation	Sample	Label Description	Supplier
AG	Autumn Gold	Medium Sweet	Taunton Cider Ltd
ARCH	Archers	Medium Sweet	Konings, Belgium
CB	Cidre Bouchê Brut	Medium Sweet	Montgommery Saintes, Calvados
CL	Crispen Light	Low Alcohol	H.P. Bulmers Ltd
DB	Diamond Blush	Cooler	Taunton Cider Ltd
DBLT	Dry Blackthorn	Dry	Taunton Cider Ltd
DD	Duc Dwal	Medium Sweet	Konings, Belgium
DW	Diamond White	Dry	Taunton Cider Ltd
IMD	West Country Scrumpy	Dry	Inch's Cider Ltd
IMS	West Country Scrumpy	Medium Sweet	Inch's Cider Ltd
K	K	Premium	W. Gaymers and Son Ltd
LF	Longfellows	Medium Sweet	Gateways Ltd
MAX	Max	Dry	H.P. Bulmers Ltd
MD	Vintage Dry	Dry	Merrydown Wine plc
MG	Vintage Gold	Medium Sweet	Merrydown Wine plc
NAT	Natch	Dry	Taunton Cider Ltd
OE	Old English	Medium Sweet	W. Gaymers and Son Ltd
POM	Pomagne	Dry	H.P. Bulmers Ltd
SAINS	Normandy Cider	Dry	Sainsburys
SBOW	Strongbow	Dry	H.P. Bulmers Ltd
SHEP	Gold Medal Farmhouse	Medium	R.J. Sheppy and Son
SS	Strongbow Super	Dry	H.P. Bulmers Ltd
WO	Original Cider	Dry	H. Weston and Son Ltd
WPECK	Woodpecker	Medium Sweet	H.P. Bulmers Ltd
WV	Vintage Cider	Dry	H. Weston and Son Ltd

Free-Choice Profiling

Assessors attended sessions (2 × 0.5 h) where they were instructed in the FCP procedure and presented with 12 and 13 samples, respectively. They were asked to smell and taste the samples and to describe, using their own terms, the flavor, aroma and mouthfeel of the ciders.

Repertory Grid Method

A dyadic presentation was used, since this appears easier to understand especially for untrained assessors (Ryle and Lunghi 1970; McEwan et al. 1989). The samples were sequentially paired, so that one sample was carried forward to the next pair, thus minimizing order effects. Assessors were asked to describe the similarities and differences between the samples in each pair, in terms of flavor, aroma and mouthfeel.

Quantitative Assessment Procedure

Individual vocabularies were entered into the computer system used for data collection (PSA-System V 1.61; Oliemans, Punter and Partners, PO Box 14167, 3508 SG Utcecht, The Netherlands) with replicates and subjective or hedonic terms deleted. Five samples were presented in individual booths at each session, and over ten sessions each sample was assessed twice. The order of presentation was balanced, so that over the whole experiment each sample was preceded by each other sample (MacFie *et al.* 1989). Water was provided for mouth-rinsing between samples. The assessors scored each descriptor on a continuous line scale, anchored at 10% and 90% by the terms 'weak' and 'strong,' respectively. A few samples were easily recognizable because of their unusual color, and so assessment was carried out under red lighting.

Replicate assessments by each panel were separated, providing four data sets each consisting of assessments of the 25 samples by 15 assessors. The four were then analyzed by GPA using Procrustes-PC V 2.1 (Oliemans, Punter and Partners, PO Box 14167, 3508 SG Utrecht, The Netherlands).

RESULTS

Free-Choice Profiling

The number of descriptors used varied from 12 to 32, with a mean of 19. The first two dimensions of the consensus configurations of the replicates are shown in Fig. 1 and 2, and accounted for 26% and 29% of original variance. The third and fourth dimensions accounted for a further 10% of variance in both replicates, but could not be readily interpreted and will not be discussed further.

Inspection of the figures showed that the two replicates produced generally similar sample configurations. Further interpretation will therefore be limited to the first replicate. Descriptive terms with correlations (positive or negative) > 0.5 with the dimensions are shown in Table 2. In Fig. 1 the sweet ciders were all towards the negative end of the first dimension, the associated descriptors being sweet and fruity (Table 2). The dry ciders were towards the positive end of the first dimension, associated with descriptors such as bitter, dry and astringent. However Strongbow, a dry cider, appeared near the middle of the plot, with the sweet ciders; it is dry compared with the same company's Woodpecker, but in comparison with the majority of products it is relatively sweet. The French ciders were grouped together towards the positive end of the second axis, associated with apple and fruity descriptors. In English cider manufacture the fermentation is allowed to continue until dryness is reached (Beech 1972), while

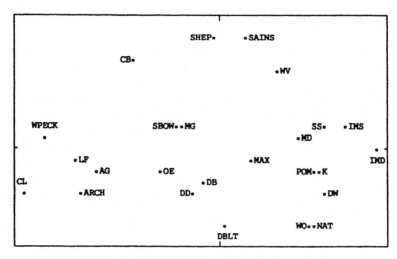

FIG. 1. CONSENSUS CONFIGURATION ON FIRST (HORIZONTAL) AND SECOND (VERTICAL) DIMENSIONS FOR FIRST REPLICATE OF PANEL USING FCP PROCEDURE
Abbreviations are shown in Table 1.

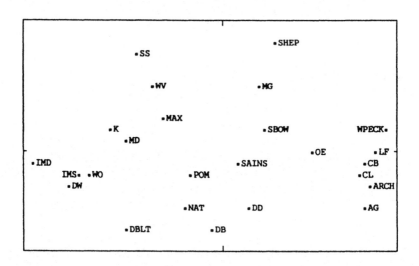

FIG. 2. CONSENSUS CONFIGURATION ON FIRST (HORIZONTAL) AND SECOND (VERTICAL) DIMENSIONS FOR SECOND REPLICATE OF PANEL USING FCP PROCEDURE
Abbreviations are shown in Table 1.

the French process involves stopping the fermentation prematurely leaving more sugars and original fruit volatiles in the product (Drilleau 1985). Other groups with similar characteristics included the premium ciders such as Max, K,

Diamond White and Diamond Blush. Inch's medium sweet and dry ciders showed similar characteristics, suggesting that similar raw materials and processes were used in their production.

TABLE 2.
INTERPRETATION OF THE FIRST FOUR PRINCIPAL AXES OF THE FIRST REPLICATE FROM GPA OF DATA FROM FCP PANEL

Principal Axis	+/- Correlation	Descriptors
1	+	Sharp, Astringent, Bitter, Dry, Acid, Acrid, Sour, Alcoholic, Chemical, Cardboard, Warming, Medicinal, Disinfectant, Wersh, Sewage, Metallic, Solvent, Beer, Catty
	–	Apples, Fruity, Apricot, Estery, Raspberry, Sweet, Syrupy, Honey, Sickly, Smooth, Flat, Mellow, Full, Bland
2	+	Apples, Fruit, Apple Peel, Grass/Silage, Burnt/Toffee, Stewed Apples
	–	Thin, Watery, Grapey

Repertory Grid Method

The number of descriptors used varied from 19 to 42, with a mean of 29. The first two dimensions of the consensus configurations of the replicates are shown in Fig. 3 and 4, and accounted for 20% and 21% of original variance. The third and fourth dimensions accounted for a further 11% of variance in both replicates, but again could not be satisfactorily interpreted.

Inspection of the figures showed that the two replicates produced generally similar sample configurations, and so further interpretation will be limited to the first replicate. Descriptive terms with correlations (positive or negative) > 0.5 with the dimensions are shown in Table 3. In Fig. 3, the sweet and dry ciders were again separated on a dimension associated with sweet/fruity and harsh terms (Table 3). The non-English ciders also formed groups, the French ciders again associated with apple, fermented and cooked apple descriptors. Similar groups can be seen, suggesting that the major differences between ciders are between producers.

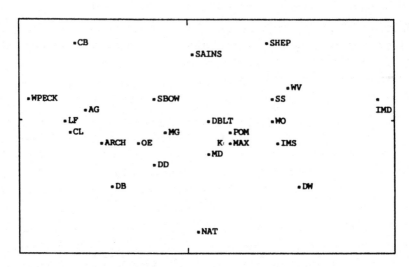

FIG. 3. CONSENSUS CONFIGURATION ON FIRST (HORIZONTAL) AND SECOND (VERTICAL) DIMENSIONS FOR FIRST REPLICATE OF PANEL USING REPERTORY GRID PROCEDURE
Abbreviations are shown in Table 1.

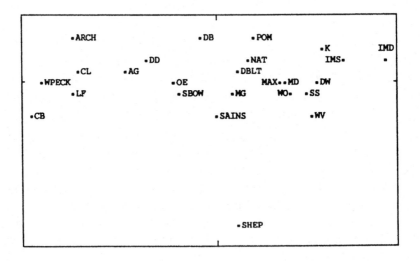

FIG. 4. CONSENSUS CONFIGURATION ON FIRST (HORIZONTAL) AND SECOND (VERTICAL) DIMENSIONS FOR SECOND REPLICATE OF PANEL USING REPERTORY GRID PROCEDURE
Abbreviations are shown in Table 1.

TABLE 3.
INTERPRETATION OF THE FIRST FOUR PRINCIPAL AXES OF THE
FIRST REPLICATE FROM GPA OF DATA FROM RGM PANEL

Principal Axis	+/- Correlation	Descriptors
1	+	Bitter, Dry, Astringent, Sour, Sharp, Perfume, Floral, Ester, Medicinal, Alcoholic, Musty/Mouldy, Catty, Solvent, Acrid, Earthy, Winey, Meths, Glue, After-taste, Sour Apples
	−	Apples, Fruity, Ester, Sweet, Syrupy, Saccharin, Caramel, Smooth
2	+	Apples, Fermented Apple, Vinyl, Cooked Apples, Solvent, Plastic, Sulphury, Smooth
	−	Sour, Vinegar, Astringent, Beer, Wine, Catty, Hops, Fizzy, Hydrogen Sulphide, Lemon Cleaning Fluid

In all cases the assessor plots showed a relatively homogeneous group, with one outlier in each panel. A final GPA was run of the complete data set, using a GENSTAT program derived from Arnold's (1987) routine, to test for segmentation between the two panels of assessors. The resulting assessor plot (Fig. 5) showed evidence of separation into the two panels, and the panel using the RGM had slightly greater residual distances from the consensus configuration.

DISCUSSION

The extent to which the first and second dimensions accounted for variation in the data varied between methods. For FCP this was 26% and 29% for the two replicates, while for the RGM the corresponding figures were 20% and 21%. This was presumably because the assessors using the RGM used a greater number of descriptors, which added error to the data without adding much further information. This would have the effect of reducing the proportion of variance accounted for by the early dimensions. The slightly higher proportion of variance accounted for by the third and fourth dimensions for the RGM suggested that it allowed the assessors to provide more useful information than could be accommodated on the first two dimensions, but given the difficulty of interpretation this could not be demonstrated.

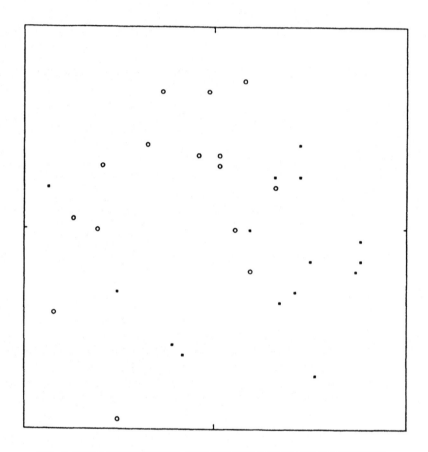

FIG. 5. FIRST (HORIZONTAL) AND SECOND (VERTICAL) DIMENSIONS OF ASSESSOR PLOT FOR ALL DATA FOR BOTH PANELS
○ = Assessors using free-choice profile procedures;
■ = assessors using repertory grid method.

The assessor plots showed some indications that the assessors were not all describing the samples in the same way, with one outlier from each panel. It seems unlikely that this was related simply to a paucity of vocabulary, as suggested by Piggott et al. (1991), since the assessors concerned used 13 and

24 terms. It is possible that these assessors either misused or misunderstood the method, or that they genuinely perceived the samples in a different way. The plot for all assessors (Fig. 5) showed clear evidence of segmentation, suggesting that the two panels were systematically different in the way they described the samples. However the first two dimensions of the consensus configurations (Fig. 1-4) were similar, and it seems that any substantial differences between the panels would only become apparent on later dimensions.

Both the methods used here allowed the assessors to generate their own personal descriptors, and no substantial difficulty of interpretation of early dimensions was encountered. McEwan *et al.* (1989) suggested that there was no real advantage in using the RGM due to the amount of time required, and this study has shown that, while the assessors using the RGM used more descriptors, there were no substantial differences in the resulting product configurations. However, some assessors tended to examine each sample in isolation, thus ignoring the principle of the repertory grid, and some assessors felt that they were being forced to produce excessive numbers of descriptors. Subsequently members of both panels reported that more descriptors were needed. Presenting all the samples at the vocabulary development stage aims to eliminate this problem, but clearly further time was required in this case to refine the vocabularies. A practice assessment run or second stage of vocabulary development would have allowed this (Gains and Thomson 1990).

No attempt was made to identify components of cider responsible for flavor characteristics, since the major purpose of this work was not to study the origins of cider flavor, but some of the descriptors used in this study can be tentatively linked with flavor compounds in cider. There are obvious, though not necessarily simple, links between sourness and acidity due in particular to malic acid, and between sweetness and sugars. Bitterness and astringency in cider are believed to originate from the phenolic compounds present (e.g., Lea 1990; Lea and Arnold 1978). Apply and fruity notes are mainly contributed by esters, especially those of low molecular weight such as butyl, pentyl and hexyl acetates. High molecular weight esters are important for winey and waxy characteristics, and perfumed notes are mainly due to the mono-, di- and tri-terpenes, which are particularly important in other fermented beverages such as wine (Williams *et al.* 1984).

CONCLUSION

The free-choice profile and repertory grid procedures both allow assessors to choose their own vocabularies. In this study the methods yielded very similar results, but the RGM elicited a greater number of descriptors and interpretation of the resulting product space was slightly easier. Neither method provided clearly superior results. The descriptive terms used could be interpreted in terms of known differences in the production processes for the ciders, and in terms of flavor compounds previously reported as present.

ACKNOWLEDGMENTS

The authors wish to thank Janice Clyne for her advice and assistance in recruiting the assessors and running the PSA-System and the assessors who came without fail to taste the samples.

REFERENCES

ARNOLD, G.M. 1987. A Generalized Procrustes macro for sensory analysis. Genstat News. *18*, 61-80.

ARNOLD, G.M. and WILLIAMS, A.A. 1986. The use of Generalized Procrustes techniques in sensory analysis. In *Statistical Procedures in Food Research*, (J.R. Piggott, ed.) pp. 233-255, Elsevier Applied Science, London.

BEECH, F.W. 1972. English cider making: Technology, microbiology and biochemistry. In *Progress in Industrial Microbiology*, (D.J.D. Hockenhul, ed.) pp. 133-213, Churchill Livingstone, London.

CAIRNCROSS, S.E. and SJÖSTROM, L.B. 1950. Flavour profiles, a new approach to flavour problems. Food Technol. *4*, 308-311.

DRILLEAU, J.F. 1985. Apple processing and cider production in France. Flussiges Obst. *52*(8), 414-418, 429, 432-433.

GAINS, N. and THOMSON, D.M.H. 1990. Sensory profiling of canned lager beers using consumers in their own homes. Food Qual. Pref. *2*, 39-47.

GUY, C., PIGGOTT, J.R. and MARIE, S. 1989. Consumer profiling of Scotch whisky. Food Qual. Pref. *1*, 69-73.

KELLY, G.A. 1955. *The Psychology of Personal Constructs*, Norton, New York.

LEA, A.G. 1990. Bitterness and astringency: The procyanidins. In *Bitterness in Foods and Beverages*, (R.L. Rousseff, ed.) pp. 123-144, Elsevier Applied Science, Amsterdam.

LEA, A.G. and ARNOLD, G.M. 1978. The phenolics of cider: bitterness and astringency. J. Sci. Food Agric. *29*, 478-483.

MacFIE, H.J., BRATCHELL, N., GREENHOFF, H. and VALLIS, L.V. 1989. Designs to balance the effect of order of presentation and first order carryover effects in hall tests. J. Sensory Studies *4*, 129-149.

McEWAN, J.A., COLWILL, J.S. and THOMSON, D.M.H. 1989. The application of two free choice profile methods to investigate the sensory characteristics of chocolate. J. Sensory Studies 3, 271-286.

McEWAN, J.A. and THOMSON, D.M.H. 1988. An investigation of the factors influencing consumer acceptance of chocolate confectionery using the repertory grid method. In *Food Acceptability*, (D.M.H. Thomson, ed.) pp. 347-362, Elsevier Applied Science, London.

ORESKOVICH, D.C., KLEIN, B.P. and SUTHERLAND, J.W. 1991. Procrustes Analysis and its applications to free-choice and other sensory profiling. In *Sensory Science Theory and Applications in Foods*, (H. Lawless and B.P. Klein, eds.) pp. 353-393, Marcel Dekker, New York.

PIGGOTT, J.R., SHEEN, M.R. and APOSTOLIDOU, S.G. 1991. Consumer perceptions of alcoholic beverages. Food Qual. Pref. 2, 177-185.

RYLE, A. and LUNGHI, M. 1970. The dyad grid: A modification of the repertory grid technique. Br. J. Psychiatry 117, 323-327.

STONE, H., SIDEL, J., OLIVER, S., WOOLSEY, A. and SINGLETON, R.C. 1974. Sensory evaluation by quantitative descriptive analysis. Food Technol. 28, 24-34.

WILLIAMS, A.A. and ARNOLD, G.M. 1985. A comparison of the aromas of six coffees characterized by conventional profiling, free-choice profiling and similarity scaling methods. J. Sci. Food Agric. 36, 204-214.

WILLIAMS, A.A. and LANGRON, S.P. 1984. The use of free choice profiling for evaluation of commercial ports. J. Sci. Food Agric. 35, 558-568.

WILLIAMS, A.A., ROGERS, C. and NOBLE, A.C. 1984. Characterization of flavour in alcoholic beverages. In *Flavour Research of Alcoholic Beverages*, (L. Nykänen and P. Lehtonen, eds.) pp. 235-253, Foundation for Biotechnical and Industrial Fermentation Research, Helsinki.

CHAPTER 6.12

DESCRIPTIVE ANALYSIS OF ORAL PUNGENCY[1]

MARGARET CLIFF and HILDEGARDE HEYMANN

At the time of original publication,
all authors were affiliated with
University of Missouri-Columbia
College of Agriculture, Food & Natural Resources
Food Science and Human Nutrition
122 Eckles Hall, Columbia, MO 65211.

ABSTRACT

Pungent spices are well-recognized for their aromatic and pungent nature; however, relatively little is known about their pungent or 'burning' responses. Therefore, this research was undertaken to characterize oral pungency of the principal irritants or red pepper (capsaicin), black pepper (piperine), cinnamon (cinnamaldehyde), cumin (cuminaldehyde), cloves (eugenol), ginger (ginger oleoresin), and alcohol (ethanol). These compounds were evaluated for four pungent qualities (burning, tingling, numbing, overall), two temporal qualities (lag time, overall duration), and three spatial qualities (longitudinal location, lateral location, localized/diffuse). The pungency of cinnamaldehyde was primarily burning and tingling. It had a quick onset and rapid decay. The pungency of eugenol had a long-lasting, predominantly numbing effect. The pungency piperine, capsaicin and ginger were primarily burning, but had different temporal and spatial responses. The pungency of ethanol was most diffuse in nature, with some burning and tingling sensations. It had the shortest perceived onset and overall duration. The pungency of cuminaldehyde was equally burning, tingling and numbing.

INTRODUCTION

Pungent spices such as red pepper, black pepper, ginger and cumin, add flavor and variety to human foods (Rozin 1990). These spices have traditionally been associated with cuisines of high cereal diets (Andrews 1984) but there is a growing demand for pungent spices in the United States. Lawless (1989)

---

[1]Contribution from the Missouri Agriculture Experiment Station. Journal Series Number 11,626.

Reprinted with permission of Food & Nutrition Press, Inc., Trumbull, Connecticut.
©Copyright 1992. Originally published in Journal of Sensory Studies *7, 279-290.*

attributes this growing demand to (1) increased multiculturalism, (2) increased popularity of ethnic restaurants among Americans of European decent, (3) increased use of 'flavored' vegetables in the diet to avoid cholesterol and saturated fats and (4) the decline in taste and smell sensitivity associated with the age-related shift of the American population.

Many spices not only induce aromatic and taste sensations, but also pungent sensation. This pungency or burning is often called the 'common chemical sense' (Moncrieff 1967) and is elicited by stimulation to the free nerve endings of the trigeminal nerve (Silver and Maruniak 1981). The free nerve endings are located in both the nasal or oral cavities, respond to a variety of compounds and induce a number of different responses (Govindarajan 1979).

Although some research has been conducted to develop a standardized methodology for evaluation of red pepper pungency (Gillette et al. 1984), scientific characterization of oral pungency has been somewhat overlooked (Lawless 1989). Spice research has primarily focused on elucidation and chemical characterization of the principal irritants (Govindarajan 1979). Some psycho-physical research has been done to evaluate intensity/concentration functions (Lawless 1984) and monitor sensitization/desensitization (Green 1989; Karrer and Bartoshuk 1991), but this research has been restricted to capsaicin. Limited research has been conducted on piperine, ginger and ethanol, and none is available on mustard, horseradish, cinnamon, and clove.

Indeed, different pungent 'qualities' have been alluded to in the literature. Todd et al. (1977) points out the qualitative differences among natural and synthetic capsaicinoids; Govindarajan (1979) notes differences between capsaicin, black pepper and ginger at higher concentrations, but neither the methodology nor the data were reported; and Lawless (1984) speculates on potential differences. Furthermore, Silver and Maruniak (1981) point out that the capability of the trigeminal nervous system for quality differentiation remains a major unresolved issue in nasal and oral chemoreception.

Therefore, this research was undertaken to quantitatively characterize the oral irritant response for the principal irritants of red pepper (capsaicin), black pepper (piperine), ginger (oleoresin), cinnamon (cinnamaldehyde), cumin (cumin aldehyde), cloves (eugenol), and alcohol (ethanol), using descriptive analysis (Stone and Sidel 1985) coupled with univariate and multivariate statistics.

MATERIALS AND METHODS

Irritants

The principal irritants, source, concentration, and purity of the compounds used are as listed in Table 1. Since the principal irritants of ginger were not

TABLE 1.
MATERIALS FOR DESCRIPTIVE ANALYSIS OF ORAL PUNGENCY

PUNGENT	PRINCIPAL IRRITANT	SOURCE	PURITY %	CONCEN. g/L
red pepper	capsaicin (8-methyl-n-vanillyl-6-nonenamide)	Sigma	98	0.002
black pepper	piperine	Aldrich	97	0.2
cinnamon	cinnamaldehyde	Aldrich	99	2.0
cumin	cuminaldehyde (4-isopropylbenzaldehyde)	Aldrich	98	15.0
cloves	eugenol	Aldrich	99	0.2
ginger	ginger oleoresin	Fritzshe		3.0
alcohol	ethanol	Fischer	95	47.5

commercially available, ginger oleoresin, containing the three pungent irritants (shogaol, gingerol, zingerone) was used (Wood 1987).

All irritants were mixed with 0.4 g/L polysorbate 80 (Aldrich Tween 80) prior to dilution with 70C double distilled water. The crystalline irritants, capsaicin and piperine, were heated with polysorbate 80 to ensure solubilization prior to dilution. Upon cooking, all solutions were made up to volume (1 L) and stored at 5C. Due to recrystallization of piperine solutions, the piperine solution was prepared prior to each evaluation and stored at room temperature.

Sensory Evaluation

Twenty-one judges, staff and students (11 females, 10 males) from the University of Missouri-Columbia, participated in the study. Judges participated in one of two round table discussion sessions for orientation and term development. At these sessions, judges were introduced to the pungent solutions and a list of tentative terms. Judges were asked to describe the perceived sensations in terms of the quality, intensity, time and location in the mouth. To mask interfering color and odor, when present, all solutions were presented in dark-blue opaque glasses and all judges wore nose-clips during the course of the evaluations. These nose-clips blocked retronasal transfer of aroma (Burdach *et al.* 1984), thereby allowing the judge to focus on the perceived mouth qualities.

The group leader served to summarize, resolve confusion, and bring the group to consensus on the final terms to be used. The final consensus of terms and their definitions are listed in Table 2.

Following group discussion, the judges participated in one practice session with the capsaicin, cinnamaldehyde, and eugenol, which were predominantly burning, tingling and numbing, respectively. During this session, judges were unaware they were being tested. If the judges were unsuccessful in characterizing the principal sensation, they were given, in each case, an individual retraining session.

At the training sessions and test sessions, judges evaluated the pungencies using a modified ASTM methodology (ASTM 1988). Judges held the entire sample (10 mL) in their mouths for 5 s, expectorated, waited 30 s and then evaluated the attributes as listed in Table 2. Judges were given three samples at each session, and were required to rinse with double distilled water and wait 5 min between samples. Judges timed their tasting, rinsing and waiting periods using stopwatches. All samples were expectorated.

Attributes were scored on 15 cm unstructured line scales. The pungent attributes (burning, tingling, numbing, and overall intensity) were scored from none to extreme; whereas the temporal attributes (lag, duration) were rated from short to long. The longitudinal and lateral location terms were scored from front

TABLE 2.
LIST OF ATTRIBUTES, ABBREVIATIONS AND DEFINITIONS
FOR CHARACTERIZATION OF ORAL PUNGENCY[1]

ATTRIBUTE	DEFINITION
1. LAG TIME	The length of time between ingestion and the **maximum** perceived pungency. (0-short; 15-long)
2. BURNING	The magnitude of the perceived hot sensation. (0-none; 15-extreme)
3. TINGLING	The magnitude of the perceived tingling sensation. (0-none; 15-extreme)
4. NUMBING	The magnitude of the perceived numbing or deading sensation, or 'lack' of feeling. (0-none; 15-extreme)
5. LONGITUDINAL LOCATION	The position of the pungent sensation from the front to the back, along the length of the tongue. (0-front; 15-back)
6. LATERAL LOCATION	The position of the pungent sensation from the middle to the side, across the width of the tongue. (0-middle; 15-side)
7. AREA	The area over which the pungent sensation is perceived as either 'localized' or 'diffuse' in nature. (0-localized; 15-diffuse)
8. OVERALL INTENSITY	The overall magnitude of the pungent sensation. (0-none; 15-extreme)
9. DURATION/ PERSISTENCE	The length of time that the pungent sensation persists. (0-short; 15-long)

[1] 15 cm unstructured linescales were used

to back and middle to side, respectively. If other areas of the mouth were affected (lips, cheeks, hard palate, soft palate, gums, throat, base of mouth, other), judges checked the appropriate box at the bottom of the scorecard. Lastly, the area attribute was scored from localized to diffuse in nature.

Experimental Design and Data Analysis

Because judges were unable to evaluate all irritants, in a single session, a balanced-incomplete-block (BIB) design (plan 11.7 Cochran and Cox 1957) was used. This allowed for day-to-day variations (block) to be considered. All irritants were evaluated in triplicate. Irritants were randomly assigned to treatments. and three irritants were evaluated per tasting session, for a total of seven sessions. Sessions were scheduled every alternate day, to avoid the possible desensitization that has been documented for capsaicin (Green 1989; Karrer and Bartoshuk 1991). At each session, judges received the three samples in random order, as assigned using SAS (SAS 1985) proc plan. All intensity scores were quantified from 0 to 15.

Panel and judge performance were first assessed using a 3-way analysis of variance (ANOVA) for the main effects of judge, irritant, and replication and all 2-way interactions. All statistical analyses were conducted using SAS (SAS 1985).

Once panel performance had been demonstrated, irritant effects were evaluated on the mean scores using the BIB software (UMC 1988). Principal component analysis was conducted, without rotation, on the mean scores using the correlation matrix (SAS 1985).

RESULTS AND DISCUSSION

As shown in Table 3, judges were highly significant sources of variation for all attributes. This reflects innate differences in sensitivity among judges (Lawless *et al.* 1984; Stevens 1990; Karrer and Bartoshuk 1991) and the unique way unstructured line scales are interpreted by judges.

Judges reproducibility was excellent, as indicated by the lack of significance of all Judge*Rep interactions. Overall panel reproducibility was also good, except for the lag time, longitudinal location, overall intensity and duration terms, as indicated by the significant F-values for replication (Table 3). This lack of reproducibility for lag time and duration was believed due to the unfamiliarity with scoring of temporal sensations under static conditions. These attributes would be best quantified using time-intensity methodology. The poor reproducibility of the overall intensity term probably reflects the complex nature of the attribute.

Judge inconsistency, across irritants, as indicated by the Judge*Irritant interactions, was significant for all attributes except longitudinal and lateral locations. However, when the variance from the irritant effects [mean square (MS) Irritant] were compared to the interaction (MS Judge*Irritant), using a more stringent mixed effect model (Goniak and Noble 1987), the new F-value for the main irritant effects remained significant, except for area of sensation. The term area of sensation was, therefore, dropped from further analyses.

When BIB analysis of variance was conducted, irritant effects were significant ($p < 0.05$) for all attributes, except lateral location ($p < 0.1$) and area. On this basis, as well as that noted above, these attributes were eliminated

TABLE 3.
F-VALUES FOR ANALYSES OF VARIANCE FOR ORAL PUNGENCY

	Judge	Irritant	Rep	Jud*Irr	Jud*Rep	Irr*Rep	MS_e
LAG TIME	8.25***	17.60***	3.07*	1.68***	0.62	1.00	7.47
BURNING	18.26***	19.17***	1.74	2.28***	1.25	2.45***	7.68
TINGING	27.05***	28.53***	2.59	2.46***	0.93	0.80	6.15
NUMBING	24.75***	66.22***	1.86	2.43***	0.84	1.21	5.76
LONG. LOC.	4.69***	7.74***	2.26***	1.85	0.83	0.87	9.77
LATERAL LOC.	8.39***	3.65**	2.00	1.05	0.83	1.30	13.34
AREA	6.39***	2.44*	2.00	1.43**	0.87	1.15	14.40
INTENSITY	24.87***	20.92***	6.24**	2.71**	1.27	1.98*	14.40
DURATION	9.53***	44.49***	8.09***	1.99***	1.22	0.90	6.04
df	20	6	2	120	40	12	240

*, **, *** significant at p<0.05, 0.01 and 0.001 respectively

from further statistical analyses. Because overall intensity was an integrated term, it was also dropped from further statistical analyses.

The mean attribute scores for the irritants are given in Fig. 1 and 2. Irritants differed significantly in their pungent character, temporal response, and spatial location or sensation. The pungency of cinnamaldehyde was primarily burning and tingling and experienced in a localized area at the tip of the tongue. Its pungency was experienced quickly and decayed rapidly, as demonstrated by the short lag time and short overall duration. In contrast, the pungency of eugenol was predominantly numbing in quality. This is consistent with its medicinal/dental use for pain relief. The pungency of piperine, capsaicin and ginger was dominantly burning, with slight tingling and numbing sensations. However, they differed in their temporal responses. The burn of ginger and piperine had intermediate lag times; whereas, the burn of capsaicin had a significantly longer lag time. Ginger and piperine had similar overall durations; however, capsaicin's overall duration was slightly shorter. This is believed due to the lower overall intensity for capsaicin (5.55) compared to that of ginger (8.71) and piperine (8.68). Had all three compounds been equi-intense, similar

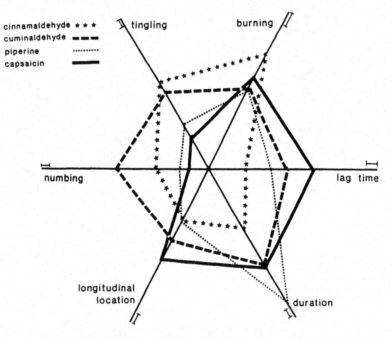

FIG. 1. COBWEB DIAGRAM OF MEAN PUNGENCY SCORES FOR CINNAMALDEHYDE, CUMINALDEHYDE, PIPERINE, AND CAPSAICIN
(n = 63; 21judges*3reps) (| - - | standard error for attribute)

durations would have been anticipated. In addition, piperine pungency was experienced more towards the front of the tongue; whereas, ginger and capsaicin were experienced farther back on the tongue. In contrast, the pungency of alcohol was most diffuse (8.75), with burning and tingling and some numbing sensations. Its short lag time and short duration are consistent with its high volatility. Finally, the pungency of cuminaldehyde differed from the other irritants. It was equipungent in burning, tingling, and numbing qualities, and these qualities were experienced most near the center of the tongue compared to the other irritants.

Using principal component analysis, the first two principal components (Fig. 3) explained 78% of the variability in the data. The first principal component (PC), accounting for 46.7% of the variance, was most positively weighted with longitudinal locations and negatively weighted with tingling. Capsaicin and ginger, located to the right on the plot, lacked the tingling quality and stimulated the back of the tongue. Whereas, alcohol and cinnamaldehyde, located to the left on the plot, were predominantly tingling and stimulated the front of the tongue.

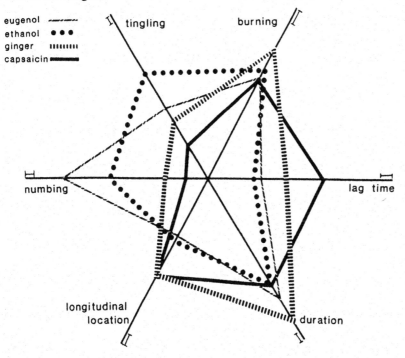

FIG. 2. COBWEB DIAGRAM OF MEAN PUNGENCY SCORES FOR EUGENOL, ETHANOL, GINGER OLEORESIN, AND CAPSAICIN
(n = 63; 21judges*3reps) (| - - | standard error for attribute)

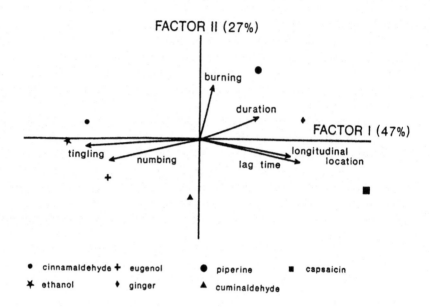

FIG. 3. PRINCIPAL COMPONENT ANALYSIS OF MEAN PUNGENCY SCORES
(n = 63; 21judges*3reps)

The second PC, accounting for another 26.5% of the variance, was most heavily weighted with the burn attribute. Piperine and ginger, located high in the plane, were predominantly burning, while those low in the plane lacked burning. Although capsaicin was also predominantly burning in pungent quality, it was not positioned high in the plane due to its lower mean intensity score. Instead, it was best characterized by a long lag time. In contrast, cinnamaldehyde was best characterized by its short lag time. Lastly, cuminaldehyde fell in the area of the plot not described well by any of the attributes; this confirms the difficulty in characterizing the pungency of this irritant compared to the other irritants.

Clearly, descriptive analysis, coupled with univariate and multivariate statistics was successful in characterizing pungent, temporal, and spatial qualities among selected irritants. This research was the first to sensorially and statistically document the capacity of the human trigeminal nervous system to differentiate nuances in the common chemical sense.

REFERENCES

ANDREWS, J. 1984. *Peppers: The Domesticated Capsicums*, University of Texas, Austin.

ASTM Committee E-18. 1988. Standard test method for sensory evaluation of red pepper heat. *American Standards on Sensory Evaluation of Materials and Products*, pp. 57-59, ASTM Philadelphia, PA.

BURDACH, K.J., KROEZE, J.H. and KOSTER, E.P. 1984. Nasal, retronasal and gustatory perception: An experimental comparison. Perception Psychophys. *36*(3), 205-208.

CLIFF, M. 1987. Temporal Perception of Sweetness and Fruitiness and their Interaction in Model System, MS Thesis, UC Davis.

COCHRAN, W.G. and COX, G.M. 1957. *Experimental Designs*, p. 472, John Wiley & Sons, New York.

GILLETTE, M.H., APPEL, C.E. and LEGO, M.C. 1984. A new method for the sensory evaluation of red pepper heat. J. Food Sci. *49*(4), 1028-1033.

GONIAK, O.J. and NOBLE, A.C. 1987. Sensory study of selected volatile sulfur compounds in white wine. Am. J. Enol. Vitic. *38*, 223-227.

GOVINDARAJAN, V.S. 1979. Pungency: The stimuli and their evaluation. In *Food Taste Chemistry*, (J.C. Boudreau, ed.) pp. 53-91, ACS Symposium Series #115, American Chemical Society. Washington, DC.

GREEN, B.G. 1989. Capsaicin sensitization and desensitization on the tongue produced by brief exposures to a low concentration. Neurosci. Lett. *107*, 173-178.

GREEN, B.G. 1991. Temporal characteristics of capsaicin sensitization and desensitization on the tongue. Physiol. Behav. *49*, 501-505.

HUTCHINSON, S.E., TRANTOW, L.A. and VICKERS, Z.M. 1990. The effectiveness of common foods for reduction of capsaicin burn. J. Sensory Studies *4*, 157-164.

KARRER, T. and BARTOSHUK, L. 1991. Capsaicin desensitization and recovery on the human tongue. Physiol. Behav. *49*, 757-764.

LAWLESS, H. 1984. Oral chemical irritation: Psychophysical properties. Chem. Senses *9*(2), 153-155.

LAWLESS, H. 1989. Pepper potency and the forgotten flavor sense. Food Technol. *11*, 52, 57-58.

LAWLESS, H., ROZIN, P. and SHENKER, J. 1985. Effects of oral capsaicin on gustatory, olfactory and irritant sensations and flavor identification in humans who regularly or rarely consume chili pepper. Chem. Senses *10*(4), 579-589.

MONCRIEFF, R.W. 1967. *The Chemical Senses*, Leonard Hill, London.

NASRAWI, C.W. and PANGBORN, R.-M. 1990. Temporal effectiveness of mouth-rinsing on capsaicin mouth-burn. Physiol. Behav. *47*, 617-623.

ROZIN, P. 1990. Getting to like the burn of chili pepper. In *Chemical Senses, Vol. 2, Irritation*, (B.G. Green, J.R. Mason and M.R. Kare, eds.) pp. 231-269, Marcel Dekker, New York.

SAS Institute. 1985. *User's Guide: Statistics*. Version 5 Ed., SAS Institute, Cary, NC.

SCOVILLE, N.L. 1912. Note on capsicum. J. Am. Pharmacol. Assoc. *1*, 453.

SILVER, W.L. and MARUNIAK, J.A. 1981. Trigeminal chemoreception in the nasal and oral cavities. Chem. Senses *6*(4), 295-305.

STEVENS, D.A. 1990. Personality variables in the perception of oral irritation and flavor. In *Chemical Senses, Vol. 2 Irritation*, (B.G. Green, J.R. Mason and M.R. Kare, eds.) pp. 217-228, Marcel Dekker, New York.

STONE, H. and SIDEL, J.L. 1985. *Sensory Evaluation Practices*, Academic Press, San Diego, CA.

TODD, P.H. JR., BESSINGER, M.G. and BIFTU, T. 1977. Determination of pungency due to capsaicin by gas-liquid chromatography. J. Food Sci. *42*, 660-666.

UMC. 1988. BIB Software. Agricultural Experiment Station Statisticians. University of Missouri-Columbia Curators.

WOOD, A.B. 1987. Determination of the pungent principles of chillies and ginger by reversed-phase high performance liquid chromatography with use of a single standard substance. Flavour Fragrance J. *2*, 1-12.

CHAPTER 6.13

A COMPARISON OF DESCRIPTIVE ANALYSIS OF VANILLA BY TWO INDEPENDENTLY TRAINED PANELS[1]

HILDEGARDE HEYMANN

*At the time of original publication,
the author was affiliated with*
Department of Food Science and Human Nutrition
University of Missouri
122 Eckles Hall
Columbia, MO 65211.

ABSTRACT

Descriptive analysis panelists create, through training, a consensus language to describe perceived differences among samples. If DA gives reliable, objective descriptions of samples, it would be expected that data obtained from independently trained panels be consistent. Two DA panels were trained independently using the same samples. Vanillin and four vanilla samples each at 3-fold, 10-fold and 20-fold concentration were evaluated. Panel J used 14 terms to describe the vanillas and panel K used 16 terms. Eleven and thirteen terms, respectively, significantly discriminated among the samples. Principal component analyses (PCAs) for the two panels were visually similar. Both separated vanillin, Bourbon, Bourbon Processed Bali, Indonesian and Indonesian Non-smoky vanillas across the first PC. Procrustes analysis of the two spaces had a 0.80 fit value. Both the PCAs and the Procrustes analysis indicated considerable overlap of similar descriptive terms. Thus it appears that DA does give reliable consistent results across independently trained panels.

INTRODUCTION

Descriptive analysis (DA) allows the sensory scientist to get an objective description of the products in terms of perceived sensory attributes. There are several descriptive analysis methods, such as Quantitative Descriptive Analysis, Sensory Spectrum and Flavor Profile (Stone *et al.* 1974; Meilgaard

---

[1]Contribution from the Missouri Agricultural Experiment Station. Journal Series Number 11,836.

Reprinted with permission of Food & Nutrition Press, Inc., Trumbull, Connecticut.
©*Copyright 1994. Originally published in* Journal of Sensory Studies **9**, 21-32.

et al. 1991; Thomson and MacFie 1983). In general, the differences among these methods reflect different sensory philosophies and approaches.

In traditional descriptive analysis (DA) the judges (usually eight to twelve people) create, through extensive training, a scientific, consensus language to describe the perceived differences among the products. This consensus language consists of a series of terms and definitions (or actual reference standards) (Stone and Sidel 1985). During the training phase the judges are trained together to ensure that they have reached consensus on the terms and that they are able to use the terms consistently. Product evaluations are performed while the judges are seated in individual, climate and light controlled booths. Judge reliability is evaluated through analysis of data from replicate judgements. DA has been used over a number of years on many products (McDaniel and Sawyer 1981; McDaniel *et al.* 1987; Heymann and Noble 1987; Noble and Shannon 1987; Guinard and Cliff 1987; Andrews *et al.* 1990; Heymann *et al.* 1990). If descriptive analysis gives reliable, objective descriptions of products, then the data obtained from two independently trained panels should be consistent. Procrustes analysis (Gower 1975; Arnold and Williams 1986; Langron 1983; Schlich 1989; Oreskovich *et al.* 1991) can be used to compare the sample configurations derived from different panels. McEwan and Hallett (1990) indicated that this technique has been successfully used to compare Norwegian and British panels evaluating milk chocolates. The objective of this study was to determine if the data derived for the same samples by two independently trained DA panels were comparable.

MATERIALS AND METHODS

Four vanilla types (Pure Bourbon, Bourbon Processed Bali, Indonesian and Indonesian nonsmoky) were each processed to 3-fold, 10-fold and 20-fold strength. Prior to evaluation 3.6 mL 3-fold, 1.2 mL 10-fold and 0.8 mL 20-fold samples were diluted with double distilled water to 1 L quantities. Vanillin at 3-fold strength (colored with unflavored caramel coloring to mimic the vanillas) was diluted by brining 3.6 mL to 1 L with double distilled water. At each session the panelist received four coded 25 mL samples served in 100 mL tulip shaped wine glasses covered with petri dish lids. A balanced incomplete block design (Cochran and Cox 1957, plan 11.22) was used to determine the samples served within a session. All panelists evaluated all samples in quadruplicate. All samples within a session were served in a randomized sequence. Judges performed all evaluations while seated in individual partitioned booths lit with white lights. All samples were expectorated and distilled water was provided as a rinse between samples.

Two panel leaders each independently trained a group of judges. Panel K was comprised of eight judges (all females ranging in age from 19 to 25 years) and panel J had ten judges (five males and five females ranging in age from 20 to 55 years). Through consensus Panel K decided to use 16 terms to describe the differences among the vanillas and panel J used 14 terms (Table 1). Each group of panelists had standards available to refer to during the evaluations (Table 1). Panel K used a 10-point scale anchored in all cases with $0 =$ no intensity and $9 =$ high intensity. Panel J used a 13 cm unstructured line scale anchored 1 cm from each end with low intensity on the left and high intensity on the right. The scale was measured from the left to the right. The two panels performed their evaluations concurrently but were independently trained.

Data Analyses

The data for each panel were analyzed as balanced incomplete block design (MAES 1987). Principal component analysis using data averaged across panelists were performed using SAS (1990). The results for the two panels (data averaged across panelists and replications) were also compared using Procrustes analysis (OPP 1989). The principal component and procrustes scores for the samples from the independent panels were correlated using SAS (1990).

RESULTS AND DISCUSSION

For panel J all terms except raisin and almond odors discriminated ($p<0.05$) among the samples (Table 2) and for panel K all terms except chocolate, kahlua and rum odor discriminated ($p<0.05$) among the samples (Table 3).

Principal component analysis (PCA) of panel J data indicated that the first principal component (PC1) explained 45.4% of the variance associated with the data space. PC2 and PC3 explained an additional 13.3% and 11.5%, respectively. The factor loadings and scores are listed in Table 4. Based on the discussion by Stevens (1986) the critical value for significance (alpha = 0.01, two-tailed) of a factor loading in this study is 0.722. PC1 contrasted tea, woody and smoky odors with marshmallow and butterscotch odors and sweet milk flavor. PC2 was positively loaded with almond odor and PC3 was negatively loaded with the rum odor. Principal component analysis (PCA) of panel K data indicated that the first principal component (PC1) explained 47.0% of the variance associated with the data space. PC2 and PC3 explained an additional 9.0% and 8.7%, respectively. The factor loadings and scores are listed in Table 5. PC1 contrasted yeasty, tobacco, smokey odors and bourbon odor and flavor

TABLE 1.
DESCRIPTIVE TERMS AND REFERENCE STANDARDS USED BY PANELS K AND J TO DESCRIBE DIFFERENCES AMONG VANILLA SAMPLES

PANEL J

Term	Type	Reference	Brand
Marshmallow	Odor	1 t marshmallow creme	Kraft
Butterscotch	Odor	1 t butterscotch topping	Kraft
Nutty	Odor	2 pecans + 2 pistachios	Planters
Tea	Odor	1 t dry tea leaves	Luzianne
Raisin	Odor	4 raisins	Dole
Prune	Odor	1 prune	Sunsweet
Woody	Odor	1 t wet oak sawdust	
Almond	Odor	5% (v/v) almond extract solution	McCormick
Rum	Odor	5% (v/v) rum extract solution	McCormick
Smokey	Odor	5% (v/v) Hickory Liquid Smoke solution	Wright's
Kahlua	Odor	1 t Kahlua Liqueur	Kahlua-Mexico
Chocolate	Odor	5% (v/v) Chocolate extract solution	McCormick
Coffee	Flavor	1 T Folgers Coffee Grounds	Folgers
Sweet milk	Flavor	1 T sweetened condensed milk	Bordens

PANEL K

Term	Type	Reference	Brand
White chocolate	Odor + Flavor	1 cube Alpine White Chocolate/100 mL water	Nestle
Butterscotch	Odor	1 butterscotch candy	Brachs
Vanillin	Odor	5% (v/v) Artificial vanilla extract solution	McCormick
Fruity	Odor	2 prunes in 100 mL water	Sunsweet
Chocolate	Odor	1 t Chocolate extract	McCormick
Rum	Odor	1 t Rum extract	McCormick
Kahlua	Odor	1 T Kahlua	Kahlua-Mexico
Bourbon	Odor + Flavor	Crown Royal Bourbon	Crown Royal
Yeasty	Odor	1 t dried yeast	Fleischmann
Earthy	Odor	Wet dirt	
Tobacco	Odor	Pipe tobacco	
Smoky	Odor	1% (v/v) Hickory Smoke Flavor solution	Wright's
Caramel	Odor	1 caramel candy in 100 mL water	Brachs
Musty	Odor	Defined as the odor on the inside of an old refrigerator	

TABLE 2.
MEAN SCORES[1] AND LEAST SIGNIFICANT DIFFERENCES (LSD) FOR PERCEIVED INTENSITY OF DESCRIPTIVE ATTRIBUTES OF VANILLA SAMPLES AS RATED BY PANEL J

Vanilla	Marshmallow	Butter	Nutty	Tea	Raisin
Vanillin	7.3 a[2]	5.0 a	1.4 d	0.6 f	1.0 a
Bourbon					
3-fold	4.4 b	2.5 bcd	1.4 d	1.2 def	1.0 a
10-fold	4.3 bc	3.0 bc	1.5 d	0.9 ef	1.2 a
20-fold	3.2 cd	3.2 b	2.1 bcd	1.1 def	1.0 a
Indonesian non-smoky					
3-fold	1.4 fg	0.9 f	3.0 ab	2.3 ab	2.0 a
10-fold	1.0 fg	0.7 f	1.5 d	1.8 abcd	0.8 a
20-fold	1.1 fg	0.7 f	3.4 a	2.5 a	1.0 a
Indonesian					
3-fold	1.2 fg	1.1 ef	2.6 abc	2.1 abc	1.3 a
10-fold	1.9 ef	1.8 de	1.7 cd	1.3 cdef	1.0 a
20-fold	0.6 g	0.8 f	3.3 a	2.2 ab	1.4 a
Bourbon processed Bali					
3-fold	3.8 bcd	2.4 bcd	2.2 bcd	1.1 def	1.1 a
10-fold	4.7 b	2.5 bcd	1.7 cd	1.5 bcde	1.5 a
20-fold	2.8 de	2.2 cd	2.1 bcd	2.2 ab	0.9 a
lsd	1.10	0.90	0.98	0.83	---

Vanilla	Prunes	Woody	Almond	Rum	Smoky
Vanillin	1.0 cde[2]	0.7 g	2.0 a	1.4 f	0.6 e
Bourbon					
3-fold	1.2 bcde	2.0 ef	2.1 a	3.2 ab	1.3 cde
10-fold	1.1 bcde	2.0 ef	1.3 a	1.5 ef	0.9 de
20-fold	0.6 e	2.2 def	1.3 a	2.0 cdef	0.9 de
Indonesian non-smoky					
3-fold	2.1 a	4.0 bc	2.0 a	2.9 abc	2.5 b
10-fold	1.6 abcd	3.2 cd	1.3 a	2.3 bcde	2.0 bc
20-fold	1.8 abc	4.7 ab	1.8 a	2.1 cdef	2.6 b
Indonesian					
3-fold	1.3 bcde	4.2 abc	1.4 a	2.9 abc	2.7 b
10-fold	0.9 de	2.8 de	1.3 a	3.1 ab	1.5 cde
20-fold	1.9 ab	5.3 a	1.8 a	2.5 abcde	3.8 a
Bourbon processed Bali					
3-fold	1.4 bcde	1.7 efg	1.8 a	3.4 a	1.2 cde
10-fold	1.3 bcde	1.6 efg	1.8 a	2.6 abcd	1.4 cde
20-fold	1.2 bcde	2.5 def	1.5 a	1.6 def	1.8 bcd
lsd	0.82	1.15	---	1.04	0.92

Vanilla	Kahlua	Chocolate	Coffee flavor	Sweet milk flavor
Vanillin	2.2 abc[2]	3.0 a	0.6 f	2.9 a
Bourbon				
3-fold	2.6 a	1.3 bcde	1.1 def	1.9 bc
10-fold	1.3 cde	1.5 bc	0.8 ef	2.2 ab
20-fold	0.8 e	1.7 b	1.0 ef	1.7 bcd
Indonesian non-smoky				
3-fold	1.8 abcd	1.4 bcd	1.9 bc	0.6 ef
10-fold	1.0 de	0.8 de	1.8 bcd	1.0 def
20-fold	1.7 bcde	0.7 de	1.9 bc	0.8 ef
Indonesian				
3-fold	1.8 abcd	1.0 cde	0.8 e	0.6 ef
10-fold	2.2 abc	0.9 cde	1.2 de	1.4 cde
20-fold	1.4 cde	0.6 e	0.8 e	0.4 f
Bourbon processed Bali				
3-fold	2.5 ab	1.3 bcde	1.3 cde	1.7 bcd
10-fold	2.2 abc	1.5 bc	1.7 bcd	2.4 ab
20-fold	1.6 cde	1.1 bcde	1.4 bcde	1.8 bc
lsd	0.89	0.74	0.72	0.79

[1] Scale ranged from 0 = no intensity to 13 = high intensity.

[2] Different superscripts within a column indicate mean values that are significantly different at p<0.05.

TABLE 3.
MEAN SCORES[1] AND LEAST SIGNIFICANT DIFFERENCES (LSD) FOR PERCEIVED INTENSITY OF DESCRIPTIVE ATTRIBUTES OF VANILLA SAMPLES AS RATED BY PANEL K

Vanilla	White choc.	Butter-scotch	Vanillin	Fruity	Chocolate
Vanillin	3.8 a	1.4 a	3.8 a	0.6 d	0.9 a
Bourbon					
3-fold	1.4 cd	0.2 bc	2.2 b	0.9 bcd	0.5 a
10-fold	2.6 b	0.6 b	1.8 bc	0.7 cd	1.4 a
20-fold	1.9 bc	0.2 bc	1.5 bcd	0.6 cd	0.4 a
Indonesian non-smoky					
3-fold	0.5 ef	0.2 bc	1.3 cde	1.4 b	1.2 a
10-fold	0.9 de	0.3 bc	0.6 ef	0.9 bcd	0.8 a
20-fold	0.1 f	0.2 bc	0.8 def	1.2 bc	1.0 a
Indonesian					
3-fold	0.5 ef	0.4 bc	0.5 f	1.3 b	0.7 a
10-fold	0.9 de	0.4 bc	0.8 def	1.1 bcd	0.7 a
20-fold	0.4 ef	0.2 bc	0.3 f	2.1 a	0.9 a
Bourbon processed Bali					
3-fold	1.3 cd	0.6 b	2.0 bc	0.8 bcd	1.1 a
10-fold	1.3 cd	0.1 c	1.4 bcde	0.6 d	1.3 a
20-fold	1.4 cd	0.2 bc	1.6 bcd	0.8 bcd	1.3 a
lsd	0.81	0.49	0.78	0.56	---

Vanilla	Rum extract	Kahlua	Earthy	Yeasty	Bourbon
Vanillin	1.3 a	0.0 a	0.3 d	0.0 g	0.3 e
Bourbon					
3-fold	1.7 a	0.4 a	1.1 bcd	1.0 cdef	4.1 abc
10-fold	1.1 a	0.9 a	0.4 d	0.5 fg	2.5 d
20-fold	1.8 a	0.3 a	1.6 ab	0.6 efg	3.2 cd
Indonesian non-smoky					
3-fold	1.6 a	0.6 a	1.5 ab	1.9 ab	4.6 ab
10-fold	1.6 a	0.4 a	1.3 abc	0.8 def	3.8 abc
20-fold	1.6 a	0.2 a	2.0 a	1.7 ab	4.1 abc
Indonesian					
3-fold	2.0 a	0.3 a	1.6 ab	2.0 a	4.7 a
10-fold	2.0 a	0.5 a	1.2 abcd	1.7 ab	4.5 ab
20-fold	1.8 a	0.3 a	1.8 ab	1.8 ab	4.6 ab
Bourbon processed Bali					
3-fold	1.2 a	0.3 a	0.5 cd	1.4 abcd	4.3 abc
10-fold	1.1 a	0.5 a	1.4 ab	1.2 bcde	3.9 abc
20-fold	1.3 a	0.2 a	1.0 bcd	1.5 abc	3.4 bcd
lsd	---	---	0.84	0.70	1.19

Vanilla	Tobacco	Smokey	Bourbon flavor	Caramel	Musty	White choc flavor
Vanillin	0.2 e	0.5 f	0.8 e	2.9 a	1.0 d	4.6 a
Bourbon						
3-fold	0.5 cde	1.8 de	2.6 bcd	2.0 bcde	1.5 bcd	2.2 cd
10-fold	0.2 e	0.7 f	2.1 d	2.7 ab	0.9 d	3.2 b
20-fold	0.3 de	0.6 f	2.3 cd	2.6 abc	1.2 cd	2.8 bc
Indonesian non-smoky						
3-fold	1.4 ab	3.7 b	3.8 a	0.6 f	2.0 ab	0.5 g
10-fold	1.0 bc	2.5 cd	3.2 abc	1.2 def	1.6 bcd	0.6 fg
20-fold	1.1 bc	3.2 bc	3.5 ab	1.1 ef	1.5 bcd	0.8 efg
Indonesian						
3-fold	1.1 bc	4.8 a	3.2 abc	0.6 f	1.9 bc	0.9 efg
10-fold	0.5 cde	0.8 f	3.9 a	1.8 cde	1.4 bcd	1.5 def
20-fold	2.1 a	4.7 a	3.4 ab	0.8 def	2.6 a	0.2 g
Bourbon processed Bali						
3-fold	0.6 cde	2.1 d	4.0 a	1.2 def	1.4 bcd	1.5 de
20-fold	0.9 bcd	0.9 ef	3.3 abc	2.0 abcd	1.4 bcd	1.6 de
20-fold	0.8 bcd	2.2 d	2.7 bcd	1.8 cde	1.2 cd	2.8 bc
lsd	0.66	0.96	1.04	0.89	0.72	0.83

[1] Scale ranged from 0 = no intensity to 13 = high intensity.

[2] Different superscripts within a column indicate mean values that are significantly different at p<0.05.

TABLE 4.
PRINCIPAL COMPONENT EIGEN VALUES, VARIABLE LOADINGS AND
SAMPLE SCORES FOR PANEL J DATA

Parameter	PC1	PC2	PC3
Eigenvalue	6.35	1.87	1.60
Variance %	45.4	13.3	11.5
Variable loadings			
Marshmallow odor	-.897	0.312	0.119
Butterscotch odor	-.854	0.116	0.232
Nutty odor	0.680	0.206	0.293
Tea odor	0.763	-.033	0.187
Raisin odor	0.298	0.421	0.450
Prune odor	0.545	0.513	0.242
Woody odor	0.901	0.024	0.114
Almond odor	0.069	0.761	-.013
Rum odor	0.266	0.206	-.789
Smokey odor	0.825	0.266	0.149
Kahlua odor	-.135	0.558	-.629
Chocolate odor	-.637	0.510	0.052
Coffee flavor	0.796	-.072	-.212
Sweet milk flavor	-.872	0.037	0.230
Sample scores			
Vanillin	-1.794	1.006	0.568
Bourbon			
3-fold	-.567	0.585	-.953
10-fold	-.837	-.521	0.690
20-fold	-.616	-.812	0.542
Indonesian non-smoky			
3-fold	0.989	0.981	0.382
10-fold	0.472	-1.091	-.082
20-fold	1.071	-.079	0.326
Indonesian			
3-fold	0.872	-.258	-.368
10-fold	0.010	-.551	-.964
20-fold	1.520	0.123	0.304
Bourbon processed Bali			
3-fold	-.343	0.504	-1.003
10-fold	-.651	0.642	0.100
20-fold	-.106	-.528	0.454

with vanillin, caramel odor and white chocolate odors and flavors. No variable was significantly loaded on either PC2 or PC3 (Stevens 1986). The PCAs for both panels separated vanillin, Bourbon, Bourbon Processed Bali, Indonesian and Indonesian Nonsmoky vanillas in this sequence across PC1. The PC1 factor scores for panel J correlated significantly with those for panel K ($r=0.954$; $p<0.001$) but the PC2 and PC3 scores were not significantly correlated ($r=0.075$ and $r=0.231$, respectively). Thus, it would seem as if the two panels differentiated among the samples in a similar fashion across the first axis but not

TABLE 5.
PRINCIPAL COMPONENT EIGEN VALUES, VARIABLE LOADINGS
AND SAMPLE SCORES FOR PANEL K DATA

Parameter	PC1	PC2	PC3
Eigen value	7.51	1.44	1.40
Variance %	47.0	9.0	8.7
Variable loadings			
White chocolate odor	-.914	0.056	0.093
Butterscotch odor	-.634	0.342	0.336
Vanillin odor	-.782	0.170	0.130
Fruity odor	0.547	0.430	0.464
Chocolate odor	-.068	-.510	0.700
Rum extract odor	0.360	0.643	0.390
Kahlua odor	-.060	-.377	-.030
Earthy odor	0.609	-.300	-.113
Yeasty odor	0.724	0.049	-.038
Bourbon odor	0.812	-.162	-.282
Tobacco odor	0.753	0.143	0.293
Smoky odor	0.791	0.175	0.299
Bourbon flavor	0.752	-.208	-.145
Caramel flavor	-.793	0.127	-.167
Musty flavor	0.648	0.259	0.287
White chocolate flavor	-.929	0.003	0.010
Sample scores			
Vanillin	-2.127	1.194	0.837
Bourbon			
3-fold	-.246	0.246	-.802
10-fold	-1.186	-.791	0.319
20-fold	-.616	0.208	-1.313
Indonesian non-smoky			
3-fold	0.941	-.231	0.678
10-fold	0.312	-.046	-.294
20-fold	0.736	-.203	0.045
Indonesian			
3-fold	0.931	0.622	0.028
10-fold	0.215	0.078	-1.033
20-fold	1.371	0.854	1.174
Bourbon processed Bali			
3-fold	-.037	-.305	0.168
10-fold	-.085	-1.199	-.145
20-fold	-.209	-.423	0.340

the second. The PCAs indicate that vanillin differs from the other samples. The Bourbon and Bourbon processed Bali samples are similar to one another and the Indonesian samples (smoky and nonsmoky) tend to be alike.

The Procrustes analysis indicated that the panel J and panel K sample spaces were similar with a 0.80 fit value and a least squares loss of 0.061 (Fig. 1). The panels disagreed most about the 10-fold Bourbon processed Bali sample since it had the largest residual (0.016); the sample residuals (indicating dissimilarity between the panels) were less than 0.010 for all other samples. Figure 2 shows the correspondence among the terms used by the two panels. The sample scores for dimension 1 of the consensus Procrustes configuration was significantly correlated with the PC1 scores for both panel J ($r=0.987$; $p<0.001$) and panel K ($r=0.984$; $p<0.01$). Dimension 2 was not significantly correlated with the sample scores on PC2 for panel J ($r=0.397$; $p>0.05$), however, dimension 2 scores were significantly correlated with the scores for PC3 for panel J ($r=0.582$; $p<0.05$). For panel K both the PC2 and PC3 scores were significantly correlated with the dimension 2 scores ($r=0.561$; $p<0.05$ and $r=0.781$; $p<0.01$, respectively).

This study indicated that DA results from independently trained panels, evaluating the same samples, are very similar. Researchers may thus, with confidence, compare and use data from different, well-trained DA panels.

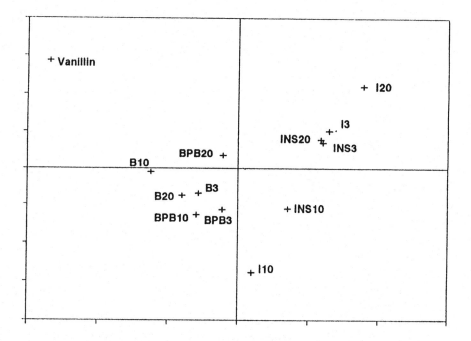

FIG. 1. PROCRUSTES CONSENSUS SAMPLE SPACE FOR VANILLAS
DERIVED BY TWO DESCRIPTIVE ANALYSIS PANELS
(B = Bourbon; BPB = Bourbon processed Bali; I = Indonesian; INS = Indonesian non-smoky; 3, 10 and 20 are 3-fold, 10-fold and 20-fold original samples).

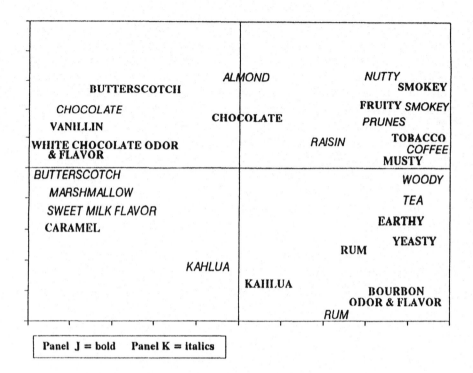

FIG. 2. POSITIONS OF ATTRIBUTES IN THE PROCRUSTES CONSENSUS SPACE DERIVED BY TWO DESCRIPTIVE ANALYSIS PANELS

ACKNOWLEDGMENTS

Vanilla samples were donated by Beck Flavors, St. Louis, MO. I would like to thank Kevin Lindsey and Jennifer Scott for training the panels.

REFERENCES

ANDREWS, J.T., HEYMANN, H. and ELLERSIECK, M.R. 1990. Sensory and chemical analyses of Missouri Seyval blanc wine. Am. J. Enol. Vitic. *41* (2), 116-120.

ARNOLD, G. and WILLIAMS, A.A. 1986. The use of generalized Procrustes technique in sensory analysis. In *Statistical Procedures in Food Research*. pp. 233-254, Elsevier Applied Science, London.

COCHRAN, W.G. and COX, G.M. 1957. *Experimental Designs*. 2nd Ed. John Wiley & Sons, New York.

GOWER, J.C. 1975. Generalized procrustes analysis. Psychometrika *40*, 33-50.

GUINARD, J.X. and CLIFF, M.A. 1987. Descriptive analysis of Pinot noir wines from Carneros, Napa and Sonoma. Am. J. Enol. Vitic. *38* (3), 211-215.

HEYMANN, H., HEDRICK, H.B., KARRASCH, M.R., EGGEMAN, M.K. and ELLERSIECK, M.R. 1990. Effect of Endpoint Temperature on Sensory and Chemical Characteristics of Fresh Pork Roasts. J. Food Sci. *55* (3), 613-617.

HEYMANN, H. and NOBLE, A.C. 1987. Descriptive analysis of commercial Cabernet sauvignon wines in California. Am. J. Enol. Vitic. *38*(1), 41-44.

LANGRON, S.P. 1983. The application of Procrustes statistics to sensory profiling. In *Sensory Quality in Foods and Beverages: Definition, Measurement and Control.* (A.A. Williams and R.K. Atkin, eds.) pp. 89-95, Ellis Horwood Limited, Chichester, U.K.

MAES. 1987. BIB-software program. Statisticians of the Missouri Agricultural Experiment Station, Columbia, MO.

MCDANIEL, M.R., HENDERSON, L.A., WATSON, JR., B.T. and HEATHERBELL, D. 1987. Sensory panel training and screening for descriptive analysis of the aroma of Pinot noir wine fermented by several strains of malolactic bacteria. J. Sensory Studies *2*, 149-167.

MCDANIEL, M.R. and SAWYER, F.M. 1981. Descriptive analysis of whisky sour formulations: magnitude estimation versus a 9-point category scale. J Food Sci. *46*, 178-181, 189.

MCEWAN, J. and HALLETT, E.M. 1990. A guide to the use and interpretation of generalized procrustes analysis. Technical Manual No. 30, Campden Food & Drink Research Assoc., Chipping Campden, U.K.

MEILGAARD, M., CIVILLE, C.V. and CARR, B.T. 1991. *Sensory Evaluation Techniques.* 2nd Ed. CRC Press, Boca Raton, FL.

NOBLE, A.C. and SHANNON, M. 1987. Profiling Zinfandel wine by sensory and chemical analysis. Am. J. Enol. Vitic. *38* (1), 1-5.

OPP. 1989. Procrustes-Pc v2.0. Oliemans, Punter & Partners, Utrecht, The Netherlands.

ORESKOVICH, D.C., KLEIN, B.P. and SUTHERLAND, J.W. 1991. Procrustes analysis and its applications to Free-Choice and other sensory profiling. In *Sensory Science Theory and Applications in Foods*, (H.T. Lawless and B.P. Klein, eds.) pp. 317-338, Marcel Dekker, New York.

SAS. 1990. Version 6.0. Statistical Analysis Systems. Cary, NC.

SCHLICH, P. 1989. A SAS/IML program for generalized Procrustes analysis. SEUGI '89. Proceedings of the SAS European Users Group International Conference, Cologne, May 9-12. pp. 529-537.

STEVENS, J. 1986. *Applied Multivariate Statistics for the Social Sciences*, pp. 344, Lawrence Erlbaum Assoc., Publ., Hillsdale, NJ.

STONE, H. and SIDEL, J.L. 1985. Sensory Evaluation Practices, Academic Press, San Diego, CA.

STONE, H., SIDEL, J., OLIVER, S., WOOLSEY, A. and SINGLETON, R.C. 1974. Sensory evaluation by quantitative descriptive analysis. Food Technol. *28*, 24-34.

THOMSON, D.M.H. and MACFIE, H.J.H. 1983. Is there an alternative to descriptive sensory assessment? In *Sensory Quality in Foods and Beverages; its Definition, Measurement and Control*, (A.A. Williams and R.K. Atkin, eds.) pp. 96-107, Ellis Horwood, Chichester, UK.

CHAPTER 6.14

MULTIVARIATE ANALYSIS OF CONVENTIONAL PROFILING DATA: A COMPARISON OF A BRITISH AND A NORWEGIAN TRAINED PANEL

EINAR RISVIK, JANET S. COLWILL, JEAN A. McEWAN and DAVID H. LYON

*At the time of original publication,
author Risvik was affiliated with
MATFORSK
Norwegian Food Research Institute
Osloveien, N-1430 AS*

and

*authors Colwill, McEwan and Lyon
were affiliated with
Department of Sensory Quality and Food Acceptability
Campden Food and Drink Research Association
Chipping Campden, Gloucestershire, GL55 6LD
United Kingdom.*

ABSTRACT

Many studies have shown that conventional profiling provides reproducible and meaningful results. However, comparison of the technique as used in different countries appears to be nonexistent. In addition, data analysis is often approached differently, and this aspect is also addressed. This paper describes a study to compare the results obtained from profiling milk chocolate samples, using trained panels in Britain and Norway. Data were analyzed using principal component analysis, generalized Procrustes analysis and partial least squares regression. Results indicate that the underlying perceptual structure of the sample spaces obtained from both panels were similar, however, the emphasis on the underlying sensory dimensions differed. Moreover, it was possible to calibrate the two profiles, which has implications for marketing products for export, as well as providing a potential tool for panel monitoring and calibration across cultures.

INTRODUCTION

A number of methods have been developed for routine use by trained sensory panels. These can be classified under the heading of difference tests and descriptive tests. The Flavor Profile Method (Caul 1957; Cairncross and Sjöström 1958), Quantitative Descriptive Analysis (Stone *et al.* 1974; Stone and Sidel 1985) and the Texture Profile (Brandt *et al.* 1963; Szczesniak 1963) are probably the most well known of the descriptive methods. Many variations of these procedures exist for describing and quantifying the sensory attributes of products, and this paper deals with the method known as conventional profiling (Williams and Arnold 1985).

The reliability of profiling in terms of providing meaningful and reproducible results is acknowledged by sensory professionals. However, a comparison of this method using different panels on the same samples has not previously been reported as far as the authors can establish.

While it is known that preferences between countries differ, it would be interesting and useful to determine whether the underlying perceptual dimensions characterizing products were stable across cultures. While one would expect differences in the attributes produced by a British versus a Norwegian trained panel, a common product map would allow easier product development for export purposes. More importantly, it is becoming increasingly desirable to standardize methods across countries. To help achieve this standardization it will be necessary to develop ways of calibrating panels over different countries. This might be achieved by distributing the same samples to each European organization involved in sensory analysis, and asking their panel to evaluate the samples according to a particular method.

To investigate the above mentioned possibilities a collaborative experiment was set up by the Campden Food and Drink Research Association (CFDRA) and MATFORSK, Norwegian Food Research Institute.

MATERIALS AND METHODS

Samples

Five British commercially formulated chocolate samples were molded into standard form and are referred to by their codes, i.e., 19, 31, 51, 53 and 59 (Colwill 1988; Risvik *et al.* 1990). The samples were formulated to represent small, but relevant, differences in their sensory characteristics. Sample 53 was made in two batches and unfortunately information as to which batch was which was not supplied. The chocolate was kept at a constant temperature of 10C to delay the aging process and to prevent the development of chocolate bloom. All samples were assessed at room temperature, which was 20C ± 1C.

Panels

The British panel comprised six of the ten trained staff at the CFDRA, while the Norwegian panel comprised ten trained staff at MATFORSK. The British panel is involved in many aspects of sensory evaluation, including difference tests, conventional and free-choice profiling, quality specifications and taint evaluation. This panel is employed as technicians on mornings five days a week, and are involved in other aspects of the department's work in between tasting. All members of the panel have undergone a six month training period, and have several years experience in sensory assessment.

The Norwegian panel works for 4 h, three days a week, and is mainly involved with descriptive work. Occasional difference tests and evaluation according to quality specifications occur. This panel was not familiar with free-choice profiling at the time of the study. Each member undergoes extensive training in sensory evaluation with spread focus on evaluation of colors using the Natural Colour System (NCS; Norwegian and Swedish standard of colors) and texture. Panel members have between 2 and 10 years experience as part of a trained profiling panel.

Criteria for successful training in both panels are high sensitivity and reproducibility of results.

Sensory

The British and Norwegian panels both used standard procedures for test design (Amerine *et al.* 1965; Meilgaard *et al.* 1987; Risvik 1985), replication, lighting (c-light), three digit codes and randomized serving orders. In addition, both panels used unstructured line scales, anchored with "low intensity" on the left and "high intensity" on the right, to quantify the attributes used to describe the samples. The anchors were in Norwegian for the MATFORSK panel.

The procedure followed by the British panel was a series of sessions where each assessor was asked to describe the attributes they perceived in a number of chocolate samples. During these sessions the panel discussed the attributes and agreed on a common (consensus) vocabulary that they could all use. No emphasis was placed on how the panel should use the measurement scale to quantify the attributes. Three replicates of the quantitative data were then collected using the SENSTEC registration system (Tecator AB, Sweden) in an open plan laboratory.

The initial Norwegian profile was generated mainly from the British profile, translated and supplemented by relevant Norwegian terms by this panel. The pretest session proceeding the profiling contained two samples (31 and 53). The assessors were introduced to the attribute list with an explanation of each attribute and an opportunity to add new attributes. Two attributes, porous and off-flavor, were added during this session. The numeric evaluations were recorded using the SENSTEC registration system with terminals in a sensory laboratory with individual booths. The pretest was instantly followed by a group discussion where the panel leader had results from the initial evaluation available to direct the discussion. The aim of this part of the training was to locate attributes that caused confusion and to promote maximum use of the measurement scale for the relevant sample space. The samples were then profiled in three replicates using SENSTEC.

Data Treatment

The data from the two profiles were analyzed using principal component analysis or PCA, generalized Procrustes analysis or PGA (Gower 1975; Arnold and Williams 1986; McEwan and Hallet 1990) and partial least squares (PLS) regression (Martens and Martens 1986). The results were compared within and across methods for the PCA and GPA, while PLS regression was used to calibrate the British data against the Norwegian data, and vice-versa. A detailed explanation of how these methods work can be found in the aforementioned references. PCA and PLS were applied using UNSCRAMBLER, while GPA was performed using a program written in the GENSTAT language.

RESULTS AND DISCUSSION

Principal Component Analysis

PCA of the Norwegian data generated a perceptual space where the first two principal component (PC's) explained 62% and 12% of the variation in the data, respectively. Leverage correction (Martens and Naes 1989) identified that only the first two PC's were significant. The two dimensional sample and attribute plots are shown in Fig. 1 and 2. The samples are separated along both dimensions, with Sample 59 showing the greatest variability in its position on the space. By referring to the attribute loading plot, reasons for the sample differences can be found. Along PC 1, samples are separated mainly according to color, with whiteness (Sample 31) and color tone (Sample 51) lying at

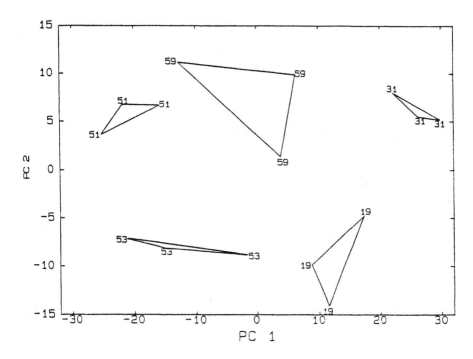

FIG. 1. SAMPLE SPACE DERIVED FROM PRINCIPAL COMPONENT ANALYSIS OF THE NORWEGIAN PROFILE DATA

opposite ends of this principal component. Attributes, such as milkiness and cocoa, are highly correlated with the color dimension and can also be found along this axis. Hardness and porosity are the most dominant characteristics on the second principal component. Other attributes are playing a role in characterizing the samples on both dimensions but these have lower loadings though could be important (or more important) in the determination of consumer acceptance.

PCA of the British data generated a perceptual space where the first two PC's explained 63% and 28% of the variation in the data, respectively. Only the first two principal components provided useful information about sample differences in spite of leverage correction, indicating that the first four PC's were significant. The two dimensional sample and attribute plots are shown in Fig. 3 and 4. The general structure of Fig. 3 is similar to that of Fig. 1 (note slight rotation of space), except that the British panel did not separate samples 19 and 53. This indicates that the Norwegian panel were sensitive to differences between those two samples, which the British panel could not detect. As before, reasons for the structure of the space can be found by examining the attribute

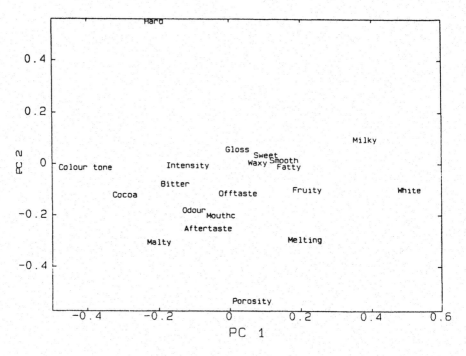

FIG. 2. ATTRIBUTE SPACE DERIVED FROM PRINCIPAL COMPONENT ANALYSIS OF THE NORWEGIAN PROFILE DATA

loading plot (Fig. 4). Along the first principal component, samples are separated according to firm, waxy, cocoa, fruity, bitter and depth of color on the left hand side and, creamy, malty and evaporated milk on the right hand side. On PC2 the attributes malty and evaporated milk lie on opposite ends on the dimension. In addition, these two attributes are correlated positively with each other on the first dimension and are also both correlated with creamy flavor. Relating this information to the sample plot (Fig. 3), it is apparent that sample 31 is high in evaporated milk flavor, while samples 19 and 53 are high in creamy and malty flavor.

Generalized Procrustes Analysis

GPA is the method routinely applied by CFDRA to analyze both conventional and free-choice profile data. It adjusts for three sources of variation inherent in sensory data. Translation of each assessor's configuration to a

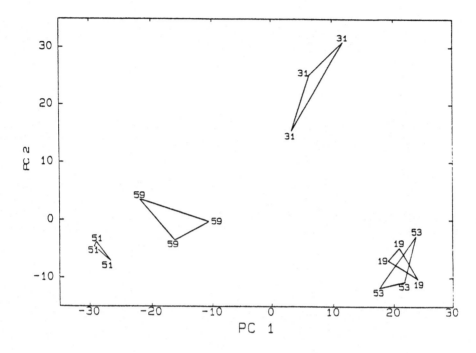

FIG. 3. SAMPLE SPACE DERIVED FROM PRINCIPAL COMPONENT
ANALYSIS OF THE BRITISH PROFILE DATA

common centroid takes account of individuals using different levels of the measurement scale, scaling adjusts for individuals using different ranges of the measurement scale, while rotation/reflection allows different interpretation of attributes to be identified.

To evaluate if the different approaches to training affected the importance of the steps "translation," "scaling" and "rotation/reflection" in the GPA, Procrustes analysis of variance (Arnold and Williams 1986) was used. This revealed that for both data sets, translation and scaling have a significant effect ($p < 0.001$) on the original configuration. The step of rotation/reflection did not significantly alter the configuration data from the Norwegian panel, but did have a slight effect ($p < 0.05$) on the British configuration data. In the authors' experience it is quite usual that the rotation/reflection step of the GPA has little or no effect on the individual configurations. This illustrates that for conventional profiling both panels were well trained in the use of the sensory terms. If either panel were not well trained, this would most likely be picked up by the GPA and give altered sample spaces. This is not seen for either of the panels.

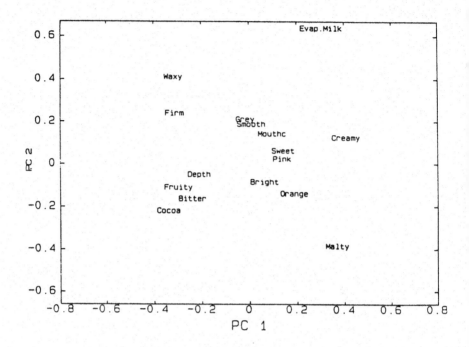

FIG. 4. ATTRIBUTE SPACE DERIVED FROM PRINCIPAL COMPONENT ANALYSIS OF THE BRITISH PROFILE DATA

Figures 5 and 6 show the consensus sample and attribute spaces for the British panel after GPA, while Fig. 7 and 8 show the corresponding plots for the Norwegian panel. Only a few attributes can be seen to change their relative positions in Fig. 6 and 8, when they are compared with the sample plots from the PCA (Fig. 4 and 2). Effects caused by uncertainty about meaning of attributes and/or effects from individual differences in use of the scale may be observed in GPA as a movement of the attribute relative to the others; this compared to a PCA loading plot. For the British panel only the relative position of brightness and orange flavor can be seen to change, although both these attributes have low loadings on the first two dimensions.

A similar phenomenon occurs on examination of the two Norwegian attribute spaces (Fig. 2 and 8) where it is apparent that the attributes gloss and fruity have increased weighting (loadings) on the second dimension (Fig. 8), and hence contribute more information to describing the sample space. However, the attribute maltiness, after GPA, has decreased weighting on the first dimension, and to some extent on the second dimension (not reflection of space along PA 1). In spite of small differences, it is evident that both the sample and attributes

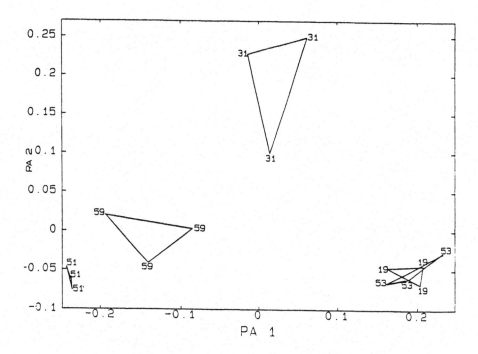

FIG. 5. CONSENSUS SAMPLE SPACE DERIVED FROM GENERALIZED PROCRUSTES ANALYSIS ON THE BRITISH PROFILE DATA

spaces have very similar structure on application of both PCA and GPA at a consensus level. It is usual with GPA to plot sample and attribute spaces for each individual assessor, as this allows agreement between assessors to be evaluated. However, within each analysis (panel), agreement between assessors was good, though these plots are not illustrated here. As an extension, comparing individual assessors from the two panels can be achieved by running a GPA on all 16 assessors. This is reported under "Analysis of Individual Differences."

Partial Least Squares Regression

PLS regression is used to calibrate one matrix of data (Y) with another matrix of data (X), by utilizing information in X in the form of factors rotated in the direction of the predicted space (Y data). The X data can be considered as a matrix of independent or predictor variables, while the Y data is a matrix of dependent or response variables. Ultimately, the model which is obtained by

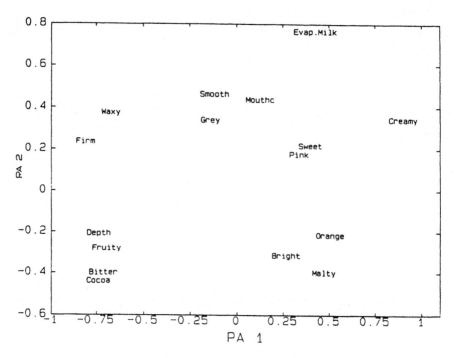

FIG. 6. CONSENSUS ATTRIBUTE SPACE DERIVED FROM GENERALIZED PROCRUSTES ANALYSIS ON THE BRITISH PROFILE DATA

application of PLS will be used to predict the Y-data given that new X-data are fed into the model. Thus, to obtain a successful prediction model, one of the important criteria must be that the information in the *X data explain a high percentage of Y data*. This point will become clearer by example in the course of the discussion.

In performing PLS with UNSCRAMBLER, it is usual to plot the X and the Y variables on the same plot. In doing so, the X variable loadings are, however, heavily dependent on the scaling of X and Y, which makes interpretation difficult as only directional information is in effect detained. For this reason, and the fact that the PCA plots reflect the X-data attribute plots, only the predicted data attribute plots (Y-data) from the PLS analysis are given in this paper.

PLS Regression: Norwegian Data (Y) and British Data (X)

In the first instance the British profile data were taken as the predictor variables, while the Norwegian data were designated the response variables. In

TABLE 1A.
USED VARIANCE PER FACTOR IN BRITISH DATA (X) TO EXPLAIN
NORWEGIAN DATA (Y), AFTER LEVERAGE CORRECTIONS

Attribute	PC1, %	PC2, %	Total after 2 factors, %
Depth of colour	81	6	87
Brightness	0	19	19
Greyness	0	46	46
Orange colour	0	62	62
Pinkness	29	0	29
Firm	28	58	86
Waxy	0	94	94
Smooth	0	44	44
Mouth coating	4	16	20
Cocoa	97	0	97
Fruity	71	13	84
Creamy	88	5	93
Evaporated milk	47	49	96
Malty	0	91	91
Sweet	42	0	42
Bitter	96	0	96
Average used variance	45	42	87

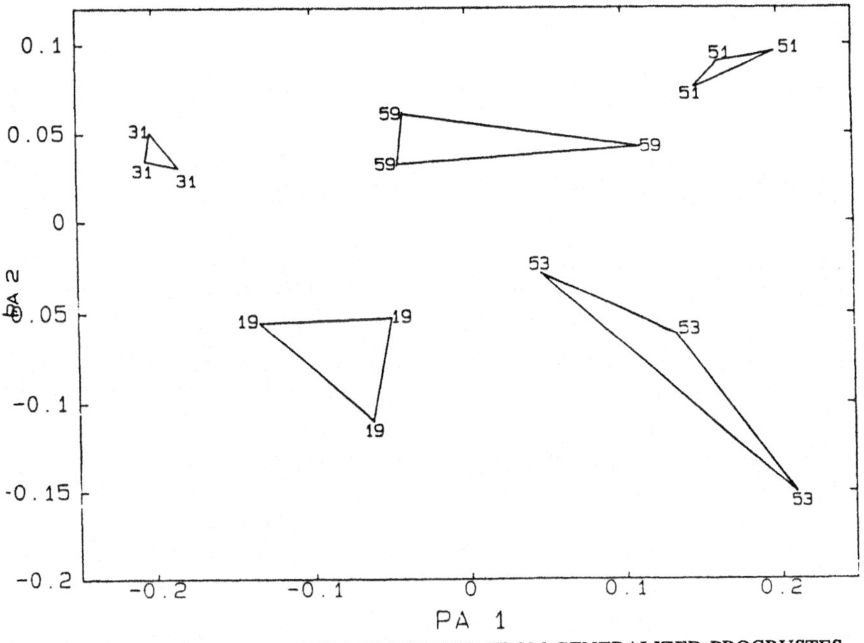

FIG. 7. CONSENSUS SAMPLE SPACE DERIVED FROM GENERALIZED PROCRUSTES ANALYSIS ON THE NORWEGIAN PROFILE DATA

TABLE 1B.

EXPLAINED VARIANCE PER FACTOR IN NORWEGIAN DATA (Y) EXPLAINED BY BRITISH DATA (X), AFTER LEVERAGE CORRECTIONS

Attribute	PC1, %	Total after PC2, %	2 factors, %
Whiteness	46	8	54
Colourtone	20	13	33
Intensity of colour	8	0	8
Gloss	0	0	0
Odour intensity	0	49	49
Hard	25	0	25
Cocoa	0	12	12
Fruity	42	0	42
Sweet	0	0	0
Bitter	0	39	39
Milk flavour	18	16	34
Malty	0	30	30
Fat	24	8	32
Aftertaste	0	12	12
Melting	53	0	53
Waxy	0	0	0
Smooth	0	0	0
Mouth coating	0	0	0
Porous	0	55	55
Offtaste	0	0	0
Average explained variance	12	13	25

other words it was desired to determine how good the British profile was in predicting the Norwegian data. Figure 9 shows the common sample space derived from PLS, which is very similar to Fig. 1 (PCA of Norwegian profile data). Thus, it would appear that a model can be set up that enables the British profile data to be used as a predictor of the sample structure apparent in the Norwegian data. As Tables 1A and 1B illustrate, about 87% of the information in the British data was used to explain only 25% of the information available in the Norwegian data. If this is looked at in more detail, the complete information from 9 attributes is used to explain a corresponding 8 attributes up to an approximate 50%. A maximum of what could be expected to be explained in this model is given in the degree of explanation in the PCA of the Norwegian data alone (72%). The gap between these two models can be evaluated as not very important.

The importance of each attribute from the British profile, which contributes to this predicted space, is better interpreted from the loading plot, since loadings are strongly related to the range of variability in the Y variables. Figure 10

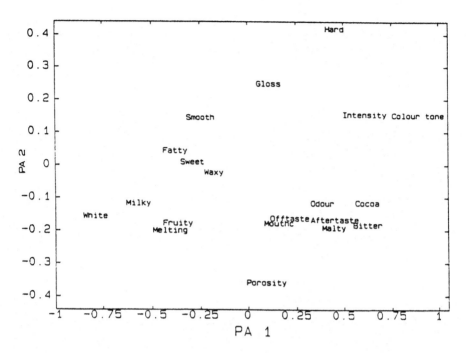

FIG. 8. CONSENSUS ATTRIBUTE SPACE DERIVED FROM GENERALIZED PROCRUSTES ANALYSIS ON THE NORWEGIAN PROFILE DATA

shows the attribute plot from the British profile after PLS. Comparing the attribute positions with those from the Norwegian PCA of Fig. 2, several interesting points emerge. Milkiness and whiteness are replaced by evaporated milk by the British panel, while color tone and cocoa of Fig. 2 are replaced by cocoa, fruity, bitter and depth of color by the British panel in describing the Norwegian sample space. This indicates, for example, that cocoa seems to be used in a similar way by the two panels, but the Norwegian panel uses two terms, whiteness and milkiness, in place of the British evaporated milk. On the second dimension, the hardness versus porosity of Fig. 2 is replaced by firmness versus malty in Fig. 10. However, care should be taken in drawing such parallels between the two data sets, since misleading conclusions could be drawn. For example, it is probably not sensible to suggest an association between porosity (Fig. 2) and malty (Fig. 10).

Returning to the percentage variation explained (Tables 1A and 1B), it is important to consider a little more what this means in practice. In effect the results indicate that only 25% of the information of the Norwegian data could be predicted utilizing just over 80% of the available information from the British

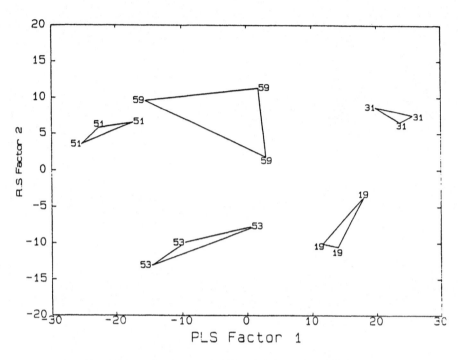

FIG. 9. COMMON SAMPLE SPACE AFTER PLS, WHERE THE PREDICTOR VARIABLES ARE THE BRITISH DATA

profile, after two PLS factors. This illustrates that there is a large component of information in the Norwegian profile, which is unique to that data. Going back to the interpretation of the first dimension of PCA and GPA (Fig. 2 and 8) it was evident that color was the main component in separating the samples, due to the training emphasis of the Norwegian panel on this aspect. This detail is not reflected in the British profile. Thus, when the British data are used as predictor variables, this color dimension will in effect be rotated "out of focus," due to the excessive information associated with the color attributes in the Norwegian (Y) data.

PLS Regression: British Data (Y) and Norwegian Data (X)

To determine whether a similar result was obtainable when using the Norwegian data to see how well it could predict the British data, a second PLS analysis was conducted. Figure 11 shows the common sample space derived from PLS, which is similar to British PCA space of Figure 3 (note reversing of

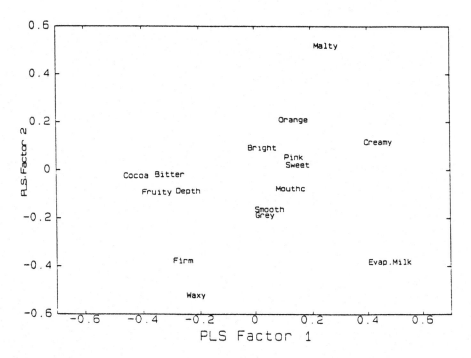

FIG. 10. ATTRIBUTE SPACE AFTER PLS, WHERE THE PREDICTOR VARIABLES ARE THE BRITISH DATA

the second dimension). Thus, it is apparent that there is information in the Norwegian profile data that can be utilized in the prediction of the British sample space. Examination of Tables 1C and 1D reveals that in two factors, about 60% of the Norwegian data was used to predict a similar percentage of the British data.

It is evident that almost all of the information available in 6 attributes are used to explain about 50% of 8 attributes and another 3 attributes to about 80%. In the PCA of the British data, 91% is explained in the two first factors, and again this shows a good recovery in the PLS.

It is worth noticing the apparent discrepancy between the used variance in Table 1C and the loadings plotted in Fig. 12. This exemplifies an important feature of the interpretation of loading plots. In Table 1C, the information is given in percentage of original information, which is a function of the actual variance in each attribute. The loadings represent the importance of each attribute X for interpretation of each dimension. In this case, dimension one accounts for 52% of the total variance, while dimension 2 only accounts for 9%.

TABLE 1C.
USED VARIANCE PER FACTOR IN NORWEGIAN DATA (X) TO EXPLAIN
BRITISH DATA (Y), AFTER LEVERAGE CORRECTIONS

Attribute	PC1, %	Total after PC2, %	2 factors, %
Whiteness	97	0	97
Colour tone	91	3	94
Intensity of colour	70	0	70
Gloss	0	0	0
Odour intensity	14	19	33
Hard	36	8	8
Cocoa	48	11	59
Fruity	37	0	37
Sweet	5	0	5
Bitter	72	16	88
Milk flavour	21	59	80
Malty	42	41	83
Fat	80	0	80
Aftertaste	4	55	59
Melting	45	22	67
Waxy	0	0	0
Smooth	30	0	30
Mouth coating	0	0	0
Porous	0	62	62
Off-taste	0	12	12
Average explained variance	52	9	61

However, loadings for each of those dimensions can be high for attributes with a low percentage variance involved. An example of this can be seen in the attribute color tone, where 3% variance shows more importance, and thus the higher loadings, than the attribute aftertaste, which has 55% variation explained. This indicates to the panel leader that the sensory scale has been used to a much wider extent for color tone than for aftertaste.

To conclude this point it is important to notice that in the interpretation of loadings it is important to include how much variance each principal component accounts for.

In Fig. 12 the position of the Norwegian attributes are shown after PLS, and this plot can be compared to Fig. 4 (British PCA), remembering that the second dimension is reversed in after PLS. As previously (Fig. 2 and 10), evaporated milk appears to be used in place of milkiness and whiteness, etc. The attribute fruity is clearly used differently by both panels lying on opposite sides of the second dimension.

TABLE 1D.
EXPLAINED VARIANCE PER FACTOR IN BRITISH DATA (Y) EXPLAINED BY NORWEGIAN DATA (X), AFTER LEVERAGE CORRECTIONS

Attribute	PC1, %	Total after PC2, %	2 factors, %
Depth of colour	19	64	63
Brightness	0	8	8
Greyness	0	29	29
Orange colour	0	66	66
Pinkness	3	6	7
Firm	0	80	80
Waxy	0	82	82
Smooth	0	17	17
Mouth coating	0	0	0
Cocoa	46	0	46
Fruity	43	23	66
Creamy	24	26	50
Evaporated milk	19	34	53
Malty	0	76	76
Sweet	6	6	12
Bitter	36	21	57
Average used variance	7	51	58

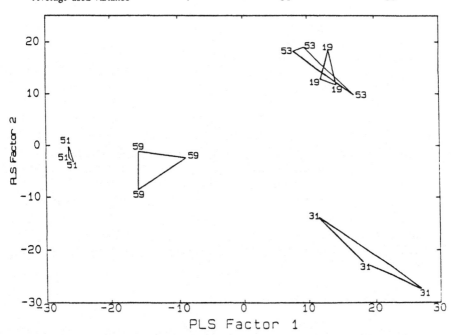

FIG. 11. COMMON SAMPLE SPACE AFTER PLS, WHERE THE PREDICTOR VARIABLES ARE THE NORWEGIAN DATA

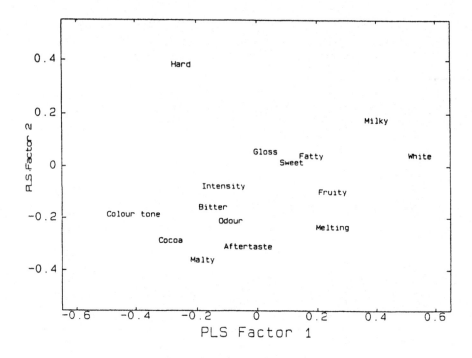

FIG. 12. ATTRIBUTE SPACE AFTER PLS, WHERE THE PREDICTOR VARIABLES ARE THE NORWEGIAN DATA

Returning to Table 1C and 1D, consider the reasons behind the variance structure of the calibration exercise. Firstly, note that the first PLS factor explains only approximately 5% of the British data utilizing 50% of the information available in the Norwegian data. On the other hand, an almost opposite phenomenon occurs when 10% of the available information in the Norwegian data are used to explain approximately 50% of the British data in the second dimension. This suggests a different focus on the underlying attribute dimensions for the two panels, which can be understood when considering the emphasis placed on the color attributes by the Norwegian panel (Fig. 2). Color is clearly less important for the British panel (Fig. 4), and the emphasis on the second dimension is most likely due to the importance of the texture and flavor attributes.

Analysis of Individual Differences

From the results discussed, it is evident that there are small differences in the underlying structure of the sample spaces derived from the two panels. It is also evident that the emphasis of the attributes used to describe the sensory dimensions differs. To investigate the data in more detail, the ability of GPA to analyze individual data was capitalized on. The individual data matrices (samples × attributes) for the British and Norwegian assessors (16 in total) were submitted to one generalized Procrustes analysis. The resultant consensus space explained 39% and 24% of the first two dimensions, and the assessor residuals were similar for assessors from both panels.

The main interest of this exercise is in comparing the sample and attribute spaces from individuals on both panels together. To this end the sample positions of each assessor were superimposed on one plot, as were the attribute positions. To illustrate this, Fig. 13 shows the sample positions of one British (C) and one Norwegian (M) assessor, while Fig. 14 gives the corresponding attribute plot. While the positions of Samples 51, 59 and 31 can be said to occupy similar positions, it is evident that perception of Samples 19 and 53 differ. This corresponds with the earlier comparison from PCA and GPA.

The attribute plot (Fig. 14) also reveals underlying differences in perception between the two panels, as reflected on the emphasis of the attributes on the two dimensions. Another point of interest is illustrated by Fig. 14, where the two typical individual's spaces have the first principal axis almost orthogonal

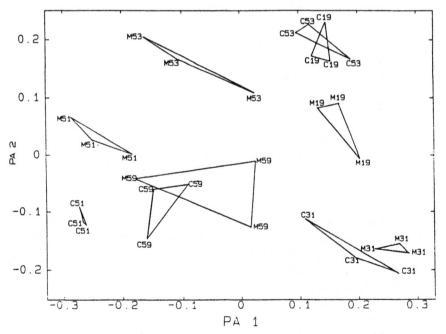

FIG. 13. SAMPLE SPACE FOR ONE BRITISH AND ONE NORWEGIAN ASSESSOR AFTER GENERALIZED PROCRUSTES ANALYSIS ON THE COMBINED DATA

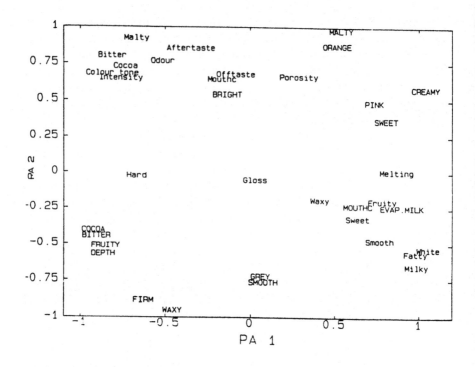

FIG. 14. ATTRIBUTE SPACE FOR ONE BRITISH AND ONE NORWEGIAN ASSESSOR
AFTER GENERALIZED PROCRUSTES ANALYSIS ON THE COMBINED DATA
(British panel is represented in capital letters)

to each other along the diagonals of the plot, with the exception of evaporated milk. This could be explained by the fact that this attribute is quite unique to the British panel, though it does coincide with the attribute directions of smoothness, whiteness, fatiness and milkiness for the Norwegian panel.

General Discussion

In reviewing the work reported in this paper one should not lose sight of the objectives stated in the Introduction. These were to establish the feasibility of profiling products for export in the country of origin, and to examine the feasibility of utilizing the methods investigated to aid the move towards standardizing methodology throughout the European community.

Considering the export problem first, several points should be considered. First, the attribute emphasis is likely to differ from country to country, both due to different training regimes and also differences in language. Indeed, within-country differences in attribute emphasis may result due to individual differences in the way the panel leader runs the discussion. It is also unlikely that two panels within one country will develop the same vocabulary of profile terms for the same set of products. However, what is important is that the relative similarities and differences between the products is similar in terms of the multidimensional product map derived. It is also important that the depth of the information used to produce the product map can be fed back into the product development function of a company to result in the same new products being produced. In the example of this paper, it is evident that 80% of the attribute information from a British panel could be used to predict only 25% of the Norwegian data, which can seem to be rather poor. However, if it was known that this 25% was the part related to acceptance, then this would be useful. This hypothesis needs to be tested in future work. In the other direction 60% of the attribute information from the Norwegian panel could predict 60% of the British information. Again, it is important to consider how this information relates to the acceptability of the product in the country of export.

Standardizing methodology throughout Europe is still at an early stage, though the multivariate methods reported offer a good starting point for the future development of this objective. Comparison of perceptual maps derived from PCA and GPA offer a first step of empirical comparison, while the more model based approach of PLS offers to take account of direction shift in the emphasis of underlying perceptual dimensions.

In summary, the research suggests that there is potential for utilizing techniques, such as partial least squares and to some extent generalized Procrustes analysis, to aid in product development for export purposes, though more work is required to look at the role of product acceptability. In terms of methodology standardization, all the multivariate methods reported are applicable. However, if successful standardization is to be achieved, more depth work is required to enable a more formal control specification to be written.

ACKNOWLEDGMENTS

Financial support from the British contribution to the study is gratefully acknowledged from the Ministry of Agriculture, Fisheries and Food. The authors are also grateful for the help by staff at CFDRA and MATFORSK.

REFERENCES

AMERINE, M.A., PANGBORN, R.M. and ROESSLER, E.B. 1965. *Principles of Sensory Evaluation of Food*, Academic Press, New York.

ARNOLD, G.M. and WILLIAMS, A.A. 1986. The Use of Generalized Procrustes Techniques in Sensory Analysis. In *Statistical Procedures in Food Research*, (J.R. Piggott, ed.) pp. 233-253, Elsevier Applied Science, London.

BRANDT, M.A., SKINNER, E.Z. and COLEMAN, J.A. 1963. Texture profile method. J. Food Sci. *28*(4), 404-410.

CAIRNCROSS, S.E. and SJÖSTRÖM, L.B. 1958. Flavour profiles: A new approach to flavour problems. Food Technol. *4*, 308-311.

CAUL, J.F. 1957. The profile method of flavour analysis. Advan. Food Res. *7*, 1-40.

COLWILL, J.S. 1988. Methods for the Collection of Consumer Acceptability and Trained Panel Sensory Data. Tech. Memorandum 506: CFDRA.

GOWER, J.C. 1975. Generalized procrustes analysis. Psychometrika *40*(1), 33-51.

MARTENS, M. and MARTENS, H. 1986. Partial least squares regression. In *Statistical Procedures in Food Research*, (J.R. Piggott, ed.) pp. 293-359, Elsevier Applied Science, London.

MARTENS, H. and NAES, T. 1989. *Multivariate Calibration*, John Wiley & Sons, New York.

McEWAN, J.A. and HALLETT, E.M. 1990. A Guide to the Use and Interpretation of Generalized Procrustes Analysis. Statistical Manual No. 1. CFDRA.

MEILGAARD, M., CIVILLE, G.V. and CARR, T.B. 1987. *Sensory Evaluation Techniques*, CRC Press, Boca Raton, Fl.

POWERS, J.J. 1984. Current practices and application of descriptive methods. In *Sensory Analysis of Foods*, (J.R. Piggott, ed.) pp. 179-242, Elsevier Applied Science, London.

RISVIK, E. 1985. *Sensory Analysis*, Tecator AB, Sweden.

RISVIK, E., McEWAN, J.A., COLWILL, J.S., ROGERS, R. and LYON, D.H. 1990. Projective mapping: A tool for sensory analysis and consumer research. (Submitted to Food Quality and Preference).

STONE, H. and SIDEL, J.L. 1985. *Sensory Evaluation Practices*, Academic Press, London.

STONE, H., SIDEL, J.L., WOOLSEY, A. and SINGLETON, R.C. 1974. Sensory evaluation by quantitative descriptive analysis. Food Technol. *28*(1), 24-34.

SZCZESNIAK, A.S. 1963. Classification of textural characteristics. J. Food Sci. *28*, 385-389.

WILLIAMS, A.A. and ARNOLD, G.M. 1985. A comparison of the aroma of six coffees characterized by conventional profiling, free choice profiling and similarity scaling methods. J. Sci. Food Agric. *36*, 204-214.

CHAPTER 7.0

COMPUTER SOFTWARE

During the last decade, we have witnessed the explosion of computer applications from computerized data collection to development of software for statistical analysis. These developments greatly enhanced the use of descriptive analysis. In this chapter, the use of various software packages for the design and statistical analysis of sensory studies will be illustrated.

T1 = LOG10(A/B);

EF = 100*(1-(EXP(2.3026*T1)));

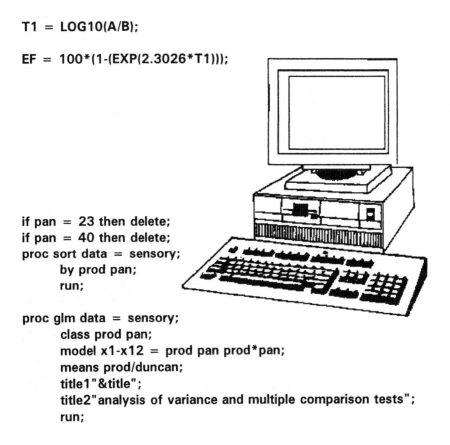

if pan = 23 then delete;
if pan = 40 then delete;
proc sort data = sensory;
    by prod pan;
    run;

proc glm data = sensory;
    class prod pan;
    model x1-x12 = prod pan prod*pan;
    means prod/duncan;
    title1"&title";
    title2"analysis of variance and multiple comparison tests";
    run;

CHAPTER 7.1

SOFTWARE PACKAGES

Statistix

The Statistix for Windows (Analytical Software) is an excellent teaching tool in the use of statistics in data analysis, especially for beginners. It is a tool for rapidly obtaining both statistical and graphical analyses of research data. In particular, the calculation of probabilities for several probability distributions encountered in practice is easily obtained. For the sensory statistician, the task of transforming proportions into normal deviates (standard normal value) in a paired comparison of samples by the Thurstone-Mosteller model is obtained by the inverse function of the standard normal distribution, $ZI(x)$, where x is a proportion. This function eliminates tedious consultation of Table A-11 provided in Gacula and Singh (1984). For example, suppose that out of 15 trained panelists, 10 indicated that Sample A is more bitter in intensity than Sample B. Hence, 10/15 gives a proportion of 0.6667. In terms of the Thurstone-Mosteller scale, the standardized distance D between Samples A and B is obtained using the TRANSFORMATIONS in the Statistix menu as follows:

$$D = ZI(A/B) = 0.4307 \qquad (Eq.\ 7.1.1)$$

The distance D would serve as the observation in the statistical analysis. See Gacula and Singh (1984) and Gacula (1993) for numerical examples in the use of the Thurstone-Mosteller model of paired comparison. This model is useful in simplifying sensory tasks in descriptive analysis, particularly when the purpose of the study is for sensory discrimination, such as in screening and ingredient substitution studies, and in a situation where the panel leader is not yet comfortable with the panel performance. The control chart technique given in Chapter 6.9 can be easily implemented using Statistix.

Design-Expert and Design-Ease

The Design-Ease (Stat-Ease, Inc.) is a software that builds factorial designs and their analyses—from test of significance to analysis of residuals. Like Statistix, both Design-Expert and Design-Ease are interactive, hence there is no need to write computer programs or codes. One of the most useful softwares for research and product development is Design-Expert. Design-Expert also builds designs for optimization work using either response surface or mixture designs.

SOFTWARE PACKAGES

These designs and their applications are given in Gacula and Singh (1984) and Gacula (1993) for sensory and consumer studies, Cornell (1981) for statistical treatment of the designs, and Box and Draper (1987) and Myers and Montgomery (1995) for statistical treatment and engineering applications. Let us illustrate the use of Design-Expert and Statistix in an optimization study with the descriptive analysis technique used in gathering the data.

Example 7.1

In this study, the first optimization work has been completed that screened out several candidate ingredients for the formula. Based on this result, three ingredients denoted by X1, X2, and X3 have been retained for the final optimization work to include overall preference using a research guidance panel. The Thurstone-Mosteller model is to be used to estimate preference scale values. Using Design-Expert, a D-optimal mixture design was built resulting in nine test formulations (Design point). The design and the corresponding responses for three intensity variables and the preference scale estimates are given in Table 7.1.1. The calculation of the preference scale values is shown in Table 7.1.2. An advantage of this scale is that we do not have to be concerned on how the panelists used the length of the scale. This is important because each pair of products was seen by different panelists, n = 28 in each pair.

TABLE 7.1.1
D-OPTIMAL MIXTURE DESIGN AND CORRESPONDING RESPONSES

Design point	Variable X1	Variable X2	Variable X3	Intensity Y1	Intensity Y2	Intensity Y3	Preference S
1	3.0	5.0	12.0	5.2, 7.0	7.9, 8.5	9.9, 11.0	-0.25
2	8.0	1.0	11.0	10.0	11.0	9.0	0.00
3	8.0	5.0	7.0	7.0, 7.8	7.0, 7.8	7.8, 8.4	-0.21
4	7.0	1.0	12.0	11.2	12.4	5.1	0.23
5	5.5	5.0	9.5	7.0, 7.5	8.8, 9.0	7.0, 8.1	-0.21
6	5.5	3.0	12.0	14.0	13.5	4.3	0.32
7	8.0	3.0	9.0	9.0, 9.8	9.5, 9.0	7.7, 8.0	-0.22
8	7.5	1.0	11.5	13.0	14.3	6.5	0.20
9	6.5	3.0	10.5	8.1	9.4	5.9	0.13

Note: A positive preference scale estimate (S) indicates the more preferred sample, whereas a negative S indicates the less preferred sample. Notice that X1 + X2 + X3 = 20%, the remaining 80% constitutes the other portions of the ingredients in the test formula.

TABLE 7.1.2
NORMAL DEVIATE Z (ROW I > COLUMN J)

	1	2	3	4	5	6	7	8	9
1	0	-0.37	1.07	0.18	0.37	0.00	0.18	0.00	0.79
2	0.37	0	-0.18	0.18	-0.79	-0.37	0.00	0.18	0.57
3	-1.07	0.18	0	0.79	0.57	1.07	0.57	-0.37	0.18
4	-0.18	-0.18	-0.79	0	-0.57	0.00	-0.57	0.00	0.18
5	-0.37	0.79	-0.57	0.57	0	0.57	-0.37	1.07	0.18
6	0.00	0.37	-1.07	0.00	-0.57	0	-1.07	0.00	-0.57
7	-0.18	0.00	-0.57	0.57	0.37	1.07	0	0.57	0.18
8	0.00	-0.18	0.37	0.00	-1.07	0.00	-0.57	0	-0.37
9	-0.79	-0.57	-0.18	-0.18	-0.18	0.57	-0.18	0.37	0
Sum	-2.22	0.04	-1.92	2.11	-1.87	2.91	-2.01	1.82	1.14
S	-0.25	0.00	-0.21	0.23	-0.21	0.32	-0.22	0.20	0.13

Note: Row i > column j indicates that, for example in a pair (1,2), sample 1 was preferred X times over sample 2, and so on. For pair (1,2), 10 panelists out of 28 preferred sample 1, thus ZI(10/28) = -0.37; conversely, ZI(18/28) = 0.37 as shown above; recall that ZI(x) is the inverse function defined earlier.

As shown in Table 7.1.2, the scale value S for each sample is simply the average of each column divided by the number of samples. The estimate of S indicates the distance between samples in a perceptual continuum centered at $S=0$, the true origin of the scale. The details of the design construction, as well as the statistical output, will not be presented because the software, being interactive, is best learned by actual use. Instead, the response surface plots and some statistical diagnostic information about the data will be discussed. In practice, effective decisions about the study can be made on response surface plots. Furthermore, information from graphs are easily conveyed to the management.

In this example, a quadratic model was used given by

$$Y1 = B_1X_1 + B_2X_2 + B_3X_3 + B_{12}X_1X_2 + B_{13}X_1X_3 + B_{23}X_2X_3 \quad \text{(Eq. 7.1.2)}$$

where B_i, $i=1,2,3$, is the pure linear effect coefficient for X_i; B_{ij} is the synergistic quadratic effect coefficient for X_iX_j, $j=1,2,3$, $i \neq j$. The term Y1 is replaced accordingly, by Y2, Y3, or S in the analysis. Obviously, the linear effect provides the linear portion of the response surface plot and the quadratic effect provides the curvilinear (non-linear) part.

First, let us analyze the overall preference response S using the quadratic model (Eq. 7.1.2). The result of the response surface plot is shown in Fig. 7.1.1. Notice the direction of the response given by five contour lines, which show the optimal lines indicated by positive value of S. The Design-Expert can further block non-optimal and optimal areas (Fig. 7.1.2a,b). Similarly, the response surface plots for Y1 and Y2 are shown in Figs. 7.1.3 and 7.1.4, respectively. Again, the contour lines show the direction of the intensity response at various levels of X1, X2, and X3. Since Y1 and Y2 are intensity

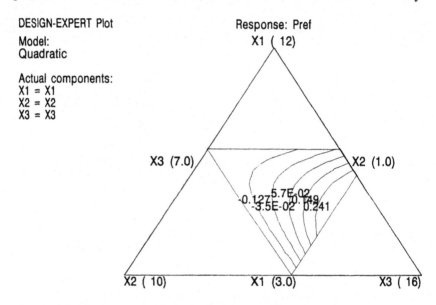

FIG. 7.1.1. CONTOUR LINES OF RESPONSE SURFACE PLOT FOR OVERALL PREFERENCE

variables, one cannot decide on the optimal intensity. However, the optimal intensity can be obtained by superimposing the intensity plots into the overall preference plots. Interactively, this can be done by "clicking the cursor" in the desired area of the optimal preference map. For example, as shown in Fig. 7.1.5a, a click gave a predicted response of Y1 = 12.24; this value corresponds to the optimal combination of X1 = 7.0%, X2 = 1.3%, and X3 = 11.7%, with a predicted overall preference score of 0.30. Thus, for variable Y1 a high intensity is desirable, which is also supported by the SAS plot of S and Y1 (Fig. 7.1.5b). A "click" on the non-optimal area (Fig. 7.1.5c) produced a negative S = −0.21, which corresponds to the following combination: X1 = 7.7%, X2 = 4.7%, and X3 = 7.5%; obviously, this is a combination to be avoided. Similar procedures can be done for Y2 and Y3.

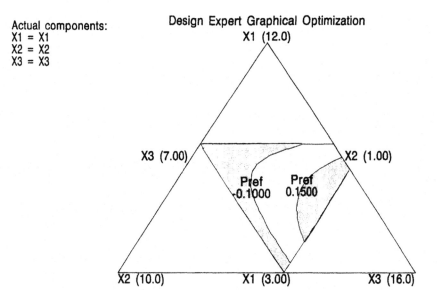

FIG. 7.1.2a. LOCATION OF OPTIMAL AREAS FOR OVERALL PREFERENCE AT VARIOUS COMBINATIONS OF X1, X2, AND X3
Positive Pref values are desirable.

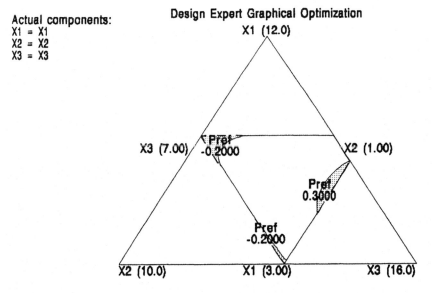

FIG. 7.1.2b. LOCATION OF OPTIMAL AREAS FOR OVERALL PREFERENCE AT VARIOUS COMBINATIONS OF X1, X2, AND X3

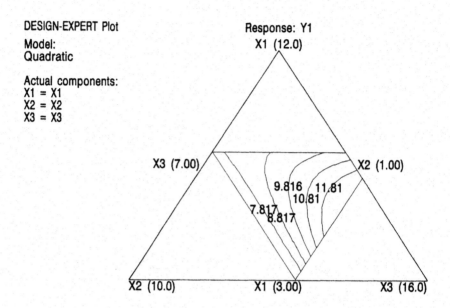

FIG. 7.1.3. CONTOUR LINES OF RESPONSE SURFACE PLOTS FOR ATTRIBUTE Y1

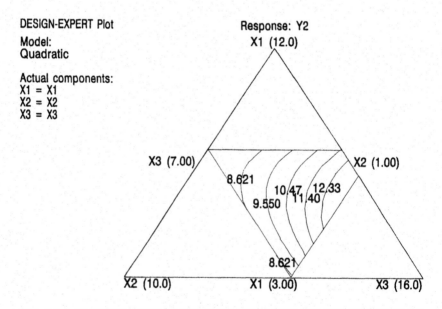

FIG. 7.1.4. CONTOUR LINES OF RESPONSE SURFACE PLOTS FOR ATTRIBUTE Y2

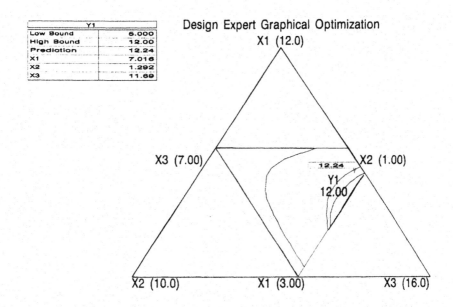

FIG. 7.1.5a. INTERACTIVE OPTIMIZATION OF OVERALL PREFERENCE AND ATTRIBUTE Y1

A useful diagnostic plot from Design-Expert is the map of the standard error (SE) of the predicted preference score (Fig. 7.1.6). In this map, the standard error of the predicted preference score in the optimal area is around 0.052. The SE provides information on the reliability of the predicted score in the optimal area. There are other diagnostic statistics which are not given in this example, and it is recommended to users to explore these useful statistics, i.e., normal probability plot of residuals, leverage, and outlier t.

Let us evaluate the data in Table 7.1.1 by regressing S on Y1-Y3 using the stepwise regression technique. Using the SAS code in Table 7.1.3a, the result is summarized in Table 7.1.3b which indicates that the most important variable for predicting overall preference is attribute Y2 followed by Y3, each respectively accounting for 77.0 and 9.4% of the variance of preference score. See Fig. 7.1.7 for the plot of S and Y2; note that the test formulas (design points) are plotted in this figure, hence it should provide a preliminary overview of potential combinations of ingredient levels.

Statistical Analysis System

The Statistical Analysis System or SAS (SAS Institute, Inc.) is a well-known software used in statistical analysis. It is a powerful system that essentially has

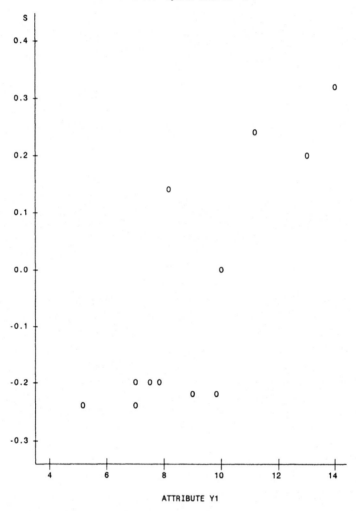

FIG. 7.1.5b. LOCATION OF DESIGN POINTS (TEST FORMULATIONS) ON THE PLOT OF OVERALL PREFERENCE S AND ATTRIBUTE Y1

the procedures for statistical analysis of all types of data. In this section, the factor analysis (FA) and the principal components analysis (PCA) will be illustrated by an example. Also, see Chapter 1.9 for a statistically-oriented

DESCRIPTIVE SENSORY ANALYSIS IN PRACTICE 697

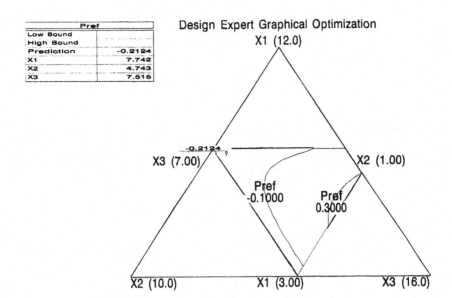

FIG. 7.1.5c. INTERACTIVE OPTIMIZATION OF OVERALL PREFERENCE

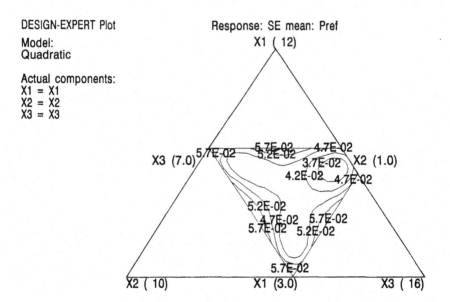

FIG. 7.1.6. STANDARD ERROR RESPONSE SURFACE PLOT FOR OVERALL PREFERENCE

TABLE 7.1.3a
SAS PROGRAM FOR STEPWISE REGRESSION ANALYSIS

```
*PROG STEPWISE;
DATA PREFER;
INFILE 'A:\PREF';
INPUT Y1-Y3 S;
%LET TITLE=EXAMPLE 7.1;
TITLE"&TITLE";
LABEL  Y1='ATTRIBUTE Y1'
       Y2='ATTRIBUTE Y2'
       Y3='ATTRIBUTE Y3';
RUN;
PROC REG DATA=PREFER;
  MODEL S=Y1-Y3 / SELECTION=STEPWISE;
  TITLE1"&TITLE";
  TITLE2"STEPWISE REGRESSION ANALYSIS OF PREFERENCE S";
  RUN;
PROC PLOT DATA=PREFER;
  PLOT S*(Y1-Y3)='*';
  TITLE1"&TITLE";
  TITLE2"PLOT OF OVERALL PREFERENCE S VS. INTENSITY";
  RUN;
```

discussion of PCA. As stated in Chapter 1.1, FA and PCA are multivariate procedures that do not distinguish between independent and dependent variables in the data. The procedures evaluate the relationships among variables. Thus, the input can be the raw data, a correlation matrix, or a covariance matrix. Except for the raw data input, other inputs should be specified in the SAS code. In our example, the raw data will be used as input stored in a SAS file called EXAMPLE. The SAS statement for factor analysis is PROC FACTOR and for principal components, PROC PRINCOMP. Refer to SAS/STAT (1990) user's manual for details.

Example 7.2

Table 7.1.4 contains a descriptive analysis of three products A, B, and C for seven sensory attributes denoted by $X1$, $X2$, $X3$, $X4$, $X5$, $X6$, and $X7$. Three judges evaluated each product. Table 7.1.5 contains the SAS program ("PROG FACTOR) for analyzing the data. These data are found in the SAS file EXAMPLE. If the data were originally entered as a Microsoft Excel format, it can be SAS readable by saving the data as a comma separated value (CSV) and the infile statement would read: INFILE'A:\EXAMPLE.CSV'DELIMITER=',';

TABLE 7.1.3b
OUTPUT OF STEPWISE REGRESSION ANALYSIS

EXAMPLE 7.1
STEPWISE REGRESSION ANALYSIS OF PREFERENCE S

Stepwise Procedure for Dependent Variable S

Step 1 Variable Y2 Entered   R-square = 0.77018064   C(p) = 6.62637827

	DF	Sum of Squares	Mean Square	F	Prob>F
Regression	1	0.42213008	0.42213008	36.86	0.0001
Error	11	0.12596222	0.01145111		
Total	12	0.54809231			

Variable	Parameter Estimate	Standard Error	Type II Sum of Squares	F	Prob>F
INTERCEP	-0.88380753	0.13740655	0.47374887	41.37	0.0001
Y2	0.08266587	0.01361529	0.42213008	36.86	0.0001

Bounds on condition number: 1,   1

Step 2 Variable Y3 Entered   R-square = 0.86435032   C(p) = 2.22338839

	DF	Sum of Squares	Mean Square	F	Prob>F
Regression	2	0.47374376	0.23687188	31.86	0.0001
Error	10	0.07434855	0.00743485		
Total	12	0.54809231			

Variable	Parameter Estimate	Standard Error	Type II Sum of Squares	F	Prob>F
INTERCEP	-0.29839710	0.24824311	0.01074255	1.44	0.2570
Y2	0.05860598	0.01427394	0.12533393	16.86	0.0021
Y3	-0.04587907	0.01741279	0.05161368	6.94	0.0250

Bounds on condition number: 1.692813,   6.771253

All variables left in the model are significant at the 0.1500 level.
No other variable met the 0.1500 significance level for entry into the model.

Summary of Stepwise Procedure for Dependent Variable S

Step	Variable Entered Removed Label	Number In	Partial R**2	Model R**2	C(p)	F	Prob>F
1	Y2 ATTRIBUTE Y2	1	0.7702	0.7702	6.6264	36.8637	0.0001
2	Y3 ATTRIBUTE Y3	2	0.0942	0.8644	2.2234	6.9421	0.0250

The PROC MEANS procedure gives the descriptive statistics for the seven variables, while the PROC GLM gives the analysis of variance. Note that GLM also provides the analysis of variance for the factor scores of products given by PROC GLM DATA=FSCORE, where FSCORE was created by PROC SCORE. In the SAS program, the PRIN method (principal component) was used to extract the underlying factors. Note that the PRIN method is not the same algorithm as that used in the PROC PRINCOMP.

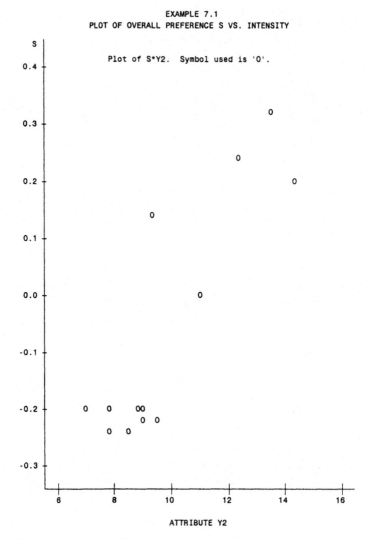

FIG. 7.1.7. LOCATION OF DESIGN POINTS (TEST FORMULATIONS) ON THE PLOT OF OVERALL PREFERENCE S AND ATTRIBUTE Y2

Let us briefly discuss important outputs of the program given in Table 7.1.6a,b,c,d. As a sensory analyst, one must first examine the "eigenvalues of the correlation matrix" (Table 7.1.6a). This shows the proportion of the variance accounted for by each factor component given by PROPORTION. As shown, 66.7% of the variation is accounted for by factor 1, 18.0% by factor 2 and 10.7% by factor 3, which totals 95.4% (CUMULATIVE). Thus, one may conclude that 3 factors would be sufficient to describe the data. The SCREE plot

TABLE 7.1.4
DESCRIPTIVE ANALYSIS DATA

Judge	Product	X1	X2	X3	X4	X5	X6	X7
1	A	3	2	3	5	6	7	5
1	B	5	6	5	4	7	6	6
1	C	5	6	6	5	6	5	7
2	A	2	3	2	6	7	7	4
2	B	6	7	7	4	7	5	5
2	C	7	7	8	5	8	4	6
3	A	3	3	2	6	5	6	5
3	B	6	6	5	5	6	5	8
3	C	7	7	7	6	7	4	8

Note: Data based on a 13-point intensity scale (0 - 12).

(Fig. 7.1.8) also provides similar information, that is three factors are sufficient; in practice, an eigenvalue equal to 1 is generally used to determine the number of factors to be retained.

Table 7.1.6b shows the "rotated factor pattern" which indicates the correlation of each variable to the factor component; this information aids the analyst in finding a descriptive name for the component. FACTOR1 is bipolar characterized by X1(0.90), X2(0.82), X3(0.78), X6(-0.92), and X7(0.92); it is bipolar because X6 has a negative relationship to FACTOR1. FACTOR2 is characterized by X5(0.95), and FACTOR3 by attribute X4(0.99).

Table 7.1.6c shows the estimates of "standardized scoring coefficients." Using Eq. (1.1.5) from Chapter 1.1, the equation for FACTOR1 is

$$Y1 = 0.22X1 + 0.17X2 + \ldots -0.26X6 + 0.40X7$$

and so on. Substitution of the observed standardized ratings (Xs) shown in Table 7.1.7 yields the factor scores. Standardization makes the sum of the mean factor scores across products equal to zero, which results in the elimination of scale effect. It should be emphasized that it is important for sensory analysts to have a general idea on how the raw data were used and analyzed, to aid in the interpretation of the results. The factor scores for each product for FACTOR1 are shown in Table 7.1.6d, where a GLM was used to obtain multiple product comparisons. Results in this table show that product B was significantly different from B and C at the 5% level. The results for the other factors are not reported here. However, the map of the products based on FACTOR1 and FACTOR2 scores is shown in Fig. 7.1.9. Note the separation of product A from the others. The analyst should investigate the reason behind this separation by examining the integrated descriptive name of the factor components.

TABLE 7.1.5
SAS PROGRAM FOR FACTOR ANALYSIS

```
*PROG FACTOR;
DATA BRISTLE;
INFILE 'A:\EXAMPLE';
INPUT JUDGE PROD $ X1-X7;
%LET TITLE=FACTOR ANALYSIS EXAMPLE 7.2;
TITLE"&TITLE";
LABEL   X1='ATTRIBUTE 1'
        X2='ATTRIBUTE 2'
        X3='ATTRIBUTE 3'
        X4='ATTRIBUTE 4'
        X5='ATTRIBUTE 5'
        X6='ATTRIBUTE 6'
        X7='ATTRIBUTE 7';
  RUN;
PROC SORT DATA=BRISTLE;
  BY PROD;
  RUN;
PROC MEANS MEAN N STD CV MAXDEC=2;
  VAR X1-X7;
  BY PROD;
  TITLE"&TITLE";
  RUN;
PROC GLM DATA=BRISTLE;
  CLASS PROD;
  MODEL X1-X7=PROD;
  MEANS PROD/DUNCAN;
  TITLE"&TITLE";
  RUN;
PROC FACTOR DATA=BRISTLE
  SCREE
  NFACTORS=3
  OUTSTAT=FACTOR
  METHOD=PRIN
  ROTATE=VARIMAX
  SCORE;
  VAR X1-X7;
  RUN;
PROC SCORE DATA=BRISTLE
  SCORE=FACTOR
  OUT=FSCORE;
  VAR X1-X7;
  RUN;
PROC GLM DATA=FSCORE;
  CLASS PROD;
  MODEL FACTOR1-FACTOR3=PROD;
  MEANS PROD/DUNCAN;
  TITLE1"&TITLE";
  TITLE2"ANALYSIS OF FACTOR SCORES";
  RUN;
PROC STANDARD DATA=BRISTLE
  MEAN=0
  STD=1
  OUT=NORMAL;
  VAR X1-X7;
  RUN;
PROC PRINT DATA=NORMAL;
  TITLE1"&TITLE";
  TITLE2"STANDARDIZED RAW DATA";
  RUN;
```

TABLE 7.1.6a
PROPORTION OF VARIANCE (EIGENVALUES) ACCOUNTED FOR BY FACTORS 1 TO 7

FACTOR ANALYSIS EXAMPLE 7.2

Initial Factor Method: Principal Components

Prior Communality Estimates: ONE

Eigenvalues of the Correlation Matrix: Total = 7 Average = 1

	1	2	3	4	5	6	7
Eigenvalue	4.6681	1.2610	0.7502	0.2141	0.0727	0.0287	0.0052
Difference	3.4071	0.5108	0.5362	0.1414	0.0439	0.0235	
Proportion	0.6669	0.1801	0.1072	0.0306	0.0104	0.0041	0.0007
Cumulative	0.6669	0.8470	0.9542	0.9848	0.9952	0.9993	1.0000

3 factors will be retained by the NFACTOR criterion.

TABLE 7.1.6b
CORRELATIONS (ROTATED FACTOR PATTERN) OF INTENSITY ATTRIBUTES TO THREE FACTORS RETAINED

Rotation Method: Varimax

Orthogonal Transformation Matrix

	1	2	3
1	0.86786	0.42633	-0.25508
2	0.47812	-0.57717	0.66202
3	-0.13502	0.69650	0.70474

Rotated Factor Pattern

	FACTOR1	FACTOR2	FACTOR3	
X1	0.89982	0.34811	-0.21921	ATTRIBUTE 1
X2	0.82243	0.41762	-0.30656	ATTRIBUTE 2
X3	0.78046	0.51220	-0.30211	ATTRIBUTE 3
X4	-0.08254	-0.13500	0.98652	ATTRIBUTE 4
X5	0.16730	0.94939	-0.13483	ATTRIBUTE 5
X6	-0.92056	-0.31904	-0.03641	ATTRIBUTE 6
X7	0.91675	-0.23068	0.05453	ATTRIBUTE 7

Let us evaluate the data in this example using the principal components analysis. The SAS code is given in Table 7.1.8, which is similar to that of factor analysis except for the PROC PRINCOMP statement. Since we know already that three factors adequately described the data, N=3 is specified in the

TABLE 7.1.6c
STANDARDIZED SCORING COEFFICIENTS FOR EACH INTENSITY ATTRIBUTE

Standardized Scoring Coefficients

	FACTOR1	FACTOR2	FACTOR3	
X1	0.22112	0.02033	-0.04113	ATTRIBUTE 1
X2	0.16760	0.07820	-0.11368	ATTRIBUTE 2
X3	0.12748	0.18061	-0.08133	ATTRIBUTE 3
X4	0.08120	0.21718	0.95910	ATTRIBUTE 4
X5	-0.20048	0.81312	0.18619	ATTRIBUTE 5
X6	-0.25605	-0.07107	-0.21256	ATTRIBUTE 6
X7	0.40167	-0.44407	0.05053	ATTRIBUTE 7

TABLE 7.1.6d
DUNCAN'S MULTIPLE COMPARISON TEST OF THREE PRODUCTS FOR FACTOR1

```
FACTOR ANALYSIS EXAMPLE 7.2
ANALYSIS OF FACTOR SCORES

      General Linear Models Procedure

Duncan's Multiple Range Test for variable: FACTOR1
```

NOTE: This test controls the type I comparisonwise error rate, not the experimentwise error rate

Alpha= 0.05   df= 6   MSE= 0.315047

Number of Means   2   3
Critical Range   1.121 1.162

Means with the same letter are not significantly different.

Duncan Grouping	Mean	N	PROD
A	0.8679	3	C
A			
A	0.2394	3	B
B	-1.1073	3	A

program. Table 7.1.9 is the first page of the output that contains the correlation coefficients among the variables. Sensory analysts should examine the correlation matrix as this will assist in the interpretation of the principal components.

DESCRIPTIVE SENSORY ANALYSIS IN PRACTICE

FACTOR ANALYSIS EXAMPLE 7.2

Initial Factor Method: Principal Components

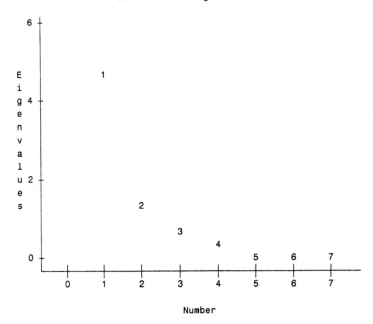

FIG. 7.1.8. SCREE PLOT OF EIGENVALUES AND NUMBER OF FACTORS

TABLE 7.1.7
STANDARDIZED SCORES FOR SEVEN INTENSITY ATTRIBUTES

FACTOR ANALYSIS EXAMPLE 7.2
STANDARDIZED RAW DATA

OBS	JUDGE	PROD	X1	X2	X3	X4	X5	X6	X7
1	1	A	-1.03030	-1.62242	-0.89443	-0.14213	-0.62994	1.37612	-0.70711
2	4	A	-1.57576	-1.11891	-1.34164	1.13707	0.50395	1.37612	-1.41421
3	7	A	-1.03030	-1.11891	-1.34164	1.13707	-1.76383	0.49147	-0.70711
4	2	B	0.06061	0.39162	0.00000	-1.42134	0.50395	0.49147	0.00000
5	5	B	0.60606	0.89513	0.89443	-1.42134	0.50395	-0.39318	-0.70711
6	8	B	0.60606	0.39162	0.00000	-0.14213	-0.62994	-0.39318	1.41421
7	3	C	0.06061	0.39162	0.44721	-0.14213	-0.62994	-0.39318	0.70711
8	6	C	1.15152	0.89513	1.34164	-0.14213	1.63785	-1.27783	0.00000
9	9	C	1.15152	0.89513	0.89443	1.13707	0.50395	-1.27783	1.41421

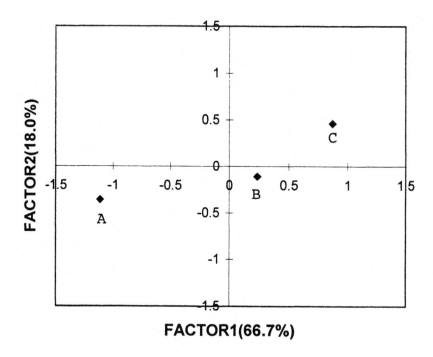

FIG. 7.1.9. MAP OF THE THREE PRODUCTS IN THE SPACE OF FACTOR1 AND FACTOR2 SCORES

Table 7.1.10 shows the "eigenvalues of the correlation matrix" that contains the variance accounted for by the three principal components—PRIN1, PRIN2 and PRIN3. Note that the eigenvalues are the same as those shown in Table 7.1.6a for factor analysis. The "eigenvector" is the resultant coefficient from the solution of Eq. (1.1.2). Thus for PRIN1, the equation for PC score is

$$Y1 = 0.46X1 + 0.45X2 + \ldots -0.43X6 + 0.32X7$$

where the Xs are the standardized ratings (Table 7.1.7) for the products of interest. The average score for each product is obtained by dividing Y1 by the number of judges. The results of these calculations for both FA and PCA are given in Table 7.1.11.

TABLE 7.1.8

SAS PROGRAM FOR PRINCIPAL COMPONENTS ANALYSIS

```
*PROG PRINCOM;
DATA BRISTLE;
INFILE 'A:\EXAMPLE';
INPUT JUDGE PROD $ X1-X7;
%LET TITLE=PRINCIPAL COMPONENT ANALYSIS EXAMPLE 7.2;
TITLE"&TITLE";
DROP JUDGE;
LABEL  X1='ATTRIBUTE 1'
       X2='ATTRIBUTE 2'
       X3='ATTRIBUTE 3'
       X4='ATTRIBUTE 4'
       X5='ATTRIBUTE 5'
       X6='ATTRIBUTE 6'
       X7='ATTRIBUTE 7';
   RUN;
PROC SORT DATA=BRISTLE;
   BY PROD;
   RUN;
PROC PRINCOMP DATA=BRISTLE
   N=3
   OUT=PRIN;
   RUN;
PROC GLM DATA=PRIN;
   CLASS PROD;
   MODEL PRIN1-PRIN3=PROD;
   MEANS PROD/DUNCAN;
   TITLE"&TITLE";
   RUN;
PROC STANDARD DATA=BRISTLE
   MEAN=0
   STD=1
   OUT=NORMAL;
   VAR X1-X7;
   RUN;
PROC PRINT DATA=NORMAL;
   TITLE1"&TITLE";
   TITLE2"STANDARDIZED RAW DATA";
   RUN;
```

The average PC scores in this table may be graphed like the plot shown in Fig. 1.1.3 (Chapter 1.1). To communicate effectively the results of the PC analysis, the sensory analyst should be able to explain the reason behind the dispersion of the products based on the integrated principal or factor components.

TABLE 7.1.9
OUTPUT FROM PROC PRINCOMP SHOWING THE CORRELATION COEFFICIENTS AMONG THE SEVEN INTENSITY ATTRIBUTES

PRINCIPAL COMPONENT ANALYSIS EXAMPLE 7.2

Principal Component Analysis

9 Observations
7 Variables

Simple Statistics

	X1	X2	X3	X4
Mean	4.888888889	5.222222222	5.000000000	5.111111111
StD	1.833333333	1.986062548	2.236067977	0.781735960

	X5	X6	X7
Mean	6.555555556	5.444444444	6.000000000
StD	0.881917104	1.130388331	1.414213562

Correlation Matrix

	X1	X2	X3	X4	X5	X6	X7	
X1	1.0000	0.9345	0.9452	-.3392	0.5068	-.9383	0.7232	ATTRIBUTE 1
X2	0.9345	1.0000	0.9289	-.4204	0.5630	-.8847	0.6231	ATTRIBUTE 2
X3	0.9452	0.9289	1.0000	-.4291	0.6339	-.8902	0.5534	ATTRIBUTE 3
X4	-.3392	-.4204	-.4291	1.0000	-.2820	0.0786	0.0000	ATTRIBUTE 4
X5	0.5068	0.5630	0.6339	-.2820	1.0000	-.4040	0.0000	ATTRIBUTE 5
X6	-.9383	-.8847	-.8902	0.0786	-.4040	1.0000	-.7037	ATTRIBUTE 6
X7	0.7232	0.6231	0.5534	0.0000	0.0000	-.7037	1.0000	ATTRIBUTE 7

TABLE 7.1.10
EIGENVECTORS OF THE THREE PRINCIPAL COMPONENTS EXTRACTED FROM THE DATA

PRINCIPAL COMPONENT ANALYSIS EXAMPLE 7.2

Principal Component Analysis

Eigenvalues of the Correlation Matrix

	Eigenvalue	Difference	Proportion	Cumulative
PRIN1	4.66813	3.40713	0.666875	0.666875
PRIN2	1.26100	0.51076	0.180143	0.847019
PRIN3	0.75024		0.107177	0.954196

Eigenvectors

	PRIN1	PRIN2	PRIN3	
X1	0.456009	0.074960	-.038703	ATTRIBUTE 1
X2	0.448950	-.045212	-.041814	ATTRIBUTE 2
X3	0.450230	-.109075	0.044402	ATTRIBUTE 3
X4	-.176262	0.615843	0.706978	ATTRIBUTE 4
X5	0.270455	-.496226	0.627642	ATTRIBUTE 5
X6	-.428423	-.249431	-.142679	ATTRIBUTE 6
X7	0.316279	0.541037	-.284028	ATTRIBUTE 7

TABLE 7.1.11
ESTIMATES OF AVERAGE SCORES OF PRODUCTS

Analysis	Factor/component	A	B	C
FA	FACTOR1(66.7%)	-1.11a	0.24b	0.87b
	FACTOR2(18.0%)	-0.36	-0.11	0.46
	FACTOR3(10.7%)	0.58a	-1.00c	0.42bc
PCA	PRIN1(66.7%)	-2.72a	0.91b	1.82b
	PRIN2(18.0%)	0.07	-0.55	0.48
	PRIN3(10.7%)	0.27	-0.70	0.44

Note: Within a row, average scores with same letter are not significantly different at the 5% level.

REFERENCES

Analytical Software. STATISTIX for Windows. Tallahassee, Florida.

BOX, G.E.P. and DRAPER, N.R. 1987. Empirical Model-Building and Response Surfaces. John Wiley & Sons, New York.

CORNELL, J.A. 1981. Experiments with Mixtures. John Wiley & Sons, New York.

GACULA, JR., M.C. 1993. Design and Analysis of Sensory Optimization. Food and Nutrition Press, Trumbull, Connecticut.

GACULA, JR., M.C. and SINGH, J. 1984. Statistical Methods in Food and Consumer Research. Academic Press, San Diego, California.

MYERS, R.H. and MONTGOMERY, D.C. 1995. Response Surface Methodology. John Wiley & Sons, New York.

SAS. SAS Institute Inc., Cary, North Carolina.

SAS/STAT. 1990. SAS/STAT User's Guide. Vols. 1 and 2, 4th Edition. SAS Institute Inc., Cary, North Carolina.

Stat-Ease, Inc. Design-Expert User's Guide. Ver. 4. Stat-Ease, Inc., Minneapolis, Minnesota.

Stat-Ease, Inc. Design-Ease Reference Manual. Ver. 3.0.7. Stat-Ease, Inc., Minneapolis, Minnesota.

INDEX

Beef flavor descriptions, 270

Canonical analysis, 283, 344
Catfish flavor descriptors, 293
Comparison of trained panels, 653, 665

Descriptive analysis
 beer, 303
 cheddar cheese, 149, 153
 chicken flavor, 275, 279
 cider, 627
 cigarette, 520
 coconut meat, 595
 coffees, 477
 comparison of scaling methods, 613
 criteria for systems of terminology, 79-82
 dairy products, 147
 detection of inconsistent panelists, 122-123
 dulce de leche, 163, 171
 fabrics, 405, 417
 farm-raised catfish, 289
 fermented milks, 185, 189, 192
 flavored dairy products, 235
 formula optimization, 2
 ice cream, 201, 203
 importance of language, 77
 meat WOF, 267
 methods, 5-6
 noodles, 617
 oral pungency, 641, 645
 paper and fabrics, 443
 reducing noise in descriptive data, 109
 restructured beef steaks, 255
 success, 1
 tea, 55, 507
 vanilla, 653
 wine, 313, 317, 335, 351 383, 459
 yoghurt, 219, 222
Descriptors for young coconut, 601

Evaluation of fabric aesthetics, 425-441
Experts versus consumers, 127-146
 expert versus non-expert, 461
 quality experts in wine, 383
 sorting wine, 391

Factor analysis, 8-12, 133-135, 155-158, 239, 278, 320, 499, 508
Flavor Profile and Profile Attribute Analysis, 5, 15
 presentation of results, 21
 training, 16-18
Free-Choice Profiling
 cider, 627
 coffee, 477
 food texture, 543
 wine, 459

Handfeel description, 445
 evaluation protocol, 446-451
Harshness of fabrics, 405
 fabric aesthetics, 417-424
 fabric properties related to harshness, 413
 one at a time evaluation, 408
 paired evaluation, 409

Intensity Variation Descriptive Methodology, 519
Interactive graphical optimization, 695, 697

INDEX

Lexicon of peanut flavor descriptors, 539

Multidimensional scaling, 501

Odor profiling, 493, 500

Panel selection and training
  beer, 303-308
  catfish, 290
  cheddar cheese, 152-153
  chicken flavor, 276-278
  control chart technique, 595
  ice cream, 202-203
  panel reproducibility and sensitivity, 295, 619-623
  paper and fabrics, 451-455
  QDA, 28-30, 35-37
  tea, 55
  terminology development, 72-73, 79-81
  wine, 336, 352-353, 372-374
  yoghurt, 221-223
Partial least squares, 179, 673
Principal components analysis, 8-12, 90-108, 168-179, 192-194, 278, 282, 378, 391-392, 429-431, 508, 650, 655, 668
  basic feature, 92
Procrustes analysis, 194, 197, 459, 480, 547, 631, 661, 670

Quantitative Descriptive Analysis
  definition, 27, 63
  panel selection, 35-35
  QDA, 5, 23-24, 63-69
  QDA plot, 32, 45-46, 48-49, 66, 603
Quantitative Flavor Profiling, 238, 245

Reference standards in training
  frame of reference, 83
  panelists, 71-76
  terminology development, 72-74
  vanilla, 656
  wine, 354-355
Repertory Grid Method, 627
Reverse engineering, 141

SAS software program
  factor analysis, 702
  principal components analysis, 707
  stepwise regression analysis, 698
Similarity scaling, 479
  INDSCAL, 480
Software packages
  Design-Ease, 3, 689
  Design-Expert, 3, 689
  SAS, 11, 695-696, 698-699
  Statistix, 7, 689
Spectrum Descriptive Analysis Method, 5, 443
Standardized flavor language, 241-243

Taste descriptive analysis: concept formation, alignment and appropriateness, 561
Test factor patterns for concordance, 53
  likelihood ratio test, 56
Thurstone-Mosteller model, 689
Training texture profile panel, 257, 261-264

CPSIA information can be obtained at www.ICGtesting.com
Printed in the USA
BVOW08*0843030615

401796BV00019B/66/P